Electronic, Magnetic, and Optical Materials

Advanced Materials and Technologies Series

Series Editor

Yury Gogotsi

Drexel University
Philadelphia, Pennsylvania, U.S.A.

Electronic, Magnetic, and Optical Materials

Pradeep Fulay

CRC Press
Taylor & Francis Group
Boca Raton   London   New York

CRC Press is an imprint of the
Taylor & Francis Group, an **informa** business

CRC Press
Taylor & Francis Group
6000 Broken Sound Parkway NW, Suite 300
Boca Raton, FL 33487-2742

© 2010 by Taylor and Francis Group, LLC
CRC Press is an imprint of Taylor & Francis Group, an Informa business

---

Library of Congress Cataloging-in-Publication Data

Fulay, Pradeep P.
    Electronic, magnetic, and optical materials / author, Pradeep Fulay.
        p. cm. -- (Advanced materials and technologies series)
    "A CRC title."
    Includes bibliographical references and index.
    ISBN 978-0-8493-9564-2 (hardcover : alk. paper)
    1. Electronics--Materials. 2. Magnetic materials. 3. Optical materials. I. Title.

TK7871.F85 2010
621.381--dc22
                                                      2009043845

Visit the Taylor & Francis Web site at
http://www.taylorandfrancis.com

and the CRC Press Web site at
http://www.crcpress.com

Contents

Preface

The society we live in today relies on and is built upon our ability to almost instantly communicate and share information at virtually any location. In the past 10 to 15 years, we have witnessed unprecedented advances in the areas of computer and information technology, health care, biotechnology, environmental sciences and engineering, and energy technology. All of these disciplines, as well as the challenges they face and the opportunities they create, are interrelated and depend on our ability to send, store, and convert information and energy.

The number of different ways in which we communicate information using the Internet, computers, and other personal communication devices such as wireless laptop computers, high-definition televisions, smart phones, electronic book readers, global positioning systems, and other technologies continues to increase very rapidly. At the very core of these technologies lie very important devices made using a number of materials that have useful electronic, magnetic, and optical properties. Thus, in this age of information technology, any scientist or engineer must have some knowledge of the basic science and engineering concepts that enable these technologies. This is especially important as we move into technologies that cut across many different disciplinary boundaries.

A good portion of information-related technologies developed to date can be traced to microelectronic devices based on silicon. Of course, the information storage and processing has also been supported by the availability of magnetic materials for data storage, optical fibers based on ultrahigh-purity silica, and advances in sensors and detectors. However, silicon has been at the core of microelectronics-related technologies and has served us extremely well for more than 50 years since the invention of the transistor on December 16, 1947. There is no doubt that silicon-based microelectronic devices that enable the fabrication of computer chips from integrated circuits will continue to serve us well for many more years to come.

The first transistor, which was based on germanium, used a bulky crystal. Now, in 2009, the state-of-the-art transistors are made at a length scale of 45 nm. A very small modern computer chip (~1.5 cm × 2 cm square) contains many "cores" and about 230 million transistors! Both the size and the cost associated with the manufacturing of the transistors have decreased significantly. The size reduction has followed the famous Moore's law, which predicted that the number of transistors on a chip will double every two years (Moore 1965). This has stayed true for nearly 50 years.

We are now at a point where we have technologies that take us into regimes that span beyond Moore's law. This means that simply scaling down transistors and devices based on these will no longer do the trick! There are opportunities to create novel electronic, optical, and magnetic devices that can compete with or enhance silicon-based electronics.

Similarly, new materials other than silicon, for example, carbon nanotubes, graphene, and gallium nitride, are also emerging! With the advent of nanowires, quantum dots, and so on, the distinction between a "device" and a "material" has been steadily fading away. Technologies such as those related to flexible electronics, organic electronics, molecular electronics, photovoltaics, fuel cells, biomedical implantable devices, tissue engineering, and new sensors and actuators are evolving rapidly. Many of these technological developments have brought the fields of electrical engineering, materials science, physics, chemistry, biomedical engineering, chemical engineering, and mechanical engineering closer together—perhaps more than ever before.

The increased interdisciplinarity and interdependencies of technologies mean that engineers and scientists whose primary training or specialization is in one discipline can benefit tremendously by learning some of the fundamental aspects of other disciplines. For example, it would be useful for an electrical engineering student not only to be trained in silicon and silicon processing but also to gain some insights into other new materials, devices, and functionalities that will likely be integrated with traditional silicon-based microelectronics. In some applications, hybrid approaches can be developed in which semiconductors such as silicon are integrated with gallium arsenide and other materials such as carbon nanotubes and graphene.

The primary motivation for this book stems from the need for an introductory textbook that captures the fundamentals as well as the applications of the electronic, magnetic, and optical properties of materials. The subject matter has grown significantly more interdisciplinary and is of interest to many different academic disciplines.

There are several undergraduate engineering and science textbooks devoted to specific topics such as electromagnetics, semiconductors, optoelectronics, fiberoptics, microelectronic circuit design, photovoltaics, superconductors, electronic ceramics, and magnetic materials. However, most of these books are geared toward specialists. I was not able to find a single introductory textbook that takes an interdisciplinary approach that is not only critical, but is also of interest to students from various disciplines. This book is an attempt to fill this gap, which is important to so many disciplines but is not adequately covered in any one of them.

I have written this book for typical junior (third-year) or senior (fourth-year) students of science, such as physics and chemistry, or engineering, such as electrical, materials science, chemical, and mechanical engineering, and I have used an interdisciplinary approach. The book is also appropriate for graduate students from different disciplines and to those who do not have a significant background in solid-state physics, electrical engineering, materials science and engineering, or related disciplines. This book builds upon my almost 20 years of experience in teaching an introductory undergraduate course on electrical, magnetic, and optical properties of materials to both engineering and engineering physics students.

I faced three major challenges, which I had anticipated. First, I had to select a few of the topics that are of central importance, in my opinion, to engineers and scientists interested in the electrical, magnetic, and optical properties of materials. This means that many other important topics are not addressed in detail or are left out altogether. For example, topics such as high-temperature superconductors, ionic conductors, and fiberoptics systems are not discussed in detail, and some are not discussed at all. I also have not developed some of the newest and "hottest" topics such as graphene or carbon nanotube-based devices, spintronics, organic photovoltaics, organic light-emitting diodes (OLEDs), and so on in detail in order to maintain this as an *introductory* textbook. Interested instructors may develop some of the topics not covered here as needed, since the book covers the fundamental framework very well.

The second challenge was maintaining a relative balance between the *fundamentals* (with respect to the underlying physics related concepts) versus the technological aspects (production of devices, manufacturing, materials processing, etc.). The approach I have adopted here was to provide as many real-world and interesting technological examples as possible whenever there was an opportunity to do so. For example, while discussing piezoelectric materials, I provide many examples of technologies such as smart materials and ultrasound imaging. I have found that the inclusion of examples of real-world technologies helps maintain a high level of interest as students relate to the topics easily (e.g., how is a Blu-ray disc different from a DVD?). Students also tend to retain what they have learned for a longer period when these real-world connections are made.

The third challenge was to determine the level at which different topics are to be covered. For example, electrical engineering students who have some exposure to circuits may find the device-related problems to be rather simple, for example, calculating limiting resistance for a light-emitting diode (LED). However, this may not be the case for many other students from other disciplines. Similarly, students who have had a class in introductory materials science may find some of the concepts related to materials synthesis, processing, and structure–property relationships to be somewhat straightforward. However, many other students may not have such a background and may be confused by the terminology used. I have therefore maintained an introductory level for all the topics covered and have tried to make the concepts as interesting as possible while challenging the students' ability and piquing their curiosity.

I take full responsibility for any mistakes or errors this book may have. Please contact me as needed so that these can be corrected as soon as possible. I welcome any other suggestions you may have as well. I intend to create a Facebook page for this book, on which I will add some valuable video content. I find that such supplemental video components and links to different Web sites can be extremely useful in promoting learning, especially for some of the device operation and manufacturing aspects.

Any book such as this is a team effort, and this book is no exception. In this regard, I wish to acknowledge the assistance of the Taylor and Francis staff members who have worked with me. In particular, I thank Allison Shatkin for helping me develop this textbook. I am thankful to Marsha Pronin for her assistance and patience in working with me on this book. I am also thankful to a number of colleagues and corporations who have provided many of the illustrations.

I am thankful to my mother, Pratibha Fulay, and father, Prabhakar Fulay, for all they have done for me and the values they taught me. I am thankful to my wife, Dr. Jyotsna Fulay, and my daughter Aarohee and son Suyash for the support, patience, understanding, and encouragement they have always provided.

Pradeep P. Fulay

REFERENCES

Moore, G. E. 1965. Cramming more components onto integrated circuits. *Electronics* 38(8).

Author

Pradeep P. Fulay is a professor of materials science and engineering in the Department of Mechanical Engineering and Materials Science at the University of Pittsburgh. He joined the University of Pittsburgh in 1989, immediately after earning a PhD in materials science and engineering from the University of Arizona, Tucson. He earned a BTech with honors and an MTech with honors from the Indian Institute of Technology in Mumbai, India, in 1983 and 1984, respectively. Dr. Fulay has authored two other textbooks, nearly 60 referred journal publications, and three U.S. patents issued in the field of materials science and engineering. Dr. Fulay's research in the areas of microwave ceramics, ferroelectric and piezoelectric materials, magnetic materials, chemical synthesis, and the processing of smart materials has received international recognition. Dr. Fulay is a fellow of the American Ceramic Society. He has also held many positions in educational and research institutions, including as the president of the Ceramic Educational Council of the American Ceramic Society and as a founding member of the Greater Pittsburgh Chapter of the Materials Research Society. Dr. Fulay has been a William Kepler Whiteford Faculty Fellow at the University of Pittsburgh. His research has been supported by several organizations, including the National Science Foundation, Ford, Alcoa, and the Air Force Office of Scientific Research (AFOSR). Dr. Fulay has also served as the program director for electronics, photonics, and device technologies in the Electrical, Communications, and Cyber Systems Division at the National Science Foundation.

1 Introduction

KEY TOPICS

- Ways to classify materials
- Atomic-level bonding in materials
- Crystal structures of materials
- Effects of imperfections on atomic arrangements
- Microstructure–property relationships

1.1 INTRODUCTION

The goal of this chapter is to recapitulate some of the basic concepts in materials science and engineering as they relate to electronic, magnetic, and optical materials and devices. We will learn the different ways in which technologically useful electronic, magnetic, or optical materials are classified. We will examine the different ways in which the atoms or ions are arranged in these materials, the imperfections they contain, and the concept of the microstructure–property relationship. You may have studied some of these—and perhaps more advanced—concepts in an introductory course in materials science and engineering, physics, or chemistry.

1.2 CLASSIFICATION OF MATERIALS

An important way to classify materials is to do so based on the arrangements of atoms or ions in the material (Figure 1.1). The manner in which atoms or ions are arranged in a material has a significant effect on its properties.

1.3 CRYSTALLINE MATERIALS

A *crystalline material* is defined as a material in which atoms or ions are arranged in a particular arrangement that repeats itself in all three dimensions. The *unit cell* is the basic unit that represents an arrangement of atoms (or ions) and repeats itself. The *crystal structure* is the specific geometrical arrangement in which atoms are arranged within the unit cell. Crystal structures are derived from *Bravais lattices*, or simply *lattices*. A lattice is simply an arrangement of points in space. Auguste Bravais, a French physicist, showed that there are only 14 independent ways of arranging points in space (Figure 1.2).

Note that the concept of lattice does *not* refer to atoms or ions. This is a mathematical idea that refers only to points in space. However, when we associate an atom or an ion, or a group of atoms or ions, with a lattice point, called the *basis*, we get crystal structures. Thus, a crystal structure is derived from a Bravais lattice, in which we have either one atom or a group of atoms associated with each lattice point. In other words:

$$\text{Crystal structure} = \text{lattice} + \text{basis} \tag{1.1}$$

We have only 14 Bravais lattices (Figure 1.2). However, there are many possible bases, which lead to hundreds of possible crystal structures. We also have many materials that exhibit the same crystal structure. However, the compositions, that is, the chemical makeup, of these materials can be very different. For example, silver (Ag), copper (Cu), and gold (Au) have the same crystal structure (see Section 1.6).

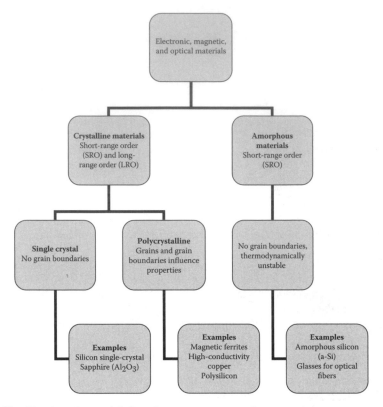

FIGURE 1.1   Classification of materials based on arrangements of atoms or ions.

Similarly, materials often exhibit different crystal structures, depending upon the temperature (T) and pressure (P) to which they are subjected. In some cases, changes in crystal structures may also result from the application of other stimuli, such as mechanical stress (σ or τ), electric field (E), or magnetic field (H), or a combination of such stimuli. The different crystal structures exhibited by a *compound* are known as *polymorphs*. For example, at room temperature and atmospheric pressure, barium titan-ate ($BaTiO_3$) exhibits a tetragonal structure (Figure 1.3). However, at temperatures slightly higher than 130°C, $BaTiO_3$ exhibits a cubic structure, and at even higher temperatures, this structure changes into a hexagonal structure. The different crystal structures exhibited by an *element* are known as *allotropes*.

A *phase* is defined as any portion of a system, including the whole, which is physically homoge-neous and bounded by a surface so that it is mechanically separable from the other portions. One of the simplest examples is water in the form of ice, which represents a phase of water. A *phase diagram* is a diagram that indicates the phases that can be expected in a given system of materi-als, which assume thermodynamic equilibrium. Thus, amorphous and crystalline materials formed under nonequilibrium conditions are not shown in a phase diagram.

For example, the phase diagram of a binary lead–tin (Pb–Sn) system is shown in Figure 1.4. Three phases, α, β, and L, are seen at different compositions and temperature ranges. The liquidus repre-sents the traces of temperature above which the material is in the liquid phase. Similarly, the solidus is the trace of temperature below which the material is completely solid. Note that although most alloys melt over a range of temperatures, some specific compositions (e.g., 61.9% tin in the lead–tin sys-tem), known as eutectic compositions, melt and solidify at a single temperature, known as the eutec-tic temperature. The lead–tin alloy has been used for soldering electronic components. Currently, lead-free substitutes (developed because of the toxicity of lead) are increasingly being used.

The electrical, magnetic, and optical properties of a material can significantly change if the crys-tal structure changes. For example, the tetragonal form of $BaTiO_3$ is *ferroelectric* and *piezoelectric*

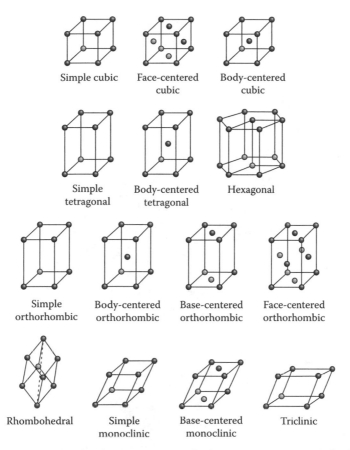

FIGURE 1.2 Bravais lattices showing the arrangements of points in space. (From Askeland, D., and P. Fulay. 2006. *The Science and Engineering of Materials.* Washington, DC: Thomson. With permission.)

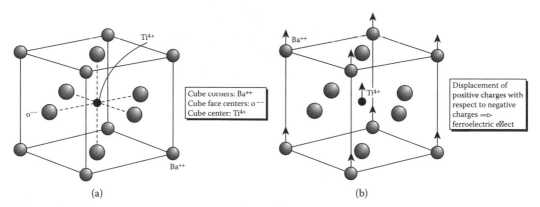

FIGURE 1.3 (a) Cubic and (b) tetragonal structures of $BaTiO_3$. (From Singh, J. 1996. *Optoelectronics: An Introduction to Materials and Devices.* New York: McGraw Hill. With permission.)

(Figure 1.3). However, the cubic form of $BaTiO_3$ is neither ferroelectric nor piezoelectric. We will learn more about these materials in later chapters.

Crystalline materials can be further classified into single-crystal and polycrystalline materials (Figure 1.1). A single-crystal material, as the name suggests, is made up of one crystal, in which the atomic arrangements of that particular crystal structure are followed, except in the external surfaces

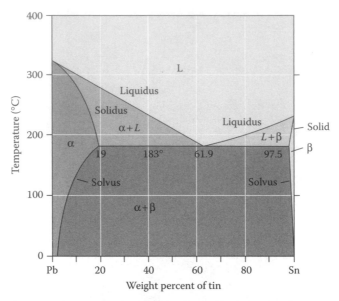

FIGURE 1.4   Lead–tin phase diagram. (From Askeland, D., and P. Fulay. 2006. *The Science and Engineering of Materials*. Washington, DC: Thomson. With permission.)

FIGURE 1.5   Single crystal of silicon. (From Askeland, D., and P. Fulay. 2006. *The Science and Engineering of Materials*. Washington, DC: Thomson. With permission.)

of the crystal. A photograph of a large single crystal of silicon (Si) is shown in Figure 1.5. Single crystals are not always large; some only a few millimeters in dimension.

Large single crystals of silicon (up to 12 inches in diameter and several feet in height) are sliced into thin wafers. These silicon wafers are then used for manufacturing integrated circuits (ICs) that are packaged into computer chips. We prefer to use the largest possible crystals for this application because it reduces the total cost of producing ICs.

In polycrystalline materials, a *long-range order* (LRO) of atoms or ions is present; that is, atoms or ions are arranged in a particular geometric arrangement within each crystal or grain. The term "grain" refers to a relatively small, single-crystal region within a polycrystalline material. The LRO can exist across relatively larger distances, ranging from a few micrometers up to centimeters for single crystals. A polycrystalline material comprises many smaller grains. The LRO in a grain ends at the boundaries of that grain. The regions or surfaces between adjacent grains are known as

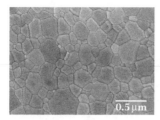 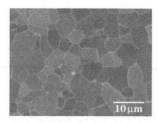

FIGURE 1.6 Microstructures of alumina ceramics. (From Kim et al. 2009. *Acta Mater* 57(5):1319–26. With permission.)

grain boundaries. The *microstructure* of a material is examined in a two-dimensional cross section in which the grain boundaries are seen as lines or curves. As we will learn in Section 1.17, grain boundaries can significantly affect some, but not all, properties.

The concept of microstructure is central to many of the ideas and technologies we will discuss throughout this book. The term "microstructure" is used to describe the arrangement of grains in a polycrystalline material. This also includes a description of the average grain size, grain-size distribution, grain shape, and whether the grains show a preferred orientation. In addition, other features such as imperfections in atomic arrangements are also often considered part of the microstructure.

The microstructure of a polycrystalline alumina (Al_2O_3) ceramic material showing grains and grain boundaries is shown in Figure 1.6. The relationships between the microstructure and properties of materials are explored in Section 1.17.

The atoms or ions of many materials do not show an LRO. Such materials are known as *amorphous materials.* We will discuss these in Section 1.18.

1.4 CERAMICS, METALS AND ALLOYS, AND POLYMERS

Another way of classifying engineered materials is based on their general behavior as metals or alloys, ceramics, or polymers/plastics. For example, stainless steels, copper, platinum (Pt), tungsten (W), and so on, are metallic materials, whereas materials such as polyethylene and Teflon are considered polymers or plastics. A *plastic* is a polymer-based material that has been formulated with one or many polymers and other additives (e.g., carbon black, conductive or dielectric particles, glass fibers, etc.). Polymers are lightweight and flexible. In recent years, significant research and development efforts have been made to develop microelectronic devices, such as transistors, solar cells, and light-emitting diodes (LEDs), based entirely on polymers or carbon-based materials, such as carbon nanotubes and graphene. You may have read or heard that some of the best television displays are based on organic LEDs (OLEDs). This field of development is based on the use of polymers for electronics and is known as *organic electronics.*

Efforts are underway aimed at integrating conventional silicon-based electronic devices with polymers to take advantage of the low density and flexibility, in addition to the electrical and optical properties, of polymers. This relatively new field of research is known as *flexible electronics.*

Materials such as Al_2O_3 and silica (SiO_2) are inorganic solids and are considered ceramics. This type of classification does not always distinguish between the details of the atomic arrangements. For example, SiO_2 is considered a ceramic material regardless of whether it is in an amorphous or crystalline form.

1.4.1 BONDING IN MATERIALS

The types of interatomic bonds that exist in materials are metallic (for metals and alloys), covalent, and ionic bonds (Figure 1.7). Most ceramics (e.g., Al_2O_3, SiO_2) tend to exhibit a mixed ionic and covalent bonding. Typically, the larger the difference between the electronegativity of the atoms or ions that form the material, the higher the ionic character of the bond.

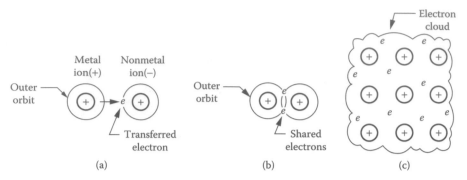

FIGURE 1.7 Different types of bonds in materials: (a) ionic, (b) covalent, and (c) metallic. (From Groover, M. P. 2007. *Fundamentals of Modern Manufacturing: Materials, Processes, and Systems*. New York: Wiley. With permission.)

In most polymers, the primary bonds (such as carbon–carbon or carbon–hydrogen) are covalent. Many technologically important semiconductors (e.g., silicon, germanium [Ge]) exhibit covalent bonds.

A secondary bond (i.e., a bond with lower energy), known as the *van der Waals bond*, is present in all materials and is caused by interactions between induced dipoles. The van der Waals forces and the resultant bonds are especially important in materials that have polar molecules, atoms, or groups (e.g., water [H_2O], hydroxyl group [OH^-], amine group [NH_2], chlorine [Cl], and fluorine [F] atoms or ions). A special type of van der Waals interaction originating from the intermolecular forces between molecules with a permanent dipole moment is known as the *hydrogen bond*. It occurs in water (hence the name "hydrogen bond") and many other solids and liquids.

The van der Waals bonds play an important role in modifying the properties of many materials, such as water, polyvinyl chloride (PVC), graphite, and clay. For example, graphite functions as a solid lubricant. On the contrary, diamond, which is also a form of carbon, is one of the hardest naturally occurring materials. In both materials, the primary bonds are covalent carbon–carbon bonds. However, graphite has a layered structure and the atoms between the layers are bonded by relatively weak van der Waals forces.

Similarly, water has a relatively high boiling point and surface tension compared to other liquids with similar molecular weight, and is denser than ice because of the presence of hydrogen bonds. PVC is more brittle because of the van der Waals forces that exist between the chlorine and hydrogen atoms between the adjacent molecular chains in this material.

1.5   FUNCTIONAL CLASSIFICATION OF MATERIALS

For electronic, magnetic, and optical applications of materials, the functional classification of materials is useful (Figure 1.8). In this classification, the *primary functionality* that a material provides is highlighted. For example, one form of iron oxide (Fe_3O_4), which is a ceramic, and another metallic material, such as Permendur (an alloy of iron [Fe] and cobalt [Co]), are classified as magnetic materials.

1.6   CRYSTAL STRUCTURES

We will now describe the crystal structures of some materials that have useful electrical, magnetic, or optical properties. Many materials that we will encounter later in this book exhibit cubic structures, comprised of *simple cubic* (SC), *face-centered cubic* (FCC), or *body-centered cubic* (BCC) arrangements of atoms or ions (Figure 1.9). In some materials, the atoms are packed in a hexagonal arrangement (Figure 1.10).

FIGURE 1.8 Functional classification of materials. (From Askeland, D., and P. Fulay. 2006. *The Science and Engineering of Materials*. Washington, DC: Thomson. With permission.)

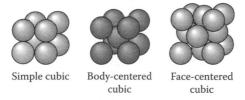

Simple cubic   Body-centered   Face-centered
                   cubic   cubic

FIGURE 1.9 Simple cubic, body-centered cubic, and face-centered cubic structures. (From Askeland, D., and P. Fulay. 2006. *The Science and Engineering of Materials*. Washington, DC: Thomson. With permission.)

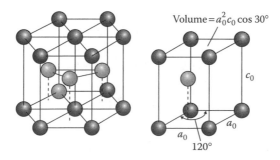

FIGURE 1.10 A unit cell for the hexagonal close-packed structure. (From Askeland, D., and P. Fulay. 2006. *The Science and Engineering of Materials*. Washington, DC: Thomson. With permission.)

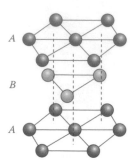

FIGURE 1.11 The packing sequence of a hexagonal close-packed structure. (From Askeland, D., and P. Fulay. 2006. *The Science and Engineering of Materials*. Washington, DC: Thomson. With permission.)

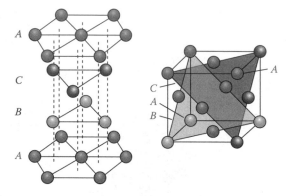

FIGURE 1.12 The packing sequence of a face-centered cubic structure. (From Askeland, D., and P. Fulay. 2006. *The Science and Engineering of Materials*. Washington, DC: Thomson. With permission.)

The hexagonal close-packed (HCP) arrangements shown in Figures 1.10 and 1.11 are the same. Similarly, the FCC structures shown in Figures 1.9 and 1.12 are also the same. Both FCC and HCP arrangements lead to the maximum possible *packing fraction* of 0.74 for atoms or spheres of the same size, but the sequence in which different layers are packed (ABCABC... for FCC vs. ABABAB for HCP) changes.

The term "packing fraction" refers to the ratio of the volume occupied by the atoms to the volume of the unit cell. It can be shown that if we have a cube-shaped box, the best we can do is fill up 74% of the space available with spheres of a given radius. The packing fraction does *not* depend upon the radius of the sphere (i.e., whether it is a basketball or a tennis ball), as long as we have spheres of the same size. When atoms are packed such that the structure exhibits the maximum possible packing fraction, the structure is referred to as a *close-packed* (CP) *structure*. Thus, a hexagonal structure with a maximum possible packing fraction of 0.74 is known as a close-HCP structure (Figure 1.11). A hexagonal structure is not necessarily closely packed.

The following example illustrates the calculation of packing fractions for cubic CP structures.

EXAMPLE 1.1: CALCULATION OF PACKING FRACTIONS

Calculate the maximum possible packing fractions for the (a) SC, (b) FCC, and (c) BCC structures. Assume that all atoms have a radius r and the unit-cell parameter is a.

SOLUTION

a. As shown in Figure 1.13, in the CP-SC structure, atoms touch along the cube edges, that is,

$$a = 2r \tag{1.2}$$

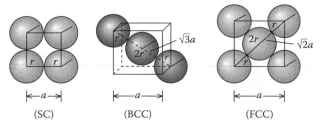

FIGURE 1.13 Packing of atoms in simple cubic (SC), body-centered cubic (BCC), and face-centered cubic (FCC) crystal structures. (From Askeland, D., and P. Fulay. 2006. *The Science and Engineering of Materials.* Washington, DC: Thomson. With permission.)

Thus, the packing fraction is

$$\text{packing fraction} = \frac{\text{volume of atoms in the unit cell}}{\text{volume of unit cell}}$$

$$= \frac{(8 \text{ atoms} \times 1/8) \times (4/3\pi \times r^3)}{a^3} = \frac{(4/3)\pi \times r^3}{8r^3} = \frac{\pi}{6} \approx 0.52 \qquad (1.3)$$

Thus, for a CP-SC structure based on atoms of a single size, regardless of whether the atoms are big or small, the packing fraction is 0.52. This means that 48% of the space in the unit cell is empty. We can introduce other smaller atoms in the voids, or so-called *interstitial sites*, found within the crystal structure (Section 1.8).

b. In the CP-FCC structure, atoms touch along the face diagonals (Figure 1.13), that is,

$$\sqrt{2}a = 4r \qquad (1.4)$$

or

$$a = 2\sqrt{2}r \qquad (1.5)$$

$$\text{FCC packing fraction} = \frac{\text{volume of atoms in the unit cell}}{\text{volume of unit cell}}$$

$$= \frac{[(6 \text{ face atoms} \times 1/2) + (8 \text{ face atoms} \times 1/8)] \times (4/3\pi \times r^3)}{a^3}$$

$$= \frac{4 \times (4/3)\pi \times r^3}{8 \times 2^{3/2} r^3} = \frac{\pi}{3\sqrt{2}} = \frac{\pi}{\sqrt{18}} \approx 0.74$$

This is the packing fraction for an FCC-CP structure. This structure has only 26% empty space. Both the CP-FCC structure and the HCP offer the same and the highest possible volume-packing fraction $(\pi/\sqrt{18})$ for spheres of uniform size.

Note that if the empty spaces or voids in the CP structure are filled with other atoms, the packing fraction will be higher. In fact, many ceramic crystal structures are rationalized using the close packing of bigger *anions* (negatively charged ions) and then stuffing the voids with smaller, positively charged *cations* (see Section 1.11).

c. Atoms in the CP-BCC structure touch along the body diagonal (Figure 1.13). Thus, the relationship between *r* and *a* is

$$\sqrt{3}a = 4r \qquad (1.6)$$

or

$$a = \frac{4r}{\sqrt{3}}$$ (1.7)

Thus, the packing fraction for a CP-BCC structure is

$$\text{BCC packing fraction} = \frac{\text{volume of atoms in the unit cell}}{\text{volume of unit cell}}$$

$$= \frac{\left[(1 \text{ cube center atom}) + (8 \text{ face center atoms} \times 1/8)\right] \times \left(4/3\pi \times r^3\right)}{a^3}$$

$$= \frac{2 \times (4/3)\pi \times r^3}{\left(4r/\sqrt{3}\right)} = \frac{\sqrt{3} \times \pi}{8} \approx 0.68$$

Thus, the BCC structure has a packing fraction of 0.68, which is in between the values for the FCC and SC structures. This structure has 32% void space available.

1.7 DIRECTIONS AND PLANES IN CRYSTAL STRUCTURES

There is a need to specify the crystallographic directions and planes in a unit cell in many applications involving the magnetic, electronic, and optical properties of materials. We use a notation known as *Miller indices* to designate specific directions and planes in cubic unit cells. In describing these, we always use a right-handed coordinate system.

1.7.1 MILLER INDICES FOR DIRECTIONS

To obtain the Miller indices of a direction that starts at point A (tail) and ends at point B (head), we subtract the coordinates of point B from those of point A. We then clear the fractions and reduce the results to the lowest integers. The results are included in square brackets (e.g., [*hkl*]). If there is a negative sign, we insert a bar on top of that index.

1.7.2 MILLER INDICES FOR PLANES

To obtain the Miller indices for planes, we start by identifying the intercepts of a plane on the three axes (i.e., *x*, *y*, and *z*). We then take the reciprocals of the intercepts and clear the fractions. If a plane is parallel to an axis, the intercept for that axis is infinity (∞).

For the Miller indices of planes, we do *not* reduce these numbers to the lowest integers. Similar to the results for directions, we use a bar above any negative index. The Miller indices of planes are written in parentheses ( ). In some cases, the intercepts of planes may not be easy to identify (e.g., a plane that passes through the origin). In this case, we can move the origin of the unit cell. In the cubic system, the direction [*hkl*] is perpendicular to the plane (*hkl*).

1.7.3 MILLER–BRAVAIS INDICES FOR HEXAGONAL SYSTEMS

In a hexagonal unit cell, the *Miller–Bravais indices* for directions and planes are designated as [*hkil*] and (*hkil*), respectively. This represents the three axes in one plane (a_1, a_2, a_3). The angle between these axes is 120°. The *c*-axis is perpendicular to this plane. Because the three axes a_1, a_2, and a_3 are in one plane, the Miller–Bravais indices corresponding to them cannot all be independent. The relationship between them is given by

$$h + k = -i$$ (1.8)

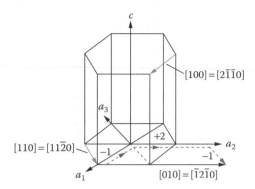

FIGURE 1.14   Typical directions in a hexagonal unit cell. The major axes are a_1, a_2, a_3, and c. (From Askeland, D., and P. Fulay. 2006. *The Science and Engineering of Materials.* Washington, DC: Thomson. With permission.)

Thus, a direction [010] in the three-axis system (i.e., [*hkl*]) will be equivalent to the direction $[\bar{1}2\bar{1}0]$. To get the Miller–Bravais indices for direction [010], we move the origin one unit cell length along a_1. Then, to get the direction, we move one step in the negative a_1 direction, two steps (unit distances) along a_2, and one step along negative a_3. Thus, the condition given by Equation 1.8 is satisfied (Figure 1.14).

We can also verify that the direction [100] is equivalent to $[2\bar{1}\,\bar{1}\,0]$.

Similar considerations apply to the Miller–Bravais indices for planes. We locate the intercepts on the three axes (a_1, a_2, a_3), in addition to that on the c-axis. Then we take the inverses of these values and clear the fractions without reducing the integers. Note that the rule $(h + k) = -i$ holds because the a-axes lie in the plane.

In the HCP structure, $[11\bar{2}0]$-type directions are close-packed. In addition, planes (0001) and (0002) are the closest-packed (i.e., they have the highest number of atoms per unit area).

1.7.4   Interplanar Spacing

In a cubic system with a lattice constant a, the distance d between a set of parallel planes with Miller indices (*hkl*) is given by the following equation:

$$d_{hkl} = \frac{a}{\sqrt{h^2 + k^2 + l^2}} \text{ (for cubic system)} \tag{1.9}$$

Examples 1.2 through 1.4 show how to obtain the Miller indices of directions and a plane, and also how to calculate interplanar distances.

EXAMPLE 1.2: MILLER INDICES FOR DIRECTIONS

What are the Miller indices for any one of the CP directions in FCC and BCC structures?

SOLUTION

In an FCC structure, the atoms touch along the face diagonal. Therefore, the face diagonals are the CP directions.

One face diagonal (marked as OA) is shown in Figure 1.15. We follow the procedure described earlier for obtaining the Miller indices of direction:

1. The coordinates of the "head" point A are 1, 1, 0.
2. The coordinates of the "tail" point B (in this case, the origin point O) are 0, 0, 0.
3. Subtracting the coordinates of the tail from those of the head, that is, 1−0, 1−0, 0−0, we get 1, 1, 0.

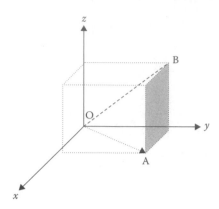

FIGURE 1.15 One of the close-packed directions in an FCC unit cell, shown as (OA). Direction (OB) is a close-packed direction for a BCC structure. Atoms are not shown.

There are no fractions to clear and no negative integers. Therefore, the Miller indices of this direction OA are [110]. Note that the opposite direction, AO, will have Miller indices of $[\bar{1}\,\bar{1}\,0]$.

In an FCC structure, all the face diagonals are close-packed. All directions along the axes are crystallographically equivalent. A family of such equivalent directions is known as the *directions of a form* and is designated as $<\ >$. For an FCC, the directions of a form for CP directions will be <110>. There would be 12 total directions in this family of CP directions for an FCC (six face diagonals and corresponding opposite directions).

You can also see that, similarly, for a BCC structure (atoms not shown in Figure 1.15), the atoms touch along the body diagonal. The Miller indices for one such direction OB (Figure 1.15) are [111]. Note that the direction OB is also a CP direction, for the BCC structure, and its Miller indices are $[\bar{1}\,\bar{1}\,\bar{1}]$.

EXAMPLE 1.3: MILLER INDICES FOR A PLANE

What are the Miller indices for the plane "P" shown in Figure 1.16?

Solution

We see that this plane, shown in Figure 1.16, intersects the x- and y-axes at a length of "1x" lattice parameter.

The plane is parallel to the z-axis, that is, it does not intersect the z-axis at all, and hence this intercept is ∞.

We follow the directions for establishing the Miller indices of a plane, as follows:

1. The intercepts on the x-, y-, and z-axes are 1, 1, and ∞, respectively.
2. The reciprocals of these are 1, 1, and 0. There are no fractions to clear. Therefore, the Miller indices of this plane are (110).
3. Similar to directions of a form, there are also *planes of a form*. These planes are equivalent and are shown in curly brackets { }. The planes of a form for {110} will include the following: (110), (101), (011), $(1\bar{1}0)$, $(10\bar{1})$, and $(01\bar{1})$.

EXAMPLE 1.4: INTERPLANAR SPACING IN MATERIALS

X-rays of a single wavelength (λ) were used to analyze a glittering sample suspected to be Au. The x-ray diffraction (XRD) analysis, in which x-rays are diffracted by different planes, showed that the (400) planes in this sample were separated by a distance of 0.717 Å. Is the sample being analyzed that of Au? Assume that the radius (r) of Au atoms is 144 pm. Assume that in this hypothetical example, no other obvious measurements, such as density, can be made. Au has an FCC structure.

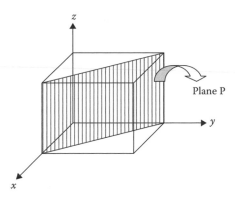

FIGURE 1.16 Plane for Example 1.3.

Solution

We are given d_{400} as 0.717 Å. From Equation 1.9, we get:

$$d_{400} = \frac{a}{\sqrt{4^2 + 0^2 + 0^2}} = 0.717 \text{ Å}$$

Therefore, the lattice constant for this sample is $a = (0.717)(4) = 2.868$ Å.

Now, it is given that the radius (r) of Au atoms is 144 pm = 144 pm × 10^{-12} m/pm × 10^{10} Å/m. This is 1.44 Å.

Because Au has an FCC crystal structure, the lattice constant (a) and the atomic radius (r) are related by Equation 1.4.

From this equation, we get $a = 2\sqrt{2} \times (1.44 \text{ Å}) = 4.07$ Å. This is the lattice constant of Au.

The sample we had has a lattice constant of 2.868 Å. Thus, the sample is not Au!

By comparing the lattice constant of 2.868 Å to that of other elements, we will see that it is close to the lattice constant of BCC Fe. Thus, the sample we have is likely BCC Fe. Because this sample glittered like Au, it may have been plated with a thin layer of Au. Thus, the lesson to be learned here is that all that glitters is not gold! In this example, we also learned that the unknown sample has a cubic structure.

1.8 INTERSTITIAL SITES OR HOLES IN CRYSTAL STRUCTURES

As mentioned in the previous section, even the CP structures have voids, known as interstitial sites. Smaller atoms or ions can enter these interstitial sites or holes in a crystal structure. Different types of interstitial sites are shown in Figure 1.17.

1.9 COORDINATION NUMBERS

The total number of nearest-neighbor atoms that surround an atom (in a lattice position or in an interstitial site) is often referred to as the *coordination number* (CN). For example, in the FCC and HCP structures, the CN is 12. This means that for any atom in a specific position of the lattice, there are 12 nearest neighbors surrounding it. In the SC structure, the CN is 6. For a BCC structure, the CN is 8. This can be easily seen by looking at the atom located at the cube center. It is surrounded by eight corner atoms. We must also realize that in a periodic structure, each atom has identical surroundings, and thus it does not matter whether we examined the atom at the cube center or at any other position in the structure.

The CNs for different structures, the corresponding relationship between the unit cell parameter (a) and the atomic radius (r), and the volume-packing fractions are shown in Table 1.1.

Coordination Number	Location of Interstitial	Radius Ratio	Representation
2	Linear	0–0.155	
3	Center of triangle	0.155–0.225	
4	Center of tetrahedron	0.225–0.414	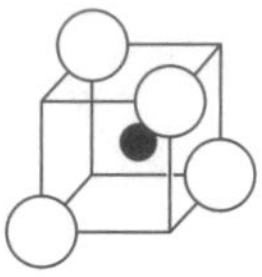
6	Center of octahedron	0.414–0.732	
8	Center of cube	0.732–1.000	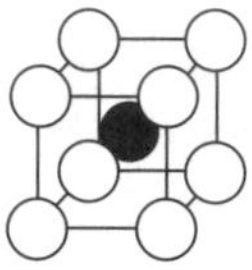

FIGURE 1.17   Different types of interstitial sites. (From Askeland, D., and P. Fulay. 2006. *The Science and Engineering of Materials*. Washington, DC: Thomson. With permission.)

TABLE 1.1
Relationships between the Unit Cell Parameter (*a*) and Radius (*r*), the Coordination Number, and the Volume-Packing Fraction for Different Unit Cells

Structure	Relationship of *a* and *r*	Atoms per Unit Cell	Coordination Number	Packing Fraction
Simple cubic	$a = 2r$	1	6	0.52
Body-centered cubic	$\sqrt{3}a = 4r$	2	8	0.68
Face-centered cubic	$\sqrt{2}a = 4r$	4	12	0.74
Hexagonal close-packed	$a = 2r$ $c \approx 1.633a$	2	12	0.74

The most common types of interstitial sites are the so-called tetrahedral and octahedral sites. A *tetrahedral site* means that an atom or ion in that site is surrounded by four other atoms or ions (CN = 4) that are located at the corners of a tetrahedron, hence the name tetrahedral site. Similarly, for an *octahedral site*, an atom or ion in this site is surrounded by six (*not* eight) atoms (CN = 6). These atoms or ions form an octahedron around the site center, hence the name octahedral site. For example, in an FCC unit cell, the center of the cube is an octahedral site. The 12 edge centers in an FCC structure are also octahedral sites. These sites are shared with a total of four neighboring unit cells. Thus, in an FCC unit cell, there are a total of $[(12 \times 1/4) + 1)] = 4$ octahedral sites.

1.10   RADIUS RATIO CONCEPT

The concept of radius ratio is useful to rationalize or to be able to guess whether a guest atom or ion is likely to enter a particular type of site. The radius ratio is the ratio of the radius of the cation to that of the anion. The radius ratio ranges and corresponding sites are shown in Figure 1.17.

The general idea behind the radius ratio concept is that we can view crystal structures of many ceramic materials as being made up of a close packing of anions. These are the negatively charged ions and are typically larger because of the extra electrons. This is conceptually followed by the "stuffing" of cations into the interstitial sites.

Note that not all the available interstitial sites in a unit cell need to be occupied. The sites to be occupied (e.g., tetrahedral or octahedral) depend upon the radius ratio, that is, in this case, r_{cation}/r_{anion}. The fraction of the sites that are occupied depends upon the stoichiometry of the compound. This is illustrated in the following discussion.

1.11 CRYSTAL STRUCTURES OF DIFFERENT MATERIALS

We will now discuss some of the more simple crystal structures exhibited by electronic, magnetic, and optical materials that are of interest to us. Many materials have far more complex crystal structures, and a discussion of these is beyond the scope of this book.

1.11.1 STRUCTURE OF SODIUM CHLORIDE

The structure of a sodium chloride (NaCl) crystal (Figure 1.18), which is shared by many ceramic materials, such as magnesium oxide (MgO), can be rationalized as follows. The structure is obtained by closely packing chlorine anions (Cl^{1-}) that have a radius of 0.181 nm. The radius of sodium cations (Na^{1+}) is 0.097 nm. This results in a radius ratio of r_{Na}^{+}/r_{Cl}^{-} of ~0.536. Therefore, we expect that sodium ions will exhibit an octahedral coordination (Figure 1.17).

Thus, we can assume this structure is obtained by creating an FCC arrangement of chlorine ions. All the octahedral sites (i.e., cube edge centers and cube center) can then be stuffed with sodium ions.

1.11.2 STRUCTURE OF CESIUM CHLORIDE

The atomic radius of a cesium ion (Cs^{+}) is 0.167 nm. The radius ratio of 0.92 for cesium chloride (CsCl) suggests an eightfold coordination. This is achieved by locating the Cs^{+} ion at the cube center. In this structure, we have eight chlorine ions at the eight corners of the cube (Figure 1.19). Each of these is shared with eight other neighboring unit cells. The number of chlorine ions per unit cell is $(8 \times 1/8) = 1$. Thus, the stoichiometry of one chlorine ion for every cesium ion is maintained, and electrical neutrality is assured.

As shown in Figure 1.19, the structure can also be represented as the chlorine ion at the center and an FCC packing of cesium ions.

Example 1.5 illustrates the use of the radius–ratio concept.

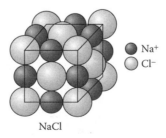

NaCl

FIGURE 1.18 Crystal structure of sodium chloride (NaCl). (From Askeland, D., and P. Fulay. 2006. *The Science and Engineering of Materials*. Washington, DC: Thomson. With permission.)

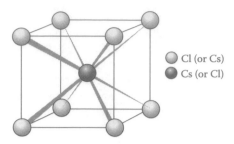

FIGURE 1.19 Cesium chloride (CsCl) structure, showing the eightfold coordination of cesium ions. (From Smart, L., and E. Moore. 1992. *Solid State Chemistry: An Introduction.* Boca Raton, FL: Chapman and Hall. With permission.)

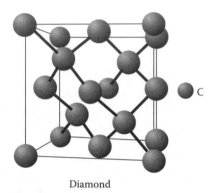

Diamond

FIGURE 1.20 Diamond cubic crystal structure. (From Askeland, D., and P. Fulay. 2006. *The Science and Engineering of Materials.* Washington, DC: Thomson. With permission.)

EXAMPLE 1.5: APPLICATION OF THE RADIUS–RATIO CONCEPT

NaCl and potassium chloride (KCl) have the same stoichiometry. The radius of the potassium ion (K^+) is 0.133 nm. The radius of the chlorine ion (Cl^+) is 0.181 nm. What will be the expected CN for K^+ ions? What will be the expected structure of KCl? Is this structure consistent with the stoichiometry?

Solution

The r_{cation}/r_{anion} for KCl is 0.133/0.181 = 0.735. This suggests that the CN for K^+ ions will be 8. This CN is possible if the K^+ ions assume the location at the cube center. Then, they are next to eight other Cl^- ions. Thus, KCl exhibits a structure similar to that of cesium chloride (CsCl) (Figure 1.19). The structure is consistent with the 1:1 stoichiometry of KCl.

1.11.3 Diamond Cubic Structure

One of the most important crystal structures of many semiconductor materials is the diamond cubic (DC) crystal structure (Figure 1.20). This is the crystal structure exhibited by semiconductors, such as silicon, germanium, and carbon in the form of diamond.

In this crystal structure, atoms are first arranged in an FCC arrangement. Note that in Figure 1.20, for the sake of clarity, the atoms are not shown touching one another in the FCC and tetrahedral arrangements. Next, the cubic unit cell is divided into eight smaller cubes known as *octants*. Then, two smaller nonadjacent octants on the top (centers of these cubes will be at a point three-fourths the unit-cell height) are selected. Similarly, two smaller nonadjacent octants at the bottom are selected (centers of these cubes will be at one-fourth of the height of unit cell). Additional atoms are placed into four of the octants. Thus, in this crystal structure, atoms exhibit what is called a "tetrahedral coordination." It is easier to visualize this tetrahedral arrangement by examining the atoms located in the centers of the octants. However, every atom in this structure ultimately has the same tetrahedral coordination.

A DC unit cell has a total of eight atoms in the unit cell. Four atoms are from the FCC arrangement [8 × (1/8) + 6 × (1/2)]. Four more atoms are derived from the inside of the unit cell. Such a description allows us to calculate the theoretical density of materials, such as silicon and germanium. This is illustrated in Example 1.6.

EXAMPLE 1.6: THEORETICAL DENSITY OF SILICON (Si)

The radius of Si in a covalent structure is 1.176 Å. What is the lattice constant of Si? What is the theoretical density of Si if its atomic mass is 28.1?

SOLUTION

If we examine the body diagonal in the DC structure exhibited by Si, we see that there are two atoms at the corner and an atom within an octant. There is space for two more atoms to be accommodated along the diagonal (Figure 1.21).

If the length of the unit cell is a, then the body diagonal is $\sqrt{3}a$. If r is the radius of the atoms in the DC structure, then from Figure 1.21, we get:

$$\sqrt{3} \times a = 8r \tag{1.10}$$

For Si:

$$a = \frac{8 \times 1.176}{\sqrt{3}} = 5.4317 \text{ Å}$$

In addition, as described before, there is an equivalent of eight atoms of Si per unit cell. Recall that one mole of an element has an Avogadro number of atoms (6.023×10^{23} atoms). Thus, in this case, 28.1 g of Si would have 6.023×10^{23} atoms. Thus, the theoretical density of Si will be

$$\text{Density} = \frac{\text{mass of 8 silicon atoms}}{\text{volume of the unit cell}} = \frac{8 \times 28.1}{6.023 \times 10^{23} \times (5.4317 \times 10^{-8} \text{ cm})^3} = 2.33 \text{ g/cm}^3$$

This value matches well with the experimentally observed values.

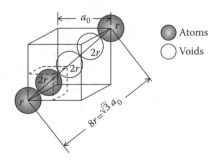

FIGURE 1.21 Schematic representation showing the arrangement of atoms and voids along a body diagonal in the diamond cubic crystal structure. (From Askeland, D., and P. Fulay. 2006. *The Science and Engineering of Materials*. Washington, DC: Thomson. With permission.)

1.11.4 Zinc Blende Structure

Zinc sulfide (ZnS) exhibits different polymorphic forms known as zinc blende and wurtzite. The structure of zinc blende is cubic (Figure 1.22), whereas that of wurtzite is hexagonal (Figure 1.23). Many compound semiconductors, such as gallium arsenide (GaAs), show the zinc blende structure.

The zinc blende structure (Figure 1.22) is similar to the DC structure (Figure 1.20). The radius of divalent zinc ions (Zn^{2+}) is 0.074 nm. The radius of the larger sulfur anion (S^{2-}) is 0.184 nm. The radius ratio of 0.402 suggests a tetrahedral coordination for Zn^{2+} ions (Figure 1.17).

We start with an FCC arrangement of S^{2-} ions. The Zn^{2+} ions enter at the four tetrahedral sites, that is, they occupy the centers of the octants inside the main unit cell (similar to the diamond-cubic unit cell).

Note that *not* all tetrahedral sites are occupied, because of the requirement of a balance of stoichiometry and electrical neutrality. In addition, each S^{2-} ion is coordinated with four Zn^{2+} ions. Note that if all ions in the zinc blende structure (Figure 1.22) were identical, we would get a DC structure (Figure 1.20).

GaAs, an important semiconductor, exhibits the zinc blende structure. The structure of GaAs can be understood by replacing the sulfur atoms in ZnS with arsenic atoms and the zinc atoms with gallium atoms. One of the polymorphs of silicon carbide (SiC) and other materials, such as indium phosphide (InP), indium antimonide (InSb), and gallium phosphide (GaP), also exhibit this type of crystal structure. Example 1.7 explores the zinc blende structure in more detail.

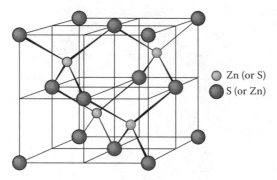

FIGURE 1.22 Schematic representation of the zinc blende structure. (From Smart, L., and E. Moore. 1992. *Solid State Chemistry: An Introduction*. Boca Raton, FL: Chapman and Hall. With permission.)

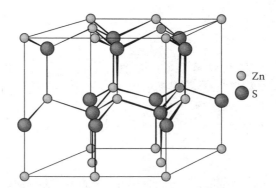

FIGURE 1.23 Structure of the wurtzite crystal. (From Smart, L., and E. Moore. 1992. *Solid State Chemistry: An Introduction*. Boca Raton, FL: Chapman and Hall. With permission.)

EXAMPLE 1.7: DENSITY OF INDIUM PHOSPHIDE (InP)

If the lattice constant (a) of InP is about 5.8687 Å, what is its theoretical density? The atomic masses of indium (In) and phosphorus (P) are 114.81 and 31, respectively.

SOLUTION

Recognize that there are four In atoms and four P atoms inside the InP unit cell (Figure 1.22). Also recall that one mole of an element has an Avogadro number, that is, 6.023×10^{23} atoms.

Thus, the theoretical density of InP will be the mass of four In atoms + mass of four P atoms divided by the volume of the cubic unit cell.

$$\text{Density of indium phosphide} = \frac{\text{mass of 4 In atoms} + \text{mass of 4 P atoms}}{\text{volume of the unit cell}}$$

$$= \frac{[4 \times (114.81) + 4 \times (31)]}{6.023 \times 10^{23} \times (5.8687 \times 10^{-8} \text{ cm})^3}$$

$$= 4.79 \text{ g/cm}^3$$

What we calculated is the so-called theoretical density. The actual density is comparable but can be a little different because the arrangement of atoms in real materials is never perfect (see Section 1.12).

1.11.5 WURTZITE STRUCTURE

According to the radius ratio, the Zn^{2+} ions have a CN of 4; that is, they are surrounded by four sulfur ions. Similarly, each sulfur ion (S^{2-}) is coordinated by four zinc ions (Zn^{2+}). The wurtzite structure is based on an HCP array of sulfide ions. Tetrahedral holes in alternate octants are occupied by the zinc ions. Many semiconductors and dielectrics, such as gallium nitride (GaN), zinc oxide (ZnO), aluminum nitride (AlN), cadmium telluride (CdTe), and cadmium sulfide (CdS), show this type of polymorph.

This structure can also be visualized by considering hexagonal close packing of zinc ions, followed by the stuffing of sulfide anions in the tetrahedral sites (Figure 1.23).

1.11.6 FLUORITE AND ANTIFLUORITE STRUCTURE

It is sometimes easier to conceptually consider that in crystal structures, the cations (i.e., smaller ions) form a CP array. For the fluorite (CaF_2) structure, we start with a CP array (FCC packing) of cations. Then, all the tetrahedral holes are filled with larger fluorine anions (F^{1-}). In this representation, the fourfold coordination of anions is clearly seen (Figure 1.24a).

To better understand this, we can extend the structure so that the cubes have fluoride ions at the corners (Figure 1.24b). In this representation, we can easily recognize the eightfold coordination of the cations. The same structure is redrawn in Figure 1.24c by shifting the origin, and in this figure, eight octants can be seen. Every other octant is occupied by a Ca^{2+} ion. The relative distances between the ion centers are shown in one of the octants in Figure 1.24d.

Ceramic materials that show a fluorite structure include CeO_2, PbO_2, UO_2, and ThO_2.

In the antifluorite structure, the positions of cations and anions are reversed (Figure 1.24c and d). Materials that show antifluorite structure include Li_2O, Na_2O, Rb_2O, K_2O, and Li_2S.

1.11.7 CORUNDUM STRUCTURE

Corundum (α-Al_2O_3) structure can be described by picturing a hexagonal close packing of oxygen anions (O^{2-}). Two-thirds of the octahedral sites are filled by the aluminum cations (Al^{3+}; Figure 1.25).

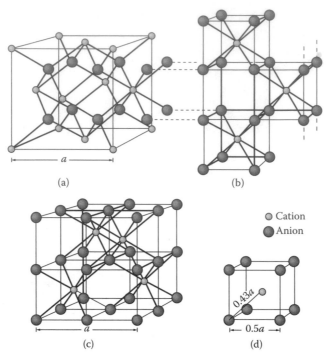

FIGURE 1.24   Structures of fluorite (a and b) and antifluorite (c and d). (From Smart, L., and E. Moore. 1992. *Solid State Chemistry: An Introduction.* Boca Raton, FL: Chapman and Hall. With permission.)

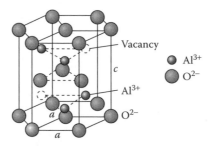

FIGURE 1.25   Crystal structure of alpha-alumina (α-Al_2O_3 or corundum). (From Askeland, D., and P. Fulay. 2006. *The science and engineering of materials.* Washington, DC: Thomson. With permission.)

1.11.8   PEROVSKITE CRYSTAL STRUCTURE

Perovskite crystal structure is one of the mixed oxide structures, that is, a structure of an oxide containing many cations. Perovskite is a calcium titanate ($CaTiO_3$) mineral that exhibits this crystal structure, which is also described as ABO_3. In this notation, the A-site cations are divalent (e.g., Ba^{2+}, Pb^{2+}, Sr^{2+}) and occupy the cube corners, as shown in Figure 1.3. The B-site cations (e.g., Ti^{4+}, Zr^{4+}) occupy the cube centers (octahedral site). The oxygen ions (O^{2-}) occupy the face center positions on the cube.

Many ceramics, with useful ferroelectric, piezoelectric, and other properties, exhibit the perovskite crystal structure. Examples include $BaTiO_3$, some compositions of ceramics in the system lead zirconium titanate (PZT), and strontium titanate ($SrTiO_3$).

The cubic form of $BaTiO_3$ is similar to a tetragonal structure and differs very slightly (~ less than 1%) in terms of its dimensions. The c-axis of tetragonal $BaTiO_3$ is ~4.01 Å. The lattice constant (cell cube length) of cubic $BaTiO_3$ is only 4.00 Å (Figure 1.3). This very slight difference of ~0.01 Å in the unit cell dimension produces very significant changes in the dielectric properties of $BaTiO_3$. For example, the cubic form of $BaTiO_3$ is not ferroelectric, whereas the tetragonal form is ferroelectric.

1.11.9 SPINEL AND INVERSE SPINEL STRUCTURES

The spinel structure is another example of a mixed oxide structure. Many magnetic materials, known as ferrites, exhibit normal or inverse spinel structures. The general formula is AB_2O_4, where A is the divalent cation (e.g., Mg^{2+}) and B is the trivalent cation (e.g., Al^{3+}). Each edge center and cube center are octahedral sites. Because the edges are shared among four unit cells, there are a total of $(12 \times 1/4) + 1 = 4$ octahedral sites. Similarly, as seen in diamond crystal structure (Figure 1.20), there are a total of eight tetrahedral sites (center of each octant) in the FCC structure.

To understand the spinel crystal structure, we will consider n formula units. In the n units, there are $8n$ tetrahedral holes and $4n$ octahedral holes. In the normal spinel structure, only one-eighth of the tetrahedral sites are filled. This is because we have n A-type atoms and $8n$ sites are available. Similarly, there are $2n$ B-type atoms and $4n$ octahedral sites. Thus, one-half of the octahedral sites are occupied.

As shown in Figure 1.26, we can break down the spinel unit cell into eight octants.

In one type of octant, called A-type octants, the A ions occupy the tetrahedral sites. These atoms are coordinated with the corner and face center anions (oxygen ions) of the unit cell. In another type of octant, known as B-type octants, the trivalent B cations are located at half of the corners of an octant. Note that out of a total of eight octants, four are A-type and four are B-type. Of the four A-type octants, only two actually contain A atoms. Similarly, of the four B-type octants, only two contain B atoms (Figure 1.26). Examples of ceramics with this crystal structure include magnesium aluminate ($MgAl_2O_4$) and zinc aluminate ($ZnAl_2O_4$).

In some materials such as Fe_3O_4 (the formula can also be written as $FeFe_2O_4$ to emphasize the divalent and trivalent forms of iron), an inverse spinel structure is observed. In this well-known magnetic material, Fe exists in both divalent (Fe^{2+}) and trivalent (Fe^{3+}) forms. The inverse spinel structure is written as $B(AB)O_4$, suggesting that half of the trivalent B-site cations occupy *tetrahedral* sites. All the A-type atoms occupy *octahedral* sites, and therefore, the structure is known as an inverse spinel structure. The other half of the B cations continue to occupy the octahedral sites. The type of ions occupying the tetrahedral and octahedral positions has a large effect on the magnetic coupling between them. This, in turn, has a significant effect on the magnetic properties of materials

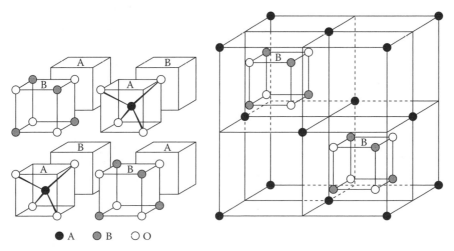

● A ◉ B ○ O

FIGURE 1.26 The spinel crystal structure. (From Smart, L., and E. Moore. 1992. *Solid State Chemistry: An Introduction*. Boca Raton, FL: Chapman and Hall. With permission.)

known as ceramic ferrites. Nickel ferrite ($NiFe_2O_4$) is another example of a material that shows inverse spinel structure.

1.12 DEFECTS IN MATERIALS

Arrangements of atoms or ions in real materials are never perfect. In some cases, atoms are missing from sites at which they are supposed to be present. This creates a defect known as a *vacancy*. The presence of vacancies can be useful in many applications of electronic ceramics because vacancies enhance the bulk or volume diffusion of specific types of ions. *Diffusion* is a process by which atoms, ions, or other species move owing to a gradient in the chemical potential (equivalent of concentration). The process of diffusion can occur through many pathways (e.g., within the grain, i.e., bulk or along the grain boundaries, surfaces, etc.). Diffusion also plays a key role in enabling semiconductor device processing as well as other materials—for example, processing steps such as sintering of metals and ceramics.

We sometimes deliberately add different atoms to a material. For example, we add boron (B) or antimony (Sb) atoms to silicon to change and better control the electrical properties of the silicon. The atoms that we add (or that are sometimes introduced inadvertently during processing) may take up the positions of the host atoms. This type of defect is known as a *substitutional atom* defect. This typically occurs if the radii of both the host and guest atoms are similar. In some other cases, the atoms we add or those introduced inadvertently may end up in the interstitial sites. This is known as an *interstitial atom* defect. These different types of point defects are shown in Figure 1.27.

When atoms of one element "dissolve" in another element as substitutional or interstitial atoms, we get what is called as a *solid solution*. This is similar to how sugar dissolves in water. Formation of solid solutions strengthens metallic materials. In Figure 1.4, α and β phases represent the solid solutions of the lead–tin system.

Solid solution formation also occurs between compounds of similar crystal structures (e.g., $BaTiO_3$ and $SrTiO_3$). The formation of solid solutions is used to "tune" the electrical and magnetic properties of ceramics, metals, and semiconductors. For example, by forming solid solutions of GaAs and aluminum arsenide (AlAs), we can produce LEDs that emit different colors.

One of the best examples of the usefulness of point defects and formation of solid solutions is the doping of Si to make an *n-type semiconductor* or a *p-type semiconductor*. As illustrated below (Figure 1.28), when a pentavalent element, such as antimony (Sb) or phosphorus, is added to Si,

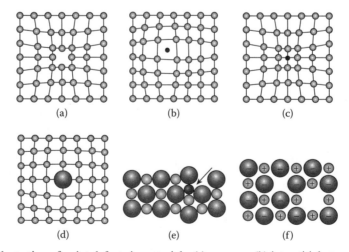

FIGURE 1.27 Illustration of point defects in materials: (a) vacancy, (b) interstitial atom, (c) small substitutional atom, (d) large substitutional atom, (e) Frenkel defect, and (f) Schottky defect. All of these defects disrupt the perfect arrangement of the surrounding atoms. (From Askeland, D., and P. Fulay. 2006. *The Science and Engineering of Materials*. Washington, DC: Thomson. With permission.)

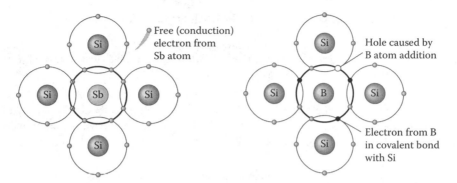

FIGURE 1.28 Creation of Sb-doped n-type and B-doped p-type silicon by the introduction of substitutional atoms. (From Askeland, D., and P. Fulay. 2006. *The Science and Engineering of Materials*. Washington, DC: Thomson. With permission.)

we create an n-type semiconductor. This is because substitutional phosphorus or antimony atoms occupy the silicon sites and thus provide an "extra" electron. At temperatures greater than ~50 K, this electron "breaks free" from the phosphorus or antimony atoms and becomes available for conduction. This leads to an n-type semiconductor (thus named because most of the charge carriers are *negatively* charged electrons).

Similarly, when atoms of trivalent elements, such as boron or aluminum, are added to silicon, there is a deficit of an electron. This is because each silicon atom contributes four electrons for covalent bonding. However, an aluminum or boron atom can contribute only three electrons from its outermost shell. This leads to the creation of a hole, which is basically a missing electron. Such semiconductors are known as p-type (because of the *positive* effective charge on a hole).

We will learn in Chapter 3 that the level of conductivity of a semiconductor can be changed by controlling the concentration of *dopant* atoms. Atoms of elements that are added purposefully and in controlled concentrations, with the assumption that they will have a useful effect, are known as dopants. Different types of semiconductors (i.e., n-type and p-type) form the fundamental basis of many electronic devices, such as transistors, diodes, and solar cells.

Note that changes in the conductivity of a semiconductor will occur regardless of whether the different atoms are added on purpose or are introduced inadvertently. Atoms of elements that find their way (usually inadvertently) into a material of interest during the synthesis or fabrication of that material are considered *impurities*. For example, when silicon crystals are grown, oxygen atoms are introduced as impurities that come from the contact of molten silicon with the quartz (SiO_2) crucibles. Typically, we want to minimize the levels of impurities in any material. This is especially the case for semiconductors, in which we want to keep the impurity levels to a minimum (parts per million [ppm] to parts per billion [ppb] for some elements).

1.13 POINT DEFECTS IN CERAMIC MATERIALS

Point defects occur in ceramic materials as well. Many ceramic materials are based on ions. It is not possible simply to remove a certain number of cations or anions from these materials without causing an imbalance in the net electrical charge. Overall, the electrical neutrality of a material has to be maintained. A *Schottky defect* is a defect in which a certain number of cations and a stoichiometrically equivalent number of anions are missing (Figure 1.27). For example, in NaCl, a Schottky defect is one in which one sodium ion (Na^+) and one chlorine ion (Cl^{1-}) are missing. For a Schottky defect in Al_2O_3, two Al^{3+} and three O^{2-} will be missing. A *Frenkel defect* is a type of point defect in which an ion (often a cation) leaves its original site and enters an interstitial site (Figure 1.27). In Section 1.14, we discuss the notation used for point defects in ceramic materials.

1.14   KRÖGER–VINK NOTATION FOR POINT DEFECTS

Defects in ceramic materials are important for many applications involving the electronic, optical, and magnetic properties of ceramics. The following rules must be observed in regard to the presence of point defects in ceramic materials:

1. *Electrical neutrality*—The material, as a whole, must be electrically neutral.
2. *Mass balance*—While introducing any defects, the mass balance must be maintained.
3. *Site balance*—In introducing defects, the overall stoichiometry of sites must be maintained.

The notation used to describe point defects in ceramics and the equations that govern their relative concentrations is called *Kröger–Vink notation*.

For example, consider an oxygen vacancy in MgO. In Kröger–Vink notation, the vacancy of oxygen is shown as $V_O^{\bullet\bullet}$. The symbol V represents a vacancy. The subscript (in this case, O) indicates the location of the defect (i.e., where the defect occurs). When an oxygen ion (O^{2-}) is missing, a negative charge of two is *missing*. This means that this defect, that is, the vacancy of oxygen, has an *effective* positive charge of two. An effective positive charge of one unit is indicated by the dot symbol ($\cdot$) that is placed as a superscript. Because we have an effective charge of +2, we use two dots in the superscript. Therefore, we describe the presence of an oxygen-ion vacancy as $V_O^{\bullet\bullet}$. Similarly, a Mg^{2+} ion vacancy will be written as V_{Mg}''. In this case, the defect will have an effective negative charge, which is shown using two dashes in the superscript.

Example 1.8 illustrates the use of Kröger–Vink notation.

EXAMPLE 1.8: YTTRIUM OXIDE (Y_2O_3)-ZIRCONIA (ZrO_2) FOR OXYGEN SENSORS AND SOLID OXIDE FUEL CELLS

a. Write down the equation that expresses the incorporation of Y_2O_3 in ZrO_2 to form a solid solution. Assume that the concentration of Y_2O_3 is small enough so that new compounds are formed.
b. What other defects are created by adding Y_2O_3 to ZrO_2? Are these defects useful?

SOLUTION

a. We assume that yttrium ions (Y^{3+}) occupy the sites of zirconium ion (Zr^{4+}). The basis for this assumption is that both are cations. We also consider the relative radii of the ions. Yttrium (Y) is a trivalent ion; when it occupies a Zr^{4+} site, there will be a deficit of one positive effective charge. Thus, this defect will have an effective charge of minus 1.

This defect is written as:

$$Y_{Zr}' \tag{1.11}$$

In ZrO_2, for each Zr^{4+} atom, there are two oxygen atoms. Therefore, to add two Y ions on two Zr^{4+} sites, we must use four oxygen sites. This is for site balance. Three oxygen ions from Y_2O_3 will enter three oxygen sites in ZrO_2.

We must use one more oxygen site in ZrO_2. However, we have no oxygen atoms left. Therefore, one oxygen site in ZrO_2 must remain vacant.

Therefore, the defect reaction for incorporation of Y_2O_3 in ZrO_2 is written as follows:

$$Y_2O_3 \xrightarrow{\;ZrO_2\;} 2Y_{Zr}' + 3O_O^x + V_O^{\bullet\bullet} \tag{1.12}$$

In this equation, the ZrO_2 above the arrow shows that Y_2O_3 (a solute) is being added to ZrO_2 (a solvent).

We check Equation 1.12 for mass balance, site balance, and electrical neutrality. We can see that for the site balance, we have used two zirconium sites and four oxygen sites (three have oxygen ions derived from yttria, and one site has an oxygen vacancy, $V_O^{\bullet\bullet}$). We also have charge balance—the defect caused by the presence of the Y ion in the Zr^{4+} site has an effective negative charge of 1, and we have two of these. This is balanced by one oxygen ion vacancy ($V_O^{\bullet\bullet}$), which has an effective charge of plus 2. We also have mass balance.

b. The point defect we did not anticipate is the oxygen ion vacancy ($V_O^{\bullet\bullet}$). For each mole of Y_2O_3 added, we get one mole of oxygen vacancies (Equation 1.12). This provides a significant amount of "room" for the oxygen ions to move around or diffuse. This is why zirconium oxide (ZrO_2) containing small amounts of Y_2O_3 (~8–10 mol %) is an oxygen ion conductor. The rate at which oxygen ions can diffuse is considerably high at higher temperatures (~1000°C).

This material is known as yttria-stabilized zirconia (YSZ). It functions as a solid electrolyte. Thus, by adding Y_2O_3 to ZrO_2 we have converted a dielectric (nonconducting) material into an ionic conductor! Applications of this material include solid oxide fuel cells (SOFCs; Figure 1.29) and oxygen sensors that are used in cars and trucks (Figure 1.30).

Another effect of adding Y_2O_3 is that when the concentration of oxygen vacancies increases, a cubic polymorph of ZrO_2 becomes stabilized at room temperature. This is why this material is known as YSZ. In some applications, the primary interest in using YSZ is for its mechanical properties and high-temperature stability. When the cubic phase is stabilized

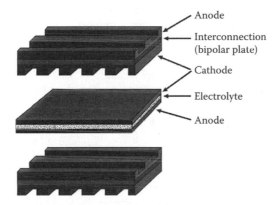

FIGURE 1.29 Schematic representation of a planar-type solid oxide fuel cell. The electrolyte is typically yttria-stabilized zirconia (YSZ). Calcium-doped $LaMnO_3$ is used as the cathode, and YSZ-containing nickel (Ni) is used as the anode. The interconnecting material is lanthanum chromite ($LaCrO_3$). (From Singhal, S. C. 2002. *Solid State Ionics* 152–153:405–10. With permission.)

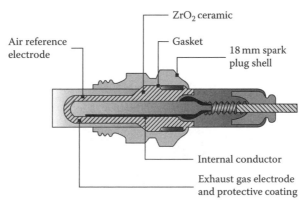

FIGURE 1.30 Schematic representation of a zirconia oxygen sensor used in cars and trucks. (Courtesy of Dynamic-Ceramic Ltd.)

by adding Y_2O_3, ZrO_2 does not show phase changes from cubic to tetragonal to monoclinic forms. This avoids the strain induced by these transformations, which would otherwise cause zirconia ceramics to break during fabrication or thermal cycling (heating and cooling). In applications such as SOFCs or oxygen gas sensors, what matters most is the ionic conductivity induced by the introduction of oxygen vacancies.

1.15 DISLOCATIONS

A *dislocation* is a line defect that represents half a plane of atoms missing from an otherwise perfect crystal structure. The two types of dislocations—an edge dislocation and a screw dislocation—are shown in Figure 1.31.

Dislocations can have a significant and deleterious effect on the properties of semiconductors and optoelectronic materials. Dislocations in semiconductors are typically formed during crystal growth or during semiconductor device processing. Many years of research have gone into ensuring that essentially dislocation-free silicon and other crystals can be grown.

In most situations, the presence of dislocations in semiconductors is considered deleterious. For example, in gallium nitride (GaN), which is a relatively new semiconductor, the presence of dislocations has a negative effect on the optoelectronic devices.

The development of LEDs and laser diodes that emit a blue or violet light was hindered by the unavailability of GaN materials with very low levels of dislocations. In recent years, superior materials-processing methods have been developed, which has in turn led to the development of GaN-based blue or violet lasers.

These blue or violet laser-emitting devices have enabled the so-called Blu-ray format for high-definition (HD) optical-data storage. The Blu-ray format provides more capacity than digital video disks (DVDs). It uses a shorter wavelength (λ) of 405 nm for optically writing the information onto a disk. This wavelength is shorter than that of the typical red lasers ($\lambda = 660$ nm) that are used for authoring DVDs. For compact discs (CDs), the wavelength of the laser used is longer—780 nm. The use of the shorter-wavelength GaN lasers means that a single-layer optical disk can hold 25 GB of data (about 9 hours of HD video content).

It is very difficult to grow defect-free GaN single crystals. Therefore, we typically grow GaN on another substrate, such as sapphire. This process is also used for the fabrication of many other semiconductor devices and is known as *epitaxy*. In this process, a strain is introduced at the interface if the lattice constants of the film and substrate do not match well.

In the case of GaN on sapphire (Al_2O_3), the mismatch is about 16% for sapphire. At the interface, the strain that exists due to lattice mismatch is often relieved by the formation of what are known as "misfit dislocations" (Figure 1.32). In GaN, dislocations known as "threading dislocations" originate at the interface and travel all the way to the surface. Dislocations in GaN, as examined by transmission electron microscopy (TEM) and high-resolution TEM (HRTEM), are shown in Figure 1.33.

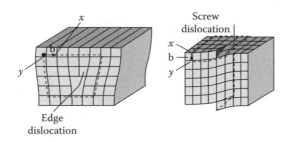

FIGURE 1.31   Illustration of an edge and a screw dislocation. (From Askeland, D., and P. Fulay. 2006. *The Science and Engineering of Materials*. Washington, DC: Thomson. With permission.)

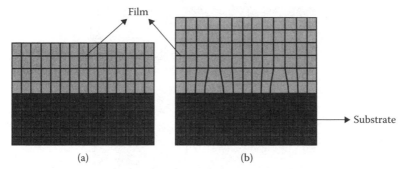

FIGURE 1.32 Illustration of a semiconductor film (a) that is coherently strained and (b) that has the lattice mismatch accommodated by misfit dislocations. (From Stach, E. A., and R. Hull. 2008. Dislocations in semiconductors. In *Encyclopedia of Materials: Science and Technology*, eds. K. H. J. Buschow et al., 2301–12. Oxford: Elsevier. With permission.)

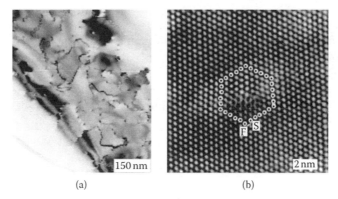

FIGURE 1.33 (a) Transmission electron microscope image of dislocations in gallium nitride deposited on sapphire (Al_2O_3). (b) The core of a dislocation as seen using an HRTEM. (Courtesy of Dr. J. Narayan, North Carolina State University.)

In some cases, the presence of dislocations away from the active regions of electrical devices can be useful. Dislocations can attract some of the impurity atoms toward them, leaving the regions in which devices are active with fewer defects. This method of segregating impurities is known as *gettering*. Impurities in a semiconductor, such as transition metal impurities including iron and nickel, are concentrated by attracting them to defects such as precipitates and dislocations.

1.16 STACKING FAULTS AND GRAIN BOUNDARIES

Similar to point defects (that are considered zero-dimensional defects) and dislocations (one-dimensional line defects), we also find area defects. One type of area defect is a *stacking fault*. We have seen that the FCC structure can be visualized by considering the stacking of atoms within the planes as ABCABCABC… (Figure 1.12). In a stacking fault, one of the planes in the expected sequence is missing. Therefore, the stacking sequence may look like ABCABABCABC…. Thus, in this stacking fault, a small region of the material shows the HCP-stacking sequence.

Other examples of area defects are domain boundaries, low-angle boundaries, and twin boundaries. A *domain* is a small region of a material in which the dielectric or magnetic polarization direction is the same. We will learn that in magnetic and ferroelectric materials, the entire material cannot be stable energetically as one large magnet or an electric dipole, respectively. To minimize the overall free energy, the material spontaneously shows the formation of multiple domains arranged in a random fashion. The formations and arrangements of domains are extremely important in

FIGURE 1.34 Microstructure of a BaTiO$_3$ ceramic. (Courtesy of Dr. Rodney Roseman, University of Cincinnati, Ohio.)

determining the properties of magnetic and ferroelectric materials. A scanning electron micrograph of the domains in ferroelectric PZT ceramic is shown in Figure 1.33.

One of the most important features of the microstructure of a polycrystalline material is the presence of grain boundaries (Figures 1.6 and 1.34). At the grain boundaries, the atomic order is disrupted. This is why grain boundaries are considered defects. Because grains are three-dimensional, the grain boundary is actually a surface in three dimensions; therefore, grain boundaries are considered three-dimensional defects.

Grain boundary is one of the most important features in terms of its effects on the properties of materials. These effects are discussed in the next section.

1.17   MICROSTRUCTURE–PROPERTY RELATIONSHIPS

1.17.1   Grain Boundary Effects

If we need to compare the electrical or other properties of two different grades of a BaTiO$_3$-based polycrystalline material, we can refer to the properties of these materials in the context of their microstructure. For example, we may conclude that a finer-grained BaTiO$_3$ formulation has a higher *dielectric constant* (*k*). We will learn in Chapter 7 that the dielectric constant is a measure of a material's ability to store an electrical charge. Grain boundary regions often also have very different electrical or magnetic properties than those of the material inside the grains.

1.17.2   Grain Size Effects

As the grain size decreases, the area of the grain boundary increases. In many applications, grain boundaries are used to control the properties of materials. Most often, a polycrystalline metallic material exhibits a higher mechanical strength than that of a coarse-grained material with essentially the same composition.

The presence of grain boundaries also has a significant effect on the electrical, magnetic, and optical properties of materials. For example, the electrical resistivity of a conductor such as copper or silver is typically higher for a polycrystalline material than that for a single crystal of the same material. This is because of the increased scattering of electrons by atoms in the grain boundary regions.

The optical properties of materials are also affected by the grain boundaries and pores (holes) in polycrystalline ceramics because these contribute to the scattering of light (Figure 1.36). In this illustration, the hexagonal regions represent the grains in a polycrystalline material.

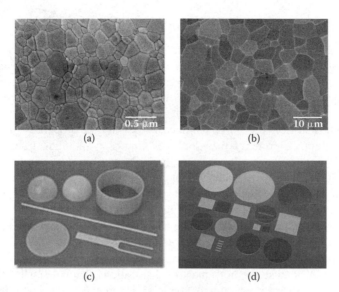

FIGURE 1.35 Effect of grain size on the optical properties of alumina ceramics. (a) Finer-sized alumina grain is opaque, (b) polycrystalline alumina with larger grain size is translucent (From Kim et al. 2009. *Acta Mater* 57(5):1319–26), (c) parts made from polycrystalline alumina with larger grain size are translucent (Courtesy of Covalent Materials Corporation, Tokyo), and (d) single-crystal sapphire substrates are transparent (Courtesy of Kyocera Corporation, Kyoto, Japan).

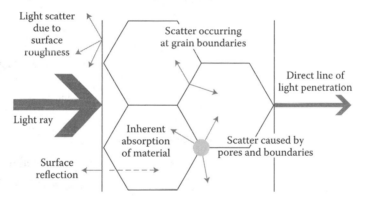

FIGURE 1.36 Processes that cause absorption and scattering of light in polycrystalline ceramics. (Adapted from Covalent Materials Corporation, Tokyo. With permission.)

Al_2O_3 ceramics can be made translucent by using special additives and processing techniques that lead to larger grain size. Polycrystalline Al_2O_3 ceramics with a fine grain size are usually opaque (Figure 1.35a). This is largely because of the scattering of light from both the pores and the grain boundaries (Figure 1.36). The microstructure of larger-grain ceramics that are optically translucent is shown in Figure 1.35b. Translucent Al_2O_3 ceramics are used in many applications, such as envelopes for high-pressure sodium vapor lamps (Figure 1.35c). With single-crystal Al_2O_3 ceramics, the material becomes essentially transparent (Figure 1.35d), because this material has very little intrinsic absorption and there are no pores or grain boundaries to scatter light.

Similar to Al_2O_3, many other transparent or translucent polycrystalline ceramic materials, such as yttrium aluminum garnet (YAG), yttrium oxide (Y_2O_3), and lead lanthanum zirconium titanate (PLZT), have also been developed for different commercial applications.

1.17.3 Microstructure-Insensitive Properties

We must emphasize two points regarding the microstructure–property relationship of materials. First, when we change the microstructure of a material, many properties can change. Thus, in the case of Al_2O_3 ceramics, as discussed before, the changes in optical properties occur concomitantly with changes in mechanical properties, such as fracture toughness. Thus, a change in the microstructure may result in changes in many properties.

The second point that needs to be emphasized is that not all properties of materials are sensitive to microstructure. A property that does not change with microstructure is known as a *microstructure-insensitive property*. For example, typically, the *Young's modulus* (Y) or *elastic modulus* (E) of a metallic or ceramic material will not change drastically with changes in the grain size. This is because Young's modulus, which is a measure of the difficulty with which elastic strain can be introduced in a material, depends on the strength of the interatomic bonds in a material. When we change the microstructure, we do not change the nature of the bonds, and hence, Young's modulus will not be affected. However, the so-called *yield stress* (σ_{YS}) of a material—a level of stress that initiates permanent or plastic deformation—typically depends strongly upon the average grain size of the material. This is because the movement of dislocations (known as slip) that causes plastic deformation is resisted by disruptions in the atomic arrangements occurring at the grain boundary regions.

We will now turn our attention to amorphous materials.

1.18 AMORPHOUS MATERIALS

In some materials, the atoms (or ions) do not exhibit an LRO. Such materials are considered *amorphous* or *noncrystalline* materials, or are simply called *glasses*. We use the term "glass" to refer to amorphous materials (metallic or ceramic) derived by the relatively rapid cooling of a melt. Amorphous materials are typically formed under nonequilibrium conditions during processing (e.g., relatively faster cooling of a melt or decomposition of a vapor). As a result, they tend to be thermodynamically unstable. Inorganic glasses based on silica (SiO_2), which are used to make optical fibers, and *amorphous silicon* (a-Si) are examples of amorphous materials.

1.18.1 Atomic Arrangements in Amorphous Materials

There exists a *short-range order* (SRO) of atoms or ions in amorphous materials. The difference between an SRO and an LRO in amorphous silicon (a-Si) and crystalline silicon (c-Si), respectively, is shown in Figure 1.37.

Note that the a-Si has a structure wherein the angles at which the silicon atom tetrahedra are connected to one another and the distances between the silicon atoms are not exactly the same throughout the structure.

A-Si is made using the so-called chemical vapor deposition (CVD) process, which involves decomposing silane (SiH_4) gas. A concentration of hydrogen atoms is also incorporated into the structure. These hydrogen atoms "pacify" some of the silicon bonds that would otherwise remain unsaturated or dangling (Figure 1.37). This is helpful for microelectronic devices based on a-Si, because the unsaturated silicon bonds would make the devices electrically inactive.

A-Si is used in making thin film transistors (TFTs). The underlying circuitry created using such a-Si–based TFTs is used to drive liquid crystal displays used in personal computers, televisions, and electronic ink (e-ink)-based bistable displays found in most state-of-the-art electronic book readers (e.g., Kindle or Sony's Reader).

Another technologically important amorphous material is SiO_2-based glass. In both c-Si and a-Si, the silicon and oxygen ions are connected in a tetrahedral arrangement. This means that each silicon ion (Si^{4+}) is always surrounded by four oxygen ions. This is the SRO. However, in

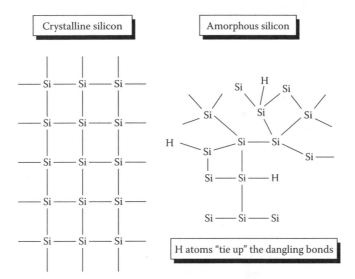

FIGURE 1.37 A schematic representation of the difference between short-range order in amorphous silicon and long-range order in crystalline silicon. (From Singh, J. 1996. *Optoelectronics: An Introduction to Materials and Devices.* New York: McGraw Hill. With permission.)

a SiO_2-based glass, the angles at which these tetrahedra are interconnected vary. Thus, the distances between silicon ions in different tetrahedra also vary. SiO_2 exhibits many equilibrium and nonequilibrium polymorphs. These have both LRO and SRO. For a given polymorph of SiO_2, the angles at which different silicon–oxygen tetrahedra are connected also remain essentially the same.

Note that a material does not have to be either crystalline or amorphous—it can exist in both forms either simultaneously or under different conditions. We often encounter materials that have both crystalline and amorphous phases. For example, in polyvinylidene fluoride (PVDF), one of the most widely-used piezoelectric materials (Chapter 8), the microstructure consists of crystalline regions embedded in an amorphous matrix. Most polymers are mixtures of crystalline and amorphous regions interspersed within the material. The processing or manufacturing methods used for creating a material have a significant effect on whether it is crystalline, amorphous, or a mixture of the two. The development of polymers for flexible and lightweight electronic devices is an active area of research and development known as *flexible electronics*.

1.18.2 APPLICATIONS OF AMORPHOUS MATERIALS

The reasons for choosing an amorphous or a crystalline form of a material for a given application vary. For example, amorphous silica is used for making optical fibers because kilometers of continuous fibers can be drawn from molten SiO_2. In principle, we could use single-crystal SiO_2 because it is optically transparent. However, this is not easy or, more important, cost-effective.

Similarly, the amorphous silica cannot be used for certain applications, such as so-called quartz clocks. Certain SiO_2 polymorphs exhibit what is described as a "piezoelectric effect," where voltage is generated across a material when it is stressed (Chapter 8). This effect is used to make so-called quartz clocks.

The properties of a material can be very different in the amorphous and crystalline states. With some technologies, we can make use of both the amorphous and crystalline forms of a material. For example, in the so-called *phase change memory* (PCM) technology, the back-and-forth switching of a material between the amorphous and crystalline states is used for data storage. Materials based on certain compositions of germanium, antimony, and tellurium (Te), known as the GST materials, are used in PCM technology. PCM technology utilizes the differences between the electrical resistivity values of the amorphous and crystalline forms of the materials.

An electrical resistor rapidly heats up the material and changes the state of the material from crystalline to amorphous (Figure 1.38). The amorphous material, in this case, exhibits a higher resistivity than the crystalline material (Figure 1.39). These changes in electrical properties are sensed by other appropriate circuitry. The amorphous state is the high-resistivity state and uses "1" as the code (data write). The crystalline state is the low-resistivity state and uses "0" as the code (data erase). The GST materials can be cycled rapidly and reversibly by heating and cooling between these "reset" amorphous and "set" crystalline phases. This is the basis for PCM technology.

A similar concept is used for rewritable CD and DVD media. For these technologies, the differences between the optical properties of materials in the amorphous and crystalline states are used. The amorphous phase is less reflective than the crystalline state.

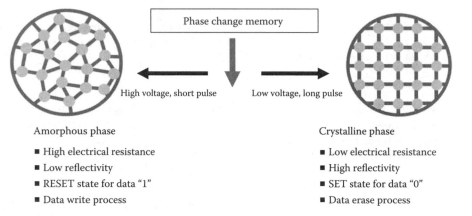

FIGURE 1.38 Phase change memory illustration. (Courtesy of Dr. Ritesh Agarwal, University of Pennsylvania.)

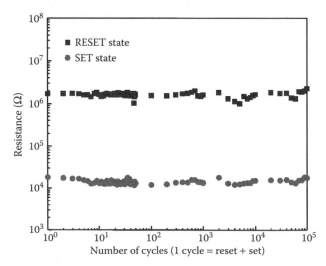

FIGURE 1.39 Resistivity of a GST-nanowire phase change memory device. (Courtesy of Dr. Ritesh Agarwal, University of Pennsylvania.)

1.19 NANOSTRUCTURED MATERIALS

In recent years, considerable research and development has occurred with respect to *nanostructured* materials. In nanostructured materials, the grain size or crystallite size is extremely small (~10 nm or less). In terms of length scales, the nanostructured form of materials falls between materials that are amorphous and crystalline. The nanostructured form causes an unusually high fraction of atoms in the material to be located at the grain boundaries (if the material is polycrystalline) or at the particle surfaces (if the material is in the form of nanoparticles). This and other phenomena (which could be quantum mechanical in nature) often lead to unique and unexpected changes in the properties of a nanostructured material.

For example, the so-called bandgap (E_g) of a semiconductor becomes larger for materials in the nanoparticle form. The increase in the bandgap of a semiconductor as a result of smaller particle size is known as "blue shift." This effect is easily observed by examining changes in the colors of dispersions of semiconductor nanoparticles, also known as quantum dots. The availability of such materials has led to the possibility of using surface effects, such as surface plasmons, because of the large surface area of nanomaterials. This has created new opportunities in the areas of nanophotonics and quantum computing.

1.20 DEFECTS IN MATERIALS: GOOD NEWS OR BAD NEWS?

The term "defect" is somewhat of a misnomer, because when defects are present, the properties of a material may actually be improved! Thus, the term "defect" just refers to the fact that the atomic arrangement in a given material is disturbed and that it is not perfect. Amorphous materials can have a very high concentration of defects. However, the presence of point defects, line defects, area defects (domains, stacking faults), and volume defects (grain boundaries) in materials is extremely important and useful.

The addition of antimony or boron to silicon (Figure 1.28) enabled the entire technology of ICs based on silicon. The presence of point defects also enables the conversion of insulating ceramic materials into semiconductors! There are plenty of examples where the presence of point defects is not just desirable but essential. One example we discussed is the creation of oxygen vacancies in zirconia to turn it into an ionic conductor. In cases of metals and alloys, the presence of point defects is also critical for the enhancement of the mechanical properties. For example, pure copper is too soft. However, when we add other alloying elements (e.g., beryllium [Be]) to form solid solutions or precipitates, the elastic modulus and yield stress of copper are improved considerably.

However, in many situations, the presence of defects can be detrimental. For example, in the case of solid solution- or precipitate-based strengthening of metals, the electrical conductivity is lowered, although some mechanical properties, such as yield strength (σ_{YS}), are improved. The effective speed (mobility) with which electrons can move in a single-crystal silicon is far greater than that in a-Si. Thus, defects can be detrimental for certain applications.

To summarize, the term "defect" does not mean that the material is defective. The presence of certain types of defects (i.e., imperfections in atomic or ionic arrangements) is beneficial and may sometimes be essential. However, certain types of defects have deleterious effects and must be avoided.

1.21 PROBLEMS

1.1 If the lattice constant of BCC Fe is 2.8666 Å, what is its density? Assume that the atomic mass of Fe is 55.85.

1.2 Show that for the DC structure, the packing fraction is about 34%.

1.3 What is the theoretical density of Ge? Assume that the atomic radius of Ge is 122.3 pm.

1.4 What is the theoretical density of GaAs? Assume that the lattice constant of GaAs is 565 pm. The atomic masses of Ga and As are 69.7 and 74.9, respectively.

1.5 Similar to the volume-packing fractions, we can calculate the area-packing fractions for planes in a unit cell. Calculate the area-packing fractions for FCC-Ni for the planes (100) and (111). Which plane is close-packed?

1.6 Sketch the following planes and directions in a cubic unit cell: [111], [$\bar{1}$10], [011], (111), and (123).

1.7 Sketch the crystal structure of a PZT ceramic with a perovskite structure. Label the different ions clearly.

1.8 Calculate the interplanar spacing for the (100), (110), and (111) planes in Si.

1.9 Show that the Miller–Bravais indices for direction [110] would be [11$\bar{2}$0].

1.10 The equation for the formation of a Schottky defect in MgO is written as

$$\text{null} \rightarrow V''_{Mg} + V^{\bullet\bullet}_{O} \tag{1.13}$$

Write down the equation for the Schottky defect formation in Al_2O_3.

1.11 In addition to Y_2O_3, we can also add MgO to ZrO_2 to stabilize the cubic crystal structure (Equation 1.12). Write the equation that will represent the incorporation of small concentrations of MgO in ZrO_2.

1.12 Why must the Al_2O_3 ceramics used in Na (sodium) vapor lamps be transparent or translucent? How does a larger grain size help with the transparency of Al_2O_3 ceramics?

1.13 Both p-Si and a-Si are used for manufacturing solar cells. For manufacturing ICs, we use single-crystal Si. Why?

1.14 Although both are based on strong covalent C–C bonds, graphite is used as a solid lubricant, whereas diamond is one of the hardest naturally occurring materials. Why is this?

1.15 What are the advantages to developing microelectronic components based on polymers?

1.16 State one example each of an application where the presence of dislocations is (a) useful and (b) deleterious.

1.17 What is the principle by which PCM technology works?

1.18 Similar to GST materials that are used in PCM, SiO_2 also exists in amorphous and crystalline forms. Compared to GST, SiO_2 is inexpensive. Will it be possible to use SiO_2 for PCM?

1.19 Why does the Blu-ray format offer much higher density for data storage in CDs and DVDs?

1.20 Conventional solar cells are made using polycrystalline silicon and offer an efficiency of about 15%. Some companies are developing solar cells based on copper indium gallium selenide (CIGS). These CIGS-based solar cells can offer efficiencies up to 40%. Why can't we just switch over to these more efficient CIGS-based solar cells?

1.21 Some companies have developed solar cells based on organic materials. The efficiency of these solar cells is about 5%. If we already have Si solar cells that operate at ~15%, why should we develop solar cells based on polymers that are not as efficient?

GLOSSARY

Allotropes: Different crystal structures exhibited by an element.

Amorphous materials: Materials in which atoms or ions do not have a long-range order.

Amorphous silicon (a-Si): An amorphous form of silicon typically obtained by the chemical vapor decomposition of silane (SiH_4) gas, which is used in thin-film transistors and photovoltaic applications.

Anions: Negatively charged ions (e.g., O^{2-}, S^{2-}).

Basis: An atom or a set of atoms associated with each lattice point. The combination of a lattice and basis defines a crystal structure.

Body-centered cubic (BCC) structure: A crystal structure in which the atoms are positioned at the corners of a unit cell with an atom at the cube center (the packing fraction is 0.68).

Bravais lattice: Any of the 14 independent arrangements of points in space.

Cations: Positively charged ions (e.g., Si^{4+}, Al^{3+}).

Close-packed structure: A crystal structure that has atoms filled in a way that achieves the highest possible packing fraction.

Coordination number (CN): The number of atoms that surround an atom on a lattice site or in an interstitial site.

Crystal structure: The geometrical arrangement in which atoms are arranged within the unit cell, described using a Bravais lattice and basis.

Crystalline material: A material in which atoms or ions are arranged in a particular three-dimensional arrangement, which repeats itself and exhibits a long-range order.

Dielectric constant (k): A measure of the ability of a material to store electrical charge.

Diffusion: A process by which atoms or ions or other species move, due to a gradient in the chemical potential (equivalent to concentration).

Domain: A small region of a ferroelectric or magnetic material in which the dielectric or magnetic polarization direction, respectively, is the same.

Dopant: Foreign atoms (e.g., boron, phosphorus) deliberately added to a semiconductor to tune its level of conductivity.

Elastic modulus (E): A measure of the difficulty with which elastic strain can be introduced into a material. This depends on the strength of the bonds within a material. It is the ratio of elastic stress to strain and is also known as the Young's modulus (Y).

Epitaxy: A process by which a thin film of one material is grown, typically over a single-crystal substrate of the same (homoepitaxy) or different (heteroepitaxy) composition. There is usually a good match between the lattice constants of the thin film and substrate.

Face-centered cubic (FCC) structure: A crystal structure in which the atoms are positioned at the corners of a unit cell and also at the six face centers (the packing fraction is 0.74).

Ferroelectric substance: A dielectric material that shows spontaneous and reversible polarization.

Flexible electronics: A field of microelectronics devoted to the development of flexible electronic components or devices, often based on polymers and/or thin metal foils.

Frenkel defect: A defect caused by an ion leaving its original position and entering an interstitial site.

Gettering: A method by which impurities in a semiconductor (such as transition metal impurities like Fe and Ni) are concentrated away from the active device regions by attracting them to defects, such as precipitates and dislocations.

Glass: An amorphous material typically obtained by the solidification of a liquid under nonequilibrium conditions. The term "glass" is typically used to describe ceramic and/or metallic amorphous materials.

Grain: A small single-crystal region in a polycrystalline material.

Grain boundaries: The regions between the grains of a polycrystalline material.

Hexagonal close-packed (HCP) structure: A crystal structure in which atoms are arranged in a hexagonal pattern such that the maximum possible packing fraction (0.74) is obtained.

Hydrogen bond: A special form of van der Waals bonds found in materials based on polar molecules or groups (e.g., water).

Impurity: Foreign atoms introduced during the synthesis or fabrication of materials such as semiconductors. Their effect is usually not desirable.

Interstitial site: A hole in the crystal structure, which may contain atoms or ions.

Interstitial atoms: Atoms or ions added to a material that occupy the interstitial sites of a given crystal structure.

Kröger–Vink notation: A notation used to indicate the point defects in materials.

Lattice: A collection of points in space. There are only 14 independent arrangements of points in space, known as the Bravais lattices.

Long-range order (LRO): Atoms or ions arranged in a particular geometric arrangement that repeats itself over relatively larger distances (~ a few micrometers up to a few centimeters). This is seen in polycrystalline and single-crystal materials.

Microstructure: A term that describes the average size and size distribution of grains, in addition to the grain boundaries and other phases (such as porosity and precipitates) and defects (e.g., dislocations).

Microstructure-insensitive property: A property that does not change substantially with changes in the microstructure of a material (e.g., Young's modulus).

Miller indices: Notation used to designate specific crystallographic directions and planes in a unit cell.

Miller–Bravais indices: A four-index system for directions and planes in the hexagonal unit cell.

n-type semiconductor: A semiconductor in which a majority of the charge carriers are negatively charged electrons (e.g., phosphorus-doped silicon).

Nanostructured materials: Materials with an ultrafine particle or grain size (~<10 nm), which have an unusually large fraction of atoms that are located at the grain boundaries (for nanocrystalline materials) or surfaces (for nanoparticles), causing them to have unusual properties.

Noncrystalline materials: Same as amorphous materials—these materials exhibit no long-range order of atoms or ions.

Octant: A smaller cube obtained by dividing the unit cell into eight smaller cubes.

Octahedral site: An interstitial site in which six atoms that form an octahedron surround the interstitial atom or ion—the coordination number is six.

Organic electronics: A field of research and development based on the use of polymeric materials, such as conductors and semiconductors, for developing flexible, lightweight microelectronic devices, such as transistors and solar cells.

p-type semiconductor: A semiconductor in which a majority of the charge carriers are positively charged holes (e.g., boron-doped silicon).

Packing fraction: The ratio of volume or space occupied by atoms to the volume of the unit cell.

Phase-change memory (PCM): A memory technology in which the amorphous and crystalline states of a material are used to store information as "1" or "0."

Phase: Defined as any portion of a system, including the whole, which is physically homogeneous and bounded by a surface so that it is mechanically separable from any other portion.

Phase diagram: A diagram indicating the different phases that can be expected in a given system of materials, assuming a condition of thermodynamic equilibrium exists.

Piezoelectric effect: The development of a voltage across a material when it is stressed.

Plastic: A polymer-based material formulated with one or many polymers and other additives (e.g., carbon black and glass fibers).

Polymorphs: Different crystal structures exhibited by a compound.

Quantum dots: Nanocrystals of semiconductors that exhibit a change in the bandgap, which in turn causes changes in their optical properties, such as color.

Radius ratio: The ratio of the radius of a cation to that of an anion.

Short-range order (SRO): Atoms and ions in amorphous and crystalline materials exhibit a short-range (up to ~ a few Å) order in the arrangement of atoms or ions in a specific geometrical fashion.

Schottky defect: A defect in which a certain number of cations and a stoichiometrically equivalent number of anions are missing, for example, the absence of one sodium ion and one chlorine ion in NaCl.

Simple cubic (SC) structure: A crystal structure in which atoms are positioned at the corners of a unit cell (the packing fraction is 0.52).

Solid solution: A new phase of a material obtained by dissolving atoms of one element into another, which strengthens metallic materials and can increase conductivity.

Stacking fault: A planar defect in which one of the planes from the stacking sequence is missing.

Substitutional atoms: Atoms or ions added to a material. They occupy sites that are usually occupied by the atoms or ions of the host material.

Tetrahedral site: An interstitial site in which four atoms that form a tetrahedron surround the interstitial atom or ion—the coordination number is four.

Unit cell: The basic unit that represents an arrangement of atoms (or ions) repeating in all three dimensions.

van der Waals bond: A secondary bond that is present in all materials, caused by interactions between induced dipoles. It is functionally important in materials with polar ions, atoms, or groups (e.g., water, PVC, PVDF). A special type of this bond is the hydrogen bond.

Vacancy: The absence of an atom or an ion from its crystallographic location within a crystal structure.

Yield stress (σ_{YS}): A level of stress that initiates a permanent or plastic deformation. This is typically a strong function of the grain size of a material.

Young's modulus: Also known as the elastic modulus, this is a measure of the difficulty with which elastic strain can be introduced in a material. It depends on the strength of bonds within the material and the ratio of the elastic stress to the strain.

REFERENCES

Askeland, D., and P. Fulay. 2006. *The Science and Engineering of Materials*. Washington, DC: Thomson.

Groover, M. P. 2007. *Fundamentals of Modern Manufacturing: Materials, Processes, and Systems*. New York: Wiley.

Han, I., R. Datta, S. Mahajan, R. Bertram, E. Lindow, C. Werkhoven, and C. Arena. 2008. Characterization of threading dislocations in GaN using low-temperature aqueous KOH etching and atomic force microscopy. *Scr Mater* 59(11):1171–3.

Kim, B.-N., K. Hiraga, K. Moritaa, H. Yoshida, T. Miyazaki, and Y. Kagawa. 2009. Microstructure and optical properties of transparent alumina. *Acta Mater* 57(5):1319–26.

Singh, J. 1996. *Optoelectronics: An Introduction to Materials and Devices*. New York: McGraw Hill.

Singhal, S. C. 2002. Solid oxide fuel cells for stationary, mobile, and military applications. *Solid State Ionics* 152–153:405–10.

Smart, L., and E. Moore. 1992. *Solid State Chemistry: An Introduction*. Boca Raton, FL: Chapman and Hall.

Stach, E. A., and R. Hull. 2008. Dislocations in Semiconductors. In *Encyclopedia of Materials: Science and Technology*, eds. K. H. J. Buschow et al., 2301–12. Oxford: Elsevier.

2 Electrical Conduction in Metals and Alloys

KEY TOPICS

- The fundamentals of what causes a material to be a conductor of electricity
- Different types of conducting materials based on metals and alloys and their "real-world" technological applications
- The classical theory of conductivity
- The band theory of solids and its use for examining the differences between conductors, semiconductors, and insulators
- The effects of chemical composition, microstructure, and temperature on the conductivity of metals and alloys
- Real world applications of conductive materials

2.1 INTRODUCTION

Metals are *conductors* of electricity; that is, they offer relatively little resistance to the flow of electricity. In metals, each atom donates one or more electrons and therefore becomes more stable. This process leads to a large "sea of electrons" (Figure 2.1). A high concentration of these donated electrons, which are free to move around within a metallic material, leads to a higher level of conductivity in metals.

Figure 2.2 shows one way to classify electronic materials based on the nature of their interatomic bonds. In Section 2.8, we will see that compared to pure metals, alloys offer a higher resistance to the flow of electricity. In contrast to metals, most ceramic materials (e.g., silica [SiO_2], zirconia [ZrO_2], alumina [Al_2O_3], and silicon carbide [SiC]) exhibit strong ionic and/or covalent bonds. Many polymers (polyethylene, polystyrene, epoxies, etc.) also primarily exhibit covalent bonds *within* the chains of atoms and van der Waals bonds *between* the chains of atoms. Therefore, ceramics and polymers are usually, but not always, electrical *insulators* and are also referred to as *dielectrics*. The use of the term "insulator" is sometimes preferred when the ability of a material to withstand a strong electric field is to be emphasized, as opposed to offering only a high electrical resistance. For example, porcelain can be described more appropriately as an insulator, rather than as a dielectric.

Covalent bonds are formed by the sharing of electrons from different atoms participating in bond formation. Therefore, the valence electrons in the covalent or ionic bonds are localized in dielectric and insulating materials, that is, they are trapped inside the bonds (Figure 2.1). In solids with strong ionic bonds, the electropositive elements donate their valence electrons to electronegative elements to form ionic bonds. In Section 3.2.1, we will see that many ceramics and some polymers can be made to exhibit a useful level of conductivity through deliberate modification of their composition and microstructure. The conducting ability of some materials (e.g., silicon [Si], germanium [Ge], and gallium arsenide [GaAs]) is in between that of insulators and conductors. These materials are known as *semiconductors* (Figure 2.2). The ranges for the values of a parameter known as the *resistivity* (ρ), discussed in Section 2.2, are also included in Figure 2.2.

Some materials, known as *superconductors*, exhibit zero electrical resistance under certain conditions (e.g., low temperatures). Examples of superconducting materials include elements such as

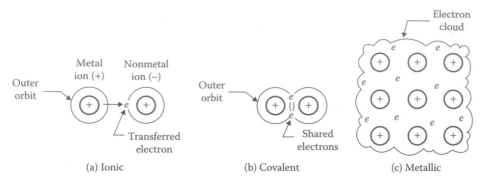

FIGURE 2.1 Illustration of (a) ionic bonds, such as those in sodium chloride and many other ceramics; (b) covalent bonds, such as those in silicon, many ceramics, and polymers; and (c) metallic bonds. (From Groover, M. P. 2007. *Fundamentals of Modern Manufacturing: Materials, Processes, and Systems.* New York: Wiley. With permission.)

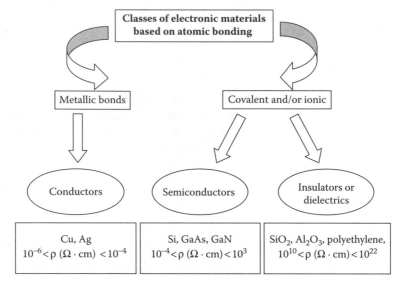

FIGURE 2.2 Classification of materials based on the nature of bonding. Typical ranges of resistivity are also shown.

mercury (Hg), intermetallic compounds such as niobium tin (Nb_3Sn), and ceramic materials such as yttrium barium copper oxide ($YBa_2Cu_3O_{7-x}$) and magnesium diboride (MgB_2). Superconductors are not shown in Figure 2.2 because the bonding in these materials is complex, and other additional mechanisms play a role in the superconductivity of these materials.

The ability of a material to conduct electricity usually also depends upon both its temperature and its microstructure. In some instances, a relatively higher temperature will promote a higher level of electrical conductivity. In other cases, such as for essentially pure metals, the ability to conduct electricity *decreases* with an increase in temperature.

Similarly, a material's structure has a significant effect on its ability to conduct electricity. For example, polymers such as polyethylene are usually considered to be nonconducting. However, some polymers are good conductors. An example of a polymeric material that shows a useful level of conductivity is polyaniline. The advantage of conducting polymers is that they are flexible and lightweight, unlike metals and semiconductors. Thus, novel and flexible electronic components can be developed using conducting and semiconducting polymeric materials.

An important point is that although we can associate a general trend in conductivity with a particular class of materials, there are always exceptions. Most polymers and ceramics are dielectrics or insulators, but not all. Similarly, most metals are good conductors, but not all.

In some cases, it is advantageous to use conductive materials based on *composites*. Composites are materials made up of two or more materials or phases, often arranged in unique geometrical arrangements to achieve desired properties. For example, silver (Ag)-filled epoxies are used in microelectronics as conductive adhesives. These composites are prepared by dispersing fine silver powder particles in an epoxy matrix. The silver provides the electrical conductivity. Epoxy provides a means of applying a paint-like material that eventually hardens into a solid plastic.

Certain aspects of electrical conductivity in materials can be explained by considering electrons to be particles and treating them using the principles of classical mechanics. However, certain other aspects related to electrical conductivity can be explained only through a quantum mechanical–based explanation, in which electrons are considered waves. We begin the discussion of these topics with Ohm's law. We will next discuss the classical and quantum mechanical theories of conductivity.

2.2 OHM'S LAW

The geometry of a *resistor* with resistance R is shown in Figure 2.3. A resistor is a component included in an electrical circuit to offer a predetermined value of electrical resistance.

Ohm's law is an empirical relationship that states the following:

$$V = IR \tag{2.1}$$

where V is applied voltage in volts (V), I is electrical current in amperes (A), and R is electrical resistance in ohms (Ω). In this chapter, unless stated otherwise, we will assume a direct current (or DC) voltage. This is in contrast to alternating current (or AC) voltage, which cycles with time.

The resistance (R) of an electrical resistor of length L and cross-sectional area A (Figure 2.3) is given by

$$R = \rho \frac{L}{A} \tag{2.2}$$

The inverse of electrical resistance is known as *conductance*, which has units of Siemens or Ω^{-1}.

In this equation, ρ (rho) is defined as the *electrical resistivity, bulk resistivity,* or *volume resistivity.* A commonly used unit for resistivity is $\Omega \cdot$ cm. The inverse of ρ is known as *conductivity* (σ).

$$\sigma = \frac{1}{\rho} \tag{2.3}$$

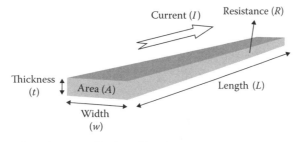

FIGURE 2.3   Geometry of a resistor used in describing Ohm's law.

A commonly used unit for conductivity is Siemens/centimeter (S/cm). When looking up the properties of materials or performing related calculations, it is important to check whether a value of conductivity or resistivity is being stated and to check what units are being used. Different sources quote values in different units. For example, resistivity values may be listed as $\Omega \cdot cm$, $\Omega \cdot m$, $\mu\Omega \cdot cm$ (micro-ohm-centimeter), $n\Omega \cdot m$ (nano-ohm–meter), and so on.

Most conductors obey Ohm's law: as the applied voltage is increased, the current increases linearly. The resistance does not depend on the voltage. There are, however, certain materials (e.g., zinc oxide–based formulations) and devices that, under certain conditions, do *not* obey Ohm's law. For these materials, resistance remains independent of the voltage up to a certain value. However, beyond a certain critical value of the *electric field* (E), the material undergoes an electrical breakdown, and the resistance decreases by orders of magnitudes. The electric field is defined as the voltage divided by the distance across which the voltage is applied. Materials that do not obey Ohm's law are sometimes described as variable resistors or *varistors*. Varistors are useful as current-limiting, surge-protection devices.

The magnitude of resistivity allows us to delineate the approximate boundaries between different classes of materials, as shown in Figure 2.2. Materials with a resistivity in the range of $\sim 10^{-6}$–10^{-4} $\Omega \cdot cm$ (1–10^2 $\mu\Omega \cdot cm$) are usually considered conductors. Insulators are materials with resistivity greater than $\sim 10^{10}$ $\Omega \cdot cm$. Resistivity values between $\sim 10^{10}$ and 10^{22} $\Omega \cdot cm$ are typical of electrically nonconducting or insulating materials. Materials that have resistivity ranging between $\sim 10^{-4}$ and 10^3 $\Omega \cdot cm$ (i.e., 10^2 10^9 $\mu\Omega \cdot cm$) are considered semiconductors. Materials with resistivities from 10^3 to 10^{10} $\Omega \cdot cm$ are classified as *semi-insulators*. The values associated with these ranges of resistivities are not rigid.

Table 2.1 shows the typical resistivity values for some metals. The values listed in Table 2.1 are based on $T_0 = 300$ K (see Equation 2.25). The values of a parameter known as the *temperature coefficient of resistivity* (α_R; TCR) are also included. We will define α_R in Section 2.8. In Table 2.1, note that silver is the element with the highest level of conductivity. Copper (Cu) and gold (Au) are also very good conductors of electricity. Note that the conductivity of tungsten (W) is higher than that of platinum (Pt).

Note that the α_R values for magnetic metals are higher than those for other metals (see Section 2.8). Examples 2.1 through 2.4 illustrate some applications of the concepts discussed in this section.

EXAMPLE 2.1: RESISTIVITY AND CONDUCTIVITY—UNITS AND CONVERSION

The resistivity of a Cu sample is 1.673 $\mu\Omega \cdot cm$.
   a. What is the value of this resistivity in $\Omega \cdot cm$? What is the value in $n\Omega \cdot m$?
   b. The conductivity of an Ag sample is listed as 62.9×10^4 S/cm. What is the resistivity in $\mu\Omega \cdot cm$?

SOLUTION

   a. Cu sample resistivity in $\Omega \cdot cm$ = 1.673 $\mu\Omega \cdot cm \times 10^{-6}$ $\Omega/\mu\Omega$ = 1.673×10^{-6} $\Omega \cdot cm$.
      Cu sample resistivity in $n\Omega \cdot m$ = 1.673×10^{-6} $\Omega \cdot cm \times 10^9$ $n\Omega/\Omega \times 10^{-2}$ m/cm = 16.73 $n\Omega \cdot m$.
   b. Conductivity of the Ag sample = 62.9×10^4 S/cm.
      Therefore, the resistivity of the Ag sample = $(62.9 \times 10^4)^{-1}$ $\Omega \cdot cm$ = 1.589×10^{-6} $\Omega \cdot cm$.
      We want the answer in $\mu\Omega \cdot cm$,
      Therefore, the resistivity of the Ag sample = 1.589×10^{-6} $\Omega \cdot cm \times 10^6$ $\mu\Omega/\Omega$ = 1.589 $\mu\Omega \cdot cm$.

EXAMPLE 2.2: ANTENNAS FOR RADIO FREQUENCY–IDENTIFICATION TECHNOLOGY

Radio frequency–identification (RFID) technology is used for applications such as automatic highway toll-payment systems (e.g., EZ-Pass™), smart cards for access control, libraries, livestock tracking, and many other inventory-control applications (Figure 2.4).

A passive RFID tag is comprised of a conductive antenna, typically made from Cu, with a small computer chip attached to it. The computer chip holds the information. In one process to manufacture antennas, we begin with a thin foil of solid Cu. A pattern is then created by etching away excess

TABLE 2.1
Conductivity/Resistivity Values for Selected Metals

Material	Conductivity (σ) S/cm	Resistivity (ρ) $\mu\Omega \cdot$ cm	Temperature Coefficient of Resistivity (α_R) ($\Omega/\Omega \cdot$ °C)
Aluminum	37.7×10^4	2.65	4.3×10^{-3}
Beryllium	25×10^4	4.0	25×10^{-3}
Cadmium	14.6×10^4	6.83	4.2×10^{-3}
Chromium	7.75×10^4 (at 0°C)	12.9 (at 0°C)	3.0×10^{-3}
Cobalt (magnetic)	16.0×10^4	6.24	5.30×10^{-3}
Copper	59.7×10^4	1.673	4.3×10^{-3}
Gold	42.5×10^4	2.35	3.5×10^{-3}
Iridium	18.8×10^4	5.3	3.93×10^{-3}
Iron (magnetic)	10.3×10^4	9.7	6.51×10^{-3}
Lead	4.84×10^4	20.65	3.68×10^{-3}
Magnesium	22.4×10^4	4.45	3.7×10^{-3}
Mercury	1.0×10^4 (at 50°C)	98.4 (at 50°C)	0.97×10^{-3}
Molybdenum	19.2×10^4 (at 0°C)	5.2 (at 0°C)	5.3×10^{-3}
Nickel (magnetic)	14.6×10^4	6.84	6.92×10^{-3}
Palladium	9.253×10^4	10.8	3.78×10^{-3}
Platinum	10.15×10^4	9.85	3.93×10^{-3}
Rhodium	22.2×10^4	4.51	4.3×10^{-3}
Silver	62.9×10^4	1.59	4.1×10^{-3}
Tantalum	0.74×10^4	13.5	3.83×10^{-3}
Tin	0.90×10^4	11.0 (at 0°C)	3.64×10^{-3}
Titanium	2.38×10^4	42.0	3.5×10^{-3}
Tungsten	18.8×10^4	5.3	4.5×10^{-3}
Zinc	16.9×10^4	5.9	4.1×10^{-3}

Unless stated otherwise, values reported are at 300 K. For some materials, the values of the coefficient of resistivity (α_R) are shown using 300 K as the reference temperature, unless stated otherwise. To convert α_R into ppm/°C, multiply the values listed by 10_6. Data from Webster, J. G. 2002. *Wiley Encyclopedia of Electrical and Electronics Engineering*, Vol. 4. New York: Wiley; and Kasap, S. O. 2002. *Principles of Electronic Materials and Devices*. New York: McGraw Hill.

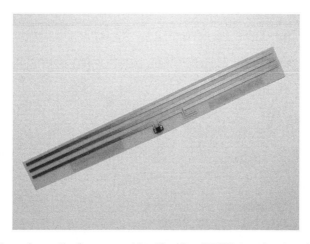

FIGURE 2.4 Typical passive radio frequency–identification (RFID) tag showing the antenna made from copper. (Courtesy of Pavel Nikitin, Intermec Technologies Corporation, USA.)

Cu to form the antenna. Other methods of manufacturing antennae include using conductive inks (made from powders of Cu or Ag) or conductive polymers. These approaches offer the possibility of lowering the cost of RFID tags.

a. An antenna is made by etching solid Cu with $\sigma = 5.8 \times 10^7$ S/m. What is its electrical resistance (R)? Assume that the antenna is in the form of a strip that is 20 cm long, 5 mm wide, and 25 μm thick.

b. What will be the resistance of a similar antenna made by screen printing *silver paste* with $\sigma = 1.6 \times 10^6$ S/m?

c. Why do you think the conductivity of Ag ink is *lower* than that of solid Cu?

Solution

a. Let us assume that the current flows along the length direction. Thus, the cross-sectional area will be 5 mm × 25 μm or 0.5 cm × 25 μm × 10^{-4} cm/μm = 125 × 10^{-5} cm². The resistance of the Cu antenna will be (Equation 2.2):

$$R_{copper} = \rho \frac{L}{A} = \frac{20 \text{ cm}}{(125 \times 10^{-5} \text{cm}^2)(5.8 \times 10^7 \text{S/m} \times 10^{-2} \text{m/cm})} = 0.02758\,\Omega$$

Note the conversions of resistivity to conductivity, and meters and micrometers to centimeters.

b. If we use Ag ink that has $\sigma = 1.6 \times 10^6$ S/m, then

$$R_{silver\ ink} = \rho \frac{L}{A} = \frac{20 \text{ cm}}{(125 \times 10^{-5} \text{cm}^2)(1.6 \times 10^6 \text{S/m} \times 10^{-2} \text{m/cm})} = 1\,\Omega$$

This value of resistance is almost 36 times higher than that calculated for an antenna made by etching solid Cu.

c. An Ag conductive ink is a liquid paint-like material obtained by dispersing Ag particles in a carrier liquid. When screen-printed Ag ink is used, the electrical contact between Ag particles is not as good as that between grains of a polycrystalline solid, even after the ink "cures" subsequent to when the solvent medium is removed. This makes it harder for the current to flow from one particle to another. Thus, although *solid silver* has the highest conductivity among all metals (Table 2.1), in this case *silver ink* has a higher resistivity than that of solid Cu.

EXAMPLE 2.3: RESISTANCE OF A LONG Cu WIRE

Calculate the resistance of 1000 m of American Wire Gauge (AWG) #18 Cu wire (Table 2.2). Assume $\sigma_{Cu} = 5.8 \times 10^7$ S/m.

Solution

From Table 2.2, for AWG #18, the wire diameter (d) is 0.040303 inches or 1.023696 mm.

Thus, the area of cross section = $\frac{\pi d^2}{4} = 8.2306 \times 10^{-7} \text{m}^2$

Therefore, the resistance of the 1000-meter-long wire will be as follows (according to Equation 2.2):

$$R = \frac{(1000 \text{ m})}{(8.2306 \times 10^{-7} \text{m}^2) \times \left(5.8 \times 10^7 \frac{S}{m}\right)} = 20.9478\,\Omega$$

TABLE 2.2
American Wire Gauge Diameter Conversion

American Wire Gauge Number	Conductor Diameter (inches)	Conductor Diameter (mm)
20	0.03196118	0.811814
18	0.040303	1.023696
16	0.0508214	1.290864
14	0.064084	1.627734
12	0.0808081	2.052526
10	0.10189	2.588006
8	0.128496	3.263798
6	0.16202	4.115308
5	0.18194	4.621276
4	0.20431	5.189474
3	0.22942	5.827268
2	0.25763	6.543802
1	0.2893	7.34822
0	0.32486	8.251444
00	0.3648	9.26592
000	0.4096	10.40384
0000	0.46	11.684

The diameter stated is that of the conductor and does not include insulation dimensions.
Source: Powerstream, http://www.powerstream.com/Wire_Size.htm

EXAMPLE 2.4: RESISTANCE, MASS, AND LENGTH (IN FEET) PER POUND OF Cu AND ALUMINUM (Al) WIRES

a. Calculate the mass and electrical resistance of 1000 feet of Al and Cu wires of AWG #3.
b. How many feet of Cu and Al wire can we get per pound of this wire? Assume that the mass of any insulation can be ignored.
   The densities of Cu and Al are 8930 and 2700 kg/m³, respectively. Assume $\sigma_{Cu} = 5.8 \times 10^7$ and $\sigma_{Al} = 3.8 \times 10^7$ S/m.

SOLUTION

a. For AWG gauge #3, the diameter is 0.22942 inches or 5.827268 millimeters (Table 2.2). Thus, the area of cross section is

$$\text{Area} = \frac{\pi d^2}{4} = 2.66698 \times 10^{-5} \text{m}^2$$

Thus, the resistance of a 1000-foot Cu wire will be (according to Equation 2.2)

$$R_{cu} = \frac{(1000 \text{ ft.}) \times 0.3048 \, \frac{m}{ft.}}{(2.66698 \times 10^{-5} \text{m}^2) \times \left(5.8 \times 10^7 \, \frac{S}{m}\right)} = 0.197 \, \Omega$$

Similarly, the resistance of a 1000-foot Al wire (AWG #3) will be

$$R_{Al} = \frac{(1000 \text{ ft.}) \times 0.3048 \, \frac{m}{ft.}}{(2.66698 \times 10^{-5} \text{m}^2) \times \left(3.8 \times 10^7 \, \frac{S}{m}\right)} = 0.309 \, \Omega$$

b. The volume of 1000 feet of AWG #3 Cu wire is

$$\text{Volume} = (2.66698 \times 10^{-5}\,\text{m}^2) \times 1000\ \text{ft.} \times 0.3048\,\text{m/ft.}$$
$$= 8.127 \times 10^{-3}\,\text{m}^3$$

The density of Cu is 8930 kg/m³; therefore, the mass of a 1000-foot Cu wire will be

$$= 8.127 \times 10^{-3}\,\text{m}^3 \times 8930\ \text{kg/m}^3$$
$$= 72.59\ \text{kg (about 160 lb.)}$$

Thus, for a Cu AWG gauge #3 wire, the length per pound will be ~1000/160 or 6.25 feet.
Because the geometry is the same, the volume of the Al wire is the same as that of the Cu wire. The density of aluminum is 2700 kg/m³. Thus, the mass of 1000 feet of Al wire of AWG gauge #3 will be 8.129 × 10⁻³ m³ × 2700 kg/m³ = 21.948 kg or ~48.27 lb.
Thus, for an Al wire of gauge AWG #3, the length (in feet) per pound will be ~1000/48.27 = 20.71.

2.3   SHEET RESISTANCE (R_s)

In some applications of microelectronics, it is customary to use a parameter known as sheet resistance (R_s) instead of resistivity (ρ). Sheet resistance is measured more easily than resistivity. As shown in Figure 2.3, let t be the thickness of the resistor. The length and width are L and w, respectively. The cross-sectional area $A = t \times w$. We can rewrite Equation 2.2 as follows:

$$R = \rho \frac{L}{t \times w} \tag{2.4}$$

In fabrication of resistors that are used in *integrated circuits* (ICs), the designer usually works with a material that has certain values of the ratio L/w. These are typically chosen such that a large number of small components fit within a given area (i.e., miniaturization). Typically, the resistivity (ρ) and thickness (t) are maintained the same for resistors in a given layer of an IC. Consider a square resistor, that is, $L = w$. According to Equation 2.4, the resistance will be equal to the sheet resistance (R_s), defined as follows:

$$R_s = \frac{\rho}{t} \tag{2.5}$$

The unit for sheet resistance (R_s) is ohms (Ω) only. We write it as $\Omega/\square$ and read it as "ohms per square" to indicate that R_s represents the resistance of a square resistor. From Equations 2.4 and 2.5, we can derive an expression for R as follows:

$$R = R_s \frac{L}{w} \tag{2.6}$$

Example 2.5 illustrates the use of the concept of sheet resistance.

EXAMPLE 2.5: Cu INTERCONNECTS FOR INTEGRATED CIRCUITS

Cu, because of its high conductivity ($\sigma = 58 \times 10^4$ S/cm), is widely used in modern-day ICs as a conductor material that provides connections between different components, known as interconnects. If the thickness of Cu is 80 nm, what is its sheet resistance (R_s)? If the width of this Cu conductor is 0.5 µm and the length is 300 µm, what is the resistance?

SOLUTION

The sheet resistance can be calculated as follows:

$$R_s = \frac{\rho}{t} = \frac{1}{\sigma t}$$

$$R_s = \frac{1}{(58 \times 10^4 \text{S/cm}) \times (80 \times 10^{-7} \text{cm})} = 0.216 \ \Omega/\square$$

From this, we can calculate the resistance as follows:

$$R = R = R_s \frac{L}{w} = 0.216 \times \frac{300 \ \mu\text{m}}{0.5 \ \mu\text{m}} = 129 \ \Omega$$

Note that in this example, the thickness of the interconnect is very small (80 nm). In Section 2.10, we will see that at such a low thickness, the resistivity of metals is actually higher than their "bulk" value.

Sheet resistance is often measured using a technique known as the *four-point probe* or *Kelvin probe* technique (Figure 2.5). In this method, we apply an external voltage that leads to a predetermined value of current (I) between two terminals. The decrease in voltage (V) across the other two probes separated by a distance s is measured. The resistivity is then calculated using the following formula:

$$\rho = 2\pi \times s \times \frac{V}{I} \tag{2.7}$$

In Equation 2.7, V is the measured voltage in volts, I is the source current in amperes, s is the spacing between probes (in cm), and ρ is the resistivity in $\Omega \cdot$cm. Note that if $t < 5$ mm, a correction factor is needed.

The sheet resistance (R_s), measured for films and wafers using the four-point probe method, is given by Equation 2.8.

$$R_s = 4.532 \times \frac{V}{I} \times k \tag{2.8}$$

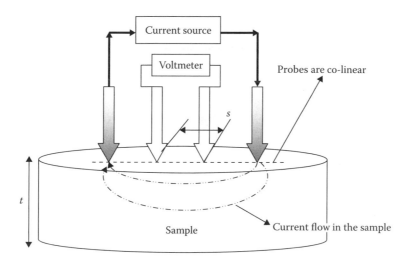

FIGURE 2.5 A schematic of the four-point probe used to measure resistivity. The spacing between probes is s, and the sample thickness is t (not to scale).

The factor 4.532 is the value of $\pi/\ln(2)$. A correction factor (k), whose value depends on the ratio of probe spacing (s) to sample diameter (d) and the ratio of wafer thickness (t) to probe spacing (s), may be needed.

For thin films and wafers (e.g., semiconductors such as germanium [Ge] or silicon [Si]), the bulk resistivity is obtained by dividing the resistivity (ρ) by the thickness (t) and is given by Equation 2.9:

$$\rho = 4.532 \times t \times \frac{V}{I} \times k \tag{2.9}$$

When measuring the resistivity values of good conductors, the current values used are limited to ~10 mA. This avoids sample heating and limits the current density at the probe tips. For measuring high-resistivity materials, lower currents are used (~1 μA). This limits the voltage measured to ~200 mV.

Using the four-point probe technique, we can measure the spatial variations in the electrical conductivity of semiconductor crystals or other materials. For example, in the fabrication of ICs, large crystals of silicon (12-inch diameter) are used as starting materials. These crystals are then sliced into silicon wafers, on which ICs are fabricated. Often, we need to measure the variation in conductivity from one wafer to another. In some cases, we need to measure the variation in conductivity, either deliberate or intrinsic to processing, across different parts of the same wafer. This measurement can be achieved using the four-point probe method.

This method works well for both conductors and semiconductors. A different technique, known as the van der Pauw method, is useful for materials with high resistivity values.

2.4 CLASSICAL THEORY OF ELECTRICAL CONDUCTION

A German scientist named Paul Drude (1863–1906) proposed the classical theory of conduction in metals. In this theory, electrons are considered particles, and an atom is considered to be a nucleus surrounded by core electrons and the valence electrons (Figure 2.6). The valence electrons are the outermost electrons, are relatively weakly bound to the nucleus, and become available for conduction. For example, for sodium (Na), there is only one valence electron per atom. Each sodium atom donates this valence electron. This leads to the formation of a "sea of electrons" (see Section 2.1), also described as "electron gas." The concentration of valence electrons in this sea of electrons is quite high. As a result, pure metals would be expected to be good conductors of electricity based on this concentration.

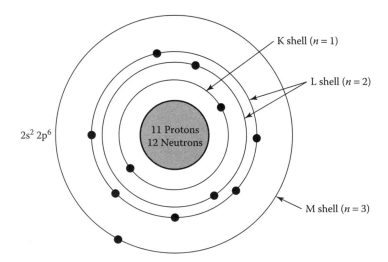

FIGURE 2.6 The structure of a sodium atom (atomic number 11), showing the nucleus surrounded by core electrons and valence electrons. (From Askeland, D. 1989. *The Science and Engineering of Materials*. 3rd ed. Washington, DC: Thomson. With permission.)

Although classical theory provides a good start for understanding the electrical properties of solids, it does not explain several features associated with electrical conduction in materials. The limitations of classical theory in explaining electrical conduction in materials are discussed in Section 2.7. These limitations are overcome using a quantum mechanical approach, which is discussed in Section 2.13. We will now begin a more detailed discussion of the classical theory of conductivity.

2.5   DRIFT, MOBILITY, AND CONDUCTIVITY

We can rewrite Ohm's law (Equation 2.1) as follows:

$$I = \frac{V}{R} \tag{2.10}$$

Instead of the electrical current (I), consider the *current density* (J), which is defined as the current per unit area. The commonly used units for current density are A/m^2 and A/cm^2.

Another way of writing Ohm's law is by applying the relationship between the current density (J) and applied electric field (E), as shown below:

$$\boldsymbol{J} = \sigma\boldsymbol{E} \tag{2.11}$$

This is known as the point form of Ohm's law. Note that $\boldsymbol{J}$ and $\boldsymbol{E}$ are vector quantities and conductivity (σ) is a tensor quantity. In this discussion, however, we treat all of these as scalar quantities. We use resistivity or conductivity instead of resistance or conductance because the resistance (R) depends upon the geometry of the resistor, whereas the resistivity (or conductivity) is an intrinsic property of a particular material. If we have a material of a certain composition and microstructure, then the resistivity value is fixed. An exception to this occurs when we examine the properties of materials or devices at a *nanoscale* (~1–100 nm length; Section 2.10).

The distinction between resistance and resistivity is also important in situations in which the resistivity of a material or a component changes across the thickness of the component. Thus, if we know the resistivity values and the geometry for different regions, we can calculate the total resistance of a component or a device.

Let us now derive the point form of Ohm's law (Equation 2.11).

The conduction electrons in metals move at very high speeds (~10^6 m/s). We can expect that as temperature increases, the kinetic energy of electrons should also increase, causing the electrons to move at higher speeds. However, as we will see in Section 2.7, this speed is largely *independent* of the temperature. When no electric field is applied, the motions of the conduction electron are in random directions and do not result in an electrical current (Figure 2.7). This is similar to ions drifting around in an electroplating solution. A less-technical analogy is that of small schoolchildren running around without the supervision of their teacher!

Now consider a voltage V applied across the length (L) of a conductor. The electric field is given by

$$E = \frac{V}{L} \tag{2.12}$$

When we apply a DC electric field to a conductor, the electrons in the material flow from the negative to the positive terminal of the voltage supply (Figure 2.7b). This current is known as the *electron current*. As a matter of convention, we state that the current or the *conventional current* flows from the positive to the negative terminal of the battery (Figure 2.7c).

The force (F) exerted on an electron by the electric field (E) is given by

$$F = qE \tag{2.13}$$

The force results in acceleration a, given by

$$F = ma \tag{2.14}$$

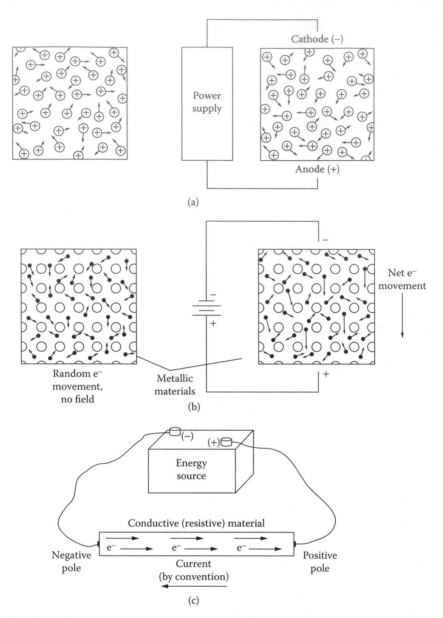

FIGURE 2.7   (a) Random motion of electrons due to thermal energy is similar to the movement of ions in an electroplating solution. (b) The drift of conduction electrons in a pure metal; note that the applied electric field (E) and drift are opposite, and overall, the electrons move within the material toward the anode. (c) The conventional current is said to flow from the positive to the negative terminal. The electron current, i.e. the actual motion of electrons, is in the opposite direction. (From Minges, M. L. 1989. *Electronic Materials Handbook*, Vol. 1. Materials Park, OH: ASM. With permission.)

Thus, the acceleration experienced by the electrons is given by

$$a = qE/m \tag{2.15}$$

Because of this externally applied driving force, that is, the electric field, conduction electrons begin to show a net movement in the direction opposite to that of the electric field (*E*). This motion of charge carriers (electrons, in this case) due to an applied electric field is known as *drift*. The drift of the carriers is characterized by an average drift velocity (v_x). This movement of conduction electrons results in an electrical current density (J_x) in the positive direction "x." Because electrons are negatively charged, the electrical currents and their drift are in opposite directions (Figure 2.7c).

If an electron accelerates and then collides with an atom after time τ, it stops. From Newton's Laws of Motion, we know that $v = u + at$, where v is final velocity, u is initial velocity, a is acceleration, and t is time. Applying this law here, assuming the initial speed is zero, we get $v = a\tau$. Thus, the average velocity of an electron between collisions will be as follows:

$$v_{\text{avg}} = \left(\frac{1}{2}\right)a\tau$$

We should not, however, consider the average time between collisions to calculate the average velocity; we should calculate the actual velocities and then obtain the average. If this is done, we can show that the average velocity or drift velocity (v_{drift}) is given by

$$v_{\text{drift}} = v_{\text{avg}} = a\tau \qquad (2.16)$$

Substituting for the acceleration a in Equation 2.16 as force divided by mass (Equation 2.15), we get

$$v_{\text{drift}} = v_{\text{avg}} = \left(\frac{qE}{m}\right)\tau = \left(\frac{q\tau}{m}\right)E \qquad (2.17)$$

According to Equation 2.17, the drift velocity of electrons is proportional to the applied electric field (E). The proportionality constant is known as the *mobility* (μ) of the charge carriers. We use the subscript "n" (to reflect the negative charge), that is, the symbol (μ_n), to describe the mobility of electrons. The term "mobility" refers to the speed with which the charge carriers drift under the influence of an external electric field (E).

$$\mu_n = \frac{q\tau}{m} \qquad (2.18)$$

The commonly used units for mobility are $\text{cm}^2 \cdot \text{s}^{-1} \cdot \text{V}^{-1}$ and $\text{m}^2 \cdot \text{s}^{-1} \cdot \text{V}^{-1}$.

The *mean free path length* (λ) of conduction electrons is defined as the average distance that the electrons travel before colliding again and is obtained by multiplying the average speed (v_{avg}) by the time between collisions (τ).

$$\lambda = v_{\text{avg}} \times \tau \qquad (2.19)$$

We can calculate the number of electrons flowing across the unit area per unit time by multiplying the drift velocity by the *carrier concentration* (n). In this case, the carriers are conduction electrons. To obtain the electrical current density, we must also multiply the carrier concentration by the charge q on each electron.

Thus, the current density (J) is given by

$$J = n \times v_{\text{drift}} \times q \qquad (2.20)$$

Substituting for v_{drift} from Equation 2.17, we get

$$J = n \times \left(\frac{q\tau}{m}\right)E \times q$$

or

$$J = \left(\frac{nq^2\tau}{m}\right) \times E \qquad (2.21)$$

This is a modified form of Ohm's law (Equation 2.11).

Therefore, in Equation 2.21, the term in brackets represents the conductivity (σ) and is given by

$$\sigma = \left(\frac{nq^2\tau}{m} \right) = \left(\frac{q\tau}{m} \right) \times n \times q = \mu_n \times n \times q$$

or

$$\sigma = \mu_n \times n \times q \qquad\qquad (2.22)$$

This equation is very important because it tells us that the conductivity of a metal will be proportional to the carrier concentration (n) and the carrier mobility (μ_n). The longer the time between the scattering events or collisions (i.e., higher τ), the *higher* the mobility of the electrons in a metal. Note that higher mobility does not automatically guarantee a high conductivity because conductivity also depends on the carrier concentration of conduction electrons.

Thus, according to the classical theory, the electrical conductivity of a metal depends on (a) the concentration of carriers carrying the electrical current, that is, how many carriers are available per unit volume of the material, (b) the mobility (μ) of the carriers, and (c) the electrical charge on each carrier. Consider a package-delivery service as an analogy. The number of packages delivered in a day depends on the following: (a) the number of delivery trucks available, (b) the speed with which the trucks can move in traffic, and (c) the number of packages each truck can carry. In some materials, the electrical current is carried by the motion of species other than electrons. We discuss these materials in the next section.

2.6 ELECTRONIC AND IONIC CONDUCTORS

In semiconductors and certain types of ceramics, the motion of species other than electrons (e.g., ions) can generate an electrical current. In this case, we also need to account for their contributions to conductivity. Materials in which a significant part of the conductivity arises from the movement of ions are known as ionic conductors. An example of a ceramic that can exhibit a good electrical conductivity is yttria (Y_2O_3)-stabilized zirconia (ZrO_2), known as YSZ. This material is used as a solid electrolyte in solid oxide fuel cells (Chapter 1). Oxygen gas sensors used in automotives and other applications are also based on YSZ. Similarly, transparent and conductive coatings based on indium–tin oxide (ITO) are used as electrodes in touch-screen displays and solar cells. Electronic conductors are defined as materials in which electrical conduction occurs mostly due to the motions of electrons or holes. A hole is an imaginary particle that signifies the absence of an electron. Thus, metals, alloys, and semiconductors are examples of electronic conductors. If the conductivity is due to movement of both ions and electrons or holes, then the material is known as a mixed conductor. We now know one more way to classify electrical conductors based on the types of charge carriers.

2.7 LIMITATIONS OF THE CLASSICAL THEORY OF CONDUCTIVITY

The classical theory, although useful, does not explain several key features associated with the electrical conductivity of materials. For example, in some materials, such as silicon or diamond (a form of carbon [C]), each with four valence electrons, the experimentally observed conductivity is very small. Furthermore, the conductivity of these materials is extremely sensitive to the presence of even very small levels of other elements. Similarly, according to the classical theory, we would expect the speed with which conduction electrons move to increase with increasing temperature, similar to the situation encountered in the kinetic theory of gases. However, it has been shown that the speed with which conduction electrons move in a material is fairly constant with the temperature. Moreover, classical theory cannot accurately predict the relationship between the *thermal conductivity* and electrical conductivity of metals. Thus, classical theory is inadequate in explaining a number of experimental observations. A more elaborate quantum mechanics–based explanation can help to better understand these observations (Section 2.13).

The following examples illustrate the applications of these concepts.

EXAMPLE 2.6: CARRIER CONCENTRATIONS IN AL AND SI

Calculate the concentration of conduction electrons in (a) Al and (b) Si. Assume that the densities of Al and Si are 2.7 and 2.33 g/cm³. Assuming that the electron mobility values are similar, what will be the expected concentration of electrons in terms of the values of conductivity of Al and Si? The atomic masses for Al and Si are ~27 and 28, respectively.

SOLUTION

a. Because Al has a valence of +3, we assume that each Al atom donates three conduction electrons.

The atomic mass of Al is 27 g; this means that the mass of 6.023×10^{23} atoms (Avogadro's number) is 27 g.

A volume of 1 cm³ is 2.7 g of Al. The number of atoms in this volume will be

$$= \frac{6.023 \times 10^{23} \times \text{density}}{\text{atomic mass}} = \frac{6.023 \times 10^{23} \text{ atoms/mole} \times 2.7 \text{ g/cm}^3}{27 \text{ g/mole}}$$

$$= 6.023 \times 10^{22} \text{ atoms/cm}^3$$

The concentration of conduction electrons will be expected to be three times the concentration of atoms because each Al atom is assumed to donate three electrons. Thus, the concentration of conduction electrons in Al will be

$$n_{Al} = 6.023 \times 10^{22} \frac{\text{atoms}}{\text{cm}^3} \times 3 \frac{\text{electrons}}{\text{atom}}$$

$$= 1.807 \times 10^{23} \text{ electron/cm}^3$$

The conduction electron concentrations for Al and other metals are listed in Table 2.3.

b. Si has a valence of +4; therefore, the concentration of valence electrons, that is, the number of valence electrons in 1 cm³ of Si, will be

$$n_{Si} = \frac{6.023 \times 10^{23} \times 2.33}{28} \times 4 \frac{\text{electrons}}{\text{atom}}$$

$$n_{Si} = 2.0 \times 10^{23} \text{ electron/cm}^3$$

TABLE 2.3

Atomic Mass, Density, Experimentally Measured Electron Mobility, and Electron Concentration for Selected High-Purity Metals

Metal	Atomic Mass (g/mol)	Density (g/cm³)	Mobility of Electrons (μ_n) (cm²/V·s)	Calculated Number of Carrier Particles (n) # (electrons/cm³)
Silver	107.868	10.5	57	5.86×10^{22}
Copper	63.546	8.92	32	8.43×10^{22} (assumes one electron per atom)
Gold	196.965	19.32	31	5.91×10^{22}
Aluminum	26.981	2.7	13	1.807×10^{23} (assumes three electrons per atom)

Data adapted from Webster, J. G. 2002. *Wiley Encyclopedia of Electrical and Electronics Engineering*, Vol. 4. New York: Wiley; and other sources.

Experimentally, we know that the electrical conductivity of Al is very high compared to that of Si. However, as we can see, the calculated concentration of valence electrons in Si is higher than that in Al. Furthermore, as we will see in Chapter 3, a very low concentration of some elements has a huge effect on the conductivity of Si. These observations show that classical theory does not explain the higher level of conductivity in some metals compared to that of Si.

EXAMPLE 2.7: CALCULATING THE CONDUCTIVITY OF Al FROM THE MOBILITY VALUES

Assume that the mobility of electrons (μ_n) in Al is 13 cm²/V·s. The carrier concentration for Al is 1.807×10^{23} electrons (assuming three electrons donated per Al atom). Calculate the expected conductivity of Al.

SOLUTION

From Equation 2.22, the conductivity of Al will be given by the expression

$$\sigma = \left(1.807 \times 10^{23} \frac{\text{electrons}}{\text{cm}^3}\right) \times \left(13 \frac{\text{cm}^3}{\text{V} \cdot \text{s}}\right) \times \left(1.6 \times 10^{-19} \frac{\text{C}}{\text{electron}}\right)$$

$$= 37.9 \times 10^4 \text{ S/cm}$$

Because this value matches well with the data in Table 2.1, our assumption that each Al atom donates three conduction electrons appears reasonable.

EXAMPLE 2.8: CARRIER CONCENTRATION IN Cu

The experimentally measured mobility of electrons in copper is 32 cm²/V·s. Calculate the carrier concentration (n) for Cu. Compare this with the values listed in Table 2.3. What does this show about the number of conduction electrons contributed per Cu atom?

SOLUTION

We use the value of Cu conductivity (σ) provided in Table 2.1.
From Equation 2.22,

$$\sigma_{Cu} = 59.7 \times 10^4 \text{ S/cm} = (n) \times \left(32 \frac{\text{cm}^2}{\text{V} \cdot \text{s}}\right) \times (1.6 \times 10^{-19} \text{C})$$

This gives us a concentration of conduction electrons $n = 1.16 \times 10^{23}$ electrons/cm³.

In Table 2.3, the value listed for the concentration of electrons for Cu is $= 8.43 \times 10^{22}$ electrons/cm³. In arriving at this value in Table 2.3, we had assumed that each Cu atom donates one conduction electron. Because we have now estimated that the actual value of conduction electrons is 1.16×10^{23} electrons/cm³, we know that each Cu atom must be donating more than one electron. The average number of electrons donated per Cu atom will be

$$\frac{1.16 \times 10^{23} \text{ electrons/cm}^3}{8.43 \times 10^{22} \text{ electrons/cm}^3} = 1.38$$

Thus, in Cu, the average number of electrons donated per atom is 1.38. This is expected because Cu does exhibit valences of +1 and +2.

EXAMPLE 2.9: RESISTIVITY OF A SEMICONDUCTOR

A particular single crystal of a semiconducting Si sample, containing a small but deliberately added level of phosphorus (P), provides conductivity with electrons as the majority carriers. Assume that the concentration of P added is 10^{18} atoms/cm³ and that each P atom provides one electron. Ignore any other possible contributions to the conductivity of this P-doped Si sample. If $\mu_n = 700$ cm²/V·s, what is the resistivity (ρ) of this semiconductor?

Because each P atom contributes only one conduction electron, the concentration of P atoms and that of conduction electrons donated by P atoms are equal; therefore, $n = 10^{18}$ electrons/cm^3.

(In the next chapter, we will learn that our assumption about the concentration of conduction electrons contributed by the P atoms, as opposed to those from the Si atoms, is correct.)

Therefore, from Equation 2.22,

$$\sigma = (10^{18}\,\text{electrons/cm}^3) \times \left(700\ \frac{\text{cm}^2}{\text{V}\cdot\text{s}}\right) \times (1.6 \times 10^{-19}\,\text{C})$$

Therefore, $\sigma = 112$ S/cm. Compared to metals (Table 2.1), this value of conductivity of Si containing P is still relatively small. The conductivity of essentially pure Si containing no added elements is even smaller.

We have been asked to calculate the resistivity (ρ), which is equal to $1/\sigma$. Therefore, $\rho = 8.9 \times 10^{-3}\ \Omega\cdot\text{cm}$. As we will see in Chapter 3, the value of the resistivity of Si is very strongly dependent on the presence of trace levels of impurities or deliberately added elements.

2.8 RESISTIVITY OF METALLIC MATERIALS

Which factors have the greatest influence on the resistivity of metallic materials? In this section, we focus on metallic materials containing small concentrations of other elements.

We must consider the three following processes (Figure 2.8). First, atoms in a metallic material vibrate. We refer to the vibrations of atoms in a material as phonons. The conduction electrons carrying the electrical current bump into or scatter off vibrations of these atoms. The process of an electron-scattering of a phonon is temperature-dependent. As the temperature increases, the number of electrons scattering off due to the vibrations of atoms also increases. Thus, the resistivity of a pure metal increases with increasing temperature. This increase in resistivity due to the scattering of electrons by lattice vibrations is linear. We refer to this temperature-dependent component as resistivity (ρ_T; Figure 2.9).

The second factor that affects the resistivity of metals is the scattering of electrons off defects, such as vacancies, dislocations, and grain boundaries (Chapter 1). If you are unfamiliar with these terms related to the microstructure of materials, it may be worthwhile to review these concepts (see Askeland and Fulay 2006). The concentrations of different defects will depend on the microstructure of the material. However, we can assume that at a given temperature or within a small temperature range, the microstructure, and therefore the concentrations of these defects (which serve as scattering centers), is essentially constant.

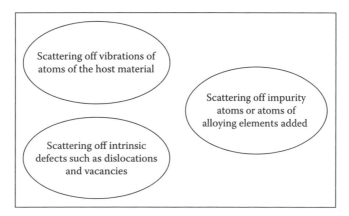

FIGURE 2.8 Schematic showing the sources of the scattering of conduction electrons in a metallic material.

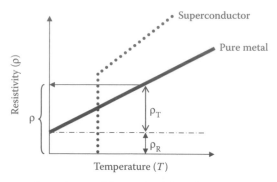

FIGURE 2.9   The temperature dependence of the conductivity of a typical metal and a superconductor.

The third factor that influences the resistivity of metallic materials is the presence of impurities, or other elements deliberately or accidentally added. Impurities cause an increase in electrical resistivity because the atoms of the added elements act as scattering centers, regardless of whether they are smaller or larger than the host atoms. Because the impurity concentrations should be independent of temperature, the effect due to the presence of impurities is generally *not* very sensitive to temperature changes.

We designate the effect of (a) microstructural defects and (b) impurities or added elements the residual resistivity (ρ_R; Figure 2.9). These are temperature-independent effects. Resistance due to these effects will remain even if we somehow eliminate the scattering of electrons due to the vibrations of atoms. The electrical resistance of most metals does *not* become zero after cooling to extremely low temperatures (Figure 2.9).

Whenever a heat treatment or any other process causes a change in the microstructure of the material, we can expect a change in the resistivity of a material. For example, *annealing* metals and alloys will probably increase their electrical conductivity because dislocations, a type of atomic-level defect in the arrangement of atoms, are annihilated. Similarly, if we bend an annealed copper wire or deform a metal or an alloy in some manner, the resistivity will increase. This is because of an increase in the dislocation density (see Section 2.11).

The conductivity of pure metals *decreases* with increasing temperature because of the increased scattering of electrons off the phonons. Adding alloying elements or the presence of impurities causes a disruption in the periodic order of atoms and introduces a strain (see Section 2.11). This causes a sudden change in the potential energy of conduction electrons as they approach the atoms of foreign elements (the alloying elements and/or impurities). The scattering of electrons from these foreign atoms then increases, thereby causing increased resistance. Therefore, we expect the conductivity of alloys to be generally lower than that of essentially pure metals. The conductivity of alloys is controlled by "extrinsic" factors, that is, the concentration and nature of the added alloying elements (see Section 2.12). Thus, in alloys, mobility is limited by the scattering of electrons off impurity atoms. We refer to the mobility that is limited by this effect as *impurity-scattering limited drift mobility*. Conductivity (σ) in essentially *pure* metals is largely limited by the scattering of electrons of the vibrations of atoms. The value of mobility that is limited by these effects is known as *lattice-scattering limited mobility*. The conductivity of pure metals limited by lattice scattering is known as *lattice-scattering limited conductivity*.

Furthermore, because the scattering of conduction electrons is controlled by a relatively large concentration of atoms of other alloying elements, the conductivity of alloys will *not* vary significantly with temperature.

For metals, we can write the resistivity (ρ) as a sum of two components ρ_T and ρ_R:

$$\rho = \rho_T + \rho_R \tag{2.23}$$

This equation is also known as *Matthiessen's rule*, where ρ_T is the temperature-dependent part of resistivity, representing the scattering of electrons due to the vibrations of the atoms. The part ρ_R represents scattering due to the presence of impurities, added alloying elements, and microstructural defects in the arrangement of atoms within a material (Figure 2.9).

If we assume that ρ_T changes linearly with temperature, we can rewrite the dependence of resistivity (ρ) with temperature as follows:

$$\rho = AT + B \tag{2.24}$$

Instead of listing the values of A and B for different metals (Equation 2.24), we will define the TCR, designated as α_R (Section 2.1).

$$\alpha_R = \frac{1}{\rho_0}\left(\frac{\rho - \rho_0}{T - T_0}\right) \tag{2.25}$$

In Equation 2.25, ρ_0 is the resistivity at some reference temperature ($T_0 = 273$ or 300 K), and ρ is the value of resistivity at some other temperature T. Note that the value of α_R will depend on what we use as the reference temperature.

From Equation 2.25, we can show that

$$\rho = \rho_0\left[1 + \alpha_R\left(T - T_0\right)\right] \tag{2.26}$$

Note that this can be rewritten as follows:

$$R = R_0\left[1 + \alpha_R\left(T - T_0\right)\right] \tag{2.27}$$

The values of α_R for different metals were listed in Table 2.1. The values of α_R for some alloys are shown in Tables 2.4 and 2.5 (see Section 2.12).

The relative change in the resistivity of some metals over a wider range of temperatures is shown in Figure 2.10. Note that nickel (Ni) and iron (Fe) are magnetic, and the change in the resistivity of these materials is not linear. In addition, aluminum melts at ~657°C; thus, there may be some changes in its microstructure as it approaches higher temperatures. If the duration of a material's exposure to higher temperatures causes microstructural changes, the resistivity of the material can change. For example, in the case of metals such as aluminum and copper, annealing the material at higher temperatures will cause the annihilation of dislocations and thus cause an increase in the conductivity.

The example below explains how the change in the resistivity of platinum can be used to make precise and accurate temperature-measurement devices.

EXAMPLE 2.10: Pt RESISTANCE–TEMPERATURE DETECTORS

A *resistance–temperature detector* (RTD) makes use of a metal wire (e.g., that of Pt; Figure 2.11). As the temperature changes, the electrical resistance changes; thus, by measuring the change in resistance, we can measure the temperature. A Pt RTD can operate at temperatures in the range of −200°C to +700°C. Assume that the α_R for the Pt material used is 3.93×10^{-3} $\Omega/\Omega \cdot$ °C. The RTD wire is designed so that the resistance at 0°C is 100 Ω. (a) What is the temperature if the resistance measured is 200 Ω? (b) Why is Pt preferred for making RTDs, in comparison to W, Cu, Ag, or Au? (c) In Europe, Pt compositions with $\alpha_R = 0.00385$ $\Omega/\Omega \cdot$ °C are often used. For $R = 200$ Ω, what is the predicted temperature if we use an RTD made from this Pt composition?

TABLE 2.4
Properties of Some Heating-Element Materials

Material	Resistivity at 20°C ($\mu\Omega \cdot cm$)	α_R at 20°C ($\Omega/\Omega \cdot °C$)	Maximum Operating Temperature (°C)	Main Applications
Nickel 80/chromium 20 (Nikorthal™-type alloys)	108	$+14 \times 10^{-3}$	1200	Furnaces, heating elements for domestic appliances
Chromium 22/aluminum 5.8/balance iron (Kanthal™-type alloys)	145	$+3.2 \times 10^{-3}$	1400	Furnaces for heat treatment
Platinum 90/rhodium 10	18.7	–	1550	Laboratory furnaces
Platinum 60/rhodium 40	17.4	–	1800	Laboratory furnaces
Molybdenum	5.7	4.35×10^{-3}	1750	Vacuum furnaces, inert atmosphere
Tantalum	13.5	3.5×10^{-3}	2500	Vacuum furnaces
Graphite	1000	-26.6×10^{-3}	3000	Furnaces requires nonoxidizing atmosphere
Molybdenum disilicide ($MoSi_2$)	40	1200×10^{-3}	1900	Glass industry, laboratory furnaces, ceramic processing
Lanthanum chromite ($LaCrO_3$)	2100	–	1800	Laboratory furnaces
Silicon carbide (SiC), also known as glowbar	1.1×10^5	–	1650	Industrial furnaces
Zirconia	–	–	2200	Laboratory furnaces, becomes an ionic conductor above ~1000°C

Source: Adapted from Laughton, M. A., and D. F. Warne, eds. 2003. *Electrical Engineer's Reference Book.* Amsterdam: Elsevier; and other sources.

TABLE 2.5
Resistivity and Temperature Coefficient Values for Selected Materials

Material	Typical Composition (wt%)	Resistivity ($\mu\Omega \cdot cm$) at 20°C	Temperature Coefficient of Resistivity (α_R) $\Omega/\Omega \cdot °C$ ($T_0 = 300$ K)
Lead (Pb)–tin (Sn) solder	Sn: 63, Pb: 37	14.7	
Brass	Cu: 60, Zn: 40	6.4	1000
	Cu: 70, Zn: 30	8.4	2000
Nichrome	Ni: 58.5, Fe: 22.5, Cr: 16, Mn: 3	100	400
Constantan	Cu: 60, Ni: 40	44.1	+2/+33
Manganin	Cu: 84, Mn: 12, Ni: 4	45	+6/−42 (12°C–100°C)
	Cu: 83, Mn: 13, Ni: 4 (Wire Alloy)	48.2	+15/−15 (15°C–35°C)
	Cu: 86, Mn: 10, Ni: 4 (Shunt Alloy)	38.3	+15/−15 (40°C–60°C)
Palladium–silver alloy	Pd: 60, Ag: 40	42–44	

Data from various sources.

SOLUTION

a.

$$\alpha_R = \frac{1}{\rho_0}\left(\frac{\rho - \rho_0}{T - T_0}\right)$$

$$\therefore 0.00393 = \frac{1}{R}\left(\frac{R - R_0}{T - T_0}\right) = \frac{1}{100}\left(\frac{R - 100}{T - 0}\right)$$

$$\therefore 0.393T = R - 100 \quad \text{or}$$

$$R = 100 + 0.393T$$

Thus, the variation in resistance is 0.393 Ω/°C.
Because the resistance measured by the RTD is 200 Ω, we rewrite as

$$200 = 100 + 0.393T \quad \text{or}$$

$$T = \frac{200 - 100}{0.393} = 254.45°C$$

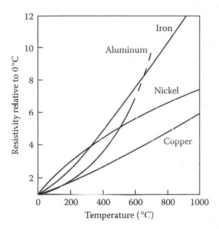

FIGURE 2.10 Relative change in resistivity with temperature (relative to 0°C) for Al, Cu, Ni, and Fe. (From Laughton, M. A., and D. F. Warne, eds. 2003. *Electrical Engineer's Reference Book*. Amsterdam: Elsevier. With permission.)

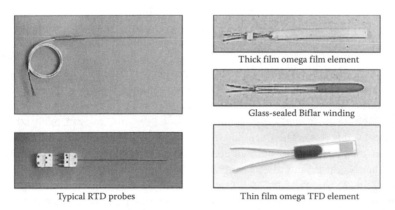

FIGURE 2.11 Commercially available resistance–temperature detector devices. (Courtesy of Omega Corporation, San Diego, CA.)

b. We could use Cu instead of Pt, but Cu has a lower resistivity. This means that we will have to use a longer wire. Cu will also oxidize at high temperatures. It can be used for high temperatures; however, it is brittle. Au and Ag exhibit much higher levels of conductivity, and we would need longer wires; moreover, these metals are expensive.

c. If we used a Pt composition with $\alpha_R = 0.00385 \; \Omega/\Omega \cdot {}^{\circ}C$, then for a wire whose resistance is $100 \; \Omega$, the change in resistance will be

$$R = 100 + 0.385T$$
$$200 = 100 + 0.385T \quad \text{or}$$
$$T = \frac{200 - 100}{0.385} = 259.7^{\circ}C$$

Note that a seemingly small change in the TCR, from 0.00393 to 0.00385 $\Omega/\Omega \cdot {}^{\circ}C$, causes a considerable change in the value of the predicted temperature. In practice, the exact coefficient of the resistivity of the RTD element must be known. Also, the data for a range of temperatures must be fitted to a polynomial and not to the equation of a straight line.

2.8.1 EFFECT OF THERMAL EXPANSION

In Example 2.10, when calculating the resistance as a function of temperature, we ignored the change in resistance of the platinum wire due to changes in its length caused by thermal expansion. If a copper wire is heated from 25°C to 75°C, the length of the copper wire will increase. The thermal expansion coefficient for copper is (α_{Cu}) ~$17 \times 10^{-6}/{}^{\circ}C$. The increase in length will cause an increase in the resistance.

Note that in some applications, this change in length—caused by a change in temperature or the presence of stress—may be important and may have to be accounted for, for example, if a type of strain gauge makes use of the change in electrical resistance to calculate the magnitude of strain.

2.9 JOULE HEATING OR I^2R LOSSES

When a material offers resistance to the flow of an electric current, the energy lost during the collisions of electrons with vibrations of atoms or other defects appears as heat. This is known as *Joule heating* or I^2R *losses*. The power dissipated by a material with resistance (R) carrying a current (I) is given by the following equation:

$$\text{Power lost} = V \times I = I^2 \times R \tag{2.28}$$

Obviously, Joule heating is not useful when the goal is to carry current with the least possible resistance. Thus, in ICs used to make computer chips, we try to minimize the I^2R losses. Heat sinks are designed and built into the computer chip packaging to conduct the heat away from the chip. The main concern regarding the use of ICs is not that the energy is wasted but that the electrical performance of components is usually reduced with increased temperatures.

In many applications, however, Joule heating can be useful. One of the most important applications of alloys such as nichrome, kanthal, and so on, and ceramic resistor materials is their use as heating elements (Table 2.4). This includes heating elements used in small consumer appliances, such as toaster ovens, hair dryers, and immersion water heaters. Incandescent light bulbs use tungsten filaments that produce light after being heated to high temperatures. Many materials, such as molybdenum disilicide ($MoSi_2$), are used to make heating elements for laboratory furnaces. Many of the metallic materials used as heating elements are oxidized by heating them in air. The oxides on the surface form a dense, adherent oxide layer (such as in SiO_2, Al_2O_3, or chromium oxide [Cr_2O_3]), which protects the base material from further oxidation. Another application where Joule heating is useful is in the manufacturing of electrical fuses. In an electrical circuit, when a certain value current is exceeded, the material in the fuse melts as a result of Joule heating. This creates an open circuit and prevents current flow.

EXAMPLE 2.11: CALCULATION OF JOULE POWER LOSSES IN A Cu BUS BAR AND THE COST OF ELECTRICITY

A Cu *bus bar*, or a conductor used in power transmission, is 300 feet long and has a cross section of ¼ × 4 inches. (a) What is the resistance of this bar at room temperature? (b) Calculate the Joule losses in kilowatts if the current is 1000 A. Assume that the current is DC. (c) What is the energy lost (in kW · h) if the current is carried for 24 hours? (d) Assuming electricity costs 4 cents/kW · h, calculate the dollar value of the energy wasted due to Joule losses on a per-year basis. Assume that the conductivity of the Cu bar is 5.8×10^7 S/m.

SOLUTION

a.

$$R_{\text{Copper bus bar}} = \rho \frac{L}{A} = \frac{300 \text{ ft.} \times 12 \text{ in./ft.} \times 2.54 \text{ cm/in.}}{\left(\frac{1}{4} \times 4\right) \text{in.}^2 \times (2.54 \text{ cm/in.})^2 \times (5.8 \times 10^7 \text{ S/m} \times 10^{-2} \text{m/cm})} = 2.44366 \times 10^{-3}\,\Omega$$

b. Now, we can calculate the Joule losses as follows:

$$\text{Power lost} = V \times I = I^2 \times R = (1000)^2 \times 0.00244366\ \Omega = 2443.66 \text{ W}$$

Thus, the power lost is 2.433 kW.
c. The energy lost in 24 hours will be

$$\text{Energy lost} = \text{power} \times \text{time} = 2.443 \text{ kW} \times 24 \text{ h} = 58.6 \text{ kW·h}$$

As a reference, the total electric power production in the United States in 2005 was ~4 × 10^{12} kW · h or 4 trillion kW · h.
d. If the cost of electricity is 4 ¢/kW · h, the total cost of electricity wasted due to Joule losses will be $2.34 each day. In a year, the cost of electricity wasted due to Joule losses will be 365 days × $5.86/day = $854.

2.10   DEPENDENCE OF RESISTIVITY ON THICKNESS

So far, we have assumed that the resistivity (ρ) of a material does not change with its geometric dimensions. However, there is one exception—if the thickness of the film is comparable to the mean free path length (λ) for the conduction electrons (see Section 2.5), then the resistivity of the material does depend on its thickness (Gupta 2003). For bulk copper (e.g., a large piece of copper or copper wire), the mean free path length (λ) for conduction electrons is ~400 Å (40 nm). As the thickness (t) of a resistor approaches the mean free path length (λ), the scattering of electrons from the film surface increases. In addition to surface scattering, the resistivity can also increase due to the presence of surface contaminants or due to higher resistivity phases formed as a result of surface oxidation or other chemical reactions. These effects tend to be more pronounced as the thickness of the film is in the nanoscale region. Therefore, a considerable scatter in the data concerning the resistivity of very thin films is often observed. For bulk, nearly pure, and annealed copper, the resistivity is ~1.7 $\mu\Omega$ · cm (Table 2.1). For copper films with a thickness of $t \sim 50$ nm, the room-temperature resistivity value is ~2.5 $\mu\Omega$ · cm. For copper films of 20 nm thickness, the copper resistivity is ~5 $\mu\Omega$ · cm.

TABLE 2.6
Conductivity of Different Materials (at 20°C) Based on the IACS Scale

Material	% IACS Conductivity	Material	% IACS Conductivity
Annealed copper	100	Nickel	25
99.999% Copper	102.5	Iron	17
Electrolytic tough-pitch copper ~99.0%	100.2–101.5	Platinum	16
Oxygen-free high-conductivity copper	101		
Silver	104	Tin	13
Aluminum	60	Lead	8

IACS = International Annealed-Copper Standard.
Data from various sources.

If we include the dependence of resistivity on thickness, then Matthiessen's rule (Equation 2.23) can be modified and rewritten as follows:

$$\rho = \rho_T + \rho_R + \rho_{thickness} \tag{2.29}$$

Copper is a very important conducting material. As a result, there is a special measuring unit based on the conductivity of copper, known as the *international annealed-copper standard* (IACS). We define 100% IACS as a conductivity of 57.4013×10^6 S/m, or 57.4013×10^4 S/cm (or $\rho = 1.74212 \times 10^{-6}$ $\Omega \cdot$ cm). This unit is based on an annealed 1-meter-long copper wire, with a cross-sectional area of 1 mm^2 and a resistance of 0.174212 Ω (Laughton and Warne 2003).

We can calculate the % IACS of any material using Equation 2.30.

$$\% \text{ IACS} = \frac{1.74212}{\text{Resistivity } (\Omega \cdot \text{m})} \times 100 \tag{2.30}$$

The conductivity of the so-called *electrolytic tough pitch* (ETP, or just TP) copper, which is 99.0% pure (~100–600 ppm oxygen), is ~100.2% IACS. The conductivity of 99.999% copper is 102.5% IACS. The IACS standard was developed in 1913 and was based on the highest-conductivity copper available at that time. Many compositions that have been developed since then have conductivity higher than 57.4013×10^4 S/cm. As a result, we have several high-purity copper samples on the IACS scale with conductivity greater than 100% (Table 2.6).

2.11 CHEMICAL COMPOSITION–MICROSTRUCTURE–CONDUCTIVITY RELATIONSHIPS IN METALS

2.11.1 INFLUENCE OF ATOMIC-LEVEL DEFECTS

We will now compare the conductivities of pure, annealed copper to that of pure, cold-worked copper. The term "cold-working" refers to a deformation process carried out at a relatively low temperature (such as bending, wire-drawing, or extrusion carried out at temperatures below the so-called recrystallization temperature). Cold-working annealed metals and alloys leads to an increase in the density of dislocations. This will cause the resistivity of the annealed metals to increase. For example, the decrease in the conductivity of sterling silver (92.5% Ag–7.5% Cu) as a function of percentage of cold work after the drawing process is shown in Figure 2.12a. The increased dislocation

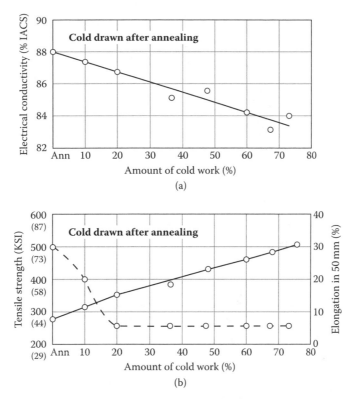

FIGURE 2.12 (a) Electrical conductivity of sterling silver (92.5% Ag–7.5% Cu) shown as % IACS; and (b) increase in tensile strength (solid line) and decrease in percentage of elongation (dotted line) as a function of % cold work. The data are for a 2.3-mm wire, which was cold-drawn after annealing. (From ASM International, eds. *Properties and Selection: Nonferrous Alloys and Special Purpose Materials*, Vol. 2. Materials Park, OH: ASM. With permission.)

density caused by the cold-working process causes a simultaneous increase in the tensile strength (Figure 2.12b). These data are for a 2.3-millimeter wire that was cold-drawn after annealing. The percentage of cold work is defined as follows:

$$\text{Percentage of cold work} = \frac{A_0 - A_f}{A_0} \times 100 \qquad (2.31)$$

where A_0 is the original cross-sectional area and A_f is the final cross-sectional area after cold-working.

In metals and alloys, the annealing heat treatment leads to the annihilation of dislocations. Because the cold-worked pure copper has a higher dislocation density, we expect its resistivity to be higher than that of annealed pure copper.

2.11.2 INFLUENCE OF IMPURITIES

The conductivity of pure metals is also very sensitive to the presence of impurities. This effect does *not* come out as a prediction of the classical theory of conductivity. The change in the conductivity of copper as a function of the different alloying elements is shown in Figure 2.13.

The impurities that have the most dramatic effect on the conductivity of copper are phosphorous, titanium (Ti), cobalt (Co), iron, arsenic (As), and antimony (Sb). Elements such as silver, cadmium (Cd), and zinc (Zn) produce relatively less increase in the resistivity. You may also know that the

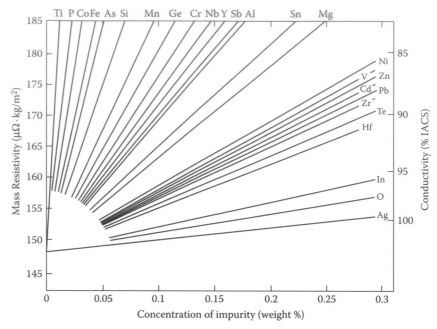

FIGURE 2.13 Effects of different impurities on the conductivity of copper. (Courtesy of Copper Development Association, New York.)

addition of alloying elements in very small concentrations usually causes an increase in the strength of a metal. This is known as solid-solution strengthening (Chapter 1).

When alloys form, up to a certain concentration of one metal often completely dissolves in another, similar to sugar or salt dissolving in water. The resultant alloy is said to be a solid solution. For example, nickel can be added to copper to form a solid solution.

Thus, silver, cadmium, and zinc are good choices for strengthening annealed copper without causing an adverse effect on its electrical conductivity. Also note that, as shown in Figure 2.14, the effect of oxygen on the resistivity of copper is smaller than expected. This is because oxygen is usually present as a fine precipitate of copper and other oxides. Oxygen does not appear as an impurity dissolved in a solid solution.

The most widely used copper-alloy conductor is known as electrolytic tough pitch (ETP) copper, which contains about 100–650 ppm (or 1 mg/kg) oxygen. Note that "ppm" is not a scientific unit. The oxygen present in copper scavenges the dissolved hydrogen and sulfur during copper refining. Moreover, oxygen reacts with the other metal impurities that are originally dissolved in copper and causes them to precipitate as oxides. By adding an optimum concentration of oxygen, we cause an *increase* in the electrical conductivity of copper (Figure 2.14). Thus, instead of having a small level of an impurity distributed throughout the material (as in a solid solution), we can concentrate that impurity in the precipitate particles of a second phase; in this way, we can increase the conductivity without changing the nominal chemical composition. In some cases, where intricate castings need to be made or joining processes like welding or brazing need to be used, we cannot make use of ETP copper. For these applications, *oxygen-free high-conductivity copper* (OFHC) is used. This copper has less than 0.001% oxygen and is more expensive. The conductivity of OFHC and ETP copper is ~101% IACS. ETP and OFHC copper are the most widely used varieties of copper. Their applications include windings for motors, generators, bus bars, and so on.

The effect of the addition of different alloying elements on the resistivity of platinum is shown in Figure 2.15. Adding alloying elements also strengthens platinum; the consequent increase in strength, as represented by increased hardness, is shown in Figure 2.16.

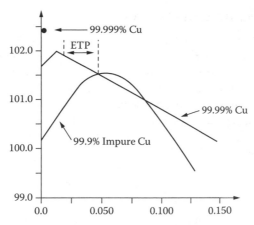

FIGURE 2.14 Effect of oxygen concentration on the conductivity of copper. (Courtesy of Copper Development Association, New York.)

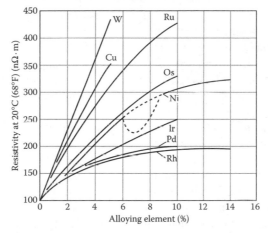

FIGURE 2.15 Effects of additions of various alloying elements on the resistivity of platinum. (From Vines, R. F. and E. M. Wise. 1941. *The Platinum Metals and Their Alloys*. New York: International Nickel Co. With permission.)

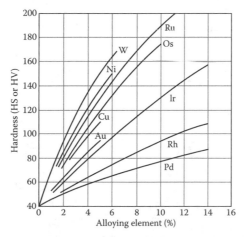

FIGURE 2.16 Increase in the hardness of platinum due to the addition of various alloying elements. (From Lampman, S. R., and T. B. Zorc, eds. 1990. *Metals Handbook: Properties and Selection: Nonferrous Alloys and Special Purpose Materials*. Materials Park, OH: ASM. With permission.)

2.12 RESISTIVITY OF METALLIC ALLOYS

Alloys are metallic materials that are characterized by the presence of more than one element. The dominant element or matrix element is what is used while referring to the alloy. For example, the term *copper alloy* means that copper is the dominant element. Other elements present in considerable concentrations are added deliberately to improve the properties of the alloy. For example, in copper–beryllium (Cu–Be) alloys, beryllium is added to increase the Young's modulus of the alloy. The electrical properties of some alloys are summarized in Table 2.5.

When alloys form a solid solution, we can relate the resistivity of the alloy to the resistivity of the pure metal using *Nordheim's rule*:

$$\rho_{alloy} = \rho_{matrix} + Cx(1-x) \tag{2.32}$$

where ρ_{alloy} is the resistivity of the alloy, ρ_{matrix} is the resistivity of the base metal without any alloying elements, C is Nordheim's coefficient, and x is the *atom fraction* of the alloying element added. The first term, that is, ρ_{matrix}, accounts for the resistivity of the base or matrix material. If an alloy system forms a solid solution over a complete range of compositions (i.e., from 0% to 100% of solute), then we expect a parabolic variation in the resistivity, as predicted by Equation 2.32. This is seen in the copper–nickel system (Figure 2.17).

Alloys of platinum and palladium (Pd) are often used as electrodes in devices such as multilayer capacitors and electrical contacts. This is another example of a system in which solid solutions are formed over the entire range of compositions. The electrical resistivity of platinum–palladium alloys as a function of the palladium concentration is shown in Figure 2.18 (Lampman and Zorc 1990).

When palladium is added to platinum or vice versa, the tensile strength increases; this effect is shown in Figure 2.19 (Lampman and Zorc 1990).

Note that most alloy systems do *not* show the formation of solid solutions over the entire composition ranges. Instead, different phases are often formed, or the solubility of one component into another is limited. In some alloys, heat treatments can lead to the "ordering" of atoms. Processes such as ordering, clustering, and precipitation of different phases can have a significant effect on the observed resistivity values. For example, in the copper–gold system, the formation of ordered phases leads to an *increase* in conductivity. Both composition and microstructure have a major effect on the resistivity of alloys.

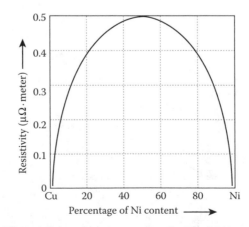

FIGURE 2.17   Variation in the resistivity of copper as a function of nickel concentration. (From Neelkanta, P. 1995. *Handbook of Electromagnetic Materials*. Boca Raton, FL: CRC Press. With permission.)

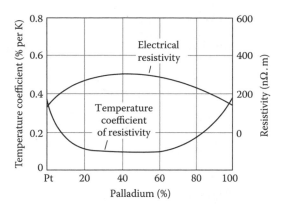

FIGURE 2.18 Electrical resistivity (in nΩ·m) and temperature coefficient of resistivity (%/K) for Pt–Pd alloys. (From Lampman, S. R., and T. B. Zorc, eds. 1990. *Metals Handbook: Properties and Selection: Nonferrous Alloys and Special Purpose Materials.* Materials Park, OH: ASM. With permission.)

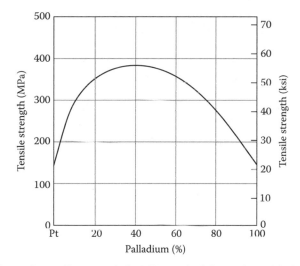

FIGURE 2.19 The change in tensile strength (in MPa on the left *y*-axis, and in ksi on the right *y*-axis) of annealed platinum–palladium alloys as a function of palladium concentration. (From Lampman, S. R., and T. B. Zorc, eds. 1990. *Metals Handbook: Properties and Selection: Nonferrous Alloys and Special Purpose Materials.* Materials Park, OH: ASM. With permission.)

EXAMPLE 2.12: RESISTIVITY OF A Cu ALLOY USING NORDHEIM'S RULE

Pure Au and Ag are too soft for most applications. As a result, Cu is added as an alloying element to strengthen these metals using solid-solution strengthening. What is the resistivity of an Au alloy containing 1.5 weight % Cu? Assume that Nordheim's coefficient (C) for Cu dissolved in Au is 450 nΩ·m (Equation 2.32).

SOLUTION

From Table 2.1, we see that the resistivity of Au is 2.35 μΩ·cm or 23.5 nΩ·m. Note that in Equation 2.32, the concentration of the alloy-forming element has to be expressed as an atom fraction, which is achieved by applying the following equation:

$$x = \frac{M_{Au}w}{(1-w)M_{Cu} + wM_{Au}}$$

where x is the atom fraction of Cu, w is the weight fraction of Cu, M_{Cu} is the atomic mass of Cu, and M_{Au} is the atomic mass of Au. In our case, the wt% of Cu is given as 1.5; thus, the weight fraction of Cu is 1.5/100 = 0.015. From the periodic table, $M_{Au} = 197$ g/mol, $M_{Cu} = 63.5$ g/mol. Thus,

$$x = \frac{197 \times 0.015}{[(1-0.015) \times 63.5] + [0.015 \times 197]} = 0.045$$

The concentration of Cu as an atom fraction is $x = 0.045$.
Therefore, using Nordheim's rule, we get:

$$\rho_{alloy} = Cx(1-x)$$
$$= 23.5 + 17.19 = 40.69 \sim 40.7 \ n\Omega \cdot m$$
$$\rho_{alloy} = 23.5 + [400 \times 0.045(1-0.045)]$$

We can see that a small concentration of Cu increases the resistivity of Au substantially. Also, note that Cu is actually a better conductor than Au. However, when we add Cu to Au as an alloying element, the Cu atoms disrupt the arrangement of the Au atoms. This increases the scattering of conduction electrons in Au and causes an increase in resistivity.

As discussed in Section 2.8, we typically expect alloys to have low electrical conductivities; moreover, the conductivity of metallic alloys does not change much with temperature (Figure 2.20).

The elements added to the alloys or those present as impurities in raw materials or processing may remain dissolved in the alloy and form a solid solution. Alternatively, impurities can react with one another or with other elements present to form various separate phases. Both the concentration and the manner in which the impurities are distributed in the microstructure influence the observed values of resistivity.

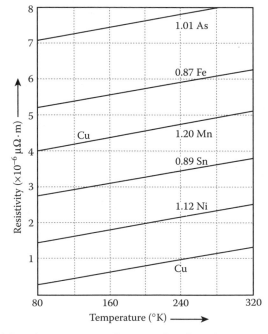

FIGURE 2.20 The resistivity of some copper alloys as a function of temperature. The data for resistivity change of copper is also shown. (From Neelkanta, P. 1995. *Handbook of Electromagnetic Materials*. Boca Raton, FL: CRC Press. With permission.)

2.13 QUANTUM MECHANICAL APPROACH TO CONDUCTIVITY

We will now turn our attention to the quantum mechanical approach to explain variations in the electrical conductivity of different solids. This approach leads to the band theory of solids. Quantum mechanics provides a powerful approach for explaining a number of features related to the conductivity of materials that are not explained by the classical theory. In quantum mechanics, electrons are treated as waves rather than as particles.

We start with a short discussion on electron quantum numbers.

The principal quantum number (n) quantizes the electron energy. Quantization means that electrons in an atom can have only certain levels of energy. The principal quantum number accounts for the Coulombic interactions between the positively charged nucleus and an electron. A value of $n = 1$ corresponds to the K shell, $n = 2$ corresponds to the L shell, and so on (Figure 2.6). Another quantum number is related to the angular momentum of the electrons. This is known as the orbital angular momentum quantum number (l) or azimuthal quantum number. For a given value of the principal quantum number n, there are subshells that are characterized by different values of the orbital angular momentum. In a description of the electronic configuration, a value of $l = 0$ corresponds to the letter for subshell "s." Similarly, a value of $l = 1$ corresponds to the letter for subshell "p," and so on. As an example, for the energy level with $n = 3$, that is, the M shell, there will be 3s, 3p, and 3d subshells. Thus, different combinations of the quantum numbers n and l represent different electron energy levels.

The different allowed values of these and other quantum numbers are shown in Table 2.7. In addition to n and l, an electron has a *magnetic quantum number (m or m_l)* that represents the component of orbital angular momentum along an external magnetic field. Finally, an electron also has a quantum number that quantizes the spin, known as the *spin quantum number (s or m_s)*. This quantum number becomes especially important in understanding the magnetic properties of materials (Chapter 9).

A complete set of quantum numbers, that is, n, l, m, and s, describes the unique quantum state of an electron. A set of quantum numbers represents what is described as a *wave function* associated with an electron. A wave function is equivalent to an orbital or an energy level. One important principle from quantum mechanics is known as the *Pauli's exclusion principle*. This principle states that no two electrons in a given system, such as an atom, can have all four identical quantum numbers. If two electrons have the same values of n, l, and m, then according to Pauli's exclusion principle, their spins must be opposite. Such electrons are said to be *spin-paired electrons*. The possible values of quantum numbers are summarized in Table 2.7.

The maximum number of electrons allowed in a shell with a given value of n is $2n^2$. Thus, for values of $n = 1$, 2, and 3, the maximum number of electrons allowed is 2, 8, and 18, respectively.

TABLE 2.7
Quantum Numbers for Electrons

Principal quantum number (n)	$n = 1, 2, 3,$
	$n = 1$ is the K shell, $n = 2$ is the L shell, and so on. The maximum number of electrons for a shell with given n is $2n^2$
Orbital angular-momentum quantum number (l) (also known as the azimuthal quantum number)	$l = 0, 1, \ldots (n-1)$
	$l = 0$ indicates the s subshell, $l = 1$ indicates the p subshell, etc. The maximum number of electrons in the various subshells are s = 2, p = 6, d = 10, f = 14, and g = 18
Magnetic quantum number (m or m_l)	$m = -l, -(l-1), \ldots, 0, (l-1), l$
Spin (s or m_s)	$s = \pm\frac{1}{2}$

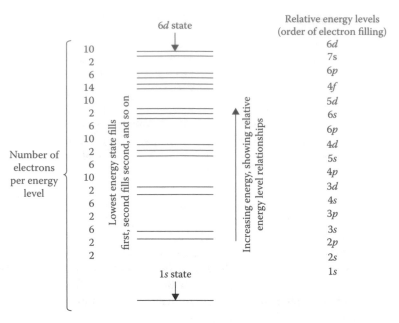

FIGURE 2.21 The order and number of electrons for different elements. (From Minges, M. L. 1989. *Electronic Materials Handbook.* Vol. 1. Materials Park, OH: ASM. With permission.)

For the s, p, d, f, and g sublevels, the maximum number of states or energy levels allowed are 2, 6, 10, 14, and 18, respectively.

The order in which the different energy levels are filled is as follows:

1s,2s,2p,3s,3p,4s,3d,4p,5s,4d,5p,6s,5d,4f,6p,7s,6d,5f.... .

Note that in this filling order, there are some inversions. For example, we use up the 4s level before 3d. A complete description of reasons for this is beyond the scope of this book. However, it is important to note that these inversions do occur. These inversions play a key role in the determination of both the electronic configuration of transition elements and the magnetic properties of ceramic ferrites and iron- and nickel-based magnetic materials.

Examples 2.13 and 2.14 show how these concepts can be used to describe the electronic configuration for different elements. Figure 2.21 contains a list of the electronic configurations for different elements.

EXAMPLE 2.13: ELECTRONIC CONFIGURATION FOR Al

Write down the electronic configuration for Al, whose atomic number (Z) is 13. Explain the meaning of the different symbols used.

Solution

For Al, Z = 13; this means that there are 13 electrons in one Al atom.

We start with $n = 1$. In this level, we can have $2(1)^2 = 2$ electrons. For $n = 1$, the only possible value of l is 0. In this s subshell, we can have only two electron energy levels. Thus, the first part of the configuration is $1s^2$ (read as "one s two"). Then, for $n = 2$ (or the L shell), the possible values of l are 0 and 1, that is, s and p sublevels. The total maximum number of electrons in this shell can be $2(2)^2 = 8$. Thus, when $n = 2$, for the s subshell, we can have two electrons; and for the p subshell, we can have six electrons. Thus, the configuration thus far will read $1s^2 2s^2 2p^6$. This accounts for a total of $2 + 2 + 6 = 10$ electrons. We have only three more electrons left. We move to the $n = 3$ level (M shell). We can have up to $2(3)^2 = 18$ electrons in this level, where the possible values of l are 0, 1, and 2 or s, p, and d subshells. We start with the s subshell, place two electrons here, and then move to the p subshell and place one more electron here.

Thus, the electronic configuration for Al will be $1s^2 2s^2 2p^6 3s^2 3p^1$.

TABLE 2.8
Electron Spin States in Iron (Fe)

$n = 1$	$n = 2$		$n = 3$		$n = 4$	
1s	2s	2p	3s	3p	3d	4s
↑	↑	↑↑↑	↑	↑↑↑	↑↑↑↑↑	↑
↓	↓	↓↓↓	↓	↓↓↓	↓	↓

An arrow pointing up (↑) means $s = +1/2$, or spin up; an arrow pointing down (↓) means $s = -1/2$, or spin down. Note, the 4s level is filled before the 3d level and two of the 3d electrons are spin-paired.

Source: Edwards-Shea, L. 1996. *The Essence of Solid-State Electronics.* Upper Saddle River, NJ: Prentice Hall. With permission.

The p subshell can hold five more electrons, but these levels will remain empty because the Al atom has only 13 electrons. Similarly, the 3d and higher energy levels, such as 4s, 4p, and 4d, will also remain empty.

The electrons in the outermost shell ($n = 3$), that is, the 3s and 3p electrons, are referred to as the valence electrons. These electrons are particularly important because they are available for both electrical conduction and chemical reactions in metallic materials.

EXAMPLE 2.14: ELECTRONIC CONFIGURATION FOR FE

Write down the electronic configuration of an Fe (Z = 26) atom.

SOLUTION

There are 26 electrons in an Fe atom. We start with $n = 1$. This orbital can take $2(n = 1)^2$ or two electrons. The only possible value of l is 0, that is, the s subshell. Thus, the electronic configuration until this level will be $1s^2$. For $n = 2$, we can have a maximum of $2(2)^2$ or eight electrons. The possible values of l are 0 and 1; or, we can have s and p subshells. Thus, now the electronic configuration will read $1s^2 2s^2 2p^6$. For $n = 3$, we can have $2(3)^2$ or 18 electrons. However, we have only $26 - 2 - 8 = 16$ electrons remaining. For $n = 3$, we can have possible values of 0, 1, and 2 for l; or s, p, and d subshells. Thus, now the electronic configuration will read as $1s^2 2s^2 2p^6 3s^2 3p^6 3d^8$.

However, according to the filling order stated earlier, the 4s shell will fill *before* the 3d shell. The 4s level will take two electrons. The balance of six electrons will enter the 3d shell.

Thus, the final electronic configuration for iron will be $1s^2 2s^2 2p^6 3s^2 3p^6 4s^2 3d^6$.

The d subshell can contain a maximum of 10 electrons, but actually contains only six. Thus, the d subshell is deficient by four electrons. Of the six electrons in the 3d sublevel, two are spin-paired, that is, they have all the same quantum numbers, except the spin quantum numbers (which are +1/2 and −1/2; Table 2.8).

The remaining four of the 3d electrons are *not* spin-paired. These unpaired electrons make it possible for an Fe *atom* to behave like a tiny bar magnet. This behavior contributes to making Fe a magnetic material. We will study this when we discuss magnetic materials (Chapter 9).

2.14 ELECTRONS IN AN ATOM

Consider the electron energy levels or associated spectrum in silicon (Figure 2.22). Note that the energy levels are shown as parts of the circles that represent electron orbits. There are two types of electron energy states. One type is that of the so-called "free states." The other type is known as the "bound state." The electrons in the free states are not bound to the atom. They have kinetic energy so high that they can easily overcome the attraction force that would have kept them bound to the nucleus. The energy at which an electron becomes free is referred to as E = 0 or E_{vacuum}. In the "free state," electrons can have any possible energy level. In the bound state, on the other hand, electrons are held together within the atom by Coulombic interactions. According to the principles

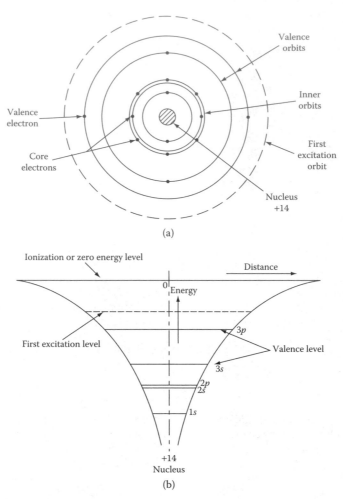

FIGURE 2.22 Electronic structure and energy levels in a Si atom: (a) The orbital model of a Si atom showing the 10 core electrons ($n = 1$ and 2), and the 4 valence electrons ($n = 3$); (b) energy levels in the Coulombic potential of the nucleus are also shown schematically. (From Streetman, B. G., and S. Banerjee. 2000. *Solid State Electronic Devices*, 5th ed. Upper Saddle River, NJ: Prentice Hall. With permission.)

of quantum mechanics, the energy states available to electron atoms are discrete or quantized. There are certain energy levels that electrons within atoms are not allowed to have. For example, electrons are not allowed in between the 1s and 2s levels, between the 2s and 3p levels, and so on.

2.15   ELECTRONS IN A SOLID

A solid material can have billions of atoms. It is important to know not just the electronic structure of individual atoms but also how different atoms interact with one another when they are in close proximity to one another. When individual atoms are brought together to form a solid material, the originally discrete bound electron energy levels or the electron wave functions begin to overlap (Figure 2.23).

This overlapping of energy levels of different atoms leads to the formation of *energy bands*, such as those in silicon (Figure 2.24).

As an example, instead of a 3s level for an atom (Figure 2.22), we now get a 3s band for an Si crystal (Figure 2.24). This concept of the formation of energy *bands* in a *solid*, as opposed to the energy *levels* in individual *atoms*, is the foundation for describing the differences in the electrical properties of insulating, conducting, and semiconducting materials.

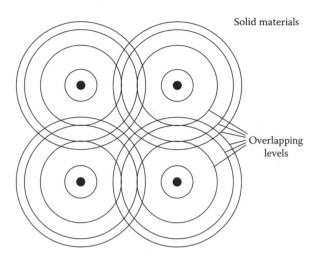

FIGURE 2.23 Overlap of the electron energy levels or wave functions as the atoms come closer. (Adapted from Edwards-Shea, L. 1996. *The Essence of Solid-State Electronics.* Upper Saddle River, NJ: Prentice Hall. With permission.)

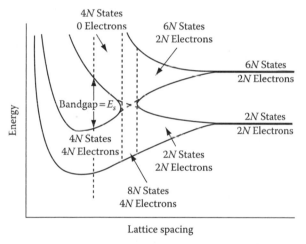

FIGURE 2.24 Formation of energy bands in silicon. (From Streetman, B. G., and S. Banerjee. 2000. *Solid State Electronic Devices*, 5th ed. Upper Saddle River, NJ: Prentice Hall. With permission.)

2.16 BAND STRUCTURE OF SOLIDS

Consider lithium (Li; Z = 3). The electronic configuration is $1s^2 2s^1$. This means that there are two electrons in the first energy level. There is only one electron in the next energy level for an atom, the 2s level. The 2s and 2p levels are collectively known as the L shell. The 2s level can accept two electrons and is thus only half-filled by a lithium atom.

Now consider a lithium crystal with N atoms. When we have N lithium atoms, there are N electrons that belong to the 2s energy band. When lithium atoms approach each other, the different 2s energy levels of the different atoms begin to overlap. This leads to the formation of a 2s band (Figure 2.25). Note that each 2s level can take two electrons; thus, for N atoms, the 2s band is capable of taking $2N$ electrons. However, there are only N 2s electrons. Thus, the 2s band is only half-full. However, this does not mean that only the lowest energy levels in the 2s band are full.

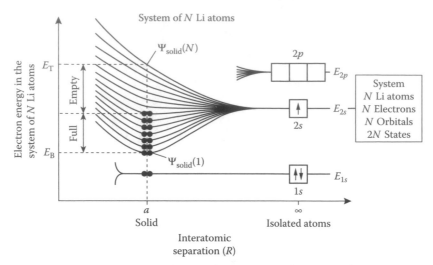

FIGURE 2.25 Formation of energy bands in lithium metal. The 2s band is only half-filled. Note that the 1s level shows very little splitting. (From Kasap, S. O. 2002. *Principles of Electronic Materials and Devices.* New York: McGraw Hill. With permission.)

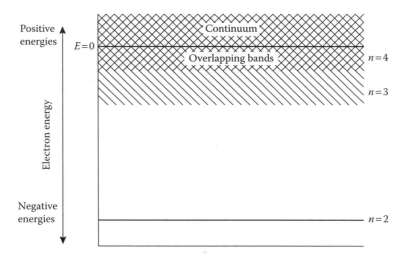

FIGURE 2.26 Schematic of a band diagram for a metal. (From Edwards-Shea, L. 1996. *The Essence of Solid-State Electronics.* Upper Saddle River, NJ: Prentice Hall. With permission.)

In the 2s band, half the energy states are empty; hence, electrons can move into these energy levels. Moreover, the higher-energy, empty energy levels for the individual atoms also begin to overlap. Thus, the 2s band is extended and overlaps with the 2p and 3s bands.

When an electric field is applied to a metal such as lithium, the electrons that gain energy have many energy levels available for them to move into. This is why alkali metals, such as sodium and lithium, are very good conductors of electricity.

Note that the 1s energy level for an atom can accommodate only two electrons. Thus, the 1s band in lithium is completely filled. Also, note that the 1s electrons are closer to the nucleus. They are not affected as much by the presence of electrons from other atoms. Hence, the 1s level does not show splitting unless we force the lithium atoms to separate by a distance less than their equilibrium separation. In addition, since the band is completely filled, the electrons in this band do not contribute to electrical conduction in lithium.

We represent these energy levels in the form of a band diagram (Figure 2.26).

A band diagram represents the electron energy levels. From an electrical properties viewpoint, the bands that involve the outermost or valence electrons are the most important in most cases. Thus, it is customary to show only the outermost bands in a band diagram. In the bottom of the band diagram is a band that is typically almost completely filled with electrons and is known as the valence band. The allowed energy band immediately above the valence band and into which electrons can move by gaining sufficient energy or momentum (either because of an applied electric field or increased temperature) is known as the conduction band. The magnitude of the energy gap between the conduction and valence bands is known as the bandgap (E_g).

Qualitative band diagrams for different types of materials are shown in Figure 2.27.

Band diagrams and the magnitudes of the bandgap energies provide an excellent way to classify materials into conductors, semiconductors, and insulators (Figure 2.28).

For magnesium (Mg), there is a partial overlap between the 3s and 3p bands (Figure 2.27). There is no apparent bandgap. When an electrical field is applied, the electrons in the 3s band accelerate and occupy the empty states in the 3p band, and possibly higher bands.

Because of the overlapping of the energy levels of different atoms, the bound energy levels become almost, but not completely, continuous. In between these "energy bands," we find regions where no electrons are allowed. Thus, there are "forbidden gaps" between the bands of allowed energy levels. We can now use this description to distinguish between solids with different electrical properties.

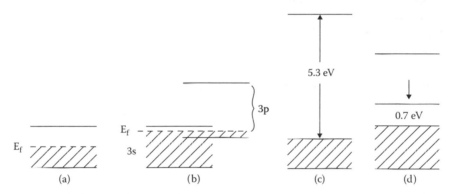

FIGURE 2.27 Band diagrams for conductors: (a) an alkali metal, (b) magnesium (Mg), a bivalent metal, (c) diamond, an insulator and (d) germanium (Ge), a semiconductor. (From Mahajan, S., and K. S. Sree Harsha. 1998. *Principles of Growth and Processing of Semiconductors*. New York: McGraw Hill. With permission.)

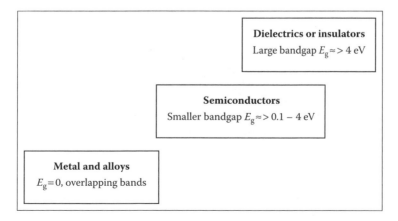

FIGURE 2.28 Classification of materials based on the values of bandgap (E_g) in electron volts (eV).

In insulators, such as diamond, or materials, such as silicon crystal at low temperatures, the valence band is completely filled. A key idea here is that when the band is completely filled, electrons cannot carry any current; hence conduction cannot take place. This is why silicon and other materials, called semiconductors, behave as *insulators* at low temperatures. In the case of semiconductors such as germanium (Ge) at low temperatures, the valence band is again completely filled. However, when the temperature increases, some electrons are able to jump across the smaller bandgap (E_g for Ge is ~0.67 eV) and enter the conduction band. This means that the electrons locked in covalent bonds between the silicon atoms can break free and jump from one bond onto another. Thus, unlike pure metals, the conductivity of semiconductor materials, such as essentially pure silicon or germanium, will *increase* as the temperature increases.

When an electrical field is applied, the electrons that enter the conduction band can move. When an electron moves from the valence band to the conduction band, we say that it has created a "hole" in the valence band. As mentioned before, a hole is an imaginary positively charged particle that represents an electron missing from a bond. We consider holes created in the valence band to be positively charged particles that can move around similarly to conduction electrons. Thus, the motions of both electrons and holes contribute to the conductivity of a semiconductor. The level of conductivity in materials such as silicon and germanium is typically lower than that of pure metals but higher than that of insulators (Figure 2.2). Therefore, we refer to materials such as germanium and silicon as semiconductors.

In insulators, the valence band is separated from the conduction band by a relatively large bandgap (Figure 2.28). For example, in diamond, the bandgap is 5.3 eV (Figure 2.27). Thus, even when the temperature is increased, the electrons cannot acquire enough energy to jump into the conduction band. Therefore, no appreciable electrical conduction occurs in diamond.

2.17 CONCEPT OF THE FERMI ENERGY LEVEL

Consider a metal such as lithium. All of the 2s electrons occupy energy levels beginning from the bottom of the 2s band (E_B) to an energy level called the Fermi level, at 0 K (E_{F0}). If we define the bottom of the valence band as zero, then the highest energy level filled is known as the Fermi energy level (E_F). The Fermi energy level of a metal is the highest energy level occupied by electrons. In semiconductors and insulators at low temperatures, the valence band is completely filled. As a result, there is no conduction. If the electrons had any levels to move into, they would be above the valence band. Consequently, the Fermi energy level for semiconductors and insulators will be expected to be somewhere above the top of the valence band. However, this is the bandgap region into which electrons are not allowed!

When the temperature increases, electrons can acquire enough energy and may be able to jump across the bandgap and into the conduction band to occupy energy levels that are allowed. The Fermi energy level (E_F) locations for a metal and a typical semiconductor are shown in Figure 2.29. The representation of E_F for a dielectric material is the same as that for a semiconductor; the only difference is that the E_g is larger.

The energy needed to excite an electron from E_F to the vacuum level, where it is essentially "freed from the solid," is equal to $q\phi$ (where q is the charge on the electron) and is known as the *work function* of a metal (Figure 2.29). The energy required to remove an electron from the bottom of the conduction bandedge to the vacuum is known as the *electron affinity* ($q\chi$) and is usually expressed in electron volts (eV). Sometimes, electron affinity is also described as χ, whose units are volts.

In the case of semiconductors, the valence band either is completely filled (at low temperatures) or is almost completely filled (at about room temperature). Thus, the probability of finding an electron in the valence band is very close to 1. The conduction band either is completely empty (at low temperatures) or is almost completely empty (at about room temperature). Thus, the probability of finding an electron in the conduction band is very small, but finite. Somewhere between the top

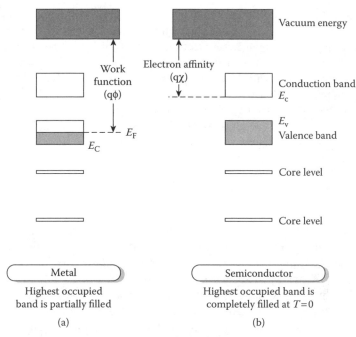

FIGURE 2.29 A schematic of band structure for (a) a typical metal and (b) a semiconductor. The work function is the energy "$q\phi$." The electron affinity for a metal is shown as the energy "$q\chi$." (Adapted from Singh, J. 2001. *Semiconductor Devices: Basic Principles*. New York: Wiley. With permission.)

edge of the valence band and the bottom edge of the conduction band, the probability of finding an electron will be 0.5 (Figure 2.29).

As noted before, certain energy levels in the bandgap are forbidden. We should define the level in the bandgap where the probability of finding an electron is half the Fermi energy level for the semiconductor. For semiconductors and insulators, the E_F lies about halfway into the bandgap, that is, between the top of the valence band and the bottom of the conduction band. We will study semiconductors in more detail in Chapter 3.

2.18 PROBLEMS

Introduction

2.1 What are the typical ranges of resistivity for metals, plastics, and ceramics?

2.2 What is the nature of bonding between atoms for most ceramics and plastics?

2.3 Compared to metals, what is the advantage in using conducting or semiconducting plastics?

2.4 Which one of these elements shows superconductivity—Ag, Au, or Al?

Ohm's Law

2.5 What is the difference between resistance and resistivity?

2.6 Do all materials obey Ohm's law? Explain.

2.7 Calculate the resistance of an AWG #20 Cu wire one mile in length.

2.8 What is the length of an AWG #16 Cu wire whose resistance is 21 Ω?

2.9 If the wire in Problem 2.8 carries a current of 5 A, what is the current density?

2.10 Al can handle current densities of ~10^5 A/cm² at about 150°C (Gupta 2003). What will be the maximum current allowed in an Al wire of AWG #18 operating at 150°C?

2.11 A circuit breaker connects an AWG #0000 Cu conductor wire 300 feet in length. What is the resistance of this wire? If the wire carries 150 A, what is the voltage decrease across this wire?

2.12 You may know that a conductor carrying an electrical current generates a magnetic field. A long wire carrying a current generates a magnetic field similar to that generated by a bar magnet. This magnet is known as an electromagnet. Consider a meter of magnetic wire AWG #2. Such wires are usually made from high-conductivity soft-drawn electrolytic Cu, and the conductor is coated with a polymer to provide insulation. What will be the electrical resistance (in ohms) of this wire?

2.13 Ground rods are used for electrical surge protection and are made from materials such as Au, Au-clad steel, or galvanized mild steel. The resistance of the actual rod itself is small; however, the soil surrounding the rod offers electrical resistance (Paschal 2001). The resistance of a ground rod is given by:

$$R = \frac{\rho}{2L\pi} \ln\left[\frac{4L}{a}\right] - 1 \qquad (2.33)$$

where R is the resistance in ohms, ρ is the resistivity of soil surrounding the ground rod (in $\Omega \cdot cm$), L is the length of ground rod in centimeters, and a is the diameter of the ground rod. (a) Assuming that the resistivity of a particular soil is $10^4 \ \Omega \cdot cm$, the length of the rod is 10 feet, and the diameter is 0.75 inches, what will be the resistance (R) of the ground rod in ohms? (b) Assuming that the ground rod is made from Cu, prove that the resistance of the rod itself is actually very small. (c) What will happen to the resistance of the metallic material as it corrodes over a period of many years?

2.14 The electrical resistance of pure metals increases with temperature. In many ceramics, the electrical current is carried predominantly by ions (such as oxygen ions in YSZ). Based on the data shown in Figure 2.30, calculate the electrical resistance of a 50-μm-thick YSZ element at 500°C and 800°C. Assume that the cross-sectional area is 1 cm² and the length is 1 meter.

2.15 Consider the material calcium oxide–stabilized ZrO_2 (Figure 2.30). Calculate the electrical resistance of a 50-μm-thick YSZ element. Assume that the cross-sectional area is 1 cm² and the length is 1 meter at 500°C and 800°C.

2.16 If a high conductivity at temperatures above 700°C was the only consideration in selecting a material for a solid oxide fuel cell electrolyte, what material would you choose (Figure 2.30)? Besides cost, what additional factors must be considered in the selection of this material?

2.17 What is unusual about the change in resistivity as a function of temperature for bismuth oxide (Bi_2O_3)?

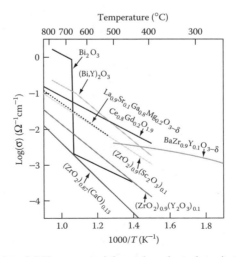

FIGURE 2.30 The conductivity of different materials used as electrolytes in the development of solid oxide fuel cells. (From Haile, S. M. 2003. *Acta Materilia* 51:5981–6000. With permission.)

Classical Theory of Electrical Conduction

2.18 Calculate the mobility of the electrons in Zn in $cm^2/V \cdot s$. Assume that each Zn atom contributes two conduction electrons. The atomic mass of Zn is 65. The resistivity is 5.9 $\mu\Omega \cdot cm$, and the density is 7.130 g/cm^3.

2.19 If the mobility of electrons in Au is 31 $cm^2/V \cdot s$, calculate the time between collisions (τ). Assume that the mass of electrons in Au is 9.1×10^{-31} kg. Calculate the mean free path length (λ) of electrons in Au if the average electron speed is 10^6 m/s.

2.20 The thermal speed of electrons is about 10^6 m/s. However, the drift velocity is rather small because electrons are scattered off by the vibrations of atoms. Calculate the drift velocity of electrons in Cu for an electric field of 1 V/m. Assume the mobility of electrons in Cu is 32 $cm^2/V \cdot s$.

2.21 If the density of Ag is 10.5 g/cm^3, what is the concentration of conduction electrons in Ag?

2.22 From the information provided in Table 2.3, calculate the expected conductivity of Ag.

2.23 Au is a face-centered cubic metal with a lattice constant of 4.080 Å. If the atomic mass of Au is 196.9655, calculate the number of conduction electrons per unit volume. Express your answer as number of electrons/cm^3. Assume that each Au atom donates one conduction electron.

2.24 A semiconductor is made so that it carries electrical current primarily from the flow of electrons. If the mobility of electrons (μ_n) is 1350 $cm^2/V \cdot s$ and the conduction electron concentration is 10^{21} cm^{-3}, what is the electrical conductivity of this material?

Joule Heating

2.25 A heating element for a flat iron is rated at 1000 W. If the iron works at 220 V, what is the resistance of this heating element?

2.26 Electronic components and devices are often tested at 125°C and −55°C to check the high- and low-temperature performances. They can then be compared with the properties observed at room temperature, 25°C. For example, using 25°C as the reference temperature and +125°C as the other temperature, α_{125} can be written as follows:

$$TCR_{125} = \alpha_{R,125} = \frac{1}{\rho_{125}} \left(\frac{\rho_{125} - \rho_{25}}{125 - 25} \right) \times 10^6 \ ppm/°C$$

Write an equation to express the temperature coefficient of resistance with $T = -55°C$ (note the negative sign) as the other temperature, using 25°C as the base or reference temperature (T_0).

2.27 In a circuit, the TCR_{125} value for a resistor is 100 ppm/°C. If the resistance (R) at 25°C is 1000 Ω, what is the resistance at 125°C?

2.28 Assume that the bus bar discussed in Example 2.11 is heated due to the Joule losses and now operates at 70°C. Calculate the resistance, power loss, and energy consumption for 24 hours and the total energy costs per year. Ignore the change in the length of the Cu bus bar because of thermal expansion.

2.29 Nichrome wire is used for cutting materials such as polystyrene (Styrofoam) into different shapes, including large facades or insulation boards. (a) Calculate the length of an AWG #20 wire that is needed to have a resistance of $R = 8 \ \Omega$. (b) What will be the resistance of this wire if it gets heated to a temperature of 200°C? See Tables 2.2 and 2.5.

Resistivity of Metallic Alloys

2.30 In the nanoscale region, why does the resistivity of thin films depend upon thickness?

2.31 What elements most affect the resistivity of high-purity Cu?

2.32 Why does the addition of oxygen in limited concentrations actually *increase* the conductivity of high-purity Cu?

2.33 Why does the resistivity of pure metals increase with temperature, whereas that of alloys is relatively stable with temperature?

2.34 Nordheim's coefficient for Au dissolved in Cu is $C = 5500$ n$\Omega \cdot$m. If the resistivity of Cu at 300 K is 16.73 n$\Omega \cdot$m, calculate the resistivity of an Au–Cu alloy containing 1 weight % Au.

Band Structure of Solids

2.35 What is the electronic configuration for an Mg atom ($Z = 12$)?

2.36 Draw a schematic of the band diagrams for a typical metal, a semiconductor, and an insulator.

GLOSSARY

Annealing: A heat treatment for metals and alloys in which a material is heated to a high temperature and then cooled slowly; after annealing, dislocations are annihilated, and the material exhibits a higher level of conductivity.

Azimuthal quantum number (l): See **Orbital quantum number**.

Band diagram: A diagram showing the electron energy levels that represent the valence and conduction bands.

Bandgap (E_g): The energy difference between the top of the valence band and bottom of the conduction band, which must be overcome to transfer an electron from the valence band to the conduction band.

Bulk resistivity (ρ): See **Resistivity**.

Bus bar: A conductor used in power transmission.

Carrier concentration: A concentration of species responsible for electrical conduction.

Cold-working: A process conducted at temperatures below the recrystallization temperature in which a metallic material is deformed or shaped, usually causing the resistivity of a material to increase.

Composites: Materials in which two or more materials or phases are blended together, sometimes arranged in unique geometrical arrangements, to achieve desired properties.

Conductance: Inverse of resistance, whose units are Siemens or Ω^{-1}.

Conduction band: The higher band on a band diagram, separated from the valence band by the bandgap, which shows the energy levels associated with the conduction electrons.

Conductivity: A microstructure-, composition-, and temperature-dependent property that conveys the ability of a material to carry electrical current and is the inverse of resistivity.

Conductors: Materials with resistivity in the range of ~10^{-6} to 10^{-4} $\Omega \cdot$cm (1 to 10^2 $\mu\Omega \cdot$cm).

Conventional current: As a matter of convention, the current that flows from the positive to the negative terminal of the battery, although the electrons themselves move in the reverse direction.

Current density (J): Current per unit cross-sectional area perpendicular to the direction of the current flow.

Dielectrics: Materials that do not allow any significant current to flow through them.

Drift: Motion of charged carriers, such as electrons, holes, or ions, in response to an electric field.

Electric field (E): Voltage divided by the distance across which the voltage is being applied.

Electrolytic tough-pitch (ETP or TP) copper: High-conductivity copper containing about 100–650 ppm (or 1 mg/kg) oxygen.

Electron affinity ($q\chi$): The energy required to remove an electron from the bottom of conduction bandedge to the vacuum, sometimes also designated as χ (in volts).

Electron current: The movement of electrons from the negative to the positive terminal of the voltage supply when a DC electrical field is applied to a material.

Electronic conductors: Materials in which a dominant part of conductivity is due to the motion of electrons or holes.

Four-point probe: A setup used for making conductivity measurements, typically involving a fixed current being applied using two outer probes, with the decrease in voltage measured using two inner probes; also known as a Kelvin probe.

Hole: An imaginary positively charged particle that represents an electron missing from a bond.

I^2R losses: The heating of a material due to resistance to electrical current, with the power dissipated (for DC voltages) given by the term I^2R (see also **Joule heating**).

IACS conductivity: Conductivity of international annealed-copper standard; 100% IACS is defined as $\sigma = 57.4013 \times 10^6$ S/m or 57.4013×10^4 S/cm (or $\rho = 1.74212 \times 10^{-6}$ $\Omega \cdot$cm), which is based on the conductivity of an annealed 1-meter-long copper wire that has a cross-sectional area of 1 mm^2 and a resistance of 0.17421 Ω.

Impurity-scattering limited drift mobility: The mobility of carrier particles in alloys, which is limited by the scattering of impurity or alloying elements and not by the thermal vibrations of the host atoms.

Indium–tin oxide (ITO): A transparent ionic conductor-based material used in touch-screen displays, solar cells, and other devices.

Insulators: Materials that neither conduct electricity (similar to dielectrics) nor easily break down electrically even in the presence of a strong electric field.

Integrated circuits (ICs): Electrical circuits typically fabricated on semiconductor substrates, such as silicon wafers, comprising resistors, transistors, and so on.

Ionic conductors: Materials in which movement of ions constitutes the major portion of the total conductivity (e.g., ITO or yttria-stabilized zirconia).

Interconnects: Conductive paths between components of an IC.

Joule heating: The heating of a material due to resistance to electrical current; for DC voltages, the power dissipated is given by I^2R (same as I^2R losses).

Kelvin probe: See **Four-point probe**.

Lattice scattering limited conductivity: The conductivity (σ) in essentially *pure* metals; it is largely limited by the scattering of electrons by the vibrations of atoms.

Lattice scattering limited mobility: The mobility of carriers in essentially pure metals; it is limited by scattering due to phonons and other defects in arrangements of atoms.

Magnetic quantum number (m or m_l): The electron quantum number representing the component of orbital angular momentum along an external magnetic field.

Mean free-path length (λ): The mean free-path length (λ) of conduction electrons, that is, the average distance that electrons travel before again colliding.

Mixed conductors: Materials in which conductivity occurs because of the movement of ions, as well as that of electrons or holes.

Mobility (μ): The speed of carriers under the influence of a unit external electric field (E).

Nanoscale: Length of scale between ~1 and 100 nm, in which unusual effects on the properties of materials are seen for materials, devices, or structures.

Nordheim's coefficient (C): See **Nordheim's rule**.

Nordheim's rule: The resistivity of a solid-solution alloy is given by the equation

$$\rho_{alloy} = \rho_{matrix} + Cx(1-x)$$

where ρ_{alloy} is the resistivity of the alloy, ρ_{matrix} is the resistivity of the base metal without any alloying elements, C is Nordheim's coefficient, and x is the *atom fraction* of alloying element added.

Orbital angular momentum quantum number (i): A quantum number related to the angular momentum of the electrons (the same as azimuthal quantum number).

Oxygen-free high-conductivity copper (OFHC): A high-conductivity copper material that has less than 0.001% oxygen. It is used when electrolytic tough-pitch copper cannot be used because of the potential welding or brazing problems.

Pauli's exclusion principle: No two electrons in a given system (such as an atom) can have all four quantum numbers identical.

Phonons: In pure metals, the vibrations of conduction electrons off of phonons leads to a resistivity component that increases with increasing temperature.

Principal quantum number (n): A quantum number that quantizes the electron energy, with values of $n = 1$, 2, and 3 corresponding to the K, L, and M shells, respectively.

Residual resistivity (ρ_R): The part of total resistivity arising from the effects of microstructural defects and impurities or added elements.

Resistivity (ρ): The electrical resistance of a resistor with a unit length and a unit cross-sectional area. This is a microstructure- and temperature-dependent property whose magnitude is the inverse of conductivity.

Resistor: A component included in an electrical circuit to offer a predetermined value of electrical resistance.

Resistance (R): The difficulty with which electrical current flows through a material. For a material with length L, cross-sectional area A, and resistivity ρ, the resistance R is given by $\rho \cdot L/A$.

Resistance-temperature detector (RTD): A temperature-measuring device based on the measurement of change in the resistivity of a metallic wire as a function of temperature.

Semiconductors: Materials that have a resistivity ranging between 10^{-4} and $10^3 \, \Omega \cdot cm$ (i.e., 10^2–10^9 $\mu\Omega \cdot cm$).

Semi-insulators: Materials with resistivity values ranging from 10^3 to $10^{10} \, \Omega \cdot cm$.

Sheet resistance (R_s): The resistance of a square resistor of a certain resistivity and thickness.

Solid solution: A solid material in which one component (e.g., Cu) is completely dissolved in another (e.g., Ni), similar to the complete dissolution of sugar in water.

Solid-solution strengthening: An effect in which the formation of a solid solution (i.e., the complete dissolution of one element into another) causes an increase in the yield stress. For example, a small concentration of Be in Cu increases the yield stress of Cu.

Spin-paired electrons: Electrons whose quantum numbers are identical, other than the spin quantum number, and which have spin directions opposite of each other to satisfy Pauli's exclusion principle.

Spin quantum number (s or m_s): An electron quantum number that quantizes the spin, its values being $\pm\frac{1}{2}$.

Superconductors: Materials that can exhibit zero electrical resistance under certain conditions.

Temperature coefficient of resistivity (TCR): A coefficient designated as α_R and defined as

$$\alpha_R = \frac{1}{\rho_0}\left(\frac{\rho - \rho_0}{T - T_0}\right)$$

where ρ is the resistivity, T is the temperature, and ρ_0 is the resistivity at the reference temperature (T_0).

Temperature-dependent component of resistivity (ρ_T): The portion of total resistivity originating from the scattering of conduction electrons off vibrations of atoms (phonons).

Valence band: The lower band on a band diagram showing the energy levels associated with the valence electrons. This band is usually completely or nearly completely filled for metals and semiconductors.

Varistors: Materials or devices with voltage-dependent resistance.

Volume resistivity: See **Resistivity**.

Wave function: A set of quantum numbers (i.e., n, l, m, and s) that represent the wave function associated with an electron.

Work function (q ϕ): The energy, in electron volts (eV), needed to excite an electron from E_F to the vacuum level, where it is essentially freed from the solid. Sometimes measured in ϕ (volts).

Yttria (Y_2O_3)-stabilized zirconia (ZrO_2) (YSZ): A zirconia ceramic doped with yttrium oxide; it has a cubic crystal structure and is an ionic conductor used in solid oxide fuel cells and oxygen gas sensors.

REFERENCES

Askeland, D. 1989. *The Science and Engineering of Materials*. 3rd ed. Washington, DC: Thomson.

Askeland, D., and P. Fulay. 2006. *The Science and Engineering of Materials*. Washington, DC: Thomson.

ASM. *Properties and Selection: Nonferrous Alloys and Special Purpose Materials*. vol. 2. Materials Park, OH: ASM.

ASM International, eds. *Properties and Selection: Nonferrous Alloys and Special Purpose Materials*. vol. 2. Materials Park, OH: ASM.

Davis, J. R. 1997. *Concise Metals Engineering Data Book*. Materials Park, OH: ASM.

Edwards-Shea, L. 1996. *The Essence of Solid-State Electronics*. Upper Saddle River, NJ: Prentice Hall.

Groover, M. P. 2007. *Fundamentals of Modern Manufacturing: Materials, Processes, and Systems*. New York: Wiley.

Gupta, T. K. 2003. *Handbook of Thick and Thin Film Hybrid Microelectronics*. New York: Wiley Interscience.

Haile, S. M. 2003. Fuel cell materials and component. *Acta Materilia* 51:5981–6000.

Kasap, S. O. 2002. *Principles of Electronic Materials and Devices*. New York: McGraw Hill.

Lampman, S. R., and T. B. Zorc, eds. 1990. *Metals Handbook: Properties and Selection: Nonferrous Alloys and Special Purpose Materials*. Materials Park, OH: ASM.

Laughton, M. A., and D. F. Warne. 2003. *Electrical Engineer's Reference Book*. Amsterdam: Elsevier.

Mahajan, S., and K. S. Sree Harsha. 1998. *Principles of Growth and Processing of Semiconductors*. New York: McGraw Hill.

Minges, M. L. 1989. *Electronic Materials Handbook*. Vol. 1 of *Electronic Packaging*. Materials Park, OH: ASM.

Neelkanta, P. 1995. *Handbook of Electromagnetic Materials*. Boca Raton, FL: CRC Press.

Paschal, J. M. 2001. *EC and M's Electrical Calculations Handbook*. New York: McGraw Hill.

Powerstream, http://www.powerstream.com/Wire_Size.htm.

Singh, J. 2001. *Semiconductor Devices: Basic Principles*. New York: Wiley.

Streetman, B. G., and S. Banerjee. 2000. *Solid State Electronic Devices*, 5th ed. Upper Saddle River, NJ: Prentice Hall.

Vines, R. F., and E. M. Wise. 1941. *The Platinum Metals and Their Alloys*. New York: International Nickel Co.

Webster, J. G. 2002. *Wiley Encyclopedia of Electrical and Electronics Engineering*, vol. 4. New York: Wiley.

3 Fundamentals of Semiconductor Materials

KEY TOPICS

- Origin of semiconductivity in materials
- Differences between elemental versus compound, direct versus indirect bandgap, and intrinsic versus extrinsic semiconductors
- Band diagrams for n-type and p-type semiconductors
- Conductivity of semiconductors in relation to the majority and minority carrier concentrations
- Factors that affect the conductivity of semiconductors
- Device applications such as light-emitting diodes (LEDs)
- Changes in the bandgap with temperature, dopant concentrations, and crystallite size (quantum dots)
- Semiconductivity in ceramic materials

3.1 INTRODUCTION

Semiconductors are defined as materials with resistivity (ρ) between ~10^{-4} and ~10^3 $\Omega \cdot$ cm. An approximate range of sensitivities exhibited by silicon (Si)-based semiconductors is shown in Figure 3.1. Elements showing semiconductivity are called *elemental semiconductors* (e.g., silicon), and the compounds that show semiconducting behavior are known as *compound semiconductors* (e.g., gallium arsenide [GaAs]). A band diagram of a typical semiconductor is shown in Figure 3.2. For most semiconductors, the bandgap energy (E_g) is between ~0.1 and 4.0 eV. If the bandgap is larger than 4.0 eV, we usually consider the material to be an insulator or a dielectric. In this chapter, we will learn that the composition of such dielectric materials can be altered so that they exhibit semiconductivity (see Section 3.21). As shown in Figure 3.2, the top of the valence band is known as the *valence bandedge* (E_v), and the bottom of the conduction band is known as the *conduction bandedge* (E_c). Recall from Chapter 2 that the band diagram shows the outermost part of the overall electron energy levels of a solid. The vertical axis shows the increasing electron energy. Thus, the magnitude of the bandgap energy (E_g) is given by

$$E_g = E_c - E_v \tag{3.1}$$

3.2 INTRINSIC SEMICONDUCTORS

Let us consider the origin of the semiconducting behavior in semiconductors such as silicon. Silicon has covalent bonds; each silicon atom bonds with four other silicon atoms, which leads to the formation of a three-dimensional network of tetrahedra arranged in a diamond cubic crystal structure (Figure 3.3).

Figure 3.4 shows a two-dimensional representation of the covalent bonds between silicon atoms. When the temperature is low (~0 K), the valence electrons shared between the silicon atoms remain in the bonds and are not available for conduction. Thus, at low temperatures, silicon behaves as an insulator (Figure 3.4). As the temperature increases, the electrons gain thermal energy. Some

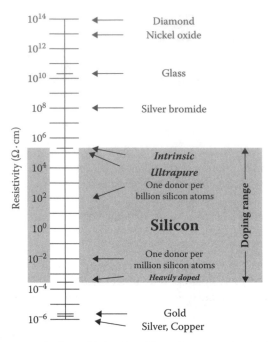

FIGURE 3.1 Approximate range of sensitivity for silicon in comparison with other materials. (From Queisser, H. J. and E. E. Haller. 1998. *Science* 281:945–50. With permission.)

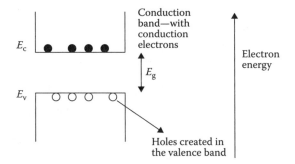

FIGURE 3.2 A band diagram for a typical semiconductor.

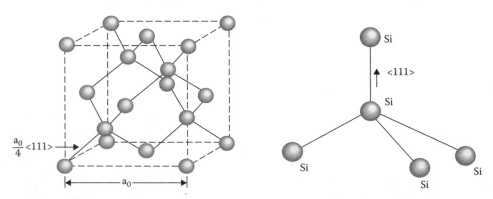

FIGURE 3.3 The diamond cubic crystal structure of silicon, showing the tetrahedral arrangement of silicon atoms. Each silicon atom is bonded to four other silicon atoms. The lattice constant is a_0. (From Mahajan, S., and K. S. Sree Harsha. 1998. *Principles of Growth and Processing of Semiconductors*. New York: McGraw Hill. With permission.)

electrons can gain sufficient energy to break away from the covalent bonds. These electrons are now free to move around and impart semiconductivity to the material (Figure 3.4).

The electrons breaking away from the bonds (Figure 3.4) can also be shown on a band diagram (Figure 3.5). At low temperatures, the valence electrons are in the covalent bonds, that is, the *valence band* is completely filled. As the temperature increases to ~>100 K, a small fraction of electrons gain enough thermal energy to make a jump across the bandgap (E_g) and into the conduction band.

When an electron breaks away from a covalent bond, it leaves behind an incomplete bond. A *hole* is an imaginary particle that represents an electron missing from a bond (Figure 3.6). On a band

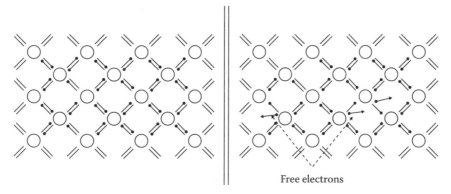

FIGURE 3.4 Two-dimensional representation of silicon bonding. (From Kano, K. 1997. *Semiconductor Fundamentals*. Upper Saddle River, NJ: Prentice Hall. With permission.)

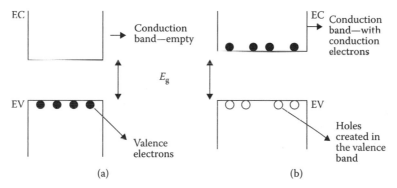

FIGURE 3.5 Typical band diagram for an intrinsic semiconductor at (a) low temperatures and (b) high temperatures.

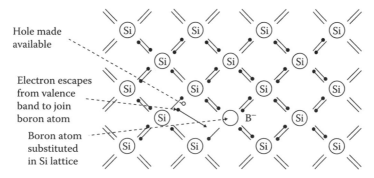

FIGURE 3.6 Creation of electron hole pairs by thermal excitation. (From Kano, K. 1997. *Semiconductor Fundamentals*. Upper Saddle River, NJ: Prentice Hall. With permission.)

diagram, a hole is an energy state left empty by an electron that moved to the conduction band. If a hole is present at site X, then another electron from a neighboring bond at site Y can move into site X. This creates a hole at site Y. Movement of an electron from site Y to X is equivalent to the movement of a hole from site X to site Y. Thus, the movement of holes also contributes to a semiconductor's electrical conductivity.

In materials such as silicon and germanium (Ge), the bandgap energy at room temperature is relatively low (for germanium and silicon, E_g is ~0.67 and ~1.1 eV, respectively). When we say the bandgap is small, we are comparing the bandgap energy with the thermal energy given by k_BT, where k_B is the Boltzmann's constant (8.617×10^{-5} eV/K or 1.38×10^{-23} J/K) and T is the temperature. At $T = 300$ K (~room temperature), the thermal energy k_BT is ~0.026 eV.

Promoting an electron into the conduction band creates an *electron–hole pair* (EHP; Figures 3.5 and 3.6). Electrons promoted into the conduction band because of thermal energy and the resulting holes that are created in the valence band are known as *thermally generated charge carriers*.

A semiconductor in which charge carriers created as a result of thermal energy are the only source for creating conductivity is known as an *intrinsic semiconductor*. Note that even if EHPs are created, the material still remains electrically neutral. In an intrinsic semiconductor, the concentration of the electrons available for conduction (n_i) is equal to that of the holes created (p_i).

$$n_i = p_i \tag{3.2}$$

Therefore, Figures 3.2 and 3.5 show the concentrations of conduction electrons and holes to be equal. The conductivity of intrinsic semiconductors is controlled by the material itself and by the temperature that changes the carrier concentrations. The word "intrinsic" emphasizes that no other "extrinsic," or foreign elements or compounds are present in significant enough concentrations to have any effect the electrical properties of an intrinsic semiconductor. Appropriately, a semiconductor whose conductivity and other electrical properties are controlled by foreign elements or compounds is known as an *extrinsic semiconductor* (see Section 3.8).

The conductivity of an intrinsic semiconductor is given by the following equation:

$$\sigma = qn_i\mu_n + qp_i\mu_p \tag{3.3}$$

In this equation, q is the magnitude of the charge on the electron or hole (1.6×10^{-19} C), and n_i and p_i are the concentrations of electrons and holes in an intrinsic material, respectively. The terms μ_n and μ_p are the mobilities of electrons and holes, respectively. Since $n_i = p_i$ for an intrinsic semiconductor, Equation 3.3 can be rewritten as

$$\sigma = qn_i(\mu_n + \mu_p) \tag{3.4}$$

For a given intrinsic semiconductor, the electron or hole concentrations depend mainly on the temperature.

Example 3.1 illustrates the calculation of the resistivity of an intrinsic semiconductor.

EXAMPLE 3.1: RESISTIVITY OF INTRINSIC GE

What is the resistivity (ρ) of essentially pure Ge? Assume that the mobilities of the electrons and the holes in Ge at 300 K are 3900 and 1900 cm²/V·s, respectively. Assume $T = 300$ K and $n_i = 2.5 \times 10^{13}$ cm⁻³.

SOLUTION

We make use of Equation 3.4

$$\sigma = (1.6 \times 10^{-19} \text{ C})\left(2.5 \times 10^{13} \frac{\text{carriers}}{\text{cm}^3} \right)(3900 + 1900) \text{ cm}^2/\text{V} \cdot \text{s}$$

Therefore,

$$\sigma = 0.0232 \text{ S/cm at 300 K}$$

The inverse of this is the resistivity (ρ) at 300 K = 43.1 $\Omega \cdot$cm.

3.3 TEMPERATURE DEPENDENCE OF CARRIER CONCENTRATIONS

As we can expect, at any given temperature the larger the bandgap (E_g) of a semiconductor, the lower the concentration of valence electrons (n_i) that can pass across the bandgap. The relationship among the carrier concentration, the bandgap, and the temperature is given by

$$n_i \propto T^{3/2} \exp\left(-\frac{E_g}{2k_B T}\right) \tag{3.5}$$

The exponential term dominates, and hence the plot of $\ln(n_i)$ with $1/T$ is essentially a straight line. Note that even though the increase is exponential, only a very small fraction of the *total number of valence electrons* actually gets into the conduction band. In Figure 3.7, a plot of n_i on a logarithmic scale is shown as a function of the inverse of the temperature. The slope of the line is essentially proportional to the bandgap (E_g). The bandgaps of germanium, silicon, and GaAs are ~0.67, 1.1, and 1.43 eV, respectively. From Equation 3.5, we expect the concentration of free electrons at a given temperature to be the highest for germanium, since it has the smallest bandgap of the three. On the other hand, GaAs will have the lowest concentration of thermally generated conduction electrons, since it has the largest bandgap of these three materials (Figure 3.7). The bandgap values and some

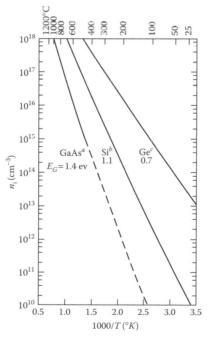

FIGURE 3.7 Intrinsic carrier concentration, plotted on a log scale as a function of the inverse of temperature for germanium, silicon, and gallium arsenide. (From Grove, A. S. 1967. *Physics and Technology of Semiconductor Devices*. New York: Wiley. With permission.)

TABLE 3.1
Properties of Selected Semiconductors

Semiconductor (Bandgap Type i: indirect, d: direct)	Bandgap (E_g) (eV) at $T = 300$ K	Mobility of Electrons (cm²/V·s) at 300 K (μ_n)	Mobility of Holes (cm²/V·s) at 300 K (μ_p)	Electric Breakdown Field (V/cm)
Carbon as diamond	5.47	800	1200	10^7
Germanium (i)	0.67	3900	1900	10^5
Silicon (i)	1.12	1500	450	3×10^5
Amorphous silicon (a-Si:H) (i)	1.7–1.8	1	10^{-2}	—
SiC (α-form) (i)	2.996	400	50	2–3×10^6
GaSb (d)	0.72	5000	850	5×10^4
GaAs (d)	1.42	8500	400	4×10^5
GaP (i)	2.26	110	75	~ 5–10×10^5
InSb (d)	0.17	8000	1250	10^3
InP (d)	1.35	4600	150	5×10^5
CdTe (d)	1.56	1050	100	—
PbTe (i)	0.31	6000	4000	—

of the other properties of semiconductors are shown in Table 3.1. The terms direct and indirect bandgap semiconductors, used in Table 3.1, are defined in Section 3.5.

3.4　BAND STRUCTURE OF SEMICONDUCTORS

In the classical theory of conductivity, electrons are considered particles. The kinetic energy (E) of an electron can be written as:

$$E = \frac{1}{2} mv^2 \tag{3.6}$$

where m is the mass of a free electron and v is the speed of the electron. Momentum (p) is defined as mass × velocity.

Therefore, we can also rewrite Equation 3.6 in terms of momentum (p) and mass as follows:

$$E = \frac{p^2}{2m} \tag{3.7}$$

Note that we have also used the symbol "p" to designate the concentration of holes.

In a quantum mechanics–based approach, an electron is considered a plane wave with a propagation constant, which is called the *wave vector* ($\mathbf{k}$). The wave function for an electron is given by:

$$\Psi_{(\mathbf{k_x}, x)} = U(\mathbf{k_x}, x) \exp(j\mathbf{k_x}x) \tag{3.8}$$

where Ψ is the wave function that is related to the probability of finding an electron, U is the function that accounts for the periodicity of the crystal structure, j is the imaginary number, $\mathbf{k_x}$ is the wave vector in x-direction along which the electron (now considered a wave) is traveling, and x is the distance.

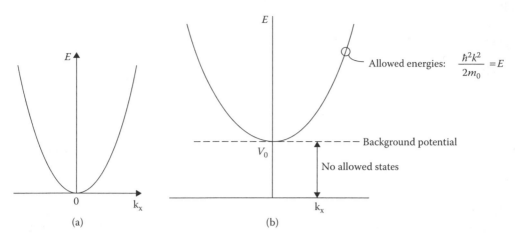

FIGURE 3.8   The band structure or energy (E) versus wave vector (**k**) for (a) a free electron; and (b) an essentially free electron in a band with starting energy V_0. (Adapted from Mahajan, S., and K. S. Sree Harsha. 1998. *Principles of Growth and Processing of Semiconductors*. New York: McGraw Hill. With permission.)

Using quantum mechanics, we can show that the electron momentum (p) is related to wave vector (**k**) by the following equation:

$$p = \hbar k \tag{3.9}$$

where $\hbar = h/2\pi = 1.054 \times 10^{-34}$ J · s, and h is the Planck's constant ($= 6.626 \times 10^{-34}$ J · s).

From Equations 3.7 and 3.9, we can write the energy of an electron as:

$$E = \frac{\hbar^2 k^2}{2m} \tag{3.10}$$

Thus, for a free electron, that is, an electron that is not experiencing any other forces due to internal or external electric or magnetic fields, the relationship between its energy (E) and wave vector (**k**) is a parabola (Figure 3.8). The plot of electron energy as a function of the wave vector (**k**) is known as the *band structure* of a material.

In a perfect crystal, if we assume that the electron is moving in a band and has a starting energy of V_0, then the energy of the electron moving in this band is given by:

$$E = \frac{h^2 k^2}{2m} + V_0 \tag{3.11}$$

In semiconductors, electrons moving in different bands are not completely free. They have built-in electric fields that are associated with other atoms, often periodically arranged in certain types of crystal structures. Thus, the electrons move around in a material as if they have a different mass, known as the *effective mass of an electron* (m_e^* or m_n^*). This is the mass that an electron would *appear* to have in a material and is different from the mass of a free electron in vacuum ($m_0 = 9.109 \times 10^{-31}$ kg). The concept of effective mass is illustrated in Figure 3.9. For convenience, we will express the effective mass as a dimensionless quantity m_e^*/m_0. This ratio is an indicator of the interactions of the electron with the atoms of the material. Similarly, holes also have an effective mass.

The significance of the effective mass is as follows: Smaller effective masses for carriers (electrons or holes) mean that the carriers can move faster, with a less apparent inertia. This means materials with smaller apparent electron or hole masses are more useful for making faster semiconductor devices. The concept of effective mass is quantum mechanical in nature, and analogies using classical mechanics must therefore be limited in scope.

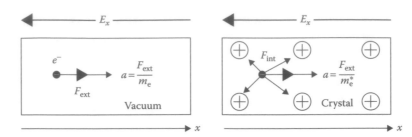

FIGURE 3.9 Illustration of the difference between the mass of electron in vacuum (m_0) and its effective mass (m_e^*). (From Kasap, S. O. 2002. *Principles of Electronic Materials and Devices*. New York: McGraw Hill. With permission.)

The effective masses of electrons and holes for some semiconductors are listed in Table 3.2. This table also lists the values of the *dielectric constant* (ε_r), which is a measure of the ability of a material to store a charge (Chapter 8). The dielectric constant is defined as the ratio of the permittivity of a material (ε) and the permittivity of the free space (ε_0).

The effective mass of electrons is related to the *E-k* curvature as follows:

$$m_e^* = \frac{\hbar^2}{\left(\dfrac{d^2E}{dk^2}\right)} \tag{3.12}$$

Since a hole is an imaginary particle that represents a missing electron in the valence band, the effective mass of the hole is the negative of the mass of the missing electron. The effective mass of holes is given by:

$$m_h^* = -\frac{\hbar^2}{\left(\dfrac{d^2E}{dk^2}\right)} \tag{3.13}$$

Note the negative sign in Equation 3.13.

The effective mass of an electron and the *E-k* curvature (Equation 3.12) is derived as follows:

The electron velocity is equal to the group velocity of the associated wave (v_g), with which the boundary of the wave propagates. The group velocity is given by the following equation:

$$v_g = \frac{d}{dk}(2\pi v) \tag{3.14}$$

where v is the frequency of the wave.

Rewrite Equation 3.14 as:

$$v_g = \frac{d}{dk}\left(\frac{2\pi h v}{h}\right) \tag{3.15}$$

Now, we replace hv with E and ($h/2\pi$) with $\hbar$ and we get:

$$v_g = \frac{1}{\hbar}\left(\frac{dE}{dk}\right) \tag{3.16}$$

The acceleration (a) of an electron is given by:

$$a = \left(\frac{dv_g}{dt}\right) \tag{3.17}$$

Substituting for v_g from Equation 3.16 into Equation 3.17, we get:

$$a = \frac{1}{\hbar}\left(\frac{d^2E}{dk^2}\right)\left(\frac{dk}{dt}\right) \tag{3.18}$$

Substituting for k with $p/\hbar$, we get:

$$a = \frac{1}{\hbar^2}\left(\frac{d^2E}{dk^2}\right)\left(\frac{dp}{dt}\right) \tag{3.19}$$

Rewrite Equation 3.19 as:

$$a = \frac{1}{\hbar^2}\left(\frac{d^2E}{dk^2}\right)\left(\frac{d(mv)}{dt}\right) = \frac{1}{\hbar^2}\left(\frac{d^2E}{dk^2}\right)F \tag{3.20}$$

From classical mechanics, we know that $F = m \times a$ or $a = F/m$. Comparing this with Equation 3.20, we get an expression for the effective mass of an electron as given in Equation 3.12. We can see from Equation 3.12 that the effective mass of an electron is inversely related to the E-k curvature. The larger the curvature, the lesser the effective mass.

For the conduction band of a semiconductor, we can write the relationship between the energy of an electron (E) and its wave vector $\mathbf{k}$ as follows:

$$E(k) = E_c + \frac{\hbar^2 k^2}{2m_e^*} \tag{3.21}$$

where E_c is the conduction bandedge energy and m_e^* is the effective mass of the electron. The effective mass of an electron depends strongly on the bandgap (E_g). The smaller the value of the bandgap (E_g), the smaller the value of m_e^* (Table 3.2 and Figure 3.10).

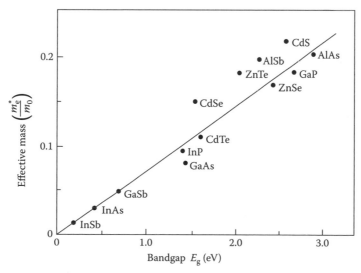

FIGURE 3.10 The relationship between electron effective mass (m_e^*) and bandgap (E_g). (From Singh, J. 2001. *Semiconductor Devices: Basic Principles*. New York: Wiley. With permission.)

TABLE 3.2
Effective Masses of Electrons and Holes and Other Properties of Different Semiconductors

Semiconductor (Bandgap Type, i: indirect, d: direct)	Bandgap (eV) ($T = 300$ K)	Mobility of Electrons at 300 K (μ_n) in cm²/V·s	Mobility of Holes at 300 K (m_p) in cm²/V·s	Effective Mass of Electrons m_n^*/m_0 (m_l^*, m_t^*)	Effective Mass of Holes m_p^*/m_0 (m_{lh}^*, m_{hh}^*)	Density (g/cm³)	Dielectric Constant (ε_r)	Lattice Constant (Å)	Melting Point (°C)
Aluminum arsenide (AlAs) (i)	2.16	1200	420	2.0	0.15, 0.76	3.60	10.9	5.66	1740
Aluminum phosphide (AlP) (i)	2.45	80	—	—	0.2, 0.63	2.4	9.8	5.46	2000
Amorphous silicon (a-Si:H)	1.7	1	10^{-2}						
Cadmium telluride (CdTe) (d)	1.56	1050	100	0.1	0.37	6.2	10.2	6.482	1098
Carbon (C) as diamond (d)	5.47	800	1200	1.4, 0.36 (at 85 K)	0.7, 2.12 (at 1.2 K)	3.51	5.7	3.566	
Gallium antimonide (GaSb) (d)	0.72	5000	850	0.042	0.06, 0.23	5.61	15.7	6.09	712
Gallium arsenide (GaAs) (d)	1.42	8500	400	0.067	0.08, 0.45	5.31	13.2	5.65	1238
Gallium phosphide (GaP) (i)	2.26	110	75	1.12, 0.22	0.14, 0.79	4.13	11.1	5.45	1467
Gallium nitride (GaN) (d)	3.47	380	—	0.19	0.60	6.1	12.2	4.5; a = 3.189; c = 5.185 (for wurtzite structure)	2530

Germanium (Ge) (i)	0.67	3900	1900	1.64, 0.082	0.04, 0.28	16	5.646	936
Indium antimonide (InSb) (d)	0.17	8000	1250	0.014	0.015, 0.4	17.7	6.474	525
Indium arsenide (InAs) (d)	0.36	$\sim 2 \times 10^4$	~500	0.027	0.025, 0.41	15	6.083	
Indium phosphide (InP) (d)	1.35	4600	150	0.077	0.089, 0.85	12.4	5.87	1070
Lead telluride (PbTe) (i)	0.31	6000	4000	0.17	0.20	30	6.452	925
Silicon (Si) (i)	1.12	1500	450	0.98, 0.19	0.16, 0.49	11.8	5.430	1415
Silicon carbide SiC (α-form) (i)	2.996	400	50	0.6	1.0	10.2	3.08	2830
Zinc sulfide (ZnS) (d)	3.6	180	10	0.23	—	8.9	5.409	1650 (vaporizes)
Zinc selenide (ZnSe) (d)	2.7	600	28	0.14	0.60	9.2	5.671	1100 (vaporizes)
Zinc telluride (ZnTe) (d)	2.25	530	100	0.18	0.65	10.4	6.101	1238 (vaporizes)

m_l^* and m_t^* are known as the longitudinal and translational effective masses, m_{hh}^* and m_{lh}^* are the light- and heavy-hole effective masses. Data are from various sources.

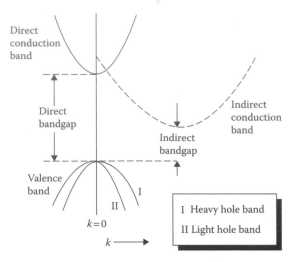

FIGURE 3.11   Bandstructure or *E-k* relationships for a direct and indirect bandgap semiconductor. (From Singh, J. 2001. *Semiconductor Devices: Basic Principles*. New York: Wiley. With permission.)

3.5   DIRECT AND INDIRECT BANDGAP SEMICONDUCTORS

The *E-k* diagrams provide one more important way to classify the semiconductors. Figure 3.11 shows the band structure of a *direct bandgap semiconductor*. In a direct bandgap semiconductor, the maximum in the valence band and the minimum in the conduction band occur at $k = 0$; an electron in the valence band can move into the conduction band if it has sufficient energy to cross the bandgap (E_g). This process does *not* require a change in the momentum of the electron.

In an *indirect bandgap semiconductor*, the valence band's maximum and the conduction band's minimum do not coincide at $k = 0$. Silicon and germanium are examples of indirect bandgap semiconductors. The actual band structures of silicon, germanium, GaAs, and aluminum arsenide (AlAs) are more complex (Figure 3.12), and their analysis is beyond the scope of this book.

In many materials based on alloys of two or more semiconductors, the bandgap can change from indirect to direct and vice versa. For example, GaAs is a direct bandgap semiconductor. When we form a solid solution with gallium phosphide (GaP; an indirect semiconductor), the bandgap of $GaAs_{1-x}P_x$ remains direct up to phosphorus (P) mole fractions of $\sim x = 0.45$–0.50. Beyond this (i.e., $x > 0.5$), the bandgap becomes indirect until we reach GaP (Figure 3.13). Another level associated with nitrogen atoms when added to GaP is shown in Figure 3.13. The significance of this will be discussed in Section 3.6.

3.6   APPLICATIONS OF DIRECT BANDGAP MATERIALS

Direct bandgap materials exhibit strong interactions with energy in the form of light waves. Because of this, direct bandgap materials are used to create *optoelectronic devices*. When an electron that has been excited into the conduction band falls back to the valence band, it recombines with a hole. If the energy of the recombination reaction is released in the form of light, this process is known as *radiative recombination*.

$$\text{Electron} + \text{hole} \xrightarrow{\text{radiative recombination}} \text{photon (light)} \qquad (3.22)$$

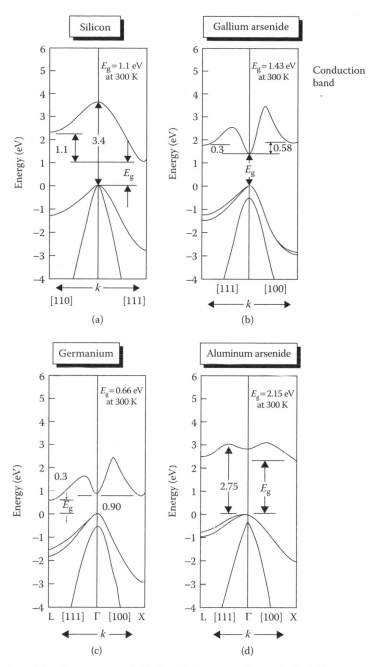

FIGURE 3.12 Actual band structures of (a) Si, (b) GaAs, (c) Ge, and (d) AlAs. (From Singh, J. 1994. *Semiconductor Devices: An Introduction*. New York: McGraw Hill. With permission.)

The radiative recombination process occurring in direct bandgap materials enables the operation of devices known as LEDs.

The electrons and the holes may also recombine, but not produce light. Instead, the energy takes the form of the vibrations of atoms known as phonons and appears as heat. This process is known as *nonradiative recombination* (Figure 3.15a).

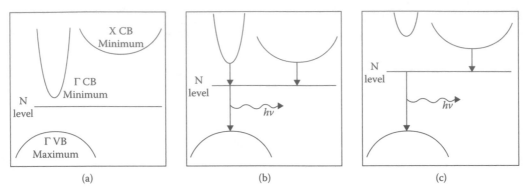

FIGURE 3.13   (a) Direct bandgap of GaAs; (b) direct bandgap in $GaAs_{0.5}P_{0.5}$; and (c) indirect bandgap of GaP. The relative level of N dopant added for optoelectronic applications is also shown. (From Schubert, F. E. 2006. *Light Emitting Diodes.* Cambridge, UK: Cambridge University Press. With permission.)

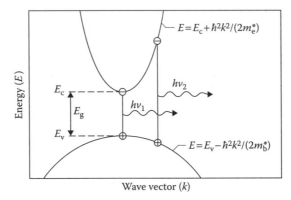

FIGURE 3.14   Vertical recombination and photon emission in a direct bandgap semiconductor. (From Schubert, F. E. 2006. *Light Emitting Diodes.* Cambridge, UK: Cambridge University Press. With permission.)

$$\text{Electron + hole} \xrightarrow{\text{nonradiative recombination}} \text{phonon (heat)} \qquad (3.23)$$

The process of radiative recombination without a change in the momentum is known as *vertical recombination* (Figures 3.14 and 3.15b). Note that radiative recombination also occurs in some indirect bandgap materials when an electron from the conduction band comes back to the valence band via a defective energy level. Nonradiative recombination occurs in direct bandgap materials as well; however, we can minimize it by using semiconductor materials that have very few point defects such as vacancies or interstitials, or other defects such as dislocations. This will improve the LED efficiency. Nonradiative recombination also occurs at the semiconductor surfaces that have incomplete or dangling bonds. Recombination dynamics is one of the factors that limits how rapidly an LED can be turned on and off.

The frequency (ν) of light emitted is related to the bandgap energy (E_g):

$$h\nu = E_g \qquad (3.24)$$

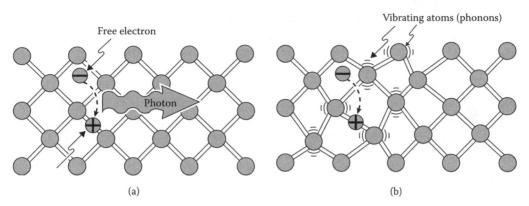

FIGURE 3.15 (a) Radiative recombination of an electron–hole pair accompanied by the emission of a photon with energy $h\nu \approx E_g$. (b) In non-radiative recombination events, the energy released during the electron–hole recombination is converted to phonons. (From Schubert, F. E. 2006. *Light Emitting Diodes*. Cambridge, UK: Cambridge University Press. With permission.)

Now, if c is the speed of light and λ is its wavelength, then

$$c = \nu \times \lambda \tag{3.25}$$

Therefore, from Equations 3.24 and 3.25:

$$E_g = \frac{hc}{\lambda} \tag{3.26}$$

This relationship forms the basis for correlating the wavelength of light emitted from LEDs. The example below illustrates how Equation 3.26 can be converted into a more useful form.

EXAMPLE 3.2: WAVELENGTH OF LIGHT EMITTED FROM LEDS

Develop a relationship between the wavelength of light (λ in micrometers) emitted from a LED and the bandgap (in electron volts).

SOLUTION

We can simplify Equation 3.26 to make it more practical for applications as follows:

$$E_g \text{ (in eV)} = \frac{(4.14 \times 10^{-15} \text{ eV} \cdot \text{s})(2.998 \times 10^8 \text{ m/s})}{\lambda}$$

$$\lambda \text{(in m)} = \frac{12.41172 \times 10^{-7}}{E_g \text{ (in eV)}}$$

$$\therefore \lambda \text{(in } \mu\text{m)} = \frac{12.41172 \times 10^{-7}}{E_g \text{ (in eV)}} \text{ m} \times 10^6 \text{ } \mu\text{m/m}$$

Therefore, the wavelength of light emitted from an LED is given by:

$$\lambda \text{(in } \mu\text{m)} = \frac{1.24}{E_g \text{(in eV)}} \mu\text{m} \tag{3.27}$$

Example 3.3 illustrates an application of optoelectronic materials.

EXAMPLE 3.3: LIGHT EMISSION FROM AN LED

An LED is made using gallium nitride (GaN; $E_g = 3.47$ eV at $T = 0$ K). What is the wavelength and the color of light emitted from this semiconductor LED?

SOLUTION

We use Equation 3.27 to calculate the wavelength:

$$\lambda(\text{in } \mu m) = \frac{1.24}{E_g(\text{in eV})} \mu m$$

$$\therefore \lambda(\text{in } \mu m) = \frac{1.24}{3.47} \mu m = 0.357 \ \mu m$$

In nanometers, this value is $0.357 \times 1000 = 357$ nm. This wavelength veers toward the blue light, which is about 475 nm.

We can adjust the composition of semiconductors, which causes the bandgap to change. In this case, indium nitride (InN) can be alloyed with GaN to form blue lasers and LEDs that emit at $\lambda = 470$ nm. Such LEDs are found in many modern-day electronic products such as CD players, videogames, and controllers.

Indirect bandgap materials such as silicon or germanium cannot normally be used for making semiconductor lasers or LEDs. This is because electrons and holes cannot combine to produce a radiative recombination in these materials. Instead, nonradiative recombination occurs, and the electron–hole recombination results in the generation of heat.

However, we cannot totally dismiss the possibility of using indirect bandgap materials. In some cases, we can use an indirect bandgap material for optoelectronic devices. For example, GaP is an indirect bandgap material. When GaP is doped with nitrogen (N), a defect level is created deep in the bandgap (Figure 3.13). This defect level then makes radiative recombination possible. The transition is between the N level and Γ VB (Figure 3.13c). We will learn in Section 3.8 that a *dopant* is an element or a compound deliberately added to enhance the electrical or other properties of a semiconductor. Since nitrogen has the same valence as phosphorous (i.e., +5), the doping of GaP with nitrogen is an example of *isoelectronic doping* (Section 3.11).

The isoelectronic doping of GaP with nitrogen provides a practical application of Heisenberg's uncertainty principle. Since the electron wave function of nitrogen atoms is highly *localized* (small Δx), the wave function is highly *delocalized* in the momentum space (large Δp). As a result, two vertical transitions can occur from the nitrogen level (Figure 3.14). One of these is radiative; an electron from the conduction band (labeled X) comes to the nitrogen impurity level, which is within the bandgap. The transition from the nitrogen level to the valence band results in emission of light (Figure 3.13). The change in the momentum of the electron as it changes from the conduction band labeled X to the conduction band labeled Γ is absorbed by the isoelectronic nitrogen atom.

3.7 MOTIONS OF ELECTRONS AND HOLES

Consider a valence band that is completely filled with valence electrons. As mentioned in Chapter 2, there is no electrical current in a completely filled band. For example, if electron 1 is moving with velocity v_1, then the effect of its motion (in terms of current generation) is nullified by the other electron (electron 2) moving with velocity v_2 in the opposite direction. Since the wave vector ($\mathbf{k}$) is proportional to the momentum (p) of the electrons (Equation 3.9), the velocities of these two electrons are opposite. If electron 1 has a wave vector $\mathbf{k_1}$, then the current generated by its motion is compensated for by electron 2 moving in the band with a wave vector $\mathbf{k_2}$.

Thus, for a completely filled band with N electrons/cm³, the motions of all electrons are canceled, and there is no net electrical current. This is the basis for the following equation for the current density of a completely filled band:

$$J = (-q)\sum_{i}^{N} v_i = 0 \tag{3.28}$$

where J is the current density, q is the magnitude of the charge on the electron, and v_i is the velocity of the ith electron. Note the negative sign in Equation 3.28, used because of the negatively charged electron.

Now consider a valence band that has a hole created by removing the jth electron. The current density in this band will be given by the sum of all current densities minus the current density due to motion of the jth electron.

$$J = \left[(-q)\sum_{i}^{N} v_i \right] - (-q)v_j \tag{3.29}$$

Note that the first term in Equation 3.29 is zero. Thus, the current density in a valence band with one hole is given by

$$J = (+q)v_j \tag{3.30}$$

We can conclude that the current in a valence band with one hole can be described as the current by the jth electron (whose motion is uncompensated) moving with velocity $-v_j$. This current contribution is the same as that of a hole with a charge of $+q$ moving in the opposite direction, that is, with a velocity of $+v_j$. The magnitude of the wave vector associated with a missing electron is equal to the wave vector of the hole created (Figure 3.16).

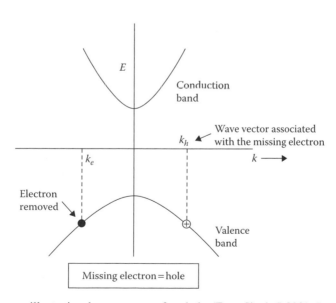

FIGURE 3.16   Diagram illustrating the wave vector for a hole. (From Singh, J. 2001. *Semiconductor Devices: Basic Principles*. New York: Wiley. With permission.)

An important observation is that during the motion of charge carriers, the current contributions of an electron moving with a certain velocity and a hole moving with opposite velocity are the same. To understand the electrical properties of the valence band, which is almost (but not completely) filled with electrons, we consider the behavior of the holes. As an analogy, consider that we look for an empty parking spot rather than parked cars while parking in a garage. To understand the electrical properties of the conduction band, we look at the behavior of the *electrons*.

In a band diagram or a band structure, we plot the electron energy so that the electron energy increases as we go up on the band diagram or band structure and the hole energy increases as we go down. As a result, the electrons excited to the conduction band seeking the minimum energy are shown at the bottom of the conduction band on the band diagram, and the holes created in the valence band minimize their energy and are shown at the top of the valence band. The band diagrams (Figures 3.2 and 3.5) therefore show the predominance of electrons at $E = E_c$ and that of holes at $E = E_v$.

3.8   EXTRINSIC SEMICONDUCTORS

In intrinsic semiconductors, the changes in the number of carriers are related exponentially to the changes in the temperature (Equation 3.5). This makes it very difficult to make practical and useful devices using intrinsic materials. We need semiconductors whose conductivity can be controlled by controlled changes in the composition. We would prefer these materials to show electrical properties that would not only be tunable with composition changes but would also be stable over a wide range of temperatures (e.g., −50°C to +150°C). In Chapter 4, we will discuss electronic devices such as transistors and diodes that can be designed and manufactured from extrinsic semiconductors. These materials show controllable variations in their conductivity due to doping and are also relatively temperature stable.

As mentioned in Section 3.2, a material in which the conductivity is largely caused and controlled by the addition of other elements or compounds is known as an extrinsic semiconductor. In extrinsic semiconductors, the charge carriers are generated by adding other elements or compounds known as dopants. A dopant is an element or a compound deliberately added to a semiconductor to influence and control electrical properties or other properties. In particular, the electrical conductivity of semiconductors is significantly affected by the presence of foreign atoms. In contrast to dopants, *impurities* are elements or compounds that are present in semiconductors either inadvertently or because of processing limitations. For example, silicon crystals grown from melt often contain dissolved oxygen (O). This impurity comes from quartz (SiO_2) crucibles used during silicon single-crystal growth. Like dopants, impurities also have a profound effect on the electrical properties of semiconductors. The concentrations of impurities are usually not completely predictable, so it is difficult to predict their effects on semiconductor properties. Thus, every effort is made in semiconductor processing to minimize the presence of impurities. In Sections 3.9 and 3.10, we describe different types of extrinsic semiconductors.

3.9   DONOR-DOPED (N-TYPE) SEMICONDUCTORS

A *donor-doped* or *n-type semiconductor* consists of an atom or an ion that provides "extra" electrons. Electrons are the *majority carriers* in these materials. Since electrons carry a negative charge, these materials are also known as n-type semiconductors.

For example, consider a silicon crystal with a low concentration of arsenic (As) forming a solid solution in which the arsenic atoms occupy the silicon sites. Each arsenic atom brings in five valence electrons. Only four of these participate in the formation of covalent bonds with other silicon atoms. The fifth electron remains bonded to the arsenic atom at low temperatures. However, as the temperature increases to ~50 to 100 K, the fifth electron from arsenic dissociates itself and is available for conduction (Figure 3.17). Thus, each arsenic atom added results in one "extra" electron available for

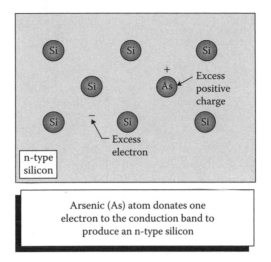

FIGURE 3.17 Illustration of arsenic-doped silicon, which is an n-type or donor-doped semiconductor. (From Singh, J. 2001. *Semiconductor Devices: Basic Principles*. New York: Wiley. With permission.)

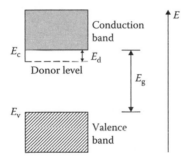

FIGURE 3.18 Band diagram for an n-type semiconductor.

conduction. Note that since the dopant atoms added are neutral, the doped semiconductor remains electrically neutral and does not acquire a net negative charge. A band diagram for an n-type semiconductor is shown in Figure 3.18. In this figure, note the donor energy level (E_d) in relation to the conduction bandedge (E_c). Examples of other n-type dopants for silicon include phosphorus and antimony (Sb). Section 3.21 discusses similar concepts through which we can create ceramic materials that show n-type semiconductivity.

3.10   ACCEPTOR-DOPED (p-TYPE) SEMICONDUCTORS

Another type of extrinsic semiconductor is an *acceptor-doped* or *p-type* semiconductor. The term "p-type" refers to the positive charge on holes that are the dominant charge carriers. In this type of material, the situation is opposite that of n-type semiconductors. We add dopant atoms or ions that create a deficit of electrons. A typical example of a p-type semiconductor is silicon doped with boron (B; Figure 3.19). Each boron atom added to silicon brings in three valence electrons. Thus, the boron atoms occupying the silicon sites have only three valence electrons that can form complete covalent bonds with three silicon atoms. The fourth boron–silicon bond is incomplete in that it is missing an electron and is thus ready to accept one. Boron-doped silicon is therefore referred to as an acceptor-doped semiconductor. Aluminum is another example of an acceptor dopant for silicon.

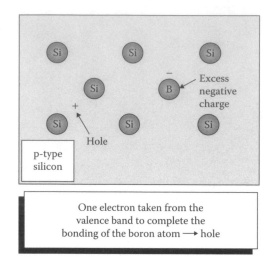

FIGURE 3.19 Illustration of boron-doped silicon, which is a p-type or acceptor-doped semiconductor. (From Singh, J. 2001. *Semiconductor Devices: Basic Principles*. New York: Wiley. With permission.)

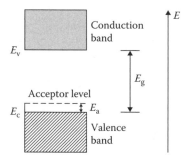

FIGURE 3.20 A band diagram for a p-type or acceptor-doped semiconductor.

The missing electron in one of the boron–silicon bonds is also described as the creation of a hole. In acceptor-doped or p-type semiconductors, the majority of carriers are holes. Hence, acceptor-doped semiconductors are also known as p-type semiconductors. The band diagram for a typical p-type semiconductor is shown in Figure 3.20. Note the position of acceptor level E_a, which is above the valence bandedge (E_v).

3.11 AMPHOTERIC DOPANTS, COMPENSATION, AND ISOELECTRONIC DOPANTS

In some cases, the same dopant element will behave as if it is both the donor and the acceptor. Such dopants are known as *amphoteric dopants*. For example, silicon in GaAs is an amphoteric dopant. If the tetravalent silicon occupies the trivalent gallium sites, it behaves as a donor. If silicon atoms occupy the pentavalent arsenic sites, it behaves as an acceptor. Therefore, is silicon a donor or an acceptor dopant when added to GaAs? The answer depends upon the concentration of silicon added to GaAs, the manner in which it is added, and the temperature during the addition. At low processing temperatures and lower concentrations, silicon acts as an acceptor, whereas at high processing temperatures and higher concentrations, it acts as a donor. We will learn about the ways to add dopants to semiconductors in the next chapter.

As mentioned in Section 3.6, the addition of isoelectronic dopants can be useful. For example, aluminum with a valency of 3 can be added to GaN. Thus, aluminum acts as neither an acceptor nor a donor. When isoelectronic dopants are added, they change the local electronic structure and in turn alter the electrical properties of the base materials. Another example of an isoelectronic dopant is germanium in silicon. Silicon–germanium (informally known as *siggy*) semiconductors have been developed. Their advantages include a higher device speed and less power consumption (Ahlgre and Dunn 2000).

In some cases, isoelectronic additions help with optical properties. As mentioned before, nitrogen-doped $GaAs_{1-x}P_x$ compositions with an indirect bandgap are used for making yellow and green LEDs (Streetman and Banerjee 2000; Figure 3.13).

There are many electronic devices that contain both types of dopants in the same volume of the material. For example, we may have a phosphorus-doped silicon crystal, and we add boron, an acceptor dopant, either to the entire crystal or parts of it. The effect of the boron addition will compensate for the effect of phosphorus doping. This is known as *compensation doping.* Note that in order for the compensated semiconductor (either the entire crystal or selected part of it) to behave as n-type or p-type, the net concentration of carriers created must exceed the concentrations of thermally generated carriers. A semiconductor containing both types of dopants behaves as n-type if the donor concentration (N_d) outweighs the acceptor concentration (N_a) and ($N_d - N_a$) $\gg n_i$. If a semiconductor has both types of dopants and if ($N_a - N_d$) $\gg n_i$ ($= p_i$), then the material behaves as a p-type semiconductor.

3.12 DOPANT IONIZATION

In an n-type or p-type material, at low temperatures, electrons and holes remain bonded to the donor or acceptor atoms, respectively. This temperature region in which the carriers remain bonded to the dopant atoms is known as the *freeze-out range* (Figure 3.21).

As the temperature increases to ~50 to 100 K, the "extra" electrons or holes bound to the donor and acceptor atoms become available for conduction. This process by which the carriers that have bonded with the dopant atoms dissociate themselves from the dopant atoms is known as *dopant ionization* (Figure 3.22).

As illustrated in Figure 3.22, when the fifth electron from the phosphorous atom becomes detached, we get a phosphorous ion that has an effective charge of +1. If a hole bound to a boron

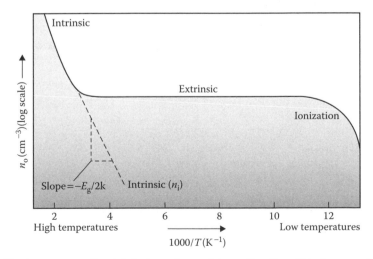

FIGURE 3.21 Carrier concentration (plotted on a log scale) as a function of the inverse of the temperature (*T*). (From Askeland, D., and P. Fulay. 2006. *The Science and Engineering of Materials*. Washington, DC: Thomson. With permission.)

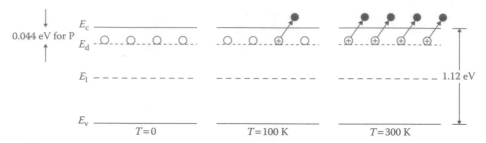

FIGURE 3.22 Dopant ionization in an n-type semiconductor. (From Kano, K. 1997. *Semiconductor Fundamentals.* Upper Saddle River, NJ: Prentice Hall. With permission.)

atom dissociates itself, we get a boron ion with an effective charge of −1. With a further increase in the temperature, the carrier concentration levels off, indicating the complete ionization. This region is labeled as "extrinsic" in Figure 3.21 since the carrier concentration is directly linked to the dopant concentration. With a further increase in the temperature, the concentrations of thermally excited carriers (n_i and p_i) are dominant. This region is therefore labeled "intrinsic," because even if the material is doped, it "behaves" as an intrinsic semiconductor.

The energy required (E_d) to remove the fifth electron from a phosphorus atom, that is, to ionize a donor atom, is given by the following equation:

$$E_d = E_c - \frac{q^4 m_e^*}{2(4\pi\varepsilon)^2 \hbar^2} \tag{3.31}$$

In Equation 3.31, E_d is the donor energy level, E_c is the conduction bandedge energy, and ε is the dielectric permittivity of the semiconductor (unit F/m). Thus, the difference ($E_d - E_c$) tells us how far below E_c the donor energy level E_d lies. This equation is an extension of Bohr's atomic model, used for predicting the binding energy of an electron to a hydrogen atom nucleus.

We rewrite Equation 3.31 to involve the ratio of the effective mass to the mass of electrons in a vacuum. The ratio of the permittivity of the semiconductor to the permittivity of the free space (ε_0) is its dielectric constant (ε_r).

In Example 3.4, we will show that

$$\frac{m_0 q^4}{2(4\pi\varepsilon_0)^2 \hbar^2} = 13.6 \text{ eV} \tag{3.32}$$

This is the energy with which an electron is bound to a hydrogen atom nucleus.

Thus, from Equations 3.31 and 3.32, we get

$$E_d = E_c - 13.6 \left(\frac{m_e^*}{m_0} \right) \left(\frac{\varepsilon_0}{\varepsilon} \right)^2 \text{ eV} \tag{3.33}$$

We can rewrite this as

$$E_d = E_c - 13.6 \frac{\left(\dfrac{m_e^*}{m_0} \right)}{\left(\dfrac{\varepsilon}{\varepsilon_0} \right)^2} \text{ eV} \tag{3.34}$$

$$E_d = E_c - 13.6 \frac{\left(\dfrac{m_e^*}{m_0}\right)}{\left(\varepsilon_r\right)^2} \text{ eV} \tag{3.35}$$

or

$$\left(E_c - E_d\right) = \frac{13.6\left(\dfrac{m_e^*}{m_0}\right)}{\left(\varepsilon_r\right)^2} \tag{3.36}$$

Note that in these equations, we use the effective mass of electrons (m_n^*) for direct bandgap semi-conductors (e.g., GaAs), whereas for indirect bandgap semiconductors such as silicon, we use the *conductivity effective mass* (m_σ^*). The examples below illustrate the calculations of dopant levels in direct and indirect bandgap materials.

The values of the dopant energy levels calculated here are approximate and use an extension of Bohr's model for the binding energy of an electron bound to the hydrogen nucleus. This is known as the hydrogen model. The actual values of the binding energy levels for different dopants used in sili-con, germanium, and GaAs depend upon the specific dopant and are shown in Table 3.3. This table shows that the binding energy for most dopants is a few millielectron volts. This is significantly lower than 13.6 eV. There are two reasons for this: (1) The effective mass of the carriers is reduced in a material; and (2) the binding potential is reduced by the crystal as described by the dielectric constant (ε_r; Equation 3.36).

Shallow dopants are those whose energy levels are close to the bandedge, for example, the dopants shown in Table 3.3. Some atoms create defect levels that are well into the bandgap; these are known as *deep-level defects*. The levels of these dopants cannot be predicted using the hydrogenic model (Equation 3.35). Deep-level dopants are introduced by atoms or ions that do not "fit" the semiconductor crystal; they severely distort the host lattice. Deep-level defects act as a trap for charge carriers. Deep-level defects can also be useful in pinning the Fermi energy level.

TABLE 3.3
Energy Levels for Some Dopants in Silicon, Germanium, and GaAs

Semiconductor	Donor Dopant	Donor Energy Level (meV)	Acceptor Dopant	Acceptor Energy Level (meV)
Si	Sb	39	B	45
	P	45	Al	67
	As	54	Ga	72
			In	160
GaAs	Si	5.8	C	26
	Ge	6.0	Be	28
	Sn	6.0	Mg	28
			Si	35
Ge	Sb	9.6	B, Al	10
	P	12	Ga	11
	As	13	In	11

Source: Singh, J. 2001. *Semiconductor Devices: Basic Principles*. New York: Wiley. With permission.

EXAMPLE 3.4: VALUE OF AN EXPRESSION RELATED TO THE BINDING ENERGY FOR AN ELECTRON IN AN H ATOM

Show that the term expressed in Equation 3.32 is equal to 13.6 eV. This is the energy with which an electron in a hydrogen (H) atom is bound to its nucleus.

SOLUTION

We use $m_0 = 9.11 \times 10^{-31}$ kg, $q = 1.6 \times 10^{-19}$ C, $\varepsilon_0 = 8.85 \times 10^{-12}$ F/m, and $\hbar = 1.05 \times 10^{-34}$ J·s in Equation 3.32.

$$\frac{m_0 q^4}{2(4\pi\varepsilon_0)^2 \hbar^2} = \frac{(9.11 \times 10^{-31} \text{ kg})(1.6 \times 10^{-19} \text{ C})^4}{2(4 \times 3.14 \times 8.85 \times 10^{-12} \text{ F/m})^2 \times (1.05 \times 10^{-34} \text{ J·s})^2} = 2.189 \times 10^{-18} \text{ J}$$

Note that 1 eV is the energy required to move the charge equal to the charge on an electron through 1 V potential difference, that is, 1 eV = 1.6×10^{-19} J.

Therefore, converting 2.189×10^{-15} J into electron volts, we get

$$(2.189 \times 10^{-18} \text{ J}) \times \frac{1 \text{ eV}}{1.6 \times 10^{-19} \text{ J}} = 13.6 \text{ eV}$$

EXAMPLE 3.5: DONOR ENERGY LEVEL CALCULATION FOR Si-DOPED GaAs

Assuming that Si acts as a donor in a GaAs sample, calculate the position of the donor energy level (E_d) relative to the conduction bandedge (E_c).

SOLUTION

We use Equation 3.35 to calculate the relative position of the donor energy level. Since GaAs is a direct bandgap semiconductor, we use the effective mass of electrons.

From Table 3.2, for GaAs $(m_n^*/m_0) = 0.067$ and the dielectric constant (ε_r) is 13.2.

$$E_d = E_c - 13.6 \frac{\left(\dfrac{m^*}{m_0}\right)}{(\varepsilon_r)^2} \text{ eV}$$

Therefore,

$$E_d = E_c - 13.6 \frac{0.067}{(13.2)^2} \text{ eV}$$

$$E_d = E_c - 5.22 \times 10^{-3} \text{ eV}$$

Thus, the donor energy level for Si in GaAs is ~5.2 meV below the conduction bandedge (Figure 3.23). Note that this calculation does not require the actual bandgap energy value. The binding energy of many dopants is significantly lower than 13.6 eV because of the lesser effective masses and dielectric constant.

EXAMPLE 3.6: COMPARISON OF DONOR ENERGY LEVEL RELATIVE TO THERMAL ENERGY

Compare the magnitude of the difference ($E_d - E_c$) calculated in Example 3.5 with the thermal energy ($k_B T$) of carriers at 300 K. (Note: Boltzmann's constant [k_B] is 8.617×10^{-5} eV/K or 1.38×10^{-23} J/K.)

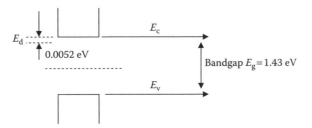

FIGURE 3.23 Illustration of the silicon donor dopant energy level in GaAs. The intrinsic Fermi energy level ($E_{F,i}$) is shown in the middle of the bandgap. Diagram is not to scale (see Example 3.5).

SOLUTION

Since we prefer to get the energy in electron volts, we use the value of k_B in electron volts. The value of $k_B T$ at 300 K is ~0.026 eV.

The energy with which the donor electron of Si in GaAs is bonded is only 0.00522 eV. Thus, the thermal energy at 300 K is almost five times greater, suggesting a nearly complete donor ionization.

EXAMPLE 3.7: DONOR ENERGY LEVEL FOR P-DOPED SI

Calculate the position of the donor energy (E_d) level relative to the conduction bandedge (E_c) for P in Si. The dielectric constant (ε_r) of Si is 11.8 (Table 3.2). The ratio of the *conductivity effective mass* of electrons in Si to m_0 is 0.26.

SOLUTION

We use the values in Equation 3.35, which have been modified to show the conductivity effective mass of the electrons in Si.

$$E_d = E_c - 13.6 \frac{\left(\dfrac{m_\sigma^*}{m_0}\right)}{(\varepsilon_r)^2} \text{ eV} \qquad (3.37)$$

Note the use of the conductivity effective mass because Si is an indirect bandgap material.

$$E_d = E_c - 13.6 \frac{0.26}{(11.8)^2}$$

Thus, the donor level for P in Si is ~0.025 eV or 25 meV below the conduction bandedge. This is illustrated in Figure 3.24.

3.13 CONDUCTIVITY OF INTRINSIC AND EXTRINSIC SEMICONDUCTORS

The conductivity of a semiconductor is given by the following equation:

$$\sigma = q \times n \times \mu_n + q \times p \times \mu_p \qquad (3.38)$$

where q is the magnitude of the charge on the electron or hole and is 1.6×10^{-19} C, n is the concentration of conduction electrons, p is the concentration of holes, and μ_n and μ_p are the mobilities of electrons and holes in the semiconductor material, respectively.

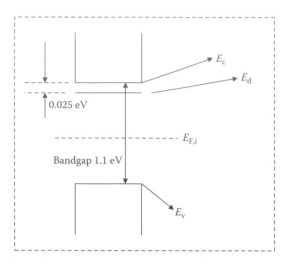

FIGURE 3.24 Position of calculated donor dopant level (E_d) in silicon relative to conduction bandedge (E_c). The dotted line in the middle of the bandgap is the Fermi energy level of the intrinsic semiconductor ($E_{F,i}$). Diagram is not to scale (see Example 3.7).

If we add one phosphorus atom to a silicon crystal and the temperature is high enough to cause donor ionization, we should get one electron in the conduction band. Let N_d be the concentration of donor atoms. Assuming that each donor dopant atom donates one electron that becomes available for conduction, we expect to get N_d number of electrons per cubic centimeter in the conduction band. Note that as the temperature increases, we expect not only the donor ionization (Figure 3.22) but also some electrons in the silicon–silicon bond to break free and provide for the intrinsic or thermally generated electrons and holes (Figure 3.2).

If n_i is the concentration of thermally generated carriers, then for an n-type semiconductor, we can write the total concentration of conduction electrons (n) as

$$n = N_d + n_i \tag{3.39}$$

In an n-type semiconductor, conduction electrons generated through the doping process dominate the conductivity and are the majority carriers, that is

$$N_d \gg n_i \tag{3.40}$$

Since the concentration of conduction electrons created by doping is significantly higher than the concentration of those generated by thermal excitation (Equation 3.40), we can assume

$$n \cong N_d \tag{3.41}$$

Recall that there is an equal concentration of holes (p_i holes/cm³) corresponding to n_i electrons/cm³. In this case, we can ignore the contribution of holes and thermally excited electrons to the total conductivity because holes are *minority carriers* in n-type semiconductors. The equation for the conductivity of an n-type semiconductor can be written as

$$\sigma = q \times N_d \times \mu_n \tag{3.42}$$

The situation is analogous for the p-type semiconductor. For this type, the concentration of holes generated due to acceptor dopants is dominant ($N_a \gg p_i$). The holes are the majority carriers, and the electrons are the minority carriers.

$$p \cong N_a \tag{3.43}$$

We can ignore the contribution of thermally generated electrons and holes to conductivity, that is

$$\sigma = q \times N_a \times \mu_p \qquad (3.44)$$

These concepts are illustrated in Example 3.8.

EXAMPLE 3.8: CONDUCTIVITY OF INTRINSIC GaAs

a. What is the resistivity of intrinsic GaAs? Assume the temperature is 300 K and $n_i = 2 \times 10^6$ electrons/cm^3 (Figure 3.7).
b. How does this value compare with the conductivity of an intrinsic Ge sample? (See Example 3.1.)

SOLUTION

a. Since this is an intrinsic semiconductor, the concentrations of electrons and holes play a role in the conductivity. From Equation 3.38

$$\sigma = q \times n \times \mu_n + q \times p \times \mu_p$$

Table 3.1 shows that the mobility values for electrons and holes for intrinsic GaAs are 8500 and 400 cm^2/V·s, respectively. The concentration of holes and electrons is the same, since each electron that moves into the conduction band creates a hole in the valence band, i.e. $n_i = p_i$.
Therefore

$$\sigma = (1.6 \times 10^{-19} \text{ C})(2 \times 10^6/\text{cm}^3)(8500 + 400) \text{ cm}^2/\text{V} \cdot \text{s}$$
$$\sigma = 2.848 \times 10^{-19} \text{ S/cm}$$

The resistivity $\rho = 3.51 \times 10^8 \ \Omega \cdot \text{cm}$.
Thus, undoped intrinsic GaAs has a relatively high resistivity. This value is actually higher than $10^3 \ \Omega \cdot \text{cm}$, which is generally considered the approximate upper limit for a material to be defined as a semiconductor (see Section 3.1).
b. As seen in Example 3.1, the resistivity of intrinsic Ge was 43.1 $\Omega \cdot \text{cm}$.
Since Ge has a smaller bandgap compared to the GaAs bandgap, it has a much higher concentration of electrons in the conduction band. Conversely, the mobilities of the carriers in Ge are lesser than those for GaAs (see Example 3.1). However, the effect of increased carrier concentrations dominates the effect of mobilities. As a result, intrinsic Ge ends up having a higher conductivity than GaAs.

3.14 EFFECT OF TEMPERATURE ON THE MOBILITY OF CARRIERS

Temperature has a major influence on the mobility of electrons (μ_n) and holes (μ_p). When the temperature is too low for dopant ionization, the electrons and holes remain bound to the dopant atoms. In this freeze-out zone, the mobility of the carriers is essentially zero. As a result, the conductivity of extrinsic semiconductors is very low at low temperatures. As the temperature increases, dopant ionization occurs. As the temperature increases above the freeze-out range, but is not too high, the extent of scattering of carriers by the vibrations of atoms decreases. In this region, the scattering of the carriers caused by the dopant atoms is dominant. The mobility of carriers limited by the scattering of the dopant or impurity atoms is known as *impurity-* or *dopant-scattering limited mobility*. If the temperature becomes too high, phonon scattering or lattice scattering (i.e., the scattering of vibrations of atoms of the host crystal) begins to dominate, and the mobility decreases again. This is known as *lattice-scattering limited mobility*. The trends in the changes in the mobility of carriers are shown in Figure 3.25.

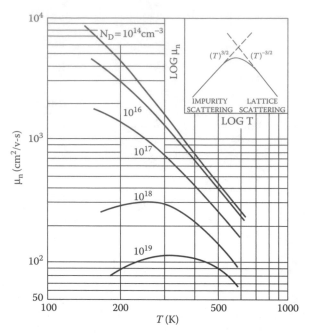

FIGURE 3.25 Electron mobility in silicon versus temperature for various donor concentrations. Insert shows the theoretical temperature dependence of electron mobility. (From Sze, S. M. 1985. *Semiconductor Devices, Physics, and Technology.* New York: Wiley. With permission.)

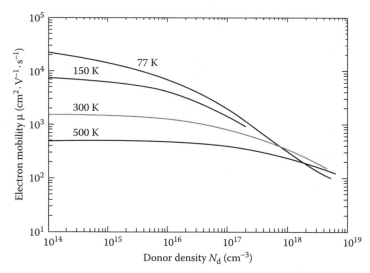

FIGURE 3.26 The variation in electron mobility for silicon at different temperatures and different donor dopant concentrations. (From Li, S. S., and W. R. Thurber. 1977. *Solid State Electron* 20(7):609–16. With permission.)

3.15   EFFECT OF DOPANT CONCENTRATION ON MOBILITY

As discussed in Section 3.14, although the dopant atoms provide carriers that contribute to conductivity, they can also act as scattering centers. This effect is important at both low temperatures and at high dopant concentrations. The variations in the electron and the hole drift mobility for silicon at different temperatures and at different dopant concentrations are shown in Figures 3.26 and 3.27. Similar trends for changes in the mobility in other semiconductors are shown in Figure 3.28.

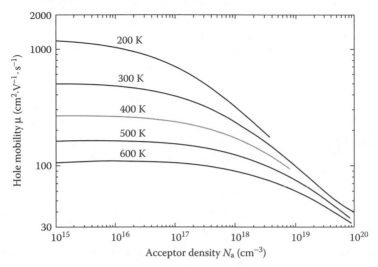

FIGURE 3.27 The variation in hole mobility for silicon at different temperatures and acceptor dopant concentrations. (From Dorkel, J. M., and P. Leturcq. 1981. *Solid State Electron* 24(9):821–5. With permission.)

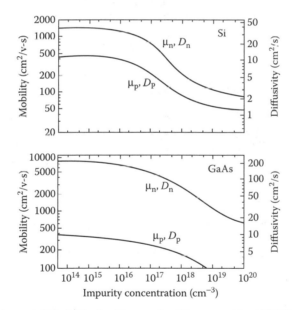

FIGURE 3.28 Mobilities and diffusivities in silicon and gallium arsenide at 300 K as a function of impurity concentration. (From Sze, S. M. 1985. *Semiconductor Devices, Physics, and Technology*. New York: Wiley. With permission.)

Example 3.9 shows how to account for these changes in mobility with the dopant concentration, while calculating the conductivity of extrinsic semiconductors.

EXAMPLE 3.9: RESISTIVITY OF N-TYPE DOPED GaAs

Calculate the resistivity of donor-doped GaAs with $N_d = 10^{13}$ atoms/cm³. Assume that $T = 300$ K and all the donors are ionized.

SOLUTION

For GaAs at 300 K, $n_i = 2.0 \times 10^6$ electrons/cm³ (Figure 3.7).

In this case, $N_d \gg n_i$; the total concentration of conduction electrons $(n) \approx N_d = 10^{13}$ electrons/cm³. From Equation 3.38:

$$\sigma = q \times n \times \mu_n + q \times p \times \mu_p$$

We ignore the contributions of the holes because the electrons are the majority charge carriers. From Figure 3.28, for GaAs with $N_d = 10^{13}$ atoms/cm³, and $\mu_n = 8000$ cm²/V·s,

$$\sigma = (1.6 \times 10^{-19} \text{ C})(10^{13})(8000)$$

$$\therefore \sigma = 1.28 \times 10^{-3} \ \Omega^{-1} \cdot cm^{-1}$$

The resistivity (ρ) is 781 $\Omega \cdot$ cm. This value is much lower than $3.51 \times 10^8 \ \Omega \cdot$ cm, which is the value of resistivity for intrinsic GaAs (Example 3.8).

3.16   TEMPERATURE DEPENDENCE OF CONDUCTIVITY

Thus, the combination of dopant ionization, changes in mobility, and thermal generation of carriers using temperature leads to changes in the conductivity of a semiconductor as a function of the temperature. The change in the resistivity of silicon as a function of the dopant concentration is shown in Figure 3.29.

3.17   EFFECT OF PARTIAL DOPANT IONIZATION

For extrinsic semiconductors near room temperature, we usually assume complete dopant ionization. However, this may not always be the case. Dopant ionization is temperature-dependent (Figures 3.21 and 3.22), and its extent depends upon the relative difference between the defect level and the nearest bandedge. The fraction of electrons still tied to the donor atoms in an n-type semiconductor is given by

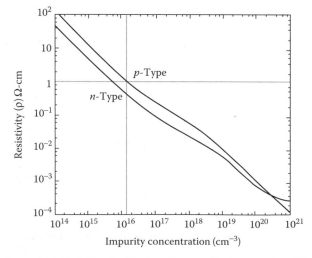

FIGURE 3.29   The change in resistivity of silicon with dopant concentration. (From Grove, A. S. 1967. *Physics and Technology of Semiconductor Devices*. New York: Wiley. With permission.)

$$\frac{n_d}{n+n_d} = \frac{1}{\left\{\dfrac{N_c}{2N_d}\exp\left(-\dfrac{(E_c - E_d)}{k_B T}\right)+1\right\}}$$ (3.45)

where n_d is the concentration of electrons still bound to donor atoms, n is the concentration of free electrons in the conduction band, $(E_c - E_d)$ is the difference between the conduction band-edge and the donor energy level, N_d is the donor dopant concentration, and N_c is the effective density of states at the conduction band. This is a parameter that indicates how many electrons can be accommodated at the energy level $E = E_c$. The actual number of electrons available in the conduction band will depend upon the distance between the donor level and the conduction bandedge.

As the temperature decreases, the exponential term in the denominator of Equation 3.45 decreases, and N_c also decreases. Overall, the right-hand side of Equation 3.45 increases. Thus, the fraction of electrons that remain bound to the donor atoms increases. Similarly, if the temperature is constant and the dopant level increases, then the denominator of the right-hand side of Equation 3.45 increases; that is, as the dopant concentration increases, the fraction of the electrons that remain bound to the donor atoms also increases.

The effective density of states at the conduction bandedge is given by

$$N_c = 2\left(\frac{m^* k_B T}{2\pi\hbar^2}\right)^{3/2}$$ (3.46)

We must be careful about the effective masses used in these calculations.

To calculate the density of states in the conduction band of *direct bandgap materials* such as GaAs, we use the effective mass of the electron as m^* in Equation 3.46; that is, we substitute the value of m_e^* for m^*.

To calculate the density of states in the conduction band of *indirect bandgap semiconductors* such as silicon, we use the *density of states effective mass* (m_{dos}^*) of electrons, defined as

$$m_{dos}^* = \left[6^{2/3}\left(m_l^* m_t^{*2}\right)^{1/3}\right]$$ (3.47)

In Equation 3.47, m_l^* and m_t^* are known as longitudinal and transverse effective masses of electrons, respectively (Table 3.2). The values m_l^* and m_t^* for the electrons in silicon are 0.98 and 0.19, respectively. This leads us from Equation 3.47 to $m_{dos}^* = 1.08$.

A detailed discussion of these parameters is beyond the scope of this book. We will use the density of states effective masses of electrons and holes in calculating the density of states, which will be discussed in Chapter 4. We will use the conductivity effective mass in calculating the response of the carriers to the electric field and the donor energies (Example 3.7).

For p-type semiconductors, there is a relationship similar to Equation 3.44. The fraction of holes that remain bound to the acceptor atoms depends upon the temperature and the distance between the acceptor level (E_a) and the valence bandedge (E_v).

$$\frac{p_a}{p+p_a} = \frac{1}{\left\{\dfrac{N_v}{2N_a}\exp\left(-\dfrac{(E_a - E_v)}{k_B T}\right)+1\right\}}$$ (3.48)

where p_a is the concentration of the holes still bound to the acceptor atoms, p is the concentration of the holes in the conduction band, $(E_a - E_v)$ is the difference between acceptor energy level and the valence bandedge, and N_a is the concentration of the acceptor dopant. In Equation 3.48, N_v is the effective density of states at the valence bandedge.

This value is given by the following equation if the semiconductor has a heavy-hole band and a light-hole band (e.g., for GaAs):

$$N_v = 2\left(m_{hh}^{3/2} + m_{lh}^{3/2}\right)\left(\frac{k_B T}{2\pi\hbar^2}\right)^{3/2} \tag{3.49}$$

In Equation 3.49, m_{hh} and m_{lh} are the effective masses of heavy and light holes, respectively (Table 3.2). The density of states effective mass of the holes in the valence band is given by

$$(m_{dos}^*)^{3/2} = \left(m_{hh}^{3/2} + m_{lh}^{3/2}\right) \tag{3.50}$$

For holes in silicon, $m_{hh}^* = 0.49$ and $m_{lh}^* = 0.16$, and this leads to $m_{dos}^* = 0.55$. For holes in GaAs, $m_{hh}^* = 0.45$ and $m_{lh}^* = 0.08$, and this leads to $m_{dos}^* = 0.47$.

The values of the density of states (at 300 K) for Ge, Si, and GaAs, calculated using Equations 3.48 and 3.1 and related equations, are shown in Table 3.4.

We usually assume that at or near 300 K, the dopants are completely ionized, and thus the fractions on the left-hand sides of Equations 3.45 and 3.48 are very small.

Example 3.10 shows how to account for partial dopant ionization.

EXAMPLE 3.10: PARTIAL DOPANT IONIZATION AND CONDUCTIVITY

A GaAs crystal is doped with Si, which acts as a donor dopant with $N_d = 10^{15}$ atoms/cm³.

   a. Calculate the fraction of electrons that is bound to the donor atoms in GaAs.
   b. What is the resistivity of this material? Assume that the donor energy (E_d) is 5.8 meV away from the conduction bandedge (E_c; Example 3.5 and Table 3.3).

SOLUTION

   a. We use Equation 3.45 to get the fraction of electrons that is still bound to the Si dopant, assumed to be a donor for GaAs. From Table 3.3, we know that the value of $(E_c - E_F)$ is 5.8 meV. Note that at 300 K, the value of $k_B T$ is 0.026 eV.
   From Table 3.4, we know that the density of states at the conduction bandedge (N_c) for GaAs at 300 K is 4.45×10^{17} cm⁻³.

$$\frac{n_d}{n + n_d} = \cfrac{1}{\left\{\cfrac{4.45 \times 10^{17}\,\text{cm}^{-3}}{2 \times 10^{15}} \exp\left(-\cfrac{(0.0058)}{0.026}\right) + 1\right\}} = 0.00586$$

TABLE 3.4
Effective Density of States for Silicon, Germanium, and GaAs ($T = 300$ K)

Semiconductor	Conduction Band Effective Density (cm⁻³) of States (N_c)	Valence Band Effective Density (cm⁻³) of States (N_v)
Si	2.78×10^{19}	9.84×10^{18}
Ge	1.04×10^{19}	6.0×10^{18}
GaAs	4.45×10^{17}	7.72×10^{18}

Thus, ~5% to 6% of the electrons remain bound to the Si atoms in this n-type GaAs sample at $T = 300$ K.

b. The actual number of the conduction electrons available is 0.00586×10^{15} and the conductivity is

$$\sigma = (1.6 \times 10^{-19}\,\text{C})(0.00586 \times 10^{15})(8000)$$

$$\therefore \sigma = 0.00715\ \Omega^{-1} \cdot \text{cm}^{-1}$$

$$\therefore \rho = 139.8\ \Omega \cdot \text{cm}$$

3.18 EFFECT OF TEMPERATURE ON THE BANDGAP

So far, we have assumed that the bandgap (E_g) of semiconductors does not change with temperature. However, the bandgap of the semiconductor does slightly change with temperature, because thermal expansion and contraction lead to a change in the interatomic distance. As we can see in Figure 3.30, as the interatomic spacing increases as a result of thermal expansion, the bandgap of silicon will decrease slightly. This trend is seen in other semiconductors as well.

Similarly, the bandgap of a semiconductor will depend upon the type and magnitude of the stress that is applied or present. In general, a tensile stress will lead to an increase in the interatomic spacing, causing a decrease in the bandgap.

The changes in the bandgap of semiconductors are given by the Varshni formula:

$$E_g = E_g\left(\text{at } T = 0 \text{ K}\right) - \frac{\alpha T^2}{T + \beta} \tag{3.51}$$

In general, the bandgap of the semiconductors slightly decreases with the increasing temperature. The *Varshni parameters* α and β for different semiconductors and their bandgaps at 0 K are listed in Table 3.5 (Schubert 2006). Examples 3.11 and 3.12 show how to account for the temperature variation in the bandgap.

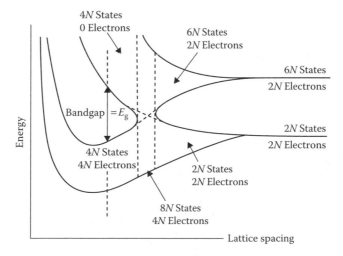

FIGURE 3.30 Energy levels in intrinsic silicon as a function of interatomic spacing. (From Streetman, B. G., and S. Banerjee. 2000. *Solid State Electronic Devices*, 5th ed. Upper Saddle River, NJ: Prentice Hall. With permission.)

TABLE 3.5
Varshni Parameters for Semiconductors

Semiconductor	E_g (at 0 K)	α (10^{-4} eV/K)	β (K)
Si	1.17	4.73	636
Ge	0.744	4.77	235
GaAs	1.519	5.41	204
GaP	2.340	6.20	460
GaN	3.470	7.70	600

Source: Schubert, F. E. 2006. *Light-Emitting Diodes.* Cambridge, UK: Cambridge University Press. With permission.

EXAMPLE 3.11: VARIATION OF THE BANDGAP WITH TEMPERATURE

Calculate the bandgap of (1) Si and (2) GaAs at 300 K using the Varshni parameters shown in Table 3.5.

SOLUTION

a. For Si, E_g at 0 K is 1.170 eV, $\alpha = 4.73 \times 10^{-4}$ eV/K and $\beta = 636$ K. Therefore:

$$E_g = E_g \text{ (at } T = 0 \text{ K)} - \frac{\alpha T^2}{T + \beta}$$

$$= 1.170 - \frac{(4.73 \times 10^{-4})(300)^2}{(300 + 636)}$$

$$= 1.124 \text{ eV}$$

b. For GaAs, the E_g at 0 K is 1.519 eV, $\alpha = 5.41 \times 10^{-4}$ eV/K and $\beta = 204$ K. Therefore:

$$E_g = 1.519 - \frac{(5.41 \times 10^{-4})(300)^2}{(300 + 204)}$$

$$= 1.422 \text{ eV}$$

Thus, the bandgaps of Si and GaAs at 300 K are 1.124 and 1.422 eV, respectively.

EXAMPLE 3.12: VARIATION IN THE WAVELENGTH OF A GAN LED

What is the wavelength of light emitted from a GaN LED operating at 300 K? Assume that the bandgap of GaN changes according to the Varshni parameters shown in Table 3.5. How does this compare with the value calculated in Example 3.3?

SOLUTION

Using the Varshni parameters for GaN (Table 3.5), we can show that the bandgap of GaN at 300 K is 3.4 eV. For this bandgap, the wavelength based on a GaN LED will be

$$\lambda \text{ (in } \mu m) = \frac{1.24}{E_g \text{ (in eV)}} \ \mu m$$

$$\therefore \lambda \text{ (in } \mu m) = \frac{1.24}{3.40} \ \mu m = 0.365 \ \mu m = 365 \text{ nm}$$

Note that in most cases, we use the room-temperature values of the bandgaps and *not* those at 0 K. When the room-temperature bandgap values are used, we often ignore the bandgap's dependence on temperature. Compared to $\lambda = 357$ nm for 0 K (Example 3.3), the wavelength of 365 nm is a little higher because of the decreased bandgap.

3.19   EFFECT OF DOPANT CONCENTRATION ON THE BANDGAP

When semiconductors are heavily doped, that is, have dopant concentrations greater than 10^{18} atoms/cm^3, the average distance between the dopant atoms becomes smaller. The electron wave functions of the dopant atoms begin to overlap. Instead of getting a single energy level associated with the dopant atoms, we get a band of energy levels associated with the dopant levels. If these dopant energy levels are shallow, that is, close to the bandedge, the overall bandgap of the entire material becomes smaller.

This reduction in the bandgap as a function of the dopant concentration is given by the following equation:

$$\Delta E_g(N) = -\frac{3q}{16\pi\varepsilon_r}\sqrt{\frac{q^2 N}{\varepsilon_r k_B T}} \tag{3.52}$$

where N is the dopant concentration, q is the electronic charge, ε_r is the dielectric constant, T is the temperature in K, and k_B is the Boltzmann's constant.

For silicon, the dielectric constant (ε_r) is 11.8, and Equation 3.52 is reduced to

$$\Delta E_g(N) = -22.5 \sqrt{\frac{N}{10^{18}}} \text{ meV} \tag{3.53}$$

This change in the bandgap for different semiconductors is shown in Figure 3.31 (Van Zeghbroeck 2004).

In some semiconductor devices, materials or parts of a material are doped so heavily that the Fermi energy level is very close to the bandedge (conduction or valence). These semiconductors are known as *degenerate semiconductors.*

3.20   EFFECT OF CRYSTALLITE SIZE ON THE BANDGAP

We have so far assumed that the materials we have considered are in a "bulk" form. This means that the size of the semiconductor crystals or the thickness of any thin films used in devices is significantly larger than the size of the atoms. In recent years, there has been a considerable interest in the nanoparticles of semiconductors, known as *quantum dots*. These semiconductor nanocrystals,

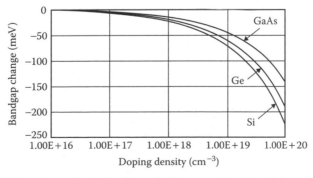

FIGURE 3.31   The relative change in the bandgap of silicon as a function of dopant concentrations. (From Van Zeghbroeck, B. 2004. *Principles of Semiconductor Devices.* Available online at http://ece-www.colorado .edu/~bart/book/. With permission.)

between ~2 and 10 nm, have many interesting optical and electrical properties. For example, when the size of a semiconductor crystal is reduced to a few nanometers, the bandgap (E_g) becomes larger compared to the bandgap of a bulk material. Thus, in a direct bandgap material, the wavelength of light absorbed or emitted is shorter. This is known as a "blue shift." In the case of cadmium selenide (CdSe), a direct bandgap semiconductor, larger crystals have a lesser bandgap energy and appear red. As the crystallite size decreases, the bandgap increases due to the so-called quantum confinement effect, and the color of the nanocrystals changes to yellow (Reed 1993).

3.21   SEMICONDUCTIVITY IN CERAMIC MATERIALS

In Chapter 2, we saw that most ceramic materials are considered to be electrical insulators. However, many ceramic materials exhibit semiconducting behavior. If the semiconductivity is due to the movement of ions, the materials are known as *ionic conductors*. We can also dope ceramics with a donor or an acceptor, similar to the way Si and GaAs are doped. If the predominant charge carriers are electrons or holes, these ceramics are known as *electronic conductors*.

Semiconducting compositions of a ceramic barium titanate ($BaTiO_3$) are used in sensors. The insulating or nonsemiconducting formulations of $BaTiO_3$ are used for manufacturing capacitors. For $BaTiO_3$, Nb^{5+} ions (in the form of niobium oxide [Nb_2O_5]) are used as a donor dopant. In this case, Nb^{5+} ions occupy titanium ion (Ti^{4+}) sites. Since each niobium ion brings in five valence electrons (compared with four of titanium), niobium functions as a donor dopant for $BaTiO_3$ (Figure 3.32). This makes $BaTiO_3$, which is otherwise considered an insulator (E_g ~3.05 eV), behave as an n-type semiconductor. As seen in Chapter 1, we need to balance the site, mass, and electrical charge while considering the doping of compounds. This balance can be expressed using the *Kröger–Vink notation* (Section 1.14).

We can create acceptor-doped or p-type $BaTiO_3$ formulations by adding manganese oxide (MnO). When we add Mn^{2+} ions in the form of MnO, they occupy the titanium (Ti^{4+}) sites (Figure 3.33). Each Mn^{2+} ion has a charge of +2 and occupies a Ti^{4+} site. This creates two holes.

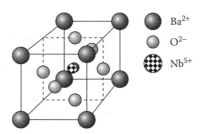

FIGURE 3.32   Donor doping of barium titanate ($BaTiO_3$) using Nb^{5+} (in the form of Nb_2O_5) ions that occupy Ti^{4+} sites results in an n-type semiconductor.

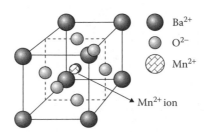

FIGURE 3.33   Illustration of acceptor-doped $BaTiO_3$, created by adding MnO. Each Mn^{2+} ion going on the titanium (Ti^{4+}) site creates two holes.

Similarly, in *solid oxide fuel cells*, strontium (Sr)-doped lanthanum manganite ($LaMnO_3$) is used as a cathode material. In this material, the divalent Sr^{2+} ions occupy trivalent lanthanum ion (La^{3+}) sites. This creates a p-type semiconductor ceramic composition.

In many solid oxide fuel cells, yttria (Y_2O_3)-doped or yttria-stabilized zirconia (ZrO_2; YSZ) is used as an ionic conductor. This is because the addition of Y_2O_3 to ZrO_2 creates oxygen ion vacancies, which facilitates the diffusion of oxygen ions, thereby creating an ionic conductor. The fuel cells make use of air and fuel gases such as natural gas. They operate at relatively high temperatures (~1000°C) and generate electric power.

3.22 PROBLEMS

3.1 Si has a density of 2.33 g/cm³. The atomic mass of Si is 28 and the valence is 4, that is, each atom has four valence electrons. Calculate the density of valence electrons in the valence band assuming all electrons are in the valence band.

3.2 The density of GaAs is 5.31 g/cm³. The atomic masses of Ga and As are 69.7 and 74.92, respectively. Show that the density of atoms, that is, the total number of Ga and As atoms per cubic centimeter in GaAs, is 4.42×10^{22}.

3.3 At 300 K, if the concentration of electrons excited to the conduction band is 1.5×10^{10} cm⁻³, what fraction of the total concentration of electrons is excited to the conduction band?

3.4 What is the resistivity of an intrinsic Si sample at 300 K? Assume $T = 300$ K and the concentration of electrons in the conduction band is $n_i = 1.5 \times 10^{10}$ cm⁻³.

3.5 What is the concentration of electrons excited to the conduction band for Si at 400 K? (See Figure 3.7.)

3.6 What is the conductivity of intrinsic Si at 400 K? Assume the mobility is proportional to $T^{-3/2}$ in this region, and use the carrier concentrations from Figure 3.7.

3.7 Silicon carbide (SiC) is used in high-temperature, high-power, and high-frequency device applications. SiC exhibits various polytypes; 4H and 6H are some of the most widely used. What is the intrinsic carrier concentration for the 6H form of SiC at 666 K, if the carrier concentration changes with temperature as shown in Figure 3.34?

3.8 The concentration of atoms in Si is $~5 \times 10^{22}$ atoms/cm³. If an Si crystal contains one part per billion (ppb) of Sb, what is the concentration of Sb expressed as atoms per cubic centimeter?

3.9 What is the resistivity of this Si containing 1 ppb of antimony? Assume $T = 300$ K and same mobility values as for intrinsic Si.

3.10 Consider an Si crystal that has 1 ppb of Al as a dopant. What is the concentration of this dopant expressed as atoms per cubic centimeter?

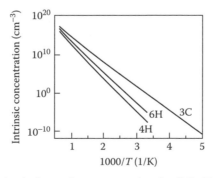

FIGURE 3.34 Changes in the intrinsic carrier concentration for SiC. (From Goldberg et al., eds. 2001. *Properties of Advanced Semiconductor Materials GaN, AlN, SiC, BN, SiC, SiGe*. New York: Wiley. With permission.)

3.11 An Si sample is doped with 10^{-4} atom % of P. Using the atomic mass of Si (28) and its density (2.33 g/cm³), show that the donor dopant concentration is 5×10^{22} atoms/cm³. From this, show that the conductivity of this n-type Si is $\sim$1200 $\Omega^{-1} \cdot$ cm⁻¹.

3.12 What is the resistivity of an Si sample doped with 1 ppb of Al? Why is this resistivity higher than that for a sample doped with 1 ppb of Sb? Use your answer to justify the trends in the data shown in Figure 3.29.

3.13 Two crystals of SiC of the so-called 4H type were grown. One was grown in an N atmosphere and the other in an argon (Ar) atmosphere. Resistivity measurements showed that one of the crystals had a very low resistivity (0.007 $\Omega \cdot$ cm; Siergiej et al. 1999). Which crystal do you think this is—the one grown in an N or in an Ar atmosphere? Explain.

3.14 The mobility of electrons in n-doped 4H-type SiC is $\sim$1000 cm²/V $\cdot$ s (Siergiej et al. 1999). What is the anticipated N-doping level in an n-type doped SiC crystal that has a resistivity of 0.007 $\Omega \cdot$ cm?

3.15 What do the terms "radiative recombination" and "nonradiative recombination" mean? Explain using a sketch.

3.16 Can radiative recombination occur in indirect bandgap materials? Explain.

3.17 Can nonradiative recombination occur in direct bandgap materials? Explain.

3.18 An LED is made using a compound that is a solid solution between InAs and GaAs. This compound can be described as $In_xGa_{1-x}As$, where x is the mole fraction of In. If the wavelength of the LED made using this compound is 1300 nm, what is the bandgap? Assume $T = 300$ K.

3.19 Amber and orange LEDs are made using AlGaInP compositions. If the wavelength of the orange LEDs is 0.6 μm, what is the bandgap of this composition? Assume $T = 300$ K.

3.20 An LED is made using AlGaAs composition with a bandgap of 1.8 eV. What is the color of this LED? What is the wavelength in nm? Assume $T = 300$ K.

3.21 The dielectric constant of Ge is 16 (Table 3.2). The conductivity effective mass of electrons in Ge is 0.12 m_0. Calculate the difference between donor ionization energy (E_d) and conduction bandedge (E_c) for Ge doped with P (see Equation 3.35).

3.22 Calculate the donor energy level position for indium arsenide (InAs), a direct bandgap material. The dielectric constant is 15. The effective mass of electrons is 0.027 m_0.

3.23 What is isoelectronic doping? Explain how this can be useful for some applications.

3.24 At high dopant concentrations ($N > 10^{18}$ atoms/cm³), the bandgap of a semiconductor becomes smaller. Derive a simplified equation (similar to Equation 3.53) for the change in bandgap energy (in meV) for Ge. Plot the data similar to that in Figure 3.31. Notice that for a given dopant level, the change in the bandgap is the maximum for the semiconductor that has the highest dielectric constant.

3.25 Derive an expression similar to Equation 3.53 for the change in bandgap as a function of dopant concentration for GaAs.

3.26 Calculate the bandgap of Si, Ge, and GaAs at 200, 400 and 500 K using the Varshni parameters shown in Table 3.5.

GLOSSARY

Acceptor: An atom or ion that provides the source of holes in an extrinsic semiconductor (e.g., boron in silicon).

Acceptor-doped semiconductor (p-type): A semiconductor doped with an element that results in an electron deficit (e.g., boron-doped silicon).

Amphoteric dopants: Dopants that can act as a donor or an acceptor (e.g., silicon in GaAs).

Band structure: The relationship between the energy (**E**) of a charge carrier and the wave vector (**k**).

Bandgap energy (E_g): The energy difference between the conduction and valence bandedges.

Compensated semiconductor: A semiconductor that contains both donor and acceptor types of dopants. The semiconductor behaves as an n-type if the donor concentration ultimately outweighs the acceptor concentration and $N_d - N_a \gg n_i$; if $N_a - N_d \gg n_i$, then the material behaves as a p-type semiconductor.

Compound semiconductor: A semiconductor based on a compound (e.g., GaAs).

Conduction bandedge (E_c): The energy level corresponding to the bottom of the conduction band.

Conductivity effective mass (m_σ^*): The mass used to calculate the donor position level in indirect bandgap semiconductors.

Dielectric constant (ε_r): A parameter related to the ability of a material to store charge. It is the ratio of the dielectric permittivity (ε) of a material to the permittivity of the free space (ε_0).

Direct bandgap semiconductor: A material in which an electron can be transferred from the conduction band into the valence band with emission of light and without a change in its momentum.

Deep-level defects: Energy level defects created deep in the bandgap, usually by atoms or ions that do not easily "fit" in the crystal structure and create a severe distortion.

Degenerate semiconductors: Heavily doped semiconductors with a Fermi energy level very close to the bandedge.

Density of states effective mass (m_{dos}^*): Defined for electrons as $m_{dos}^* = \left[6^{2/3} \left(m_l^* m_t^{*2} \right)^{1/3} \right]$ and for holes as $(m_{dos}^*)^{3/2} = \left(m_{hh}^{3/2} + m_{lh}^{3/2} \right)$. It is used for the calculation of the density of states or the position of the Fermi energy level.

Dielectric constant (ε_r): A measure of the ability of a material to store charge. It is the ratio of the permittivity of a material (ε) to the permittivity of the free space (ε_0).

Donor-doped semiconductor (n-type): A semiconductor in which the dopant atoms bring in "extra" electrons that become dissociated from the dopant atoms and provide semiconductivity.

Donor: Atoms or ions that are the source of conduction electrons in an extrinsic semiconductor (e.g., phosphorus in silicon).

Dopant: An element or a compound added to a semiconductor to enhance and control its electrical or other properties.

Dopant ionization: The process by which the electrons from the dopant atoms or holes from the acceptor atoms dissociate themselves from their parent atoms.

Effective density of states (N): A parameter that represents the effective allowed number of energy levels at a given bandedge conduction (N_c) or (N_v).

Effective mass of an electron (m_e^*): The mass that an electron would appear to have as a result of the attractive and repulsive forces an electron experiences in a crystal. It is different from the mass of an electron in a vacuum (m_0).

Electron hole pair (EHP): The process by which an electron in the valence band is promoted to the conduction band, creating a hole in the valence band.

Elemental semiconductor: An element that shows semiconducting behavior (e.g., silicon).

Extrinsic semiconductor: A donor-doped or acceptor-doped semiconductor. The conductivity of an extrinsic semiconductor is controlled by the addition of dopants.

Freeze-out range: The low-temperature range in which the electrons and the holes remain bonded to the dopant atoms and are not available for conduction.

Hole: A missing electron in a bond that is treated as an imaginary particle with a positive charge q.

Impurities: Elements or compounds present in semiconductors either inadvertently or because of processing limitations. Their presence is usually considered undesirable.

Impurity- or dopant-scattering limited mobility: The value of mobility at low temperatures (but above the carrier freeze-out) limited by the conduction electrons scattering off of the dopant or impurity atoms.

Indirect bandgap semiconductor: A semiconductor in which an electron that has been excited into the conduction band transfers into the conduction band with a change in its momentum.

Intrinsic semiconductor: A material whose properties are controlled by thermally generated carriers. It does not contain any dopants.

Isoelectronic dopants: Dopants that have the same valence as the sites they occupy (e.g., germanium in silicon or nitrogen in GaP–GaAs materials). This does not create an n-type or a p-type behavior.

Kröger–Vink notation: Notation used for expressing point defects in materials.

Lattice-scattering limited mobility: The limited mobility of carriers at high temperatures by the scattering of vibrations of atoms known as phonons.

Majority carriers: Carriers of electricity that dominate overall conductivity; for example, in an n-type semiconductor, electrons created by donor atoms would be the majority carriers.

Minority carriers: Carriers of electricity that do not dominate the overall conductivity; for example, in an n-type semiconductor, holes are the minority carriers, and in an ionic conductor, both electrons and holes are the minority carriers.

Nonradiative recombination: A process by which electrons and holes recombine, transferring the energy to vibrations of atoms or phonons. The energy appears as heat, and the process occurs in both direct and indirect bandgap materials.

n-type semiconductor: A semiconductor in which the majority carriers are electrons; also known as a donor-doped semiconductor.

Optoelectronic devices: Devices such as light-emitting diodes (LEDs), usually based on direct bandgap materials that have a strong interaction with light waves.

p-type semiconductor: A semiconductor in which the majority carriers are holes; also known as an acceptor-doped semiconductor.

Quantum dots: Nanocrystalline semiconductors whose bandgap is larger than that of a bulk material.

Radiative recombination: The process by which an electron that has been excited to the conduction band falls back to the valence band, recombining with a hole and causing emission of light. It may involve a defect energy level.

Semiconductors: Elements or compounds with resistivity (ρ) between 10^{-4} and 10^3 ($\Omega \cdot cm$). This range is approximate.

Shallow-level dopants: Dopants with energy levels only a few millielectron volts from the bandedge.

Solid oxide fuel cells: Fuel cells that make use of ceramic oxides, such as yttria-doped zirconia, as an electrolyte.

Thermally generated charge carriers: Electrons and holes that are generated as electrons jump from the valence band into the conduction band as a result of thermal energy.

Valence bandedge (E_v): The energy level corresponding to the top of the valence band.

Varshni formula: A formula that shows the variation in the bandgap with temperature (Equation 3.51).

Vertical recombination: The process of radiative recombination without a change in the momentum.

Wave vector (k): In the quantum mechanics-based approach, an electron that is considered a plane wave with a propagation constant.

REFERENCES

Ahlgre, D. C., and J. Dunn. 2000. SiGe comes of age. *MicroNews-IBM* 6(1):1.

Askeland, D., and P. Fulay. 2006. *The Science and Engineering of Materials*. Washington, DC: Thomson.

Dorkel, J. M., and Ph. Leturcq. 1981. Carrier mobilities in silicon semi-empirically related to temperature, doping, and injection level. *Solid State Electron* 24(9):821–5.

Goldberg, Yu., M. E. Levinshtein, and S. L. Rumyantsev, eds. 2001. *Properties of Advanced Semiconductor Materials GaN, AlN, SiC, BN, SiC, SiGe*. New York: Wiley.

Grove, A. S. 1967. *Physics and Technology of Semiconductor Devices*. New York: Wiley.

Kano, K. 1997. *Semiconductor Fundamentals*. Upper Saddle River, NJ: Prentice Hall.

Kasap, S. O. 2002. *Principles of Electronic Materials and Devices*. New York: McGraw Hill.

Levinshtein, M., S. Rumyantsev, and M. Shur, eds. 1999. *Handbook Series on Semiconductor Parameters*. London: World Scientific.

Li, S. S., and W. R. Thurber. 1977. The dopant density and temperature dependence of electron mobility and resistivity in *n*-type silicon. *Solid State Electron* 20(7):609–16.

Mahajan, S., and K. S. Sree Harsha. 1998. *Principles of Growth and Processing of Semiconductors*. New York: McGraw Hill.

Queisser, H. J., and E. E. Haller. 1998. Defects in semiconductors: Some fatal, some vital. *Science* 281:945–50.

Reed, M. A. 1993. Quantum dots. *Scientific American* (January):118–23.

Schubert, F. E. 2006. *Light Emitting Diodes*. Cambridge, UK: Cambridge University Press.

Siergiej, R. R., R. C. Clark, S. Sriram, A. K. Agarwal, R. J. Bojko, A. W. Mors, V. Balakrishna, et al. 1999. Advances in SiC materials and devices: An industrial point of view. *Mater Sci Eng (B)* B61–62:9–17.

Singh, J. 1994. *Semiconductor Devices: An Introduction*. New York: McGraw-Hill.

Singh, J. 2001. *Semiconductor Devices: Basic Principles*. New York: Wiley.

Streetman, B. G., and S. Banerjee. 2000. *Solid State Electronic Devices*. Upper Saddle River, NJ: Prentice Hall.

Sze, S. M. 1985. *Semiconductor Devices, Physics, and Technology*. New York: Wiley.

Van Zeghbroeck, B. 2004. *Principles of Semiconductor Devices*. http://ece-www.colorado.edu/~bart/book/.

4 Fermi Energy Levels in Semiconductors

KEY TOPICS

- The concept of Fermi energy level
- Fermi–Dirac distribution for electron energy
- Electron and hole concentrations
- Fermi energy positions for n-type and p-type materials
- Invariance of Fermi energy levels

4.1 FERMI ENERGY LEVELS IN METALS

We have seen in Chapter 1 that in metals, the outermost band is usually nearly half-filled. Let us define the energy at the bottom of the partially filled band as $E = 0$ when $E = E_B$ (Figure 4.1).

In metals at around 0 K, the highest energy level at which electrons are present is known as the Fermi energy level (E_F). Above this level, the states are available for electrons, but no electrons occupy these states. Thus, in metals at around 0 K, the probability of finding an electron at $E > E_F$ is zero. The Fermi energy values for different metals are shown in Table 4.1. This table also shows the value of the *work function* (ϕ), which for metals refers to the energy needed to remove the outermost electron from $E = E_F$. For example, the E_F value for copper (Cu) is 7.0 eV (Table 4.1). Thus, as shown in Figure 4.1, the work function (ϕ) of copper is 4.65 eV.

The Fermi energy level for metals at 0 K is given by

$$E_{F,\text{metal at 0 K}} = E_{F,0} = \left(\frac{\hbar^2}{2m_e}\right)(3\pi^2 n)^{2/3} \quad \text{(for metals)} \tag{4.1}$$

where m_e is the rest mass of the electron $\hbar = h/2\pi$, h is the Planck's constant, and n is the free electron concentration in the metal.

Example 4.1 illustrates the calculation of the Fermi energy level for a metal.

EXAMPLE 4.1: FERMI ENERGY OF A METAL AT $T = 0$ K

In Cu, if we assume that each Cu atom donates one electron, the free electron concentration is 8.5×10^{22} cm^{-3}. (a) What is the value of $E_{F,0}$? (b) If the concentration of conduction electrons estimated by conductivity measurements is 1.5×10^{23} cm^{-3}, then find the number of electrons donated per Cu atom.

SOLUTION

a. We first convert the concentration of electrons per cubic centimeter to per cubic millimeter and then substitute the rest mass of electrons (9.1×10^{-30} kg) and the value of $\hbar$ as 1.05×10^{-34} J·s.

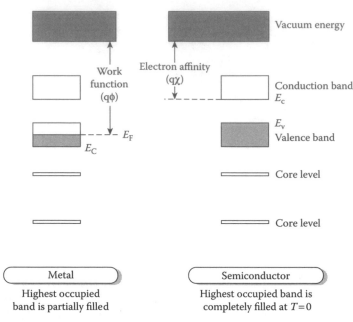

FIGURE 4.1 Fermi energy level representation for a metal and a semiconductor.

TABLE 4.1
Fermi Energy Levels ($E_{F,0}$) and Work Functions (ϕ) for Different Metals

Metal	Work Function (ϕ) (eV)	Fermi Energy Level at 0 K ($E_{F,0}$)
Silver	4.26	5.5
Aluminum	4.28	11.7
Gold	5.1	5.5
Copper	4.65	7.0
Lithium	2.3	4.7
Magnesium	3.7	7.1
Sodium	2.75	3.2

Source: Kasap, S. O. 2002. *Principles of Electronic Materials and Devices*. New York: McGraw Hill.

From Equation 4.1,

$$E_{F,0} = \frac{(1.05 \times 10^{-34}\ \text{J} \cdot \text{s})^2}{2 \times (9.1 \times 10^{-30}\ \text{kg})}\left(3\pi^2 \times (8.5 \times 10^{28}\ \text{m}^{-3})\right)^{2/3} = 1.135 \times 10^{-18}\ \text{J}$$

$$= \frac{1.135 \times 10^{-18}\ \text{J}}{1.6 \times 10^{-19}\ \text{J/eV}} = 7\ \text{eV}$$

b. From the conductivity measurements, the estimated electron concentration in Cu is ~1.15×10^{29} m^{-3}. Thus, the average number of electrons contributed by a Cu atom is larger than 8.5×10^{28}.

The number of electrons donated per Cu atom is given by

$$\frac{1.15 \times 10^{29}\ \text{m}^{-3}}{8.5 \times 10^{28}\ \text{m}^3} = 1.36$$

This value seems reasonable. In many compounds, Cu exhibits a valence of +1 and +2, for example Cu_2O and CuO.

The Fermi energy of metals does vary slightly with temperature and is given by

$$E_F(T) = E_{F,0}\left[1 - \frac{\pi^2}{12}\left(\frac{k_B T}{E_{F,0}}\right)^2\right] \tag{4.2}$$

Note that since the $E_{F,0}$ for metals is much higher than the $k_B T$ (Table 4.1). As a result, the E_F of metals does not change significantly with temperature. The Fermi energy level of semiconductors, especially those that are extrinsic, does change appreciably with temperature (see Section 4.6).

An electron with energy $E = E_F$ represents the highest-energy electron in a metal. This is the significance of the Fermi energy level in metals. If v_F is the speed of an electron with energy $E_{F,0}$, then

$$\frac{1}{2}mv_{F,0}^2 = E_{F,0} \tag{4.3}$$

From the typical values of $E_{F,0}$ for metals, we can see that at $T = 0$ K, the highest speed of electrons with energy $E_{F,0}$ is $\sim 10^6$ m/s (see Equation 4.3). As mentioned in Chapter 3, if electrons are considered classical particles, then the speed of electrons at 0 K should be zero. The *effective speed of electrons* (v_e) or the *root mean speed of electrons* in a metal is given by

$$\frac{1}{2}mv_e^2 \cong \frac{3}{5}E_{F,0} \tag{4.4}$$

The effective speed of electrons (v_e) in a metal is relatively insensitive to temperature and depends on $E_{F,0}$.

4.2 FERMI ENERGY LEVELS IN SEMICONDUCTORS

At low temperatures, the conduction band of semiconductors is empty, and the valence band is completely filled. Thus, the probability of finding an electron at $T = 0$ K at the valence bandedge ($E = E_v$) is 1. As the temperature increases, some electrons make a transition across the bandgap into the valence band. The probability of finding an electron at or above the conduction bandedge ($E \geq E_c$) is zero. At higher temperatures (e.g., 300 K), the probability of finding an electron in the valence band is still 1, but there is a small (but nonzero) probability of finding an electron in the conduction band as well, that is, $E \geq E_c$. Thus, as we go from $E = E_v$ to $E = E_c$ and higher in the conduction band, the probability of finding an electron begins at 1 and eventually approaches zero well into the conduction band. In general, the probability function for a system of classical particles is given by the Boltzmann function:

$$f(E) = A\exp\left(\frac{-E}{k_B T}\right) \tag{4.5}$$

where A is a constant, k_B is the Boltzmann constant, and T is the temperature.

Electrons in materials are also constrained by the Pauli exclusion principle, which states that no two electrons can have the same energy level (i.e., same set of quantum numbers). With this restriction on the occupancy of states, the probability of finding an electron occupying energy level E is given by the so-called *Fermi–Dirac distribution function*, which is defined as

$$f(E) = \frac{1}{\left[1 + \exp\left(\dfrac{E - E_F}{k_B T}\right)\right]} \tag{4.6}$$

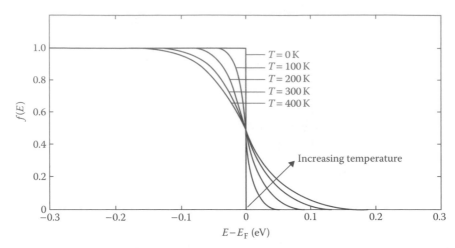

FIGURE 4.2 Fermi–Dirac function at different temperatures. E_F remains constant. (From Pierret, R. F. 1987. *Advanced Semiconductor Fundamentals*. Reading, MA: Addison-Wesley. With permission.)

where E is the energy of the electron, $f(E)$ is the probability of an electron occupying the state with energy E, k_B is the Boltzmann constant, T is the temperature, and E_F is the *Fermi energy level*.

The Fermi energy level is defined as the energy level for which the probability of finding an electron is 0.5 (Figure 4.2). Note that this level is in the middle of the bandgap, and based on the band diagram of a semiconductor, no electrons are allowed between $E = E_v$ and $E = E_c$. Thus, the Fermi–Dirac function gives a numerical value for any value of energy (E). However, this value of the function corresponds to the probability of finding an electron only if the energy level is allowed.

At a given temperature (e.g., $T = T_1$), as E increases, $f(E)$ decreases, regardless of whether $E - E_F$ is positive or negative. The probability of finding an electron decreases, as shown in Figure 4.2. This probability becomes zero only as T approaches infinity.

In Figure 4.2, we can see that at $T = 0$, $f(E) = 1$ if $E < E_F$. This means that every available level below E_F is filled at $T = 0$ K. This situation is the same as that for metals at $T = 0$ K. Also, note that at $T = 0$ K, $f(E) = 0$ if $E > E_F$. This means that every energy level below E_F is empty. At the Fermi energy level, $E = E_F$, $f(E) = 0.5$. However, the Fermi–Dirac function gives us only a mathematical value. Since no electrons are allowed in the bandgap of an intrinsic semiconductor, the actual number of electrons in the bandgap is zero.

When $E > E_F$, as the temperature (T) increases, the $\exp[(E - E_F)/k_B T]$ decreases and the value of $f(E)$ increases. This means that the probability of finding an electron at higher energy levels becomes higher as the temperature increases.

When $E < E_F$, as the temperature increases, the term $\exp[(E - E_F)/k_B T]$ increases and the value of $f(E)$ decreases. That is, the probability of finding an electron at lower energy levels becomes lesser as the temperature increases, because chances are better that the electrons have moved up to higher energy levels.

An electron creates a hole in the valence band when it is promoted to the conduction band. Because a hole can be considered to be a missing electron, the probability of finding a hole is given by

$$f(E)_{\text{hole}} = 1 - f(E) = 1 - \frac{1}{1 + \exp\left(\dfrac{E - E_F}{k_B T}\right)} \tag{4.7}$$

Another interesting situation to consider is one in which $(E - E_F)$ is far greater than k_BT. These energy levels are closer to either the valence or conduction bandedges. If $(E - E_F) \gg k_BT$, then $\exp[(E - E_F)/k_BT] \gg 1$ and Equation 4.6 can be simplified as:

$$f(E) = \frac{1}{\left[\exp\left(\dfrac{E - E_F}{k_BT}\right)\right]} = \exp-\left(\frac{E - E_F}{k_BT}\right) \tag{4.8}$$

Note the negative sign in Equation 4.8. This equation is a form of the Boltzmann equation for classical particles. Thus, the "tail" of the Fermi–Dirac distribution (i.e., when E is much greater than E_F) can be described by the Boltzmann distribution equation. The importance of this is as follows: In the conduction band of a semiconductor, when the number of states available is significantly greater than the number of electrons, the chance of two electrons occupying the same energy level (i.e., having the same quantum numbers) is not very high. Therefore, we need not worry about the limitation on electrons having the same quantum number imposed by the Pauli exclusion principle for the electrons in the conduction bandgap. We can use the Boltzmann equation (Equation 4.8) to calculate the concentration of electrons in the conduction band and the concentration of holes in the valence band (Section 4.3).

4.3 ELECTRON AND HOLE CONCENTRATIONS

The conductivity of a semiconductor depends upon the carrier concentrations and their mobilities. In this section, we will learn to calculate the carrier concentrations. If we know the probability of finding an electron at a given energy level in the conduction band, and if we know the number of states available, we can then calculate the total number of electrons in the conduction band. The following analogy may help.

Consider a high-rise hotel building. We want to know how many guests are in the building at a given time. Assume that we know how many rooms there are on each floor. If we know the probability of finding guests in their rooms, then we can calculate the total number of people in the hotel. We can use the same logic to calculate the number of electrons in a conduction band.

Consider the band diagram for a typical intrinsic semiconductor (Figure 4.3). We also plot the Fermi–Dirac function, which describes the probability of finding an electron, and the density of states function. Note that the vertical axis is the electron energy (E).

The electron energies in the conduction band range from $E = E_c$ to $E = \infty$. If at any energy level E, the density of states (i.e., the number of energy levels available) is given by $N(E)$, the product

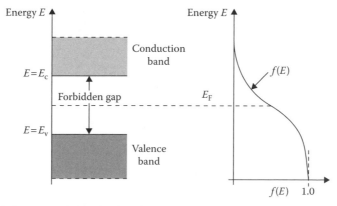

FIGURE 4.3 Band diagram and Fermi–Dirac function for an intrinsic semiconductor. (From Pierret, R. F. 1988. *Semiconductor Fundamentals Volume.* Reading, MA: Addison-Wesley. With permission.)

$f(E)N(E)$ gives us the number of electrons at any energy level in the conduction band. We integrate the product $f(E)N(E)$ over the range $E = E_c$ to $E = \infty$ to get the total concentration of electrons in the conduction band. When the energy of an electron becomes $E = E_c + q \cdot \chi$, the electron becomes free (Figure 4.1). Therefore, we should use $E_c + q \cdot \chi$ as the upper limit of integration. However, as E tends to ∞, the function $f(E)$ approaches zero rapidly. Therefore, we will use $E = \infty$ as the top limit for the integral.

$$n = \int\limits_{E=E_c}^{E=\infty} N(E)f(E)dE \tag{4.9}$$

The concentration of electrons in the conduction band is shown in Figure 4.3 as the shaded area under the curve, representing the integral shown in Equation 4.9.

We will now use the *effective density of states approximation*. We assume that most, if not all, electrons reside at the bottom of the conduction band at $E = E_c$. This is reasonable, because electrons in the conduction band tend to minimize their energies. The effective density of states at the conduction bandedge (N_c) is given by

$$N_c = 2\left(\frac{m_n^* k_B T}{2\pi\hbar^2}\right)^{3/2} \tag{4.10}$$

Thus, Equation 4.9 now becomes

$$n = 2\left(\frac{m_n^* k_B T}{2\pi\hbar^2}\right)^{3/2} f(E_c) \tag{4.11}$$

Note that because we assumed that most electrons in the conduction band reside at the bottom, we multiplied the effective density of states by the value of the probability function at $E = E_c$.

From the definition of the Fermi–Dirac function (Equation 4.6), we get

$$f(E_c) = \frac{1}{\left[1 + \exp\left(\dfrac{E_c - E_F}{k_B T}\right)\right]} \tag{4.12}$$

Since $(E_c - E_F) \gg k_B T$, the exponential term is much larger than 1, and hence $f(E_c)$ is written as

$$f(E_c) = \frac{1}{\left[\exp\left(\dfrac{E_c - E_F}{k_B T}\right)\right]} = -\exp\left(\frac{E_c - E_F}{k_B T}\right) \tag{4.13}$$

Substituting the value of $f(E_c)$ from Equation 4.13 into Equation 4.11, we get

$$n = 2\left(\frac{m_n^* k_B T}{2\pi\hbar^2}\right)^{3/2} \exp-\left(\frac{E_c - E_F}{k_B T}\right) \tag{4.14}$$

By keeping the first term as N_c, we get

$$n = N_c \exp-\left(\frac{E_c - E_F}{k_B T}\right) \tag{4.15}$$

The negative sign outside is shown so that the term inside is positive, because E_c is above E_F. We can rewrite Equation 4.14 as follows to eliminate the negative sign in the exponential term:

$$n = 2\left(\frac{m_n^* k_B T}{2\pi\hbar^2}\right)^{3/2} \exp\left(\frac{E_F - E_c}{k_B T}\right) \tag{4.16}$$

In Sections 4.5 and 4.6, we will see that we can calculate the concentration of electrons for intrinsic as well as extrinsic semiconductors using Equation 4.16 if we know the position of the E_F relative to E_c.

We can now derive a similar expression for calculating the concentration of holes in the valence band. Remember that the conduction band is nearly empty and has few electrons, the concentration of which we have calculated (Equation 4.16). Now the valence band is nearly completely filled with valence electrons in the covalent bonds of the semiconductor atoms. When some of these bonds break, the electrons move to the conduction band. This creates holes in the valence band. The concentration of holes (p) in the valance band is given by

$$p = \int_{E=0}^{E=E_v} N(E)[1 - f(E)]\, dE \tag{4.17}$$

Notice that the density of states $N(E)$ is now multiplied by $1 - f(E)$, which is the probability of finding a hole. We again use the effective density of states at the valence bandedge ($N(E_v)$); that is, we assume that most holes remain at the top of the valence band. Recall that electron energy increases as we move up the band diagram, whereas the hole energy increases as we go down the band diagram. The effective density of states at $E = E_v$ is given by

$$N_v = 2\left(\frac{m_p^* k_B T}{2\pi\hbar^2}\right)^{3/2} \tag{4.18}$$

Rewriting Equation 4.17, we get

$$p = 2\left(\frac{m_p^* k_B T}{2\pi\hbar^2}\right)^{3/2} (1 - f(E_v)) \tag{4.19}$$

From the Fermi–Dirac function

$$1 - f(E_v) = 1 - \frac{1}{\left[\exp\left(\dfrac{E_v - E_F}{k_B T}\right) + 1\right]} = \frac{1 + \exp\left(\dfrac{E_v - E_F}{k_B T}\right) - 1}{\exp\left(\dfrac{E_v - E_F}{k_B T}\right) + 1} \tag{4.20}$$

$$\therefore 1 - f(E_v) = \frac{\exp\left(\dfrac{E_v - E_F}{k_B T}\right)}{1 + \exp\left[\dfrac{(E_F - E_v)}{k_B T}\right]} \tag{4.21}$$

Note the denominator of the exponential term now contains $(E_F - E_v)$ and not $(E_v - E_F)$.

Again, if $(E_F - E_v) \gg k_B T$, then the Fermi–Dirac function simplifies to a Boltzmann function since $\exp[(E_F - E_v)/k_B T] \gg 1$. The denominator in Equation 4.21 becomes 1.

Thus, Equation 4.21 now becomes

$$1 - f(E_v) = \exp\left(\frac{E_v - E_F}{k_B T}\right)$$

$$\therefore 1 - f(E_v) = -\exp\left(\frac{E_F - E_v}{k_B T}\right) \tag{4.22}$$

The concentration of holes in the valence band from Equation 4.19 now becomes

$$p = 2\left(\frac{m_p^* k_B T}{2\pi\hbar^2}\right)^{3/2}\left[-\exp\left(\frac{E_F - E_v}{k_B T}\right)\right] \tag{4.23}$$

For the sake of convenience, we can represent the first term as N_v and can write Equation 4.23 as

$$p = N_v \exp-\left(\frac{E_F - E_v}{k_B T}\right) \tag{4.24}$$

The negative sign outside is shown so that the term inside is positive, because E_F is above E_v. This equation gives us the concentration of holes in either an intrinsic semiconductor or an extrinsic semiconductor. If we know the difference between the valence bandedge and the Fermi energy level, temperature, and the effective mass of holes, we can calculate the concentration of holes (p).

Let us multiply the equations for electron and hole concentrations. From Equations 4.15 and 4.24, we get

$$n \times p = N_c N_v \exp-\left(\frac{E_c - E_F}{k_B T}\right)\exp-\left(\frac{E_F - E_v}{k_B T}\right) = N_c N_v \exp-\left(\frac{E_c - E_v}{k_B T}\right) \tag{4.25}$$

Since $(E_c - E_v) = E_g$, we get

$$n \times p = N_c N_v \exp\left(\frac{-E_g}{k_B T}\right) \tag{4.26}$$

Thus, for a given semiconductor, the product of the concentration of electrons and the concentration of holes at a given temperature is constant. This is known as the *law of mass action for semiconductors*. It is similar to the ideal gas law: pressure × volume is a constant. For example, for silicon (Si) ($E_g \approx 1.1$ eV at 300 K), the value of $n \times p$ is constant, whether we have intrinsic, n-type, or p-type silicon. For an n-type silicon crystal, created by doping with antimony (Sb), the electron concentration in the conduction band is higher compared to n_i. The increase in the concentration of electrons in an n-type semiconductor is compensated by a decrease in the concentration of holes in the valence band. Similarly, for a p-type semiconductor, holes are the majority carriers, that is, $p \gg n$. To maintain a constant value of $n \times p$, the concentration of electrons (n) decreases.

4.4   FERMI ENERGY LEVELS IN INTRINSIC SEMICONDUCTORS

For an intrinsic semiconductor, $n_i = p_i$. Applying Equations 4.15 and 4.24 to an intrinsic semiconductor, we get

$$N_c \exp-\left(\frac{E_c - E_{F,i}}{k_B T}\right) = N_v \exp-\left(\frac{E_{F,i} - E_v}{k_B T}\right) \tag{4.27}$$

In this equation, $E_{F,i}$ is the Fermi energy level of an intrinsic semiconductor. Taking a natural logarithm of both sides, we get

$$E_{F,i} = \frac{1}{2}(E_c + E_v) + \frac{1}{2}k_BT \ln\left(\frac{N_v}{N_c}\right) \qquad (4.28)$$

Substituting for N_c and N_v from Equations 4.10 and 4.18, respectively, we get

$$E_{F,i} = \frac{1}{2}(E_c + E_v) + \frac{3}{4}k_BT \ln\left(\frac{m_p^*}{m_n^*}\right) \qquad (4.29)$$

The first term in Equation 4.29 represents the middle of the bandgap. Thus, the Fermi energy level of an intrinsic material ($E_{F,i}$) is very close to the middle of the bandgap (Figure 4.4).

$$E_{F,i} - E_{midgap} = \frac{3}{4}k_BT \ln\left(\frac{m_p^*}{m_n^*}\right) \qquad (4.30)$$

For intrinsic semiconductors with a larger effective mass of holes, the Fermi energy level is slightly above the middle of the bandgap. When the effective mass of holes is higher, the value of N_v is higher. Thus, the $E_{F,i}$ moves away from the valence band and is therefore slightly above the middle of the bandgap (Figure 4.5). For intrinsic materials with a larger effective electron mass, the $E_{F,i}$ is slightly below the middle of the bandgap. The effective masses also change with the temperature,

FIGURE 4.4 Fermi energy level position for a typical intrinsic semiconductor. (From Pierret, R. F. 1987. *Advanced Semiconductor Fundamentals*. Reading, MA: Addison-Wesley. With permission.)

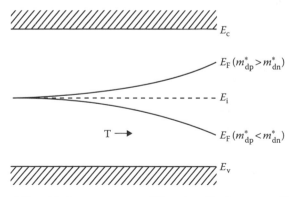

FIGURE 4.5 Variation of $E_{F,i}$ with the temperature. When the effective mass of holes is higher, the Fermi energy level is in the upper half of the bandgap. (From Li, S. S. 2006. *Semiconductor Physical Electronics*. New York: Springer. With permission.)

as does the bandgap (E_g). The variation of $E_{F,i}$ as given by Equation 4.30 is shown in Figure 4.5. Note that the dependence of E_g on the temperature, as given by the Varshni parameters, is not shown in the figure. The slight deviation that causes the $E_{F,i}$ to shift from the middle of the bandgap as a result of the difference between the effective mass of holes and electrons is illustrated in Example 4.2.

EXAMPLE 4.2 INTRINSIC FERMI ENERGY LEVEL FOR Si

Calculate the locations of the Fermi energy level for intrinsic Si located relative to the middle of the bandgap. Assume that the effective masses for electrons and holes in Si are 1.08 and 0.56, respectively.

Solution

From Equation 4.30, we get

$$E_{F,i} - E_{midgap} = \frac{3}{4}(0.026 \text{ eV}) \ln\left(\frac{0.56}{1.08}\right) = -12.8 \text{ meV}$$

Thus, the intrinsic Fermi energy level for Si is 12.8 meV below the middle of the bandgap. Compared to 550 meV (~half of the bandgap [E_g] of ~1.1 eV), this deviation is much smaller.

4.5 CARRIER CONCENTRATIONS IN INTRINSIC SEMICONDUCTORS

For intrinsic semiconductors, we can derive the electron and the hole concentrations using Equation 4.16 and Equation 4.23, respectively, as follows:

$$n_i = N_c \exp-\left(\frac{E_c - E_{F,i}}{k_B T}\right) \tag{4.31}$$

$$p_i = N_v \left[\exp-\left(\frac{E_{F,i} - E_v}{k_B T}\right)\right] \tag{4.32}$$

As mentioned previously, the negative sign is included so that the terms inside the brackets remain positive, because E_c is above $E_{F,i}$ and $E_{F,i}$ is above E_v. In Equations 4.31 and 4.32, n_i and p_i are the electron and hole concentrations, the subscript i indicates the intrinsic semiconductor, $E_{F,i}$ is the Fermi energy level for an intrinsic semiconductor, and N_c and N_v are the effective densities of states for the conduction and valence bandedges, respectively.

We now multiply Equation 4.31 by Equation 4.32 to get

$$n_i \times p_i = N_c N_v \exp-\left(\frac{E_c - E_{F,i}}{k_B T}\right)\exp-\left(\frac{E_{F,i} - E_v}{k_B T}\right) = N_c N_v \exp-\left(\frac{E_c - E_v}{k_B T}\right)$$

or since $E_c - E_v = E_g$, the bandgap energy

$$n_i \times p_i = N_c N_v \exp-\left(\frac{E_g}{k_B T}\right) \tag{4.33}$$

This is the same as Equation 4.26, which confirms that the $n \times p$ value is a constant for a given semiconductor at a given temperature. It does not matter whether the semiconductor is extrinsic or intrinsic.

We have now shown that $n \times p$ value remains constant for a given semiconductor at a given temperature:

$$n \times p = n_i \times p_i \qquad (4.34)$$

This is a very important equation, since it allows us to make a connection between the properties of intrinsic and extrinsic semiconductors.

Another way to rewrite Equation 4.33 is as follows:

$$n_i = \sqrt{N_c N_v} \, \exp{-\left(\frac{E_g}{2k_B T}\right)} \qquad (4.35)$$

We can see quantitatively from Equation 4.35 that the concentration of thermally generated carriers changes with the temperature.

We discussed the values of n_i and p_i for different semiconductors in Chapter 3. The intrinsic carrier concentrations are strongly temperature-dependent and are inversely related to the bandgap (E_g). The values of electron or hole intrinsic concentrations for silicon, germanium (Ge), and gallium arsenide (GaAs) are shown in Figure 4.6.

As we can see in Figure 4.6, the value of n_i for Si at 300 K is ~1.5×10^{10} electrons/cm^3.

The bandgap of the semiconductors (E_g) also changes with the temperature. This dependence is given by the Varshni parameters. For silicon, the change in densities of states N_c and N_v

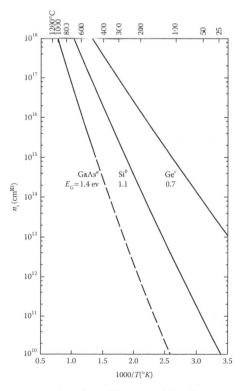

FIGURE 4.6 Intrinsic carrier concentrations for gallium arsenide, silicon, and germanium. (From Grove, A. S. 1967. *Physics and Technology of Semiconductor Devices*. New York: Wiley. With permission.)

TABLE 4.2
Density of States (N_c, N_v), Bandgap (E_g), and Intrinsic Carrier Concentration (n_i), Values for Si at Different Temperatures

T (K)	N_c (cm^{-3})	N_v (cm^{-3})	E_g (eV)	n_i (cm^{-3})
200	1.52×10^{19}	1.43×10^{19}	1.1483	5.03×10^4
250	2.15×10^{19}	2.20×10^{19}	1.1367	7.59×10^4
300	2.86×10^{19}	3.10×10^{19}	1.1242	1.07×10^{10}
350	3.65×10^{19}	4.13×10^{19}	1.1104	3.92×10^{11}
400	4.51×10^{19}	5.26×10^{19}	1.0968	6.00×10^{12}
450	5.43×10^{19}	6.49×10^{19}	1.0832	5.11×10^{13}
500	6.46×10^{19}	7.81×10^{19}	1.0695	2.89×10^{14}

(Equations 4.10 and 4.18) and the variation in the bandgap (E_g) with a temperature in the range 200–500 K are given by the following simplified equations (Green 1990; Bullis and Huff 1994):

$$N_c = 2.86 \times 10^{19} \left(\frac{T}{300} \right)^{1.58} \tag{4.36}$$

$$N_v = 3.10 \times 10^{19} \left(\frac{T}{300} \right)^{1.82} \tag{4.37}$$

The Varshni formula that describes the temperature dependence of the bandgap for silicon is

$$E_g \text{(in eV)} = 1.206 - 2.73 \times 10^{-4} T \tag{4.38}$$

In Equations 4.36 through 4.38, the density of states is expressed as states per cubic centimeter and E_g is expressed in electron volts. Note that these Varshni parameters are for the temperature range 200–500 K and are slightly different from the values of 1.206 and 2.73×10^{-4} for a wider temperature range, as discussed in Chapter 3.

For silicon, a more precise equation for the change in n_i is

$$n_i = 1.4514 \times 10^{20} \exp\left(\frac{-6997.4}{T} \right)\left(\frac{T}{300} \right)^{1.715} \tag{4.39}$$

The values of N_c, N_v, E_g, and n_i from Equations 4.36 through 4.39, calculated as a function of the temperature, are shown in Table 4.2. The value of n_i for silicon at 300 K is 1.07×10^{10} electrons/cm^3, whereas the normally quoted value is ~1.5×10^{10} electrons/cm^3. The use of these higher n_i values leads to the prediction of a higher current in silicon-based devices.

4.6 FERMI ENERGY LEVELS IN N-TYPE AND P-TYPE SEMICONDUCTORS

In Section 4.4, we saw that the Fermi energy level of an intrinsic semiconductor lies close to the middle of the bandgap. Now we will find the location of E_F for extrinsic materials. Since $n \times p$ is constant for a given semiconductor at a given temperature,

$$n \times p = n_i \times p_i \tag{4.40}$$

Since the intrinsic semiconductor has the same number of electrons and holes (i.e., $n_i = p_i$), we can rewrite Equation 4.40 as

$$n \times p = n_i^2 = p_i^2 \qquad (4.41)$$

Recall from Equation 4.15 that

$$n = N_c \exp{-\left(\frac{E_c - E_F}{k_B T} \right)}$$

Substitute for N_c from Equation 4.31 and rewrite Equation 4.15 as

$$n = N_c \exp{-\left(\frac{E_c - E_F}{k_B T} \right)} = \frac{n_i}{\exp{-\left(\dfrac{E_c - E_{F,i}}{k_B T} \right)}} \left[\exp{-\left(\frac{E_c - E_F}{k_B T} \right)} \right]$$

or

$$\therefore n = n_i \exp{\left(\frac{E_F - E_{F,i}}{k_B T} \right)} \qquad (4.42)$$

This is also an important equation because it relates the conduction band electron concentration in an *extrinsic semiconductor* to its E_F and to the properties of the same semiconductor in an *intrinsic* form. Since the product $n \times p$ is a constant, we can also calculate the concentration of holes in an n-type semiconductor.

Similarly, in a p-type semiconductor, the concentration of holes (p) is related to the density of states available in the valence band and the Fermi energy level by Equation 4.24:

$$p = N_v \exp{-\left(\frac{E_F - E_v}{k_B T} \right)}$$

We substitute the value of N_v from Equation 4.32:

$$p_i = N_v \left[\exp{-\left(\frac{E_{F,i} - E_v}{k_B T} \right)} \right]$$

Therefore,

$$p = p_i \exp{-\left(\frac{E_F - E_{F,i}}{k_B T} \right)} \qquad (4.43)$$

To remove the negative sign, and since $n_i = p_i$, we rewrite Equation 4.43 as

$$p = n_i \exp{\left(\frac{E_{F,i} - E_F}{k_B T} \right)} \qquad (4.44)$$

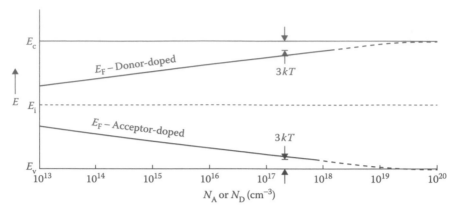

FIGURE 4.7   Variation in E_F as a function of the dopant concentration of silicon at 300 K. (Note: It assumes $n_i = 10^{10}$ cm^{-3}.) (From Pierret, R. F. 1988. *Semiconductor Fundamentals Volume*. Reading, MA: Addison-Wesley. With permission.)

Note that in Equation 4.44, $p \approx N_a$. Thus, we can calculate the position of the Fermi energy level relative to the center of the bandgap. As discussed in Section 4.4, the Fermi energy level for an intrinsic semiconductor ($E_{F,i}$) lies close to the center of the bandgap (Figures 4.4 and 4.5).

To summarize, we now have a set of two important equations that allows us to relate the dopant concentration of an extrinsic semiconductor (n-type or p-type) to their Fermi energy levels.

$$n = n_i \exp\left(\frac{E_F - E_{F,i}}{k_B T} \right)$$

$$p = n_i \exp\left(\frac{E_{F,i} - E_F}{k_B T} \right)$$

(4.45)

From Equation 4.44, we can conclude that as the doping level increases, the Fermi energy level of an n-type semiconductor will move closer to the conduction bandedge. *The Fermi energy level of n-type semiconductors lies in the upper half of the bandgap.* Similarly, as the doping level increases, the Fermi energy level of a p-type semiconductor will move closer to the valence bandedge. *The Fermi energy level of p-type semiconductors lies in the lower half of the bandgap.* Figure 4.7 shows the variation of E_F of Si as a function of the dopant concentrations (Pierret 1988). Note that this was calculated assuming $n_i = 10^{10}$ cm^{-3}.

4.7   FERMI ENERGY AS A FUNCTION OF THE TEMPERATURE

For an intrinsic semiconductor, the $E_{F,i}$ varies slightly with the temperature (Equation 4.29) because of the change in the effective masses of holes and electrons with the temperature (Figure 4.5).

When an extrinsic semiconductor is at very low temperatures, a carrier freeze-out occurs, causing it to behave as an intrinsic semiconductor. Thus, at very low temperatures, E_F is equal to $E_{F,i}$. As the temperature increases, donor ionization occurs. The Fermi energy level then moves to the location given by Equation 4.45. At higher temperatures, the thermally generated carriers begin to dominate. The semiconductor behaves as if it is intrinsic, and the E_F goes back to close to the center of the bandgap, that is, near the $E_{F,i}$ (Figure 4.8).

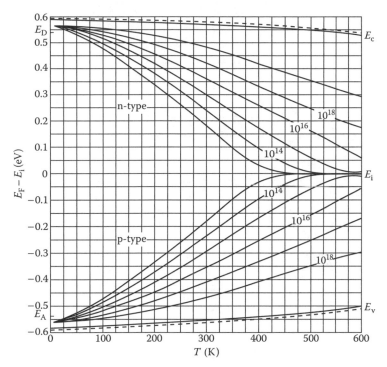

FIGURE 4.8   Position of E_F of silicon for different temperatures and dopant concentrations. (From Pierret, R. F. 1987. *Advanced Semiconductor Fundamentals*. Reading, MA: Addison-Wesley. With permission.)

As we will see in Section 4.9, when the dopant concentration is very high, the Fermi energy level moves very close ($\sim$ within $3\,k_BT$) to a bandedge. Such materials are called *degenerate semiconductors*. We cannot use Equation 4.45 to calculate the position of E_F for these materials.

The calculation of E_F for degenerate semiconductors is discussed in Section 4.9. The calculation of the location of E_F relative to $E_{F,i}$ for *nondegenerate semiconductors* is shown in Examples 4.3 and 4.4.

EXAMPLE 4.3: FERMI ENERGY LEVEL IN N-TYPE Si

An Si crystal is doped with $10^{16}/cm^3$ Sb atoms. At 300 K:

a. What is the concentration of electrons (n)?
b. What is the concentration of holes (p) in this n-type Si?
c. Where is the Fermi energy level (E_F) for this material relative to the Fermi energy level for intrinsic Si (E_i)?

SOLUTION

a. The donor dopant (Sb) concentration (N_d) is 10^{16} atoms/cm^3. The intrinsic electron concentration (n_i) in Si at 300 K (Figure 4.6) is 1.5×10^{10} electrons/cm^3 (Figure 4.6). Since $N_d \gg n_i$, we have assured $n \approx N_d$. We assume that the complete dopant ionization occurs at 300 K, and therefore the electron concentration (n) $\approx N_d = 10^{16}$ electrons/cm^3.

b. From Equation 4.41,

$$n \times p = n_i^2 = p_i^2$$

Therefore,

$$10^{16} \times p = (1.5 \times 10^{10})^2 = 2.25 \times 10^{20}$$

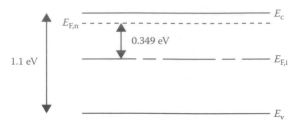

FIGURE 4.9   A band diagram and the Fermi energy level location for n-type silicon (see Example 4.3).

Thus, the concentration of holes in this n-type semiconductor is $p = 2.25 \times 10^4$ holes/cm^3. When compared to the electron concentration of 10^{16} (100,000 trillion), there are only 22,500 holes/cm^3. Therefore, holes are considered a minority carrier in an n-type semiconductor. We had seen this concept in Chapter 3 but had not calculated the actual concentration of minority carriers.

c. Now, we calculate the relative position of E_F relative to $E_{F,i}$ using Equation 4.42:

$$n = n_i \exp\left(\frac{E_F - E_{F,i}}{k_B T}\right)$$

or

$$E_F - E_{F,i} = k_B T \ln\left(\frac{n}{n_i}\right) \tag{4.46}$$

Therefore,

$$E_F - E_{F,i} = (0.026 \text{ eV}) \ln\left(\frac{10^{16}}{1.5 \times 10^{10}}\right) = 0.349 \text{ eV}$$

Note that we used $k_B T = 0.026$ eV.

Thus, the Fermi energy level for this n-type semiconductor is 0.349 eV *above* $E_{F,i}$. The band diagram for this n-type Si is shown in Figure 4.9. In this diagram, the Fermi energy level is designated as $E_{F,n}$.

EXAMPLE 4.4: FERMI ENERGY LEVEL IN B-DOPED Si

An Si crystal is doped with 10^{17}/cm^3 boron (B) atoms. At 300 K:

a. What is the concentration of holes (p)?
b. What is the concentration of electrons (n) in this p-type Si?
c. Where is the Fermi energy level (E_F) for this material relative to the Fermi energy level for intrinsic Si (E_i)?

SOLUTION

a. The acceptor dopant (B) concentration (N_a) is 10^{17} atoms/cm^3. We assume that complete dopant ionization occurs at 300 K. Therefore, the hole concentration (p) $\approx N_a = 10^{17}$ holes/cm^3.
b. From Equation 4.41,

$$n \times p = n_i^2 = p_i^2$$

Therefore,

$$n \times 10^{17} = (1.5 \times 10^{10})^2 = 2.25 \times 10^{20}$$

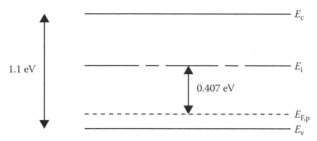

FIGURE 4.10 A band diagram for boron-doped silicon (see Example 4.4).

Thus, the concentration of electrons in this p-type semiconductor is 2.25×10^3 electrons/cm^3. Therefore, electrons are considered a minority carrier in a p-type semiconductor.

c. Now, we calculate the relative position of E_F for this p-type semiconductor relative to the $E_{F,i}$ using Equation 4.44:

$$p = n_i \exp\left(\frac{E_{F,i} - E_F}{k_B T}\right)$$

or

$$E_{F,i} - E_F = k_B T \ln\left(\frac{p}{n_i}\right) \tag{4.47}$$

Therefore,

$$E_{F,i} - E_F = (0.026 \text{ eV}) \ln\left(\frac{10^{17}}{1.5 \times 10^{10}}\right) = 0.407 \text{ eV}$$

or

$$E_F = E_{F,i} - 0.407 \text{ eV}$$

Thus, the Fermi energy level (E_F) for this p-type semiconductor is 0.407 eV *below* the $E_{F,i}$. The band diagram for this n-type Si is shown in Figure 4.10.

4.8 FERMI ENERGY POSITIONS AND THE FERMI–DIRAC DISTRIBUTION

It is clear from Equation 4.45 that the E_F of a semiconductor moves closer to the bandedge as the doping level increases. This is also reflected in the Fermi–Dirac distribution function. For n-type semiconductors (Figure 4.11b), the E_F is closer to the conduction bandedge, which means that the Fermi function is also pushed up on the diagram. Similarly, for a p-type semiconductor (Figure 4.11c), the Fermi energy level is closer to the valence bandedge, and thus the Fermi energy function is pushed down. Note that for p-type semiconductors, the value of $f(E_v)$ is not 1, because there are holes in the valence band. The value of $f(E_v)$ is slightly less than 1, and this $1 - f(E_v)$ is the probability of finding a hole in the valence band.

4.9 DEGENERATE OR HEAVILY DOPED SEMICONDUCTORS

As mentioned in Section 4.6, with very high levels of doping, the Fermi energy level of a semiconductor comes very close to the bandedge. When $|E_F - E_c|$ or $|E_F - E_v|$ is $\leq 3\ k_B T$, the semiconductor

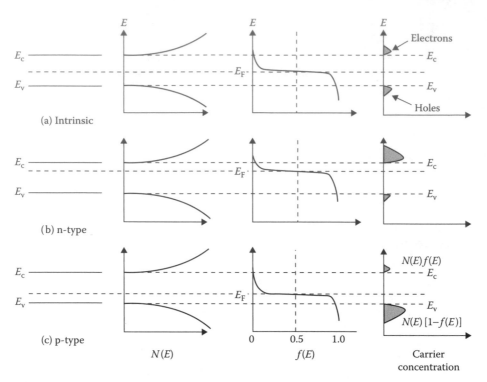

FIGURE 4.11 The Fermi energy level, the Fermi–Dirac distribution function, and the related band diagram for (a) an intrinsic, (b) an n-type, and (c) a p-type semiconductor. (From Streetman, B. G., and S. Banerjee. 2000. *Solid State Electronic Devices*. Upper Saddle River, NJ: Prentice Hall. With permission.)

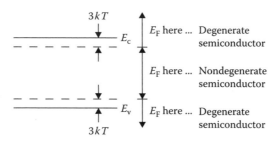

FIGURE 4.12 Definition of degenerate or nondegenerate semiconductors. (From Pierret, R. F. 1988. *Semiconductor Fundamentals Volume*. Reading, MA: Addison-Wesley. With permission.)

is known as a degenerate semiconductor (Figure 4.12). The position of the Fermi energy level for degenerate semiconductors cannot be calculated using Equation 4.45.

In degenerate semiconductors, the dopant concentration is high, and the distance between dopant atoms is very small. As we discussed in Chapter 3, there is no longer a discrete energy level (E_d or E_a) at such high dopant concentrations. Instead, we get a band of energies near the bandedge, and the bandgap effectively becomes narrow. If the doping levels are too high, then the Fermi energy level of an n-type material actually lies in the conduction band. The Fermi energy level of a p-type semiconductor lies in the valence band. Tunnel diodes can be made by using such degenerate semiconductors. In these devices, a phenomenon known as electron tunneling occurs. For silicon at 300 K, a donor dopant concentration (N_d) $\approx \geq 1.6 \times 10^{18}$ atoms/cm^3 and an acceptor dopant concentration

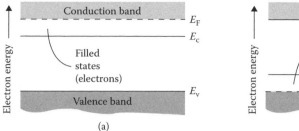
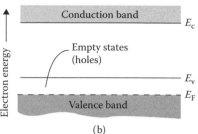

FIGURE 4.13 Fermi energy levels for (a) degenerate n-type and (b) degenerate p-type semiconductors. (From Neaman, D. 2006. *An Introduction to Semiconductor Devices*. New York: McGraw Hill. With permission.)

$(N_a) \approx \geq 9.1 \times 10^{17}$ atoms/cm³, result in a degenerate semiconductor (Pierret 1988). Terms such as "highly" or "heavily" doped and "n⁺–material" or "p⁺–material" are also used to describe degenerate semiconductors (Figure 4.13).

We use the *Joyce–Dixon approximation* (Equation 4.48) to calculate E_F for degenerate semiconductors:

$$E_F = E_c + k_B T \left[\ln \frac{n}{N_c} + \frac{1}{\sqrt{8}} \frac{n}{N_c} \right] \quad \text{(for n-type)}$$

$$E_F = E_v - k_B T \left[\ln \frac{p}{N_v} + \frac{1}{\sqrt{8}} \frac{p}{N_v} \right] \quad \text{(for p-type)}$$

(4.48)

Example 4.5 illustrates the use of the Joyce–Dixon approximation.

EXAMPLE 4.5: FERMI ENERGY LEVEL FOR A DEGENERATE SEMICONDUCTOR

A GaAs n-type sample is doped so that the concentration of electrons is 10^{17}. What is the value of the E_F for this sample relative to E_c? Assume that $T = 300\,\text{K}$ and that the semiconductor is degenerate.

SOLUTION

The density of states for GaAs at 300 K is $4.45 \times 10^{17}\,\text{cm}^{-3}$ and $k_B T = 0.026\,\text{eV}$.

$$E_F - E_c = (0.026\,\text{eV}) \left[\ln \frac{10^{17}}{4.45 \times 10^{17}} + \frac{1}{\sqrt{8}} \left[\frac{10^{17}}{4.45 \times 10^{17}} \right] \right]$$

Therefore, $E_F - E_c$ is 0.039 eV. Note that this value is within $3\,k_B T$ of E_c.

4.10 FERMI ENERGY LEVELS ACROSS MATERIALS AND INTERFACES

We have discussed in detail how to calculate the location of E_F for intrinsic and extrinsic semiconductors. The significance of the location of E_F will become clearer in Chapter 5 when we discuss the formation of the *p-n junction* and devices based on the same. When electrically different materials (e.g. a p-type semiconductor and a metal) are brought together or when electrically different interfaces are created within the same material, the equilibrium Fermi energy level remains invariant. There is no discontinuity or gradient that can exist in the equilibrium Fermi energy level. As an example, consider two materials A and B with Fermi energy levels $E_{F,A}$ and $E_{F,B}$ (Figure 4.14). It

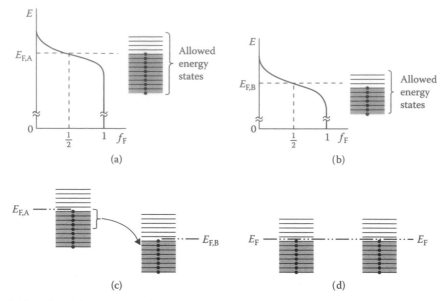

FIGURE 4.14 Materials A and B with different Fermi energy levels before and after making a contact. (a) material A in thermal equilibrium, (b) material B in thermal equilibrium, (c) materials A and B at the instant they are placed in contact, and (d) materials A and B in contact at thermal equilibrium. (From Neaman, D. 2006. *An Introduction to Semiconductor Devices*. New York: McGraw Hill. With permission.)

does not matter whether these materials are the same or different semiconductors or whether they are n-type or p-type. Materials A and B can also be metals.

We usually create a p-type material in a crystal that is uniformly donor-doped or vice versa, or we deposit a metal on a semiconductor crystal. When we bring materials A and B with different Fermi energy levels together (Figure 4.14), electrons from material A (which has a higher Fermi energy level and an overall higher energy) begin to flow to material B (Figure 4.14c), until the electrons in both materials have the same energy distribution (Figure 4.14d). The flow of electrons from material A to B will be proportional to the number of energy states filled with electrons in material A and the number of empty energy states available in material B.

The rate of transfer of electrons from material A to B is $\propto [N_A(E)f_A(E)] \times [N_B(E)(1-f_B(E))]$.

Similarly, the rate of transfer of electrons from material B to material A is $\propto [N_B(E)f_B(E)] \times [N_A(E)(1-f_A(E))]$.

When equilibrium is reached following the contact between the materials with different Fermi energy levels, there is no net current flow. At equilibrium, the rate of the transfer of electrons from material A to material B is equal to that of the transfer of electrons from material A to material B. Therefore,

$$[N_A(E)f_A(E)] \times [N_B(E)(1-f_B(E))] = [N_B(E)f_B(E)] \times [N_A(E)(1-f_A(E))]$$

This simplifies to

$$f_A(E) = f_B(E)$$

or

$$1 + \exp\left(\frac{E - E_{F,A}}{k_B T}\right) = 1 + \exp\left(\frac{E - E_{F,B}}{k_B T}\right)$$

This is true if $E_{F,A} = E_{F,B}$.

Therefore, the Fermi energy level remains constant across materials A and B, or

$$\frac{dE_F}{dx} = 0 \tag{4.49}$$

In this discussion, if materials A and B were n-type and p-type semiconductors, respectively, then we would create a p-n junction by bringing these materials together. In Chapter 5, we will discuss the use of this important concept, called *invariance of Fermi energy*, in formation of p-n junctions.

4.11 PROBLEMS

4.1 A metal has an electron concentration of 10^{22} atoms/cm³. What is the value of the Fermi energy level at 0 K (use Equation 4.1)?

4.2 What is the maximum speed of an electron in Cu at 0 K? Does this speed increase significantly with the temperature? Explain.

4.3 What is the effective speed of electrons in Cu at 0 K? Does this speed depend strongly on the temperature? Explain.

4.4 What is the E_F at $T = 300$ K for the metal discussed in Problem 4.1?

4.5 How is the Fermi energy level of a semiconductor different from both the donor and acceptor levels? Show using an illustration.

4.6 A semiconductor has $E_F = 0.26$ eV below the conduction bandedge. What is the probability that the state 0.026 eV ($k_B T$) above the conduction bandedge is occupied by an electron? What type of semiconductor is this material—n-type or p-type?

4.7 An energy level is located at 0.3 eV above the E_F. Calculate the probability of occupancy for this state at 0, 300, and 600 K.

4.8 What is the probability that an energy state 0.4 eV below E_F is empty for the temperatures of 0, 300, and 600 K?

4.9 Show that the Fermi–Dirac function is symmetric around E_F, that is, that the probability of a state ΔE above E_F is the same as the probability of finding a hole at a state E below E_F.

4.10 The reduced effective density of states mass for electrons and holes in GaAs are ~0.067 and 0.50. Use Equation 4.30 to calculate the exact position of the Fermi energy level for intrinsic GaAs at 300 K. How is this location different from that for Ge at 300 K?

4.11 Use Equation 4.15 to express the concentration of electrons (n) in the conduction band. Then, use a similar equation to express the concentration of electrons n_d in the donor state E_d and density of states N_d.

4.12 The reduced effective density of states mass for electrons and holes in Ge are ~0.56 and 0.40 at 300 K. Use Equation 4.30 to calculate the exact position of the Fermi energy level for intrinsic Ge at 300 K.

4.13 Use Equation 4.39 to calculate the intrinsic electron concentration in Si at 350 and 400 K.

4.14 What is the resistivity of the intrinsic Si sample discussed in Problem 4.13? Use the mobility values from Table 3.1 in Chapter 3, and note that the mobility does depend on the temperature.

4.15 Derive an equation equivalent to Equation 4.39 that describes the variation in the concentration of thermally excited electrons in Ge.

4.16 Derive an equation equivalent to Equation 4.39 that describes the variation in the concentration of thermally excited electrons in GaAs.

4.17 An Si sample is doped so that $n = 2 \times 10^5$ cm⁻³. What is the hole concentration for this material? Is this an n-type or a p-type Si? Assume that $T = 300$ K.

4.18 What is the conductivity (σ) of the sample described in Problem 4.17?

4.19 An Si crystal is doped with 10^{15} cm^{-3} of P atoms. Calculate the E_F of this crystal relative to the center of the bandgap, which can be assumed as the location of $E_{F,i}$.

4.20 A GaAs crystal is doped such that the E_F is 0.2 eV above the E_v. Is this an n-type or a p-type GaAs? What are the electron and hole concentrations in this material? Assume that $E_g = 1.43$ eV and $T = 300$ K.

4.21 What is the resistivity (ρ) at 300 K of the sample discussed in Problem 4.20? Use the appropriate mobility values from Table 3.1 in Chapter 3.

4.22 An Si crystal is first uniformly doped with a donor dopant, and then the process of doping the entire crystal with an acceptor is started. Sketch the variation of the E_F of this semiconductor, starting out as an n-type material as a function of the increasing acceptor concentration. Where is the E_F when the donor and acceptor concentrations are equal?

4.23 An Si crystal is doped with 10^{16} cm^{-3} of Sb atoms (see Example 4.3), and then the entire crystal is doped with 3×10^{17} atoms/cm^3 of B. What is the new location of the Fermi energy level for this crystal containing both B and Sb?

4.24 What is the resistivity of Si (at 300 K) discussed in Problem 4.23 after Sb-doping and B-doping? Use the appropriate values of mobility and consider the dependence of mobility on the dopant concentration.

4.25 As stated in Section 4.9, for Si at 300 K, a donor dopant concentration (N_d) $\approx \geq 1.6 \times 10^{18}$ atoms/cm^3 or an acceptor dopant concentration (N_a) $\approx \geq 9.1 \times 10^{17}$ atoms/cm^3 result in a degenerate semiconductor. Verify that this is correct using the definition of a degenerate semiconductor.

4.26 What is the minimum donor dopant concentration for GaAs at 300 K, at which we consider it degenerate? (Hint: The solution for $[E_F - E_c]$ must be less than $3\,k_BT$.)

4.27 A sample of a GaAs crystal is doped so that $N_d = 10^{17}$ donor atoms/cm^3 (see Example 4.5). Calculate the Fermi energy position using Equation 4.45, in which we assume that the sample is nondegenerate. Compare the result with that used in Example 4.5, where we used the Joyce–Dixon approximation.

GLOSSARY

Degenerate semiconductors: Semiconductors that are heavily doped so that the Fermi energy moves very close (within $3\,k_BT$) to a bandedge.

Effective density of states approximation: The assumption that most electrons or holes reside at the bottom of the conduction band or top of the valence band, respectively. Thus, N_c or N_v can be used in calculations of the concentrations of electrons and holes in the conduction and valence bands, respectively.

Effective speed of electrons (v_e): or the **root mean speed of electrons** in a metal is given by $(1/2)mv_e^2 \cong (3/5)E_{F,0}$.

Fermi–Dirac distribution function: The function that gives the probability of occupancy of a state by an electron and is defined as

$$f(E) = \frac{1}{\left[1 + \exp\left(\dfrac{E - E_F}{k_BT}\right)\right]}$$

Fermi energy (E_F): The energy level at which the probability of occupancy of a state is 0.5.

Invariance of Fermi energy: When materials with different Fermi energy levels are brought in contact with each other under equilibrium conditions, the Fermi energy level remains continuous across the interface between the different materials.

Joyce–Dixon approximation: An equation used for calculating the Fermi energy level for degenerate semiconductors.

Law of mass action for semiconductors: For any given semiconductor, the value of $n \times p$ is a constant at a given temperature.

p-n junction: The electrical junction between a p-type and an n-type semiconductor. This is the basis for many devices, including diodes and transistors.

Work function ($q\phi$): The energy required to remove an electron from E_F to a vacuum.

REFERENCES

Bullis, W. M., and H. R. Huff. 1994. Silicon for microelectronics. In *Encyclopedia of Advanced Materials*, vol. 4, eds. D. Bloor, S. Mahajan, M. C. Flemings, and R. J. Brook. Oxford, UK: Pergamon Press.

Green, M. A. 1990. Intrinsic concentrations, effective density of states, and effective mass in silicon. *J Appl Phys* 67:2944–54.

Grove, A. S. 1967. *Physics and Technology of Semiconductor Devices*. New York: Wiley.

Kasap, S. O. 2002. *Principles of Electronic Materials and Devices*. New York: McGraw Hill.

Li, S. S. 2006. *Semiconductor Physical Electronics*. New York: Springer.

Neaman, D. 2006. *An Introduction to Semiconductor Devices*. New York: McGraw Hill.

Pierret, R. F. 1987. *Advanced Semiconductor Fundamentals*. Reading, MA: Addison-Wesley.

Pierret, R. F. 1988. *Semiconductor Fundamentals Volume*. Reading, MA: Addison-Wesley.

Streetman, B. G., and S. Banerjee. 2000. *Solid State Electronic Devices*. Upper Saddle River, NJ: Prentice Hall.

5 Semiconductor p-n Junctions

KEY TOPICS

- Formation of a p-n junction
- Concept of built-in potential
- Band diagram for a p-n junction
- Diffusion and drift currents in a p-n junction
- What makes the p-n junction useful?
- Current–voltage (*I*–*V*) curve for a p-n junction
- Some devices based on p-n junctions

5.1 FORMATION OF A p-n JUNCTION

Consider a hypothetical experiment in which we have two crystals of a semiconductor: one is n-type and the other is p-type. The band diagrams for these before the materials are brought together are shown in Figure 5.1.

When we bring the two semiconductors together to form a p-n junction, electrons start flowing from the n-side to the p-side to minimize the concentration gradient. This will continue until a new common Fermi energy level (E_F) is established. Since electrons are negatively charged, the direction of the current associated with their motion is opposite to the direction of their actual motion. The direction of the current associated with the diffusion of electrons is from the p-side to the n-side of the p-n junction (Figure 5.2).

A p-doped material has a much higher concentration of holes on the p-side than that on the n-side. Thus, when a p-n junction is formed, the holes diffuse from the p-side to the n-side. Since holes are positively charged particles, the current induced due to their movement is in the same direction as their motion (Figure 5.2).

Note that a concentration gradient of dopant atoms exists because there is a significant concentration of donor atoms on the n-side and almost no donor atoms on the p-side. However, there is no diffusion of donor atoms or ions, as their mobility is very low at or near room temperature. Thus, here we consider the diffusion of holes and electrons only.

Since diffusion is concentration gradient driven, we expect that the flow of electrons and holes induced will continue until the concentrations of electrons and holes become equal on both sides of the p-n junction. For example, if we join a crystal of copper (Cu) and a crystal of nickel (Ni) and then heat it to a high temperature (e.g., 500°C) to promote diffusion (Figure 5.3), then after a few hours, the copper atoms will diffuse into the nickel crystal and vice versa. This process of interdiffusion will continue until the concentrations of copper and nickel atoms are equal on both sides of the original interface.

We will now discuss the difference between the interdiffusion of copper and nickel atoms (Figure 5.3) and the diffusion of holes and electrons in the formation of a p-n junction (Figure 5.2). When electrons in the n-type material begin to diffuse out into the p-type material, they leave behind positively charged donor ions. For example, when a donor such as phosphorus (P) is added as a neutral atom, it donates an electron and becomes ionized, turning into a P^{1+} ion. Similarly, when holes from the p-side begin to diffuse onto the n-side of the p-n junction, they leave behind negatively charged acceptor ions (e.g., boron; B^{1-} ion).

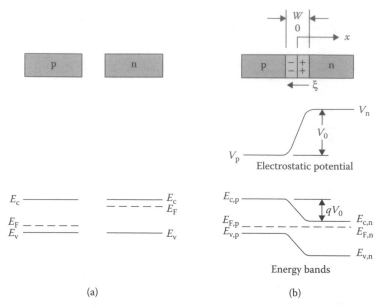

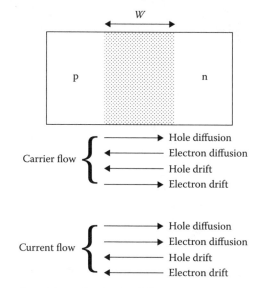

FIGURE 5.1 Band diagrams for the n- and p-semiconductors (a) prior to and (b) after the formation of a p-n junction. (From Streetman, B. G., and S. Banerjee. 2000. *Solid State Electronic Devices*, 5th ed. Upper Saddle River, NJ: Prentice Hall. With permission.)

FIGURE 5.2 The directions of particle motion for the diffusion and drift of electrons and their corresponding currents for a p-n junction under equilibrium. (From Edwards-Shea, L. 1996. *The Essenece of Solid State Electronics*. Upper Saddle River, NJ: Prentice Hall. With permission.)

Thus, as electrons diffuse from the n-type into the p-material, they leave behind a region of positively charged donor ions, shown with + signs in Figure 5.1b. Similarly, as holes diffuse from the p-side to the n-side, they create a region comprised of negatively charged acceptor ions (Figure 5.1b). Thus, at the p-n junction, there is a region known as the *depletion region*, so named because it is depleted of electrons and holes. It is also called the *space-charge region* or *space-charge layer*, which refers to the positive and negative charges present in this region.

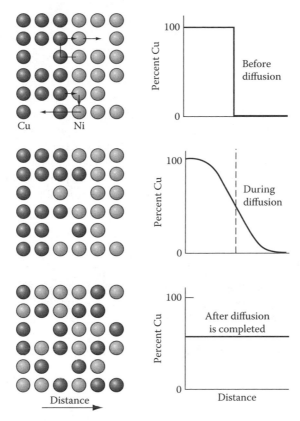

FIGURE 5.3 Illustration of the interdiffusion of copper and nickel atoms. (From Askeland, D., and P. Fulay. 2006. *The Science and Engineering of Materials.* Washington, DC: Thomson. With permission.)

Across the width of the depletion region (*w*), an *internal* or *built-in electric field* (*E*) is set up because of the presence of positively charged donor ions and negatively charged acceptor ions (Figure 5.1). This electric field (*E*) is directed from the positively charged donor ions to the negatively charged acceptor ions. If an electron tries to diffuse from the n-side to the p-side, it begins to "see" the acceptor ions. The motion of such electrons toward the p-side is opposed by the negatively charged acceptor ions (N_a^-). Similarly, if a hole tries to diffuse from the p-side to the n-side, it begins to experience the repelling force of the positively charged donor ions (N_d^+). We will use the symbols N_d and N_a for the concentrations of donor and acceptor atoms or ions, respectively. Thus, the built-in electric field stops the diffusion of electrons from the n-side to the p-side and the diffusion of holes from the p-side to the n-side.

The voltage corresponding to the internal electric field is known as the *contact potential* or *built-in potential* (V_0) and is expressed in volts. This potential difference can appear when a p-n junction between different materials is formed. As we can see from Figure 5.1, the built-in electric field is directed toward the −*x*-direction. In the depletion region, the electric field is related to the potential as follows:

$$E(x) = -\frac{dV(x)}{dx} \tag{5.1}$$

The electrostatic potential (*V*) is higher on the n-side, which is a part of the depletion layer with a net positive charge (Figure 5.1b). The contact potential (V_0) is the difference between V_n and V_p, where V_n and V_p are the electrostatic potentials in the n- and p-neutral regions, respectively. Thus,

$$V_0 = V_n - V_p \tag{5.2}$$

We plot the electron energy, which is related to the electric potential by a factor of $-q$, on the band diagram. Therefore, $E_{c,n}$, $E_{F,n}$ and $E_{v,n}$, that is, different energy levels on the n-side, appear lower on the p-n junction band diagram than the corresponding levels on the p-side (Figure 5.1b).

The development of the contact potential, also known as the *diffusion potential*, makes the p-n junction interesting and useful for device applications. We will see in Section 5.10 that by applying a *forward bias*, that is, by connecting the positive terminal of an external voltage supply to the p-side, this built-in potential can be overcome. If this happens, the p-n junction begins to conduct as the diffusion of electrons and holes resumes. Conversely, if we apply a *reverse bias* by connecting the negative terminal of an external voltage supply to the p-side, then we add to the built-in potential barrier (see Section 5.9). Thus, the p-n junction will be able to carry very little current. A p-n junction is a "tunable device" and is used to create diodes, transistors, and so on.

5.2 DRIFT AND DIFFUSION OF CARRIERS

The built-in or internal electric field (E) created in the p-n junction stops further diffusion of electrons and holes. It also plays another important role by setting up drift currents (Figure 5.2). The term "drift" refers to the motion of charge carriers under the influence of an internal or external electric field. When thermally generated *electrons* from the p-side (where they are minority carriers) experience the internal electric field, they are driven toward the positively charged region of donor ions. The internal electric field causes electrons on the p-side to drift toward the n-side (Figure 5.2). Similarly, thermally generated *holes* on the n-side drift toward the negatively charged space-charge region on the p-side. The internal electric field causes these drift currents. The total drift current in a p-n junction is known as the *generation current* (see Section 5.11). It is an important part of the current–voltage (I–V) characteristics of a p-n junction. Note that the directions of the diffusion of electrons and the drift of electrons are opposite. The current due to electron diffusion is from p-side to n-side. The electron drift current is from the n-side to the p-side. The directions of the diffusion of majority carriers and the drift of minority carriers on the n-side and p-side are shown in Figure 5.2.

For a p-n junction at equilibrium (Figure 5.2), the drift and diffusion current densities (J) cancel out, since a p-n junction at the equilibrium carries no electrical current.

Therefore,

$$J_p(\text{diffusion}) + J_p(\text{drift}) = 0 \tag{5.3}$$

and

$$J_n(\text{diffusion}) + J_n(\text{drift}) = 0 \tag{5.4}$$

Subscripts p and n refer to the motion of holes and electrons, respectively.

5.3 CONSTRUCTING THE BAND DIAGRAM FOR A p-n JUNCTION

In Chapter 4, we learned how to calculate the relative position of the Fermi energy level for a semiconductor (Examples 4.2 and 4.3) and the invariance of the Fermi energy level (Section 4.10). We will now draw the band diagram for a p-n junction using this information and graphically calculate the value of qV_0, and hence V_0. This is illustrated in Example 5.1.

EXAMPLE 5.1: ESTIMATION OF CONTACT POTENTIAL FROM THE BAND DIAGRAM

Consider a p-n junction in Si. Assume that the n- and p-sides have a dopant concentration of $N_d = 10^{16}$ and $N_a = 10^{17}$ atoms/cm³, respectively. (a) Calculate the Fermi energy position for the n-type and p-type semiconductors, and draw the band diagrams for the n-type and p-type semiconductors before they are joined. (b) Draw the band diagram for this p-n junction and estimate its contact potential (V_0). Assume that $T = 300$ K, $n_i = 1.5 \times 10^{10}$ electrons/cm³ and $E_g = 1.1$ eV.

SOLUTION

a. As we have seen in Chapter 4,

$$n = n_i \exp\left(\frac{E_F - E_{F,i}}{k_B T}\right) \tag{5.5}$$

Assuming a complete donor ionization, that is, $n \approx N_d = 10^{16}$ atoms/cm³, we get

$$E_{F,n} - E_{F,i} = (0.026 \text{ eV}) \ln\left(\frac{10^{16}}{1.5 \times 10^{10}}\right) = 0.349 \text{ eV}$$

Thus, the Fermi energy level for this n-type semiconductor is 0.349 eV above $E_{F,i}$. This is shown in Figure 5.4.

Similarly, for the p-side,

$$E_{F,i} - E_F = k_B T \ln\left(\frac{p}{n_i}\right) \tag{5.6}$$

Substituting $N_a = 10^{17}$ atoms/cm³ $= p$ and the value of n_i,

$$E_{F,i} - E_{F,p} = (0.026 \text{ eV}) \ln\left(\frac{10^{17}}{1.5 \times 10^{10}}\right) = 0.407 \text{ eV}$$

or

$$E_{F,p} = E_{F,i} - 0.407 \text{ eV}$$

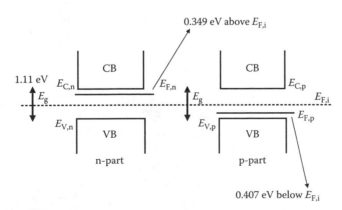

FIGURE 5.4   Band diagrams for n-type and p-type regions (see Example 5.1).

b. Now, we draw the band diagram for the p-n junction (Figure 5.5). First, we recognize that a common Fermi energy level is established after the n-type and p-type semiconductors are joined. Therefore, we draw a horizontal line that represents this common E_F (Figure 5.5). On the n-side of this junction, but far away from the boundary of the interface between the two materials, the common Fermi energy level E_F is 0.349 eV above the $E_{F,i,n}$. Using a suitable scale, we draw a line 0.349 eV below $E_{F,i,n}$ and label it $E_{F,n}$. This represents the intrinsic Fermi energy level position on the n-side. Since this is almost at the middle of the bandgap, we draw the conduction bandedge $E_{c,n}$ at 0.55 eV (1/2 of E_g) above $E_{F,i,n}$. We next draw $E_{v,n}$ 0.55 eV below the $E_{F,i,n}$.

On the p-side of this junction, but far away from the boundary of the interface between the two materials, the $E_{F,p}$ is 0.407 eV below the dotted $E_{F,i,p}$ line. Thus, we draw a line 0.407 eV above the common energy level $E_{F,p}$ and label it as $E_{F,i,p}$. This represents the intrinsic Fermi energy level position on the p-side. We then draw the conduction bandedge on the p-side $E_{c,p}$ at 0.55 eV (1/2 of E_g) above $E_{F,i,p}$. We next draw $E_{v,p}$ 0.55 eV below the $E_{F,i,p}$ line. Note that this automatically puts the bandedges on the p-side higher than their counterparts on the n-side.

The steps needed to complete the p-n junction band diagram are summarized in Figure 5.5. We now join the conduction bandedges and valence bandedges on both sides as shown in Figure 5.6. We will see in Section 5.7 that the variation of electron energy across the depletion region is parabolic.

From this band diagram, we can see that

$$qV_0 = \text{the height of the energy barrier} = (E_{c,p} - E_{c,n}) \qquad (5.7)$$

Looking at the p-side of the diagram,

$$E_{c,p} = 0.407 \text{ eV} + 0.55 \text{ eV} \qquad (5.8)$$

Looking at the n-side of the diagram,

$$E_{c,n} = 0.55 \text{ eV} - 0.349 \text{ eV} \qquad (5.9)$$

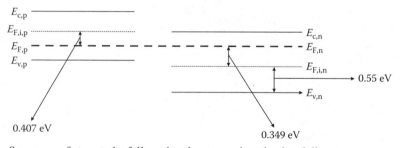

FIGURE 5.5   Summary of steps to be followed to draw a p-n junction band diagram.

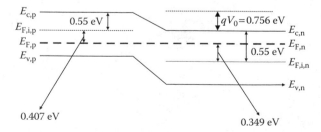

FIGURE 5.6   A band diagram for the p-n junction in Example 5.1.

Therefore, subtracting Equation 5.9 from Equation 5.8, we get

$$qV_0 = \text{the height of the energy barrier setup} = E_{c,p} - E_{c,n} = 0.407 \text{ eV} + 0.55 \text{ eV} - 0.55 \text{ eV} + 0.349 \text{ eV}$$
$$= 0.756 \text{ eV}$$

Thus, the value of the contact potential for this p-n junction (V_0) is 0.756 V (Figure 5.6). Note that the contact potential is $V_n - V_p$ (Figure 5.1).

5.4   CALCULATION OF CONTACT POTENTIAL

If the doping level on the n-side increases, then $E_{F,n}$ will be farther away from $E_{F,i,n}$ (Figure 5.5) and the $E_{c,n}$ and $E_{v,n}$ will move down to center themselves around $E_{F,i,n}$. If nothing changes on the p-side, then the value of qV_0 will be expected to increase. Similarly, if the doping level on the p-side is increased, then $E_{F,i,p}$ will move up relative to $E_{F,p}$ (Figure 5.4). Then, $E_{v,p}$ and $E_{c,p}$ will also move up to center around $E_{F,i,p}$. If nothing changes on the n-side, qV_0 will increase.

We will now derive an equation that quantitatively describes how the value of V_0 changes with the dopant concentrations on the n- and p-sides of the p-n junction.

According to Fick's first law of diffusion, the flux of diffusing species is proportional to the negative of the concentration gradient. The negative sign means that there will be a movement of species from a region of higher concentration to a region of lower concentration. In this case, we want to calculate the *electrical charge* flowing per unit time, not just the number of electrons or holes. Therefore, we will multiply the flux of species by q, the magnitude of the charge of an electron or a hole.

Thus, the diffusion current density (J) due to the motion of the holes is

$$J_{p,\text{diffusion}} = -q \times D_p \times \frac{dp(x)}{dx} \tag{5.10}$$

We have taken the direction from p to n as the positive x-direction (Figure 5.2). In this equation, D_p is the diffusion coefficient for holes.

The drift current due to the movement of the holes from the n-side, where they are the minority carriers, to the p-side is given by

$$J_{p,\text{drift}} = q \times \mu_p \times p(x) \times E(x) \tag{5.11}$$

In Equation 5.11, $p(x)$ is the hole concentration along x-direction and $E(x)$ is the built-in electric field. We know the hole concentration on both the p-side and the n-side, where the holes are minority carriers.

The current induced by the diffusion of majority carriers and the current caused by the drift of minority carriers are in opposite directions and cancel each other out (Equation 5.3). Therefore, we get

$$q \times D_p \times \frac{dp(x)}{dx} = q \times \mu_p \times p(x) E(x) \tag{5.12}$$

Simplifying,

$$\frac{dp(x)}{dx} \frac{1}{p(x)} = \frac{\mu_p}{D_p} E(x) \tag{5.13}$$

Since we want to calculate the value of V_0, the contact potential, we change the electric field (E) to electrostatic potential (V) by substituting for $E(x)$ from Equation 5.1 into Equation 5.13:

$$\frac{dp(x)}{dx} \frac{1}{p(x)} = -\left[\frac{\mu_p}{D_p}\right] \frac{dV(x)}{dx} \tag{5.14}$$

Using the so-called *Einstein relation* (not derived here) and applying it to holes,

$$\frac{\mu_p}{D_p} = \frac{q}{k_B T} \tag{5.15}$$

From Equations 5.14 and 5.15, we get

$$\left[\frac{dp(x)}{dx}\right]\left[\frac{1}{p(x)}\right] = -\left[\frac{q}{k_B T}\right]\frac{dV(x)}{dx} \tag{5.16}$$

We know the concentrations of holes on both the p-side and the n-side in the neutral regions, that is, the regions away from the p-n junction that do not have any built-up net charge. We assume a one-dimensional model; that is, we assume that the carriers will diffuse and drift along x-direction (+ or −) only. We now integrate Equation 5.16 from the p-side to the n-side:

$$-\frac{q}{k_B T}\int_{V_p}^{V_n} dV = \int_{p_p}^{p_n} \frac{dp}{p} \tag{5.17}$$

In Equation 5.17, V_p and V_n are the electrostatic potentials on the p-side and n-side of the neutral regions, where there is no built-up net charge. The hole concentrations in the neutral regions on the p-side and n-side are p_p and p_n, respectively. The subscripts indicate the side of the p-n junction. Simplifying Equation 5.17,

$$-\frac{q}{k_B T}(V_n - V_p) = \ln(p_n) - \ln(p_p) = \ln\left(\frac{p_n}{p_p}\right) \tag{5.18}$$

Note that the potential difference $V_n - V_p$ is V_0, the contact potential.

$$-\left(\frac{q}{k_B T}\right)V_0 = \ln\left(\frac{p_n}{p_p}\right) \tag{5.19}$$

Eliminating the negative sign, we get

$$V_0 = \frac{k_B T}{q}\ln\left(\frac{p_p}{p_n}\right) \tag{5.20}$$

For a *step junction*, we move abruptly from the p-side, with N_a acceptors per cubic centimeter, to the n-side with N_d donors per cubic centimeter. We rewrite Equation 5.20 as

$$V_0 = \frac{k_B T}{q}\ln\left(\frac{N_a N_d}{n_i^2}\right) \tag{5.21}$$

To get Equation 5.21, we used $p_p = N_a$ and $p_n \cdot N_d = n_i^2$. This form is useful in calculating the contact potential (V_0) associated with a p-n junction.

We can rewrite Equation 5.20 as

$$\frac{p_p}{p_n} = \exp\left(\frac{q V_0}{k_B T}\right) \tag{5.22}$$

Since $p_p \times n_p = p_n \times n_n = n_i^2$, we can write

$$\frac{p_p}{p_n} = \frac{n_n}{n_p} = \exp\left(\frac{q V_0}{k_B T}\right) \tag{5.23}$$

We can see from Equation 5.23 that if the dopant concentration on the p-side increases, then qV_0 will increase. We also saw this in the calculation of V_0 from the p-n junction band diagram in Section 5.3 and Figure 5.6.

From Equation 5.22, we substitute for p_p and p_n in terms of the density of states (N) and the difference between E_F relative to the valence bandedge on each side:

$$\frac{p_p}{p_n} = \frac{N_v \exp-\left(\dfrac{E_{F,p} - E_{v,p}}{k_B T}\right)}{N_v \exp-\left(\dfrac{E_{F,n} - E_{v,n}}{k_B T}\right)} = \exp\left(\frac{qV_0}{k_B T}\right) \tag{5.24}$$

In Equation 5.24, the additional subscript for E refers to the side of the junction. Thus, $E_{F,p}$ is the Fermi energy level on the p-side, $E_{v,n}$ is the valence bandedge on the n-side, and so on. Rearranging Equation 5.24,

$$\exp\left(\frac{qV_0}{k_B T}\right) = \frac{\exp-\left(\dfrac{E_{F,p} - E_{v,p}}{k_B T}\right)}{\exp-\left(\dfrac{E_{F,n} - E_{v,n}}{k_B T}\right)}$$

Note that the Fermi energy is constant for a p-n junction under equilibrium; that is, $E_{F,n} - E_{F,p} = 0$, so we get

$$\exp\left(\frac{qV_0}{k_B T}\right) = \exp\left(\frac{E_{F,n} - E_{F,p}}{k_B T}\right)\exp\left(\frac{E_{v,p} - E_{v,n}}{k_B T}\right) = \exp\frac{E_{v,p} - E_{v,n}}{k_B T}$$

Therefore,

$$qV_0 = E_{v,p} - E_{v,n} \tag{5.25}$$

The contact potential barrier energy (qV_0) is the difference between the valence and conduction bandedges on each side of the p-n junction. This was shown in Figure 5.6, the band diagram for a p-n junction. Example 5.2 examines the application of these equations.

EXAMPLE 5.2: CALCULATION OF CONTACT POTENTIAL (V_0)

A step junction in Si is such that n- and p-sides have dopant concentrations of $N_d = 10^{16}$ and $N_a = 10^{17}$ atoms/cm³, respectively (see Example 5.1). Calculate the contact potential (V_0), assuming that $T = 300$ K and $n_i = 1.5 \times 10^{10}$ electrons/cm³.

SOLUTION

From Equation 5.21,

$$qV_0 = \frac{k_B T}{q} \ln\left(\frac{N_a N_d}{n_i^2}\right) = (0.026 \text{ eV}) \ln\left(\frac{10^{17} \text{atoms/cm}^3 \times 10^{16} \text{atoms/cm}^3}{(1.5 \times 10^{10} \text{electrons/cm}^3)^2}\right)$$

$$qV_0 = 0.757 \text{ eV}$$

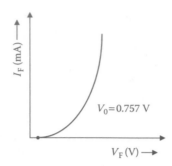

FIGURE 5.7 The *I–V* curve for the p-n junction in Example 5.2 with a forward bias.

Thus, the value of potential energy barrier set up by the built-in electric field for this p-n junction is 0.757 eV and the corresponding voltage (V_0) is 0.757 V, which is the same as in Example 5.1.

Using forward bias (the "p-side to positive," a useful mnemonic), we overcome this built-in potential, and the diffusion current starts to flow again. This assumes that the voltage drop is very small in the neutral regions, where there is no space charge accumulated. For a forward $V_F > 0.757$ V, this Si p-n junction begins to conduct an electrical current (I_F) (Figure 5.7).

5.5   SPACE CHARGE AT THE P-N JUNCTION

The depletion region has an extremely small concentration of electrons or holes. The charge density for the depletion region on the n-side of the junction is $q \times N_d$, where N_d is the concentration of donor ions. Assume that the cross-sectional area of the p-n junction is A and the width of the depletion region on the n-side is $x_{n,0}$. The subscript n refers to the n-side, and the subscript 0 stands for a p-n junction under equilibrium, that is, no external voltage or bias is applied. Thus, the volume of the depletion region on the n-side is $A \times x_{n,0}$. In this region, the total positive electrical charge is

$$Q_+ = q \times A \times x_{n,0} \times N_d \tag{5.26}$$

Similarly, if $x_{p,0}$ is the width of the penetration of the depletion region into the p-side and the cross-sectional area is A, then the magnitude of the negative charge on the p-side of the depletion region is

$$Q = q \times A \times x_{p,0} \times N_a \tag{5.27}$$

The donor and acceptor ion charges accumulated on the n- and p-sides must be equal for an electrical neutrality of the entire p-n junction, that is, $Q_+ = Q_-$ or

$$q \times A \times x_{n,0} \times N_d = q \times A \times x_{p,0} \times N_a \tag{5.28}$$

Therefore,

$$\frac{x_{n,0}}{N_a} = \frac{x_{p,0}}{N_d} \tag{5.29}$$

The width of penetration of the depletion region varies *inversely* to the dopant concentration. For example, if the p-side of the semiconductor is doped more heavily than the n-side (as in Example 5.1), the penetration of the depletion region on the p-side will be smaller. If one side is heavily doped, it needs a lesser volume of that material to compensate for the charge on the other

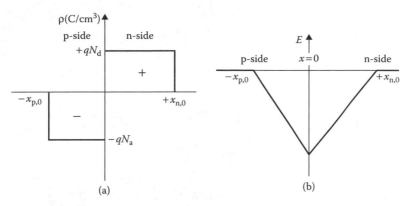

FIGURE 5.8 (a) Space charge density and (b) electric field variation at a p-n junction. (From Neaman, D. 2006. *An Introduction to Semiconductor Devices.* New York: McGraw Hill. With permission.)

side (Figure 5.8). If one side is lightly doped (in this case, the n-side), then we need more volume of that material to compensate for the charge on the other side. This means a higher penetration depth in the one-dimensional model.

5.6 ELECTRIC FIELD VARIATION ACROSS THE DEPLETION REGION

Using *Poisson's equation*, we can calculate the electric field and the electron energy variation across the depletion region. According to the one-dimensional form of this equation, the gradient of the electric field is related to the charge density and the dielectric permittivity of the material (ε):

$$\frac{d^2\phi(x)}{dx^2} = -\frac{\rho(x)}{\varepsilon} = -\frac{dE(x)}{dx} \quad \text{Poisson's equation} \tag{5.30}$$

where ϕ is the electric potential, ρ is the charge density, ε is the dielectric permittivity of the material, E is the electric field, and x is the distance. Note that $\varepsilon = \varepsilon_0 \times \varepsilon_r$, where ε_r is the dielectric constant and ε_0 is the permittivity of the free space. The dielectric constant (ε_r) is a measure of the ability of a dielectric material to store an electrical charge.

We assume that the doping on the n- and p-sides is uniform, and hence the charge density is as shown in Figure 5.8a.

Applying Poisson's equation to the n-side region in the depletion zone ($x = 0$ to $x = x_{n,0}$),

$$\frac{dE}{dx} = \frac{qN_d}{\varepsilon} \quad (\text{for } 0 \le x \le x_{n,0}) \tag{5.31}$$

We can obtain the electric field (E) variation by integrating Equation 5.31.

$$E = \int \frac{qN_d}{\varepsilon} dx = \frac{qN_d}{\varepsilon} x + C_1 \tag{5.32}$$

Note that when $x = x_{n,0}$, $E = 0$, that is, outside the depletion layer, there is no net built-up charge and the electric field becomes zero. Using this condition in Equation 5.32, we get

$$0 = \frac{qN_d}{\varepsilon} x_{n,0} + C_1$$

$$\therefore C_1 = -\frac{qN_d}{\varepsilon} x_n \tag{5.33}$$

Using this value of the constant of integration (C_1) in Equation 5.32, the electric field variation across the n-side of the depletion region is given by

$$E(x) = -\left(\frac{qN_d}{\varepsilon}\right)(x_{n,0} - x) \quad 0 \le x \le x_{n,0} \tag{5.34}$$

The maximum in the electric field occurs at $x = 0$. Its magnitude is given by

$$|E_{max}| = \left(\frac{qN_d}{\varepsilon}\right)x_{n,0} \tag{5.35}$$

Similarly, we can show that on the p-side of the junction, in the depletion region,

$$E(x) = -\left(\frac{qN_a}{\varepsilon}\right)(x_{p,0} + x) \quad -x_{p,0} \le x \le 0 \tag{5.36}$$

The variation in the electric field across the depletion layer width (w), that is, from $-x_{p,0}$ to $x_{n,0}$, is shown in Figure 5.8.

As we can see from Poisson's equation, the electric field variation is in the form of a straight line for a uniform charge density. With a linear variation in the electric field, the electric potential will have a parabolic change.

5.7   VARIATION OF ELECTRIC POTENTIAL

We use Poisson's equation to compute the electrostatic potential (ϕ) across the p-n junction. We can obtain the electrostatic potential by integrating the electric field across the distance over which the potential appears.

$$\phi(x) = -\int E(x)dx \tag{5.37}$$

For the p-side of the depletion region, substituting for $E(x)$ from Equation 5.36 into Equation 5.37, we get

$$\phi(x) = -\int\left(\frac{-qN_a}{\varepsilon}\right)(x_{p,0} + x)\,dx \tag{5.38}$$

Therefore,

$$\phi(x) = \frac{qN_a}{\varepsilon}\left(\frac{x^2}{2} + x_{p,0}x\right) + C_2 \tag{5.39}$$

Since the electrostatic potential is zero at $x = -x_{p,o}$, we can calculate the integration constant C_2 in Equation 5.39 as follows:

$$C_2 = \frac{qN_a}{2\varepsilon}x_{p,0}^2 \tag{5.40}$$

Substituting the value of C_2 in Equation 5.39, we get

$$\phi(x) = \frac{qN_a}{2\varepsilon}(x + x_{p,0})^2 \quad (-x_{p,0} \le x \le 0) \tag{5.41}$$

We will now calculate the electrostatic potential (ϕ) on the n-side of the depletion layer by integrating the electric field from Equation 5.34 as follows:

$$\phi(x) = -\int \left(-\frac{qN_d}{\varepsilon}\right)(x_{n,0} - x)\, dx \tag{5.42}$$

Therefore,

$$\phi(x) = \left(\frac{qN_d}{\varepsilon}\right)\left(x_{n,0}x - \frac{x^2}{2}\right) + C_3 \tag{5.43}$$

The electrostatic potential is continuous across the p-n junction, that is, at $x = 0$, the value of potential (ϕ) can also be calculated using Equation 5.41 and is equal to the value given by Equation 5.43. Therefore, we evaluate the integration constant C_3 using the following equation:

$$C_3 = \left(\frac{qN_a}{2\varepsilon}\right)x_{p,0}^2 \tag{5.44}$$

Substituting this value of C_3 into Equation 5.43, we get

$$\phi(x) = \left(\frac{qN_d}{\varepsilon}\right)\left(x_{n,0}x - \frac{x^2}{2}\right) + \frac{qN_a}{2\varepsilon}x_{p,0}^2 \quad (0 \leq x \leq x_{n,0}) \tag{5.45}$$

This variation in the electrostatic potential across the p-n junction is shown in Figure 5.9 (Neaman 2006).

The value of the contact potential (V_0) can be calculated by evaluating this equation at $x = x_{n,0}$. Therefore,

$$V_0 = \phi(x = x_{n,0}) = \frac{q}{2\varepsilon}\left[N_d x_{n,0}^2 + N_a x_{p,0}^2\right] \tag{5.46}$$

Note that the electrostatic potential (ϕ; unit is volts) and the electron energy (unit is electron volts) shown on the band diagram are related by the following equation:

$$\text{Electron energy} = -q \times \phi \tag{5.47}$$

The electrostatic potential is higher on the n-side (Figure 5.9), which means that the electron energy is lower on the n-side. One way to visualize the higher electrostatic potential on the n-side is to note that this is the side of the p-n junction with positively charged donor ions left behind (Figure 5.1).

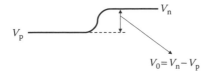

FIGURE 5.9 Electrostatic potential variation across the p-n junction.

5.8 WIDTH OF THE DEPLETION REGION AND PENETRATION DEPTHS

The semiconductor p-n junction is electrically neutral as a whole (Equations 5.28 and 5.29). Substituting for x_p from Equation 5.29 into Equation 5.46 and solving for x_n, we get

$$x_n = \left[\left(\frac{2\varepsilon V_0}{q} \right) \left(\frac{N_a}{N_d} \right) \left(\frac{1}{N_a + N_d} \right) \right]^{1/2} \tag{5.48}$$

Note that we dropped the subscript 0 in $x_{n,0}$, and that the p-n junction is not biased.

Similarly, we can solve for x_p by substituting for x_n from Equation 5.29 into Equation 5.46:

$$x_p = \left[\left(\frac{2\varepsilon V_0}{q} \right) \left(\frac{N_d}{N_a} \right) \left(\frac{1}{N_a + N_d} \right) \right]^{1/2} \tag{5.49}$$

Now, the width of the depletion layer is

$$w = x_p + x_n \tag{5.50}$$

Substituting for x_p and x_n,

$$w = \left[\left(\frac{2\varepsilon V_0}{q} \right) \left(\frac{N_a + N_d}{N_a N_d} \right) \right]^{1/2} \tag{5.51}$$

Thus, for a p-n junction with known doping levels, we can calculate the built-in potential (V_0) using Equation 5.21 or 5.46. We can calculate the depletion layer width using Equation 5.51, and the maximum electric field in a p-n junction at $x = 0$ using Equation 5.35. Examples 5.3 and 5.4 illustrate the calculation of the contact potential (V_0), the depletion layer width (w), and the maximum electric field.

EXAMPLE 5.3: CALCULATION OF THE DEPLETION REGION WIDTH

A step junction in Si is such that the n- and p-sides have a uniform dopant concentration of $N_d = 10^{16}$ and $N_a = 10^{17}$ atoms/cm³, respectively (see Examples 5.1 and 5.2). (a) On which side will the depletion layer penetration be smaller? Why? (b) Calculate the penetration depths on the n-side and p-side. (c) Calculate the total depletion layer width (w).

SOLUTION

a. For this p-n junction, the depletion layer penetration depth on the p-side will be smaller because of the relatively higher acceptor dopant concentration.
b. From Equations 5.48 and 5.49,

$$x_n = \left[\left(\frac{2 \times 11.8 \times 8.85 \times 10^{-14} \text{ F/cm} \times 0.757 \text{ V}}{1.6 \times 10^{-19} \text{ C}} \right) \left(\frac{10^{17} \text{ atoms/cm}^3}{10^{16} \text{ atoms/cm}^3} \right) \left(\frac{1}{10^{17} + 10^{16}} \right) \right]^{1/2}$$

$$x_n = 2.98 \times 10^{-5} \text{ cm} \quad \text{or} \quad 298 \text{ nm}$$

The space charge width on the p-side (x_p) is given by

$$x_p = \left[\left(\frac{2 \times 11.8 \times 8.85 \times 10^{-14}\ \text{F/cm} \times 0.755\ \text{V}}{1.6 \times 10^{-19}\ \text{C}}\right)\left(\frac{10^{16}\,\text{atoms/cm}^3}{10^{17}\,\text{atoms/cm}^3}\right)\left(\frac{1}{10^{17}+10^{16}}\right)\right]^{1/2}$$

$x_p = 2.98 \times 10^{-6}$ cm   or   0.0298 μm   or   29.8 nm

Because of the higher dopant concentration, the penetration depth on the p-side (x_p) is smaller.

c. The total width of the depletion layer (w) will be = 298 + 29.8 = 327.8 nm.

EXAMPLE 5.4: CALCULATION OF THE MAXIMUM ELECTRIC FIELD

A p-n junction in Si has $N_d = 10^{16}$ atoms/cm³ and the acceptor doping level is $N_a = 5 \times 10^{17}$ cm⁻³.
a. Calculate the built-in potential (V_0) and (b) the maximum electric field in the depletion region.

SOLUTION

a. From Equation 5.21,

$$qV_0 = k_B T\ \ln\left(\frac{N_a N_d}{n_i^2}\right) = (0.026\ \text{eV})\ \ln\left(\frac{5 \times 10^{17}\ \text{atoms/cm}^3 \times 10^{16}\,\text{atoms/cm}^3}{(1.5 \times 10^{10}\,\text{electrons/cm}^3)^2}\right) = 0.795\ \text{eV}$$

The built-in voltage (V_0) is 0.795 V.
b. The maximum electric field occurs at $x = 0$, the metallurgical boundary of the p-n junction. From Equation 5.35, we get
   When $x = 0$, we get

$$E(x = 0) = -\frac{(1.6 \times 10^{-19}\ \text{C})(10^{16}\ \text{atoms/cm}^3)x_n}{8.85 \times 10^{-14}\ \text{F/cm} \times 11.8} \tag{5.52}$$

We calculate x_n value from Equation 5.48:

$$x_n = \left[\left(\frac{2\varepsilon V_0}{q}\right)\left(\frac{N_a}{N_d}\right)\left(\frac{1}{N_a + N_d}\right)\right]^{1/2}$$

$$x_n = \left[\left(\frac{2 \times 11.8 \times 8.85 \times 10^{-14}\ \text{F/cm} \times 0.795\ \text{V}}{1.6 \times 10^{-19}\ \text{C}}\right)\left(\frac{5 \times 10^{17}\ \text{atoms/cm}^3}{10^{16}\ \text{atoms/cm}^3}\right)\left(\frac{1}{5 \times 10^{17} + 10^{16}}\right)\right]^{1/2}$$

$x_n = 3.19 \times 10^{-5}$ cm   or   0.319 μm

Therefore, the magnitude of the electric field is

$$E_{max}(x = 0) = \left|\frac{1.6 \times 10^{-19} \times 10^{16}}{11.8 \times 8.85 \times 10^{-14}}\right|(3.19 \times 10^{-5}\,\text{cm}) = 4.91 \times 10^4\ \text{V/cm}$$

$$E_{max} = 4.91 \times 10^4\ \text{V/cm}$$

This is a fairly large electric field at ~50,000 V/cm. The depletion region has almost no free electrons or holes, and therefore there is very little drift current despite this large built-in electric field.

5.9   REVERSE-BIASED p-n JUNCTION

The real utility of the p-n junction is that its ability to conduct can be changed with doping and by application of an external voltage. Consider a reverse-biased p-n junction. This means that we apply the n-side of the junction to the positive terminal of an external direct current (DC) voltage supply (Figure 5.10).

Applying a reverse bias, that is, connecting the n-side (which has the positive space-charge region) to the positive terminal, causes the electrostatic potential on the n-side (V_n) to increase (Figure 5.10b).

Recall that electron energy is related to electrostatic potential by $-q$. Thus, with a reverse bias, the Fermi energy level on the n-side is decreased. The band diagram for a reverse-biased p-n junction is shown in Figure 5.10c.

The internal electric field and the external electric field follow the same direction, from the n-side to the p-side. The depletion layer width (w) for a reverse-biased junction is higher than that for a p-n junction with no bias (Figure 5.11a).

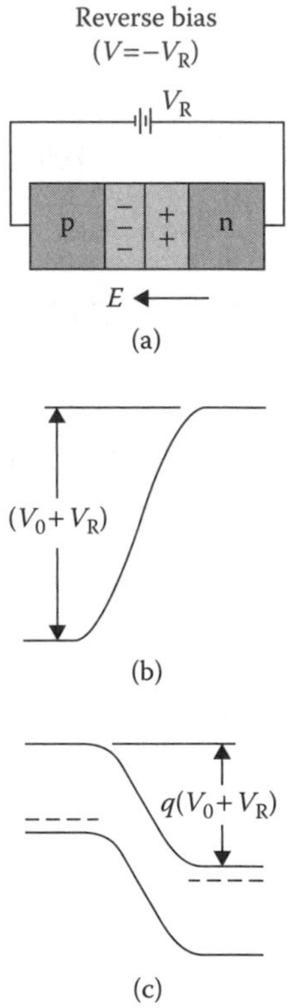

Electrons attempting to diffuse from the n-side to the p-side under a concentration gradient will encounter a larger energy barrier of height $q(V_0 + V_R)$, which can be visualized easily on a band diagram (Figure 5.10c). The magnitude of the diffusion current due to the diffusion of electrons from the n-side to the p-side becomes very small. Similarly, holes attempting to diffuse from the p-side to the n-side, again under a concentration gradient, will encounter a larger potential barrier that can be easily visualized on the electrostatic potential diagram (Figure 5.10b). The magnitude of the total diffusion current due to the movement of holes and electrons is negligible because of the increased energy barrier (Figure 5.11c).

We can calculate the width of the depletion region (w) and the penetration depths in the p- and n-regions of a p-n junction by substituting ($V_R + V_0$) for V_0 in Equations 5.48, 5.49, and 5.51 derived in Section 5.8.

Therefore, for a reverse-biased p-n junction,

Reverse bias
$(V = -V_R)$

(a)

$(V_0 + V_R)$

(b)

$q(V_0 + V_R)$

(c)

FIGURE 5.10 (a) A reverse-biased p-n junction showing the directions of built-in and applied electric fields, (b) the variation in the electrostatic potential and (c) a band diagram. (From Streetman, B. G., and S. Banerjee. 2000. *Solid State Electronic Devices*, 5th ed. Upper Saddle River, NJ: Prentice Hall. With permission.)

$$x_n = \left[\left(\frac{2\varepsilon(V_0 + V_R)}{q} \right) \left(\frac{N_a}{N_d} \right) \left(\frac{1}{N_a + N_d} \right) \right]^{1/2} \tag{5.53}$$

$$x_p = \left[\left(\frac{2\varepsilon(V_0 + V_R)}{q} \right) \left(\frac{N_d}{N_a} \right) \left(\frac{1}{N_a + N_d} \right) \right]^{1/2} \tag{5.54}$$

The depletion layer width under the reverse bias can be calculated by adding x_n and x_p values or by using the following equation:

$$w = \left[\left(\frac{2\varepsilon(V_0 + V_R)}{q} \right) \left(\frac{N_a + N_d}{N_a + N_d} \right) \right]^{1/2} \tag{5.55}$$

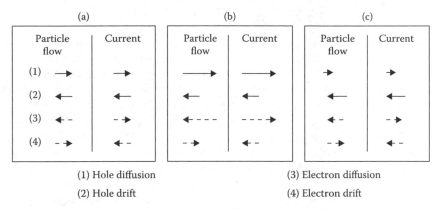

(1) Hole diffusion   (3) Electron diffusion

(2) Hole drift   (4) Electron drift

FIGURE 5.11   The directions of particle motion for the diffusion and drift of electrons and their corresponding currents for p-n junction with (a) zero bias, (b) forward bias, and (c) reverse bias. (From Askeland, D., and P. Fulay. 2006. *The Science and Engineering of Materials*. Washington, DC: Thomson. With permission.)

5.10   DIFFUSION CURRENTS IN A FORWARD-BIASED P-N JUNCTION

In a forward-biased p-n junction, the p-side is connected to the positive terminal of an external power supply. This increases the electrostatic potential on the p-side. Due to the built-in potential, the applied electric field will now oppose the internal electric field (Figure 5.11a). This means that the total height of the potential barrier will be reduced (Figure 5.12b).

Electrons, the majority carriers on the n-side, can now diffuse more easily to the p-side. In the p-region, these injected electrons move toward the positive terminal, where they are collected. While making their way through the p-region, some of the electrons will recombine with the holes. The positive terminal of the power supply compensates for the holes lost as a result of the recombination. The current due to the electron diffusion is thus maintained by the electrons from the n-side. The negative terminal of the battery provides these electrons.

Similarly, we can see on the electrostatic potential diagram (Figure 5.12c) that the energy barrier for hole diffusion is reduced. This means that more holes, the majority carriers on the p-side, diffuse onto the n-side. The holes injected in the n-side diffuse toward the negative terminal of the power supply. Some holes will end up recombining with the electrons on the n-side. The negative terminal of the power supply replaces the electrons lost to the recombination. The current is sustained as more holes continue to diffuse over to the n-side (Figure 5.11b). The positive terminal of the power supply supplies these holes.

The motion of the majority carriers produces a *forward current* (I_F). Under a sufficient forward bias, the total diffusion current increases and the p-n junction begins to conduct. The depletion layer width (w) decreases under a forward bias because the net electric field in the depletion layer is decreased (Figure 5.11b). We can obtain the new values of the penetration depths and the depletion layer width by substituting V_0 by ($V_0 - V_f$) in Equations 5.48, 5.49, and 5.51 for these quantities.

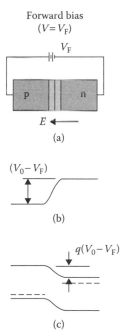

FIGURE 5.12 (a) The bias direction for a forward-biased p-n junction, (b) electrostatic potential variation, and (c) a band diagram. (From Streetman, B. G., and S. Banerjee. 2000. *Solid State Electronic Devices*, 5th ed. Upper Saddle River, NJ: Prentice Hall. With permission.)

5.11   DRIFT CURRENTS IN A p-n JUNCTION

In a p-n junction at room temperature, there are thermally-created electron–hole pairs (EHPs) in both the neutral regions and the depletion regions of the p-n junction. The thermally generated carriers in the *neutral region* first diffuse to the depletion region under a concentration gradient. For example, there are thermally generated holes in the n-side neutral region ($x > x_n$). However, the concentration of holes near the space-charge region is essentially zero at $x = x_n$. Thus, the holes flow from the neutral region to the boundary of the depletion layer due to the concentration gradient. These holes then drift toward the n-side under the influence of the electric field present in the depletion region. The *hole diffusion length* (L_p) is the distance that thermally generated holes on the neutral n-side can travel from the transition region to the n-side through the electric field in the depletion region. Similarly, the *electron diffusion length* (L_n) is the average distance that an electron can travel before recombining with a hole. Carriers generated within the length of the diffusion distance will successfully pass to the depletion layer without recombination. Holes or electrons generated at a distance larger than the diffusion distance (for that carrier) will most likely recombine and thus will not contribute to this so-called "reverse current." All thermally generated carriers therefore do not end up contributing to generation current.

The first source for the drift current (I_{Shockley}) is induced by the diffusion of thermally generated *minority carriers* in the neutral region, followed by their drift across the depletion region. This current is given by the Shockley equation:

$$I_{\text{Shockley}} = \left[\left(\frac{qD_p}{L_p N_d} \right) + \left(\frac{qD_n}{L_n N_a} \right) \right] n_i^2 \tag{5.56}$$

In the Shockley equation, q is the magnitude of the charge on the electron or hole, D is the diffusion coefficient, L is the diffusion length, and subscripts p and n refer to the holes and electrons, respectively. Recall from Chapter 4 that the intrinsic carrier concentration (n_i) increases exponentially with the temperature and is inversely related to the bandgap (E_g).

The second source for the drift current is from the thermally generated carriers *inside* the depletion region. The internal electric field separates these carriers, and they drift toward the neutral regions. This aspect of the reverse current is given by

$$I_{\text{generation,depletion}} = \frac{qWn_i}{\tau_g} \tag{5.57}$$

In Equation 5.57, τ_g is the *mean thermal generation time*. This is the average time needed to thermally create an EHP. The total current due to the drift of thermally generated minority carriers, whether created in the neutral region or in the space-charge region, is called the *generation current* (I_0).

Thus, the total generation current (I_0) is given by combining Equations 5.56 and 5.57:

$$I_0 = \left[\left(\frac{qD_p}{L_p N_d} \right) + \left(\frac{qD_n}{L_n N_a} \right) \right] n_i^2 + \frac{qW_i}{\tau_g} n_i \tag{5.58}$$

From Equation 5.58, we can see that either the first term with n_i^2 or the second, thermal-generation term with n_i is controlling the generation current. In Figure 5.13, the reverse current for a germanium (Ge) diode is shown. In this case, we show a *dark current* to avoid any current due to the photogeneration of EHPs.

The total generation current is controlled at low temperatures (<240 K) by the second term, thermal generation. At lower temperatures, the slope of the line is proportional to $E_g/2$ because $n_i \propto \exp(-E_g/2kT)$. At higher temperatures, the diffusion term dominates, and the current is

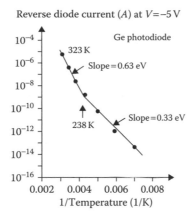

Reverse diode current (A) at $V = -5$ V

FIGURE 5.13 Generation current (I_0) for a germanium diode with $1/T$. The effect of photogenerated carriers is excluded, i.e., this is a dark current. (From Kasap, S. O. 2002. *Principles of Electronic Materials and Devices*. New York: McGraw Hill. With permission.)

controlled by the n_i^2 term. The slope of the line shown in Figure 5.13 is nearly proportional to the bandgap of the semiconductor.

The *generation current* (I_0) is *independent* of the applied bias; this is because the electric field in the depletion region affects how rapidly the carriers will overcome the potential barrier, but not the total amount of electrical charge moving across per unit of time. Although the forward current (I_F), which is due to the diffusion of majority carriers, is strongly affected by the type and amount of bias, the magnitude of the generation current (I_0) is independent of the applied voltage (Figure 5.11).

The generation current depends upon the temperature because the carrier concentration due to thermal excitation also depends on the temperature. Similarly, the absorption of light energy leads to the creation of EHPs. This also causes an increase in the generation current.

Thus, the total current (I) flowing through the p-n junction is given by

$$I = I_{\text{generation}} + I_{\text{diff}} \tag{5.59}$$

When there is no applied voltage (Figure 5.11a), the total drift current and the current due to the diffusion of majority carriers are equal and opposite, that is,

$$\left| I_{\text{diff}} \right| = \left| I_{\text{generation}} \right| \quad \text{for} \quad V = 0 \tag{5.60}$$

When the p-n junction is reverse-biased (Figures 5.11c and 5.14c), the diffusion of majority carriers is almost negligible, and the total current is equal to the generation current.

$$\left| I_{\text{generation}} \right| = I_0 \quad \text{for reverse bias} \tag{5.61}$$

Therefore, the generation current (I_0) is also known as the *reverse-bias saturation current* (I_s). The generation current is usually in the range of $10^{-14} - 10^{-12}$ A. The magnitude of the generation current depends upon its doping levels, the size of the cross-sectional area of the junction, temperature, and so on (Equation 5.58). It does not depend on the reverse-applied voltage.

For a p-n junction under a forward bias (Figures 5.11b and 5.14b), the diffusion current ($I_{\text{diffusion}}$) changes exponentially with the applied voltage and is given by

$$I_{\text{diffusion}} = I_0 \left(\frac{qV}{kT} \right) \tag{5.62}$$

Thus, the total current under a forward bias is given by the difference between the diffusion currents:

$$I = I_0 \left[\exp\left(\frac{qV}{k_B T} \right) - 1 \right] \tag{5.63}$$

Equation 5.63 is also known as the *ideal diode equation*. The current–voltage (*I–V*) curve for a p-n junction is shown in Figure 5.17.

If the applied bias (*V*) is large compared to the magnitude of ($k_B T/q$), then the exponential term will be much larger than 1, and Equation 5.63 can be rewritten as

$$I \approx I_0 \exp\left(\frac{V}{(q/k_B T)} \right) \tag{5.64}$$

At room temperature (*T* = 300 K), the value of ($q/k_B T$) is ~0.026 V, hence

$$I \approx I_0 \exp\left(\frac{V(\text{in V})}{(0.026 \text{ V})} \right) \quad \text{at} \quad T = 300 \text{ K} \tag{5.65}$$

The p-n junction behaves like a one-way electrical valve. It functions as a rectifier by allowing current to flow through under a forward bias but not under a reverse bias.

A summary of p-n junction band diagrams and changes in the electric field in the depletion layer width, as well as electrostatic potential variation under zero, forward, and reverse bias, is shown in Figure 5.14.

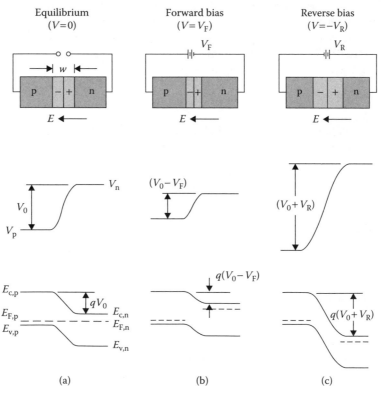

FIGURE 5.14 Summary of band diagrams and the changes in the electric field, electrostatic potential, and depletion layer width of a p-n junction under (a) zero, (b) forward, and (c) reverse bias. (From Streetman, B. G., and S. Banerjee. 2000. *Solid State Electronic Devices*, 5th ed. Upper Saddle River, NJ: Prentice Hall. With permission.)

5.12 DIODE BASED ON A p-n JUNCTION

The electrical symbol for a diode is shown in Figure 5.15.

A diode can serve several useful functions, including rectification. For an alternating current voltage (AC) input, the diode will produce the output voltage shown in Figure 5.16.

Two diodes can be used to create a full-wave rectification of AC voltage. The p-n junction is also used as the basic building block for solar cells and devices known as transistors. Millions of transistors are connected to form miniature circuits, which are then used to create semiconductor chips that are used in computers and other electronic equipment.

An actual *I–V* characteristic curve for a p-n junction is shown in Figure 5.17. Note that under a forward bias, the current (*I*) is in

FIGURE 5.15 Symbol for a p-n junction diode in an electrical circuit.

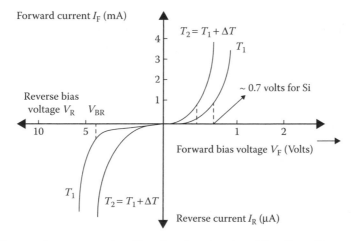

FIGURE 5.16 Half-wave rectification using a diode.

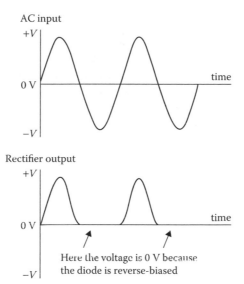

FIGURE 5.17 I-V curves for representative silicon based p-n junctions. The effect of increased temperature on lowering of breakdown and knee voltage is shown. Note the current scale for forward and reverse bias is in milli- and micro-amperes, respectively. Similarly, the voltage scales for forward and reverse bias are different. (From Floyd, T. 2006. *Electric Circuit Fundamentals*, 7th ed. Upper Saddle River, NJ: Prentice Hall. With permission.)

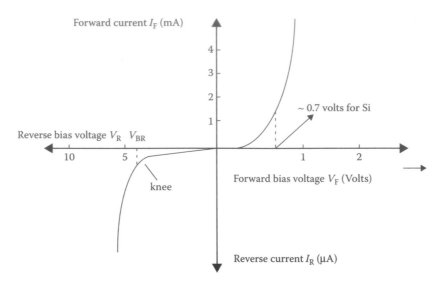

FIGURE 5.18 I-V curve for a representative silicon based p-n junctions. Note the current scale for forward and reverse bias is in milli and micro-amperes, respectively. Similarly, the voltage scales for forward and reverse bias are different. (From Floyd, T. 1998. *Electronics Fundamentals: Circuits, Devices, and Applications*, 4th ed. Upper Saddle River, NJ: Prentice Hall. With permission.)

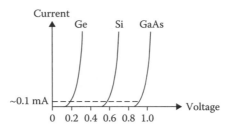

FIGURE 5.19 Schematic of *I–V* curves for different semiconductors showing different expected knee voltages (current axis not to scale). (From Kasap, S. O. 2002. *Principles of Electronic Materials and Devices*. New York: McGraw Hill. With permission.)

milliamperes (mA). Under a reverse bias, the current is in microamperes (μA). When the forward-bias voltage is less than this *knee voltage* (a voltage close to the built-in potential V_0), a small current is set up across the p-n junction. As the voltage increases, the current also increases. However, this increase is nonlinear; that is, it does not follow Ohm's law (Figure 5.18).

For silicon (Si), the knee voltage is ~0.7 V. Recall that this is the typical built-in potential (V_0) for silicon p-n junctions (Example 5.1). As the temperature increases from 25°C to $25 + \Delta T$, the knee voltage decreases slightly. At an elevated temperature the forward current (I_F) at any given voltage is higher than that at 25°C. As expected, more minority carriers are generated on both sides of the p-n junction at higher temperatures. There is thus a very slight increase in the reverse-bias current at higher temperatures. The sketch in Figure 5.17 is not to scale and exaggerates this difference in the change in the magnitude of the generation current.

For germanium-based p-n junctions, the knee voltage is ~0.3 V. A schematic of *I–V* curves expected for p-n junctions for different semiconductors are shown in Figure 5.19.

EXAMPLE 5.5: CURRENT IN AN Sɪ ᴘ-ɴ JUNCTION DIODE

The built-in potential (V_0) of an Si p-n junction diode is 0.8 V. Assume that the reverse-bias saturation current (I_0) is 10^{-13} amperes (A). A forward bias of 0.5 V is then applied.

a. What is the new height of the potential barrier seen on the electrostatic potential diagram?
b. What is the current flowing through this diode with a forward bias of 0.5 V?
c. What is the current if the forward bias changes to 0.6 V?

Compare the different current values with each other.

SOLUTION

a. When the applied voltage is 0.5 V, the new height of the barrier seen on the electrostatic potential will be $(V_0 - V_F) = (0.8 - 0.5) = 0.3$ V (Figure 5.14b).
b. Since the forward bias applied ($V = 0.5$ V) is much larger than kT/q (0.0259 V), that is, $qV \gg kT$, the exponential term is much larger than 1, and we use Equation 5.65. Therefore,

$$I_F \approx (10^{-13}\,\text{A})\exp\left(\frac{0.5\ \text{V}}{0.0259\ \text{V}}\right) = 2.42 \times 10^{-5}\,\text{A}$$

This is about 24 μA.
c. When the forward bias is 0.6 V, the current is given by

$$I_F \approx (10^{-13}\,\text{A})\exp\left(\frac{0.6\ \text{V}}{0.0259\ \text{V}}\right) = 1.15 \times 10^{-3}\,\text{A}$$

This current of about 1 mA is about 40 times greater than the current when the bias of 0.5 V was applied. This shows that the effective resistance of the p-n junction drops by ~40 times with a bias increase of only 0.1 V.

The p-n junction under forward bias has a *dynamic resistance* (r_d') that changes with the applied voltage (Figure 5.20). Most materials (e.g. metals, alloys, semiconductors, and so on) show a resistance that is constant and is not voltage-dependent.

The dynamic resistance of a p-n junction can be calculated from the slope of the I–V curve at a given value of voltage:

$$r_d' = \frac{dV_{\text{applied}}}{dI_F} \tag{5.66}$$

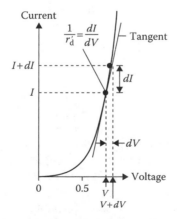

FIGURE 5.20   A graphical illustration of the dynamic resistance of a p-n junction. (From Kasap, S. O. 2002. *Principles of Electronic Materials and Devices.* New York: McGraw Hill. With permission.)

A p-n junction circuit uses a resistor (R_L), known as the limiting resistor, to limit the total current and protect the p-n junction from damage due to an excessive current. When such a resistor is used in series with the diode, the forward current is given by

$$I_F = \frac{(V_{applied} - V_{knee})}{R_L} \tag{5.67}$$

If a limiting resistor is not used, the p-n junction can be damaged by Joule heating.

If the dynamic resistance of the p-n junction is accounted for, then the forward current is given by

$$I_F = \frac{(V_{applied} - V_{knee})}{(R_L + r'_d)} \tag{5.68}$$

The application of this equation is illustrated in Example 5.6.

EXAMPLE 5.6: FORWARD CURRENT IN A DIODE

An Si p-n junction functions as a diode and is connected to a resistor $R_L = 1000\ \Omega$. Assume that the knee voltage is 0.7 V.

a. What is the forward current if a 4.5 V forward bias is applied?
b. What is the forward current, assuming that the diode offers a dynamic resistance of 10 Ω at the selected I_F value?

SOLUTION

a. From Equation 5.67,

$$I_F = \frac{(4.5 - 0.7)\ V}{1000\ \Omega} = 3.8\ mA$$

b. When the dynamic resistance of the diode (r'_d) must be accounted for, we add that resistance to the R_{limit}.
   From Equation 5.68,

$$I_F = \frac{(4.5 - 0.7)\ V}{(1000 + 10)\ \Omega} = 3.762\ mA$$

As can be expected by adding the dynamic resistance, the magnitude of I_F decreases.

5.13   REVERSE-BIAS BREAKDOWN

A reverse bias is sometimes so high that it causes a *reverse-bias dielectric breakdown* of the p-n junction (Figure 5.21). When this happens, the current flowing through the p-n junction increases rapidly. Critical breakdown voltages (V_{br}) for silicon diodes generally start at ~60 V. The term "breakdown" is a bit misleading; if the p-n junction breaks down electrically, this does not mean that the device is permanently damaged. The breakdown of a p-n junction, which should be distinguished from a broken or defective p-n junction, is useful in some applications.

A limiting resistor (R_L) is used to limit the flow of current during a reverse-bias breakdown. If such a current-limiting resistor is not used, the p-n junction will be damaged because of overheating caused by the excessive current. This will occur in both forward and reverse bias.

The reverse-bias breakdown can occur by two mechanisms. In the *avalanche breakdown* mechanism, a high-energy conduction electron collides with a silicon–silicon bond and knocks off a

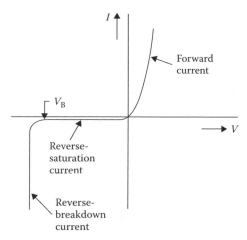

FIGURE 5.21 *I–V* characteristic of p-n junction including a reverse-bias breakdown. (From Mahajan, S., and K. S. Sree Harsha. 1998. *Principles of Growth and Processing of Semiconductors*. New York: McGraw Hill. With permission.)

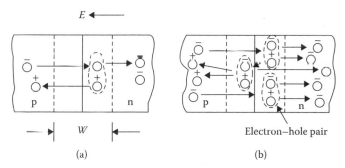

FIGURE 5.22 Illustration of the impact ionization process. (a) A single ionizing collision with a host lattice atom by an incoming electron and (b) multiplication of carriers due to multiple collisions. (From Mahajan, S., and K. S. Sree Harsha. 1998. *Principles of Growth and Processing of Semiconductors*. New York: McGraw Hill. With permission.)

valence electron from that bond, which makes this electron free. This process is known as *impact ionization* (Figure 5.22).

From a band diagram viewpoint, the impact ionization process sends an electron from the valence band to the conduction band, creating a hole in the valence band. The first high-energy electron continues, knocking off another electron from one more silicon–silicon covalent bond and continuing the process. The impact ionization process is shown on the band diagram in Figure 5.23.

In the mechanism known as an *avalanche breakdown*, multiple collisions from a single high-energy electron can create many EHPs.

A p-n junction with one side heavily doped is known as a one-sided junction. If the p-side is heavily doped, we refer to this as a p^+-n junction. The depletion layer of a p^+-n is mainly on the n-side. The charge density distribution for this junction is shown in Figure 5.24. When this junction breaks down electrically, the breakdown will occur on the n-side at a location where the electric field is at a maximum.

Figure 5.24 shows the critical breakdown field as a function of background doping for a *one-sided p-n junction*, in which one of the sides is very heavily doped. If the p-side is heavily doped, the junction is referred to as p^+-n; if the n-side is heavily doped, the junction is shown as p-n^+. The charge density expected for a p^+-n junction is shown in Figure 5.25.

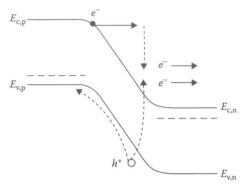

FIGURE 5.23 Impact ionization process illustrated on a band diagram. (From Streetman, B. G., and S. Banerjee. 2000. *Solid State Electronic Devices*. Upper Saddle River, NJ: Prentice Hall. With permission.)

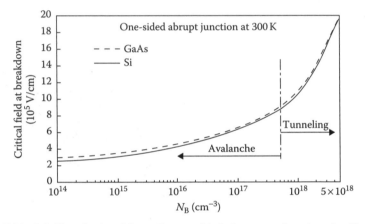

FIGURE 5.24 Critical field at the breakdown for one-sided abrupt p-n junctions in silicon and GaAs (at $T = 300$ K). (From Mahajan, S., and K. S. Sree Harsha. 1998. *Principles of Growth and Processing of Semiconductors*. New York: McGraw Hill. With permission.)

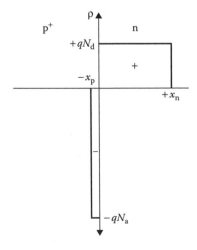

FIGURE 5.25 Schematic of the charge density for a p^+-n junction. (From Neaman, D. 2006. *An Introduction to Semiconductor Devices*. New York: McGraw Hill. With permission.)

The breakdown voltage for a p⁺-n junction is given by

$$V_{\text{BD}} = \frac{\varepsilon E_{\text{critical}}^2}{2qN_d} \tag{5.69}$$

where N_d is the donor dopant concentration and E_{critical} is the critical field that causes breakdown (Singh 2001). The breakdown field for a p⁺-n junction will depend upon the level of doping. This change for the silicon and gallium arsenide (GaAs) p-n junctions is shown in Figure 5.24.

Example 5.7 shows the calculation of the breakdown voltage by avalanche breakdown mechanism.

EXAMPLE 5.7: BREAKDOWN VOLTAGE FOR A p-n Si DIODE

In an Si diode, $N_a = 10^{19}$ atoms/cm³ and $N_d = 5 \times 10^{15}$ atoms/cm³. If the critical breakdown field is 4×10^5 V/cm, what is the breakdown voltage for this diode at $T = 300$ K?

SOLUTION

The p-side is heavily doped, thus the breakdown will occur in the depletion layer that extends to the n-side. We apply Equation 5.69:

$$V_{\text{br}} = \frac{(11.8)(8.85 \times 10^{-14}\,\text{F/cm})(4 \times 10^5\,\text{V/cm})^2}{2(1.6 \times 10^{-19}\,\text{C})(5 \times 10^{15}\,\text{atoms/cm}^3)} = 103.5\,\text{V}$$

For diodes that have a higher breakdown field, the breakdown voltage will be higher. Thus, such diodes as those made from silicon carbide (SiC), which has a high breakdown voltage, are useful for high-temperature, high-power applications.

As the doping level increases, another mechanism of breakdown can be applied. In the *Zener tunneling mechanism*, electrons tunnel across the p-n junction instead of climbing over the energy barrier. This is commonly seen in heavily doped p-n junctions. The breakdown occurs across very thin depletion regions and thus typically occurs at low voltages (~0.1–5 V). Heavier doping means a smaller depletion region width (Equations 5.48 and 5.49). With Zener tunneling, electrons from the p-side valence band can tunnel across into the conduction band on the n-side because the reverse bias pushes down the E_c on the n-side and aligns it with the valence bandedge (E_v) on the p-side. The Zener tunneling mechanism is also known as the Zener effect or band-to-band tunneling (Figure 5.26).

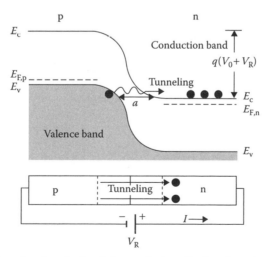

FIGURE 5.26 Zener effect showing the breakdown of a heavily doped p-n junction. (From Kasap, S. O. 2002. *Principles of Electronic Materials and Devices*. New York: McGraw Hill. With permission.)

The probability of Zener tunneling (T) is given by

$$T \approx \exp\left(-\frac{4\sqrt{2m_e^*}\,E_g^{3/2}}{21q\hbar E}\right) \tag{5.70}$$

where m_e^* is the reduced effective mass of the electron, E_g is the bandgap, and E is electric field (Singh 2001). The Zener breakdown is important in heavily doped junctions and in narrow-bandgap materials. The value of probability (T) generally needs to be ~10^{-6} for tunneling to initiate this breakdown process. Example 5.8 illustrates the calculation of the tunneling probability.

EXAMPLE 5.8: ZENER TUNNELING IN InAs

Calculate the probability of tunneling in indium arsenide (InAs) if the applied electric field (E) is 3×10^5 V/cm. Assure that the m_e^* for electrons in InAs is 0.02 m_0^*. The bandgap of InAs is 0.4 eV.

SOLUTION

From Equation 5.70, the probability of tunneling is

$$T = \exp\left(\frac{-4\sqrt{2(0.02 \times 9.1 \times 10^{-31}\text{ kg})}(0.4\text{ eV} \times 1.6 \times 10^{-19}\text{ J/eV})^{3/2}}{3(1.6 \times 10^{-19}\text{ C})(1.05 \times 10^{-34}\text{ J·s})(3 \times 10^4\text{ V/cm} \times 10^2\text{ cm/m})}\right)$$

Note the use of SI units, including the conversion of the electric field from V/cm to V/m. $T = 2.82 \times 10^{-4}$, which is greater than 10^{-6}. Therefore, Zener tunneling is important for this value of the electric field in InAs.

5.14 ZENER DIODES

The breakdown by either Zener or avalanche mechanism is the basis for the *Zener diode* (Figure 5.27a), a device that is typically operated under a reverse bias. The *I–V* curve for a Zener diode including the reverse-breakdown region is shown in Figure 5.27b. The normal operating region for a regular p-n junction diode is also shown in this figure for comparison (Figure 5.27c).

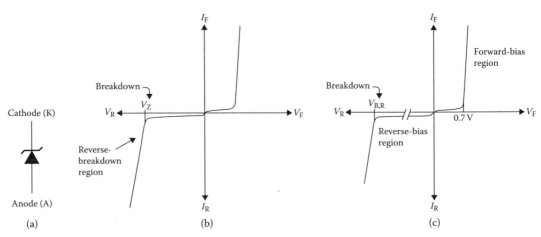

FIGURE 5.27 (a) Symbol for a Zener diode. The anode (+) is on the left and cathode (−) is on the right. The positive terminal of the power supply is connected to the cathode of the Zener diode to operate under a reverse bias. (b) Typical *I–V* curve for a Zener diode. (c) Normal operating region for a regular p-n junction diode. (From Floyd, T. 1998. *Electronics Fundamentals: Circuits, Devices, and Applications*, 4th ed. Upper Saddle River, NJ: Prentice Hall. With permission.)

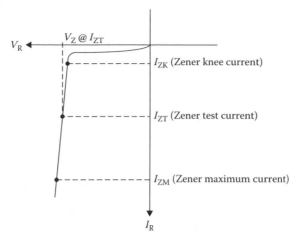

FIGURE 5.28   Close-up of a reverse-bias portion of the *I–V* curve for a Zener diode. (From Floyd, T. 1998. *Electronics Fundamentals: Circuits, Devices, and Applications*, 4th ed. Upper Saddle River, NJ: Prentice Hall. With permission.)

Since the Zener diode has an avalanche or Zener breakdown after a certain applied voltage is reached, the voltage across this diode remains essentially constant. Therefore, Zener diodes are used as voltage regulators (Example 5.9). Under a forward bias, the Zener diode will function like a regular diode, with the typical knee voltage of ~0.7 V for silicon diodes.

The voltage at which a Zener diode will begin to conduct current under a reverse bias is known as the *Zener voltage*. Diodes can be designed to have a Zener voltage ranging from a few volts to a few hundred volts.

In practice, the Zener *I–V* curve is not completely vertical (Figure 5.28). We define ΔV_Z as the change in voltage across the reverse-biased Zener diode as the current changes from a Zener test current (I_{ZT}) to a higher value, up to the Zener maximum current (I_{ZM}). The Zener diode impedance or resistance (Z_Z) is given by

$$Z_Z = \frac{\Delta V_Z}{\Delta I_Z} \tag{5.71}$$

In Equation 5.71, $\Delta I_Z = I_{ZM} - I_{ZT}$.

Example 5.9 shows how a Zener diode under a reverse bias functions as a voltage regulator.

EXAMPLE 5.9: ZENER DIODE VOLTAGE REGULATOR

Consider a voltage regulator supply connected to a 200 Ω current-limiting resistance (R_L) and a reverse-biased Zener diode. The knee current (I_{ZK}; Figure 5.28) is 0.2 mA and the maximum current (I_{ZM}) is 100 mA. Assume that the voltage across the Zener diode is always 10 V. What is the range of input voltages that can be regulated using this Zener diode?

SOLUTION

The job of this circuit (Figure 5.29) is to provide a 10 V constant output even though the input voltage of the variable power supply will change. We need to find the range of voltages that can be controlled using this Zener diode and the 200 Ω resistor.

If the applied voltage is too low, the Zener diode will not conduct. When the applied voltage is just above the Zener voltage, the current is close to the knee current of 0.2 mA. This current

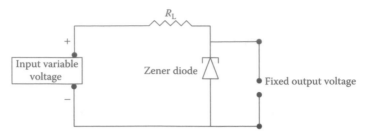

FIGURE 5.29   A Zener diode as a voltage regulator (see Example 5.9).

flows through the Zener diode and the 200 Ω resistor. The voltage drop across the resistor is given by Ohm's law:

$$V_{resistor} = I \times R = I_{ZK} \times R = 0.2 \text{ mA} \times 200 \text{ } \Omega = 40 \text{ mV}$$

The voltage drop across the Zener diode is always 10 V. The total input voltage will be 10 V + 0.04 V = 10.04 V. This is the minimum input voltage that can be controlled. If the applied voltage is lower than this, then the Zener diode will not conduct.

   If we now increase the voltage beyond 10.04 V, the Zener diode will conduct a current. The maximum current this diode is designed for is 100 mA. The voltage drop across the resistor for this current is given by

$$V_{resistor} = I \times R = I_{ZM} \times R = 100 \text{ mA} \times 200 \text{ } \Omega = 20 \text{ V}$$

Once again, the voltage drop across the Zener diode is only 10 V. Under these conditions, the total voltage must be = 20 + 10 = 30 V.

   This combination of a Zener diode and a current-limiting resistor would provide a constant 10 V output even as the input voltage changes between 10.04 and 30 V. If the voltage exceeds 30 V, it exceeds the maximum allowed current of 100 mA and will damage the Zener diode.

5.15   PROBLEMS

5.1   A p-n junction is formed in Ge so that the donor and acceptor concentrations are 10^{16} and 10^{17} atoms/cm^3, respectively. Ge has a small bandgap ($E_g \sim 0.67$ eV). Based on this, do you expect the built-in potential to be smaller or greater than Si ($E_g \sim 1.1$ eV)? What is the built-in potential (V_0) for this p-n junction? Use the intrinsic carrier concentrations for Ge from Figure 3.7 and assume $T = 300$ K.

5.2   A p-n junction is formed in GaAs so that the donor and acceptor concentrations are 10^{16} and 10^{17} atoms/cm^3, respectively. Show that the built-in potential for this p-n junction is 1.22 V. Use the intrinsic carrier concentrations for GaAs from Figure 3.7 and assume that $T = 300$ K.

5.3   A p-n junction is such that the acceptor and donor dopant levels are 0.16 eV from the nearest bandedge. The doping level on both sides is 10^{16} atoms/cm^3. What is the built-in potential for this junction?

5.4   An Si p-n junction at 300 K is P-doped on one side such that $N_d = 5 \times 10^{16}$ atoms/cm^3. On the acceptor side, the B dopant level is 10^{16} atoms/cm^3. Calculate the contact potential, maximum electric field, width of the depletion region, and the penetration depths on each side. It may be easier to set up a spreadsheet to solve this and other problems.

5.5   For the junction described in Example 5.3, calculate the maximum electric field in V/cm.

5.6   For the p-n junction discussed in Equation 5.4, calculate the value of the penetration depth on the p-side and the width of the depletion layer (w).

5.7 What is the forward current in a diode as discussed in Example 5.6, if the forward bias is 0.4 and 0.7 V?

5.8 What is the breakdown field for an Si $p^+ = n$ junction where the $N_d = 5 \times 10^{16}$ atoms/cm^3? Assume that the p-side is very heavily doped and the breakdown therefore occurs on the n-side. Use the data in Figure 5.25.

5.9 What is the breakdown voltage for the diode discussed in Example 5.8, if $N_d = 2 \times 10^{17}$ atoms/cm^3?

5.10 What is the breakdown voltage for a diode in a diamond p-n junction? Assume that the breakdown field is 10^7 V/cm and $N_d = 10^{16}$ atoms/cm^3. How does this value compare with the breakdown voltage for a similar diode made from Si?

5.11 What donor doping level will be needed for a Si p^+-n diode, if the breakdown voltage is 30 V? Use the data in Figure 5.25 and assume that the p-side is heavily doped so that the breakdown occurs on the n-side.

5.12 What is the minimum voltage that can be regulated using a Zener diode with a knee current (I_{ZK}) of 2 mA and a V_Z of 15 V?

5.13 True or false: The Zener diode operates on a breakdown mechanism by avalanche or tunneling. Explain.

GLOSSARY

Avalanche mechanism: A high-voltage breakdown mechanism occurring in diodes in which a conduction electron with high energy scatters a valence electron and transfers it into the conduction band, creating a hole in the valence band. This is different from the Zener mechanism, which occurs at lower voltages.

Band-to-band tunneling: A tunneling mechanism in which electrons from the valence band of the p-side tunnel flow into the conduction band on the n-side. See also **Zener tunneling**.

Built-in electric field: See **Internal electric field**.

Built-in potential (V_0): See **Contact potential**.

Contact potential (V_0): This is the potential difference that appears when a junction is formed between different materials (also known as the built-in potential). On the band diagram, this is expressed as energy qV_0. We need to apply a voltage greater than V_0 or provide energy greater than qV_0 to cause the p-n junction to conduct.

Dark current: A current measured for a nonilluminated p-n junction in order to avoid any current due to the photogeneration of electron–hole pairs, as opposed to thermal generation.

Depletion region: A region at the electrical interface between the n- and p-regions of a p-n region that is depleted of charge carriers, that is, electrons and holes. A built-in electric field exists over this region.

Diffusion: The motion of electrons, holes, and ions from a region of a high concentration and chemical potential to a region of a low concentration.

Diffusion potential: See **Contact potential**.

Drift: The motion of charge carriers under the influence of an internal or external electric field.

Dynamic resistance (r'_d): The voltage-dependent resistance of a p-n junction under a forward bias when the applied voltage is less than the voltage near the knee of the I–V curve.

Einstein relation: An equation that describes the relationship between the mobility of species and their diffusion coefficient.

$$\frac{\mu}{D} = \frac{q}{k_B T}$$

Electron diffusion length (L_n): The average distance an electron can diffuse before recombining with a hole.

Forward bias: A voltage applied to a p-n junction in which the positive terminal of an external voltage supply is connected to the p-side of the p-n junction. This bias can overcome the built-in electric field and cause the p-n junction to conduct.

Forward current (I_F): The current in a p-n junction under a forward bias.

Generation current (I_0): The total current due to the drift of thermally generated carriers under the influence of an electric field in the depletion region. These go from the n-side to the p-side.

Hole diffusion length (L_p): The average distance a hole can diffuse before recombining with an electron.

Ideal diode equation: The equation that describes I–V characteristics of an ideal p-n junction (Equation 5.63):

$$I = I_0 \left[\exp\left(\frac{qV}{k_B} \right) - 1 \right]$$

Impact ionization: A process in which a high-energy electron collides with other electrons, breaking bonds in a material and causing one of the valence electrons to move into the conduction band, which creates a hole. This process eventually leads to an avalanche breakdown in a p-n junction.

Internal electric field: The electric field developed by positively charged donor ions, which are left behind as the electrons diffuse from the n-side to the p-side, and negatively charged acceptor ions, which are left behind as the holes diffuse from the p-side to the n-side.

Knee voltage: The voltage on the I–V curve for a diode, at which the forward current begins to increase exponentially. This value is very close to the built-in potential (V_0) for the p-n junction.

Mean thermal generation time (τ_g): The average time needed to thermally create an electron–hole pair.

One-sided p-n junction: A p-n junction in which one side is very heavily doped. This side is indicated with a + superscript (e.g., a p^+-n junction is where the p-side is very heavily doped).

Poisson's equation: An equation relating the gradient of an electric field to the charge density and the dielectric permittivity of the material (ε), given by Equation 5.30:

$$\frac{d^2\phi(x)}{dx^2} = -\frac{\rho(x)}{\varepsilon} = -\frac{dE(x)}{dx}$$

Reverse bias: A voltage applied to a p-n junction so that the negative terminal of the external voltage supply is connected to the p-side of the p-n junction. This reverse bias adds to the built-in electric field and makes the p-n junction nonconducting.

Reverse-bias dielectric breakdown: A high value of reverse-bias applied voltage, which causes a high level of current in the p-n junction.

Reverse-bias saturation current (I_s): A current due to the drift of thermally generated carriers in a p-n junction, which is the same as the generation current (I_0). It is *independent* of the applied voltage.

Shockley equation: An equation that describes one source of the generation current in a p-n junction originating from the diffusion of minority carriers that are generated in the neutral regions of the p-n junction (Equation 5.56):

$$I_{\text{Shockley}} = \left[\left(\frac{qD_p}{L_p N_d} \right) + \left(\frac{qD_n}{L_n N_a} \right) \right] n_i^2$$

Space charge layer: See **Depletion region**.
Space charge region: See **Depletion layer**.
Step junction: A p-n junction in which the transition from the p-side to the n-side is abrupt.
Zener diode: A diode based on a Zener or avalanche breakdown occurring in a p-n junction.
Zener breakdown mechanism: A low-voltage breakdown mechanism in which a p-n junction breaks down when electrons tunnel across the p-n junction instead of climbing over the energy barrier. Electrons from the p-side valence band tunnel across into the conduction band on the n-side because the reverse bias pushes down the E_c on the n-side and aligns it with the valence bandedge (E_v) on the p-side. This is seen more often in heavily doped junctions with a smaller bandgap.
Zener tunneling: The tunneling seen in the Zener breakdown mechanism, also known as band-to-band tunneling.
Zener voltage: The voltage at which a Zener diode begins to carry current under a reverse bias (typical values ~1–200 V).

REFERENCES

Askeland, D., and P. Fulay. 2006. *The Science and Engineering of Materials*. Washington, DC: Thomson.
Edwards-Shea, L. 1996. The *Essenece of Solid State Electronics*. Upper Saddle River, NJ: Prentice Hall.
Floyd, T. 1998. *Electronics Fundamentals: Circuits, Devices, and Applications*, 4th ed. Upper Saddle River, NJ: Prentice Hall.
Floyd, T. 2006. *Electric Circuit Fundamentals*, 7th ed. Upper Saddle River, NJ: Prentice Hall.
Kasap, S. O. 2002. *Principles of Electronic Materials and Devices*. New York: McGraw Hill.
Mahajan, S., and K. S. Sree Harsha. 1998. *Principles of Growth and Processing of Semiconductors*. New York: McGraw Hill.
Neaman, D. 2006. *An Introduction to Semiconductor Devices*. New York: McGraw Hill.
Singh, J. 2001. *Semiconductor Devices: Basic Principles*. New York: Wiley.
Streetman, B. G., and S. Banerjee. 2000. *Solid State Electronic Devices*, 5th ed. Upper Saddle River, NJ: Prentice Hall.

6 Semiconductor Devices

KEY TOPICS

- Metal–semiconductor junctions
- Schottky and ohmic contacts
- Solar cells
- Light-emitting diodes (LEDs)
- The operation of a bipolar junction transistor (BJT)
- Principle of the field-effect transistor (FET)

6.1 METAL–SEMICONDUCTOR CONTACTS

In semiconductor device fabrication, metals are often deposited as electrodes onto the semiconductor surfaces. The deposition of a metallic material onto a semiconductor results in an *ohmic contact* or it can result in another type of contact known as a *Schottky contact*. The ohmic contact between a metallic material and a semiconductor is such that there is no energy barrier to block the flow of carriers in either direction at the metal–semiconductor junction. In some cases, however, the contact between a metal and a semiconductor is rectifying; this type of contact is known as the Schottky contact.

To better understand the origin of Schottky and ohmic contacts, recall that the *work function* of a material is the energy required to remove an electron from its Fermi energy level and set it free (Figure 6.1). Note that while there are electrons at $E = E_F$ for a metal, there are no electrons at $E = E_F$ for a semiconductor.

Therefore we define the *electron affinity of a semiconductor* ($q\chi_S$) as the energy needed to remove an electron from the conduction band and set it free (Figure 6.1). It is *not* the lowest energy required to remove an electron, because when an electron is removed from the conduction band, another electron must move from the valence band to conduction band, which requires additional energy.

For semiconductors, we usually refer to the values of electron affinity, and not the work function, since electrons are present in the conduction band but not at $E = E_F$. The work function and electron affinity values for different metals and semiconductors are shown in Table 6.1. The values of the electron affinity and the work function are often expressed as the potential in volts (V) or as energy in electron volts.

6.2 SCHOTTKY CONTACTS

A Schottky contact is a rectifying contact between a metal and a semiconductor. It can function as a diode, known as the *Schottky diode* (Figure 6.2).

The Schottky contact is formed if $\phi_M < \phi_S$ for a p-type semiconductor or $\phi_M > \phi_S$ for an n-type semiconductor. The subscripts "m" and "s" stand for metal and semiconductor, respectively.

6.2.1 BAND DIAGRAMS

Consider a metal–semiconductor system in which the work function of the metal (ϕ_M) is greater than that for an n-type semiconductor (ϕ_S; Figure 6.1a).

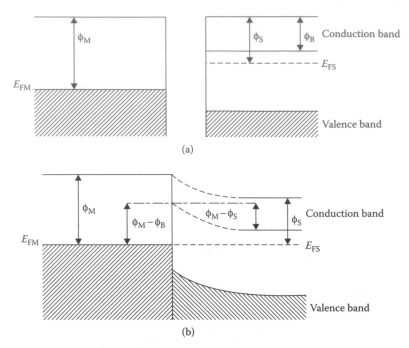

FIGURE 6.1 Energy diagrams for a junction between a metal and an n-type semiconductor ($\phi_M > \phi_S$), (a) before contact, (b) after contact the Fermi levels agree ($E_{FM} = E_{FS}$). (From Solymar, L. and D. Walsh. 1998. *Electrical Properties of Materials*, 6th ed. Oxford, UK: Oxford University Press. With permisson.)

TABLE 6.1
Some Metal Work Function ($q\phi_M$) and Semiconductor Electron Affinity ($q\chi_s$) Values

Metal	Work Function ($q\phi_M$) (eV)	Semiconductor	Electron Affinity ($q\chi_s$)(eV)
Silver (Ag)	4.26	Germanium (Ge)	4.13
Aluminum (Al)	4.28	Silicon (Si)	4.01
Gold (Au)	5.1	Gallium arsenide (GaAs)	4.07
Chromium (Cr)	4.5	Aluminum arsenide (AlAs)	3.5
Molybdenum (Mo)	4.6		
Nickel (Ni)	5.15		
Palladium (Pd)	5.12		
Titanium (Ti)	4.33		
Tungsten (W)	4.55		

Anode

Cathode

FIGURE 6.2 Symbol for a Schottky diode.

We will construct a band diagram for this metal–semiconductor junction using the principle of the invariance of Fermi energy. We will follow steps similar to those used in Chapter 5 for creating a band diagram for a p-n junction. At this metal–n-type semiconductor junction, where $\phi_M > \chi_S$, electrons flow from the higher-energy states of the semiconductor conduction band to the lower-energy states of the metal. This creates a positively charged depletion region in the n-type semiconductor. A negative surface charge builds up on the metal. Since the free electron concentration in a metal is very high, this charge is within an atomic distance from the surface. A built-in electric field is directed from the n-type semiconductor to the metal. The associated built-in potential (V_0) prevents the further flow of electrons from the n-type semiconductor to the metal (Figure 6.3).

As we can see in Figure 6.3, there is also a barrier ($q\phi_B$) to the flow of electrons from the metal to the n-type semiconductor:

$$q\phi_B = q\phi_M - q\chi_S \tag{6.1}$$

This barrier, called the *Schottky barrier* ($q\phi_B$), prevents any further injection of electrons into the n-type semiconductor from the metal. When a metal is deposited onto a semiconductor, we may think that the electrons from the metal flow into the semiconductor easily. However, this is not always the case.

The magnitude of the built-in potential (qV_0) for transferring an electron from the conduction band of the semiconductor to the Fermi energy level metal is given by

$$qV_0 = q\phi_B - qV_n \tag{6.2}$$

where qV_n is the energy difference between E_c and E_F of the n-type semiconductor (Figure 6.3).

Substituting for the Schottky barrier from Equation 6.1 ($q\phi_B$) as ($q\phi_M - q\chi_S$) in Equation 6.2, we get

$$qV_0 = q\phi_M - q\chi_S - qV_n$$
$$\therefore qV_0 = q\phi_M - q\chi_S - (q\phi_S - q\chi_S)$$
$$\therefore qV_0 = q\phi_M - q\phi_S$$

Thus, the built-in potential barrier (qV_0) is equal to the difference between the two work functions of the materials forming the Schottky contact.

$$qV_0 = q\phi_M - q\phi_S \tag{6.3}$$

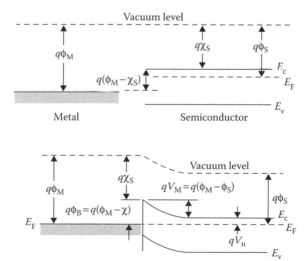

FIGURE 6.3 Formation of a Schottky contact between a metal and n-type semiconductor ($q\phi_M > q\phi_S$). (From Sze, S. M. 2002. *Semiconductor Devices, Physics, and Technology*, 2nd ed. New York: Wiley. With permission.)

6.2.2 Surface Pinning of the Fermi Energy Level

It may seem from Equation 6.1 that for different metals deposited on a given semiconductor (e.g., silicon [Si]), the barrier height $q\phi_B$ changes with the work function of the metal ($q\phi_M$). However, in practice the Schottky barrier height does not change appreciably for different metals deposited onto a given semiconductor (Table 6.2). As shown in Table 6.2, the Schottky barrier height of ~0.8 eV is essentially independent of the metal deposited for n-type silicon. For n-type gallium arsenide (GaAs), the Schottky barrier height is ~0.9 eV.

Many metals deposited on silicon are thermodynamically unstable, so that when the semiconductor–metal contact is exposed to high temperatures during processing, the metals react with the silicon and form a *silicide* intermetallic compound. For example, platinum (Pt) reacts with silicon and forms platinum silicide (PtSi). This lowers the Schottky barrier by ~0.06 eV, from 0.90 for platinum to 0.84 eV for PtSi. Tantalum silicide ($TaSi_2$) and titanium silicide ($TiSi_2$) are the preferred materials for forming Schottky contacts in silicon semiconductor processing. The values of the Schottky barrier for some silicides are listed in Table 6.2.

The Schottky barrier is independent of the metal or alloy used to create it because of the *surface pinning* of the semiconductor's Fermi energy level in the interface region (Figure 6.4).

At the semiconductor surface or its interface with another metal or material, additional energy levels are introduced into the otherwise forbidden bandgap. The physical interface between the semiconductor and its surroundings (another metal, surrounding atmosphere, and so on) is not perfect; there are dangling or incomplete bonds, which means the atoms at the surface or interface are not fully coordinated with the other atoms. For example, inside a single crystal of silicon, an atom of silicon should be coordinated with four other silicon atoms in a tetrahedral fashion. However, this is not the case for silicon atoms at the surfaces or interfaces with other metals or materials. The interface between a metal and a semiconductor is not sharp at an atomistic level. There is a very small region at the interface in which it is unclear whether the material is a metal or a semiconductor. Nanoscale oxide particles, intermetallic compounds, and so on may be present at this interface. These surface atoms have incomplete bonds and other imperfections and introduce a large

---

TABLE 6.2
Schottky Barrier Heights (in Electron Volts) for Metals and Alloys on Different Semiconductors

Metal/Alloy	n-Si	p-Si	n-GaAs
Aluminum (Al)	0.72	0.8	0.80
Titanium (Ti)	0.5	0.61	
Chromium (Cr)	0.61		
Tungsten (W)	0.67		
Nickel (Ni)	0.61		
Molybdenum (Mo)	0.68		
Gold (Au)	0.79	0.25	0.9–0.95
Silver (Ag)			0.88
Platinum (Pt)			0.86–0.94
Platinum silicide (PtSi)	0.85	0.2	
Nickel silicide (NiSi)	0.7	0.45	
Nickel silicide ($NiSi_2$)	0.70		
Tantalum silicide ($TaSi_2$)	0.59		
Titanium silicide ($TiSi_2$)	0.60		

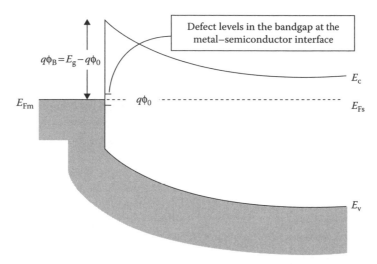

$q\phi_B = E_g - q\phi_0$

E_{Fm}

$q\phi_0$

Defect levels in the bandgap at the metal–semiconductor interface

E_c

E_{Fs}

E_v

FIGURE 6.4 Pinned Fermi energy level ($q\phi_0$) in semiconductors. (From Singh, J. 2001. *Semiconductor Devices: Basic Principles*. New York: Wiley. With permission.)

number of energy states or available defect-related energy levels into the interface region of the semiconductor's bandgap (Figure 6.4).

Thus, although theoretically there are no energy levels allowed in the bandgap, some energy levels occur in the bandgap near the surface or interface region in the bulk of the semiconductor. These states effectively "pin" the Fermi energy level of the semiconductor in the interface region.

Surface pinning means that the Fermi energy level in the surface region of the semiconductor is at a fixed level (ϕ_0) that does not change with the addition or removal of electrons in the rest of the semiconductor (via doping). This also means that the position of the semiconductor's conduction bandedge is fixed in the interfacial region. Thus, regardless of the metal deposited, the Fermi energy level of the metal must align with the pinned Fermi energy level (ϕ_0; Figure 6.4). When the Fermi energy level is pinned, the Schottky barrier height ($q\phi_B$) for the injection of electrons from a metal into the semiconductor is given by

$$q\phi_B = E_g - q\phi_0 \tag{6.4}$$

The Schottky barrier of different metals deposited on a given semiconductor is essentially constant (Table 6.2).

6.2.3 CURRENT–VOLTAGE CHARACTERISTICS FOR SCHOTTKY CONTACTS

We now consider the ideal current–voltage (*I–V*) curve for a Schottky contact between an n-type semiconductor and a metal, such that $q\phi_M > q\chi_S$. When there is no applied bias, some electrons on the semiconductor side will have a high enough energy to overcome the built-in potential barrier (qV_0) and flow onto the metal side. This process of *thermionic emission* creates the thermionic current. (IMS) If the n-type semiconductor is heavily doped, the depletion layer is thin, and it is possible for electrons to tunnel from the n-side semiconductor to the metal. Thermionic emission and tunneling both create a flow of electrons from the semiconductor to the metal. The resultant *conventional current* (I_{MS}) is directed from the metal to the semiconductor (Figure 6.5). This current is balanced by the conventional current resulting from the flow of electrons from the metal to the semiconductor (I_{SM}). These currents cancel each other out in a Schottky contact under equilibrium.

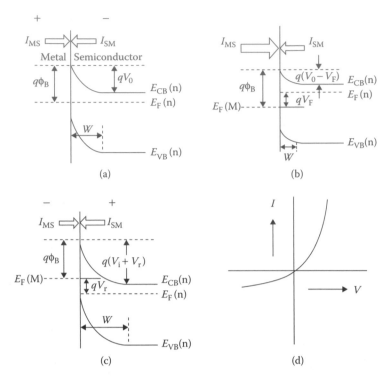

FIGURE 6.5 Current flows in a Schottky contact: (a) no bias; (b) forward bias; (c) reverse bias; (d) the *I–V* curve. (From Mahajan, S., and K. S. Sree Harsha. 1998. *Principles of Growth and Processing of Semiconductors*. New York: McGraw Hill. With permission.)

A forward bias is applied to this Schottky contact by connecting the positive terminal of a power supply to the metal side. The applied voltage is opposite the internal field, directed from the n-type semiconductor to the metal. The potential barrier for the flow of electrons from the semiconductor to the metal is reduced from qV_0 to $q(V_0 - V_F)$. There is therefore an increased flow of electrons from the semiconductor to the metal. This means that the current directed from the metal to the n-type semiconductor (I_{MS}) increases. This is represented in Figure 6.5b by the relatively thicker and longer arrow for I_{MS}.

Note that since the Schottky barrier does not change much for a given semiconductor (Table 6.2), the *current due to the motion of electrons* from the metal to the semiconductor (I_{SM}) does not change. Thus, the value of I_{SM} does not change under an applied bias. The net result is that under a forward bias, the overall conventional current flow from the metal to the semiconductor (I_{SM}) increases.

Under a reverse bias, that is, when the metal side is connected to the negative terminal of a DC power supply, the potential barrier for the flow of electrons from the semiconductor toward the metal increases from qV_0 to $q(V_0 + V_r)$. This causes the resultant I_{MS} to decrease, whereas the value of I_{SM} remains unchanged (Figure 6.5c). The resultant *I–V* curve for a Schottky diode is shown in Figure 6.5d.

We have considered a Schottky contact formed between an n-type semiconductor and a metal (Figure 6.6a and b). This type of rectifying contact can occur between a p-type semiconductor and a metal when $\phi_M < \phi_S$ (Figure 6.6).

The current through a Schottky diode is given by the following equation:

$$I = I_S \left[\exp\left(\frac{qV}{\eta k_B T} \right) - 1 \right] \tag{6.5}$$

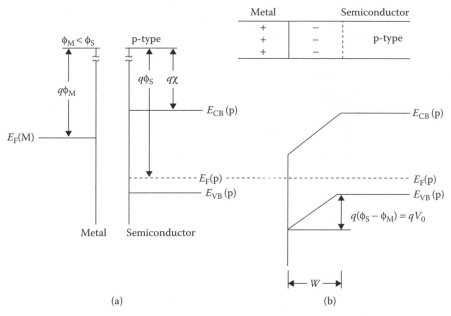

FIGURE 6.6 Energy-band diagrams of a metal/p-type semiconductor contact with $\phi_M < \phi_S$: (a) two materials isolated from each other and (b) at thermal equilibrium after the contact is made. (From Mahajan, S., and K. S. Sree Harsha. 1998. *Principles of Growth and Processing of Semiconductors.* New York: McGraw Hill. With permission.)

where I_S is the reverse saturation current, V is the applied bias, and η is the ideality factor for a Schottky diode. The value of η is between 1 and 2 and is closer to 1 for a Schottky diode. The reverse-bias saturation current (I_S) is given by

$$I_S = A \times \left(\frac{m^* q k_B^2}{2\pi^2 \hbar^3} \right) \times T^2 \times \exp\left(-\left\{ \frac{q\phi_B}{k_B T} \right\} \right) \tag{6.6}$$

where A is the area through which the Schottky current flows.

The term $\left(\dfrac{m^* q k_B^2}{2\pi^2 \hbar^3} \right)$ is known as the *effective Richardson constant* (R^*).

$$R^* = \left(\frac{m^* q k_B^2}{2\pi^2 \hbar^3} \right) = 120 \left(\frac{m^*}{m_0} \right) A \ \mathrm{cm}^{-2} \ \mathrm{K}^{-2} \tag{6.7}$$

where m^* is the carrier effective mass and m_0 is the carrier rest mass.

Instead of the saturation current (I_S), we can write Equation 6.7 in the form of a saturation current density (J_S) as

$$J_S = \left(\frac{m^* q k_B^2}{2\pi^2 \hbar^3} \right) \times T^2 \times \exp\left(-\left\{ \frac{q\phi_B}{k_B T} \right\} \right) \tag{6.8}$$

The values of the effective Richardson's constant (R^*), predicted from Equation 6.7, are high. Values from more detailed calculations are shown in Table 6.3.

The Schottky diode I–V curve (Figure 6.5) appears very similar to that for a p-n junction diode. However, there are important differences in the magnitude of the currents, which are illustrated in Example 6.1 and Figure 6.7.

TABLE 6.3
Effective Richardson Constants for Semiconductors

Carriers and Semiconductor	Effective Richardson Constant (R*) $(A \cdot cm^{-2} \cdot K^{-2})$
Electrons in silicon (Si)	110
Holes in silicon (Si)	32
Electrons in GaAs	8
Holes in GaAs	74

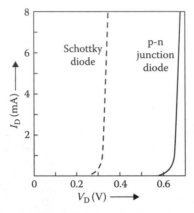

FIGURE 6.7 Comparison of the *I–V* curves for a p-n junction and a Schottky diode. (From Neaman, D. 2006. *An Introduction to Semiconductor Devices*. New York: McGraw Hill. With permission.)

EXAMPLE 6.1: SCHOTTKY DIODE CURRENT

A Schottky diode made from an Si and tungsten (W) junction has a saturation current density (J_s) of 10^{-11} A/cm^2. (a) What is the current density if the applied forward bias (V) is 0.3 V? (b) If the cross-sectional area of this diode is 5×10^{-4} cm^2, what is the. value of the current (I) in mA?

SOLUTION

a. We rewrite Equation 6.5 as

$$J = J_S \left[\exp\left(\frac{qV}{\eta k_B T} \right) - 1 \right] \tag{6.9}$$

We assume that under a forward bias of 0.3 V, the exponential term is much larger than 1. This is because $(k_B T/q) \sim 0.026$ V at 300 K. We also assume $\eta = 1$. Therefore,

$$J = J_S \left[\exp\left(\frac{V}{\left(\frac{k_B T}{q} \right)} \right) \right] = 10^{-11} \, \text{A/cm}^2 \times \exp\left(\frac{0.3\,\text{V}}{0.026\,\text{V}} \right)$$

$$J = 10.25 \, \text{A/cm}^2$$

b. The magnitude of the current is $I = 10.25$ A/cm^2 × $(5 × 10^{-4}$ cm$^2) = 5.12$ mA. This is a large forward current. A typical Si p-n junction diode, with a built-in potential of ~0.7 V, hardly carries any current, for a forward voltage of 0.3 V.

Schottky diodes have a lower voltage drop compared to p-n junction diodes because they are made from a metal–semiconductor junction, not a p-n junction. A forward current of ~1 mA can be achieved for forward voltages as small as ~0.1–0.4 V.

6.2.4 ADVANTAGES OF SCHOTTKY DIODES

The Schottky diode is considered a *majority device*, that is, minority carriers do not play an important role. Therefore, Schottky diodes exhibit faster switching times. The lower forward voltage means that a Schottky diode does not dissipate as much power. Because of these advantages, Schottky diodes are used in high-speed computer circuits.

The junction capacitance of a Schottky diode is lower compared to that of a typical p-n junction based on the same semiconductor. Since the reverse saturation current is high, the voltage and current ratings of Schottky diode for a forward bias are lower.

Silicon Schottky diodes have relatively smaller breakdown voltages. The silicon diodes work well up to a breakdown voltage of ~100 V. The resistance of the diode increases significantly in silicon diodes that have larger breakdown voltages. For any given forward voltage drop, the value of the current decreases significantly with higher breakdown voltage diodes. This limits the use of silicon Schottky diodes to rectify relatively lower voltages (Figure 6.8).

We can use semiconductors with higher breakdown voltages to control currents and voltages in high-powered electronics. For example, there is considerable interest in using silicon carbide (SiC) Schottky diodes. The forward current characteristics for a form of silicon carbide (known as 4H-SiC) are shown in Figure 6.9.

Compared to silicon diodes, these diodes have higher Schottky barrier height of ~1.1 eV. They can also be designed for higher breakdown voltages and still offer a lower resistance. Thus, compared to silicon, Schottky diodes made using SiC are better suited for high-power applications that involve higher voltages and currents.

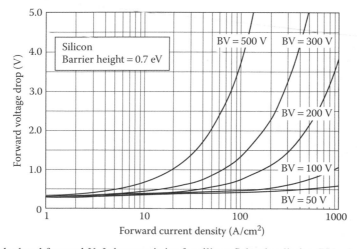

FIGURE 6.8 Calculated forward *V–I* characteristics for silicon Schottky diodes. BV = breakdown voltage. (From Baliga, B. J. 2005. *Silicon Carbide Power Devices*. Singapore: World Scientific. With permission.)

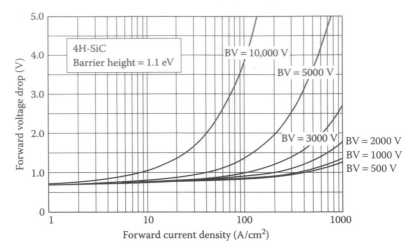

FIGURE 6.9 Calculated forward *V–I* characteristics for silicon carbide Schottky diodes. BV = breakdown voltage. (From Baliga, B. J. 2005. *Silicon Carbide Power Devices*. Singapore: World Scientific. With permission.)

6.3 OHMIC CONTACTS

Ohmic contacts are necessary in many semiconductor devices, such as solar cells and transistors. Ohmic contacts are formed when $\phi_M < \phi_S$ for an n-type semiconductor or $\phi_M > \phi_S$ for a p-type semiconductor. These contacts are nonrectifying, meaning that there is no energy barrier to block current flow in either direction. The use of the word "ohmic" does not, however, mean that the resistance of the contact is constant with voltage.

6.3.1 BAND DIAGRAM

The band diagrams for a metal–semiconductor junction forming an ohmic contact are shown in Figure 6.10.

Ohmic contacts can be formed on semiconductor surfaces using different strategies such as those shown in Figure 6.11. The usual approach is to make tunneling possible using a Schottky barrier of a lower height and then doping to reduce the depletion layer width.

Aluminum (Al) metallization is often used to create an ohmic contact on silicon. As we can see from Table 6.2, aluminum can form a Schottky contact on silicon by itself. However, when an evaporated aluminum film on p-type silicon is heated to ~450–550°C, the aluminum diffuses into the silicon and creates a heavily doped (p^+) layer at the interface between the aluminum and p-type silicon. Silicon can also diffuse out into aluminum and form an aluminum–silicon alloy. The contact between the p^+ layer of silicon and the aluminum–silicon alloy is ohmic in nature. Similarly, gold (Au) containing a small concentration of antimony (Sb), when deposited on n-type silicon, can create an ohmic contact by diffusing some of the antimony into the n-type silicon and creating an n^+ layer at the interface. Electrons can tunnel through this barrier and create an ohmic contact. The current–voltage (*I–V*) characteristics of an ohmic contact and a Schottky contact are compared in Figure 6.12.

6.4 SOLAR CELLS

A *solar cell* is a p-n junction–based device that generates an electric voltage or current upon optical illumination. Solar cells are useful in alternative energy technologies that are "greener" and utilize resources such as energy from the sun and wind. The field of research and development converting light energy into electricity is known as *photovoltaics*.

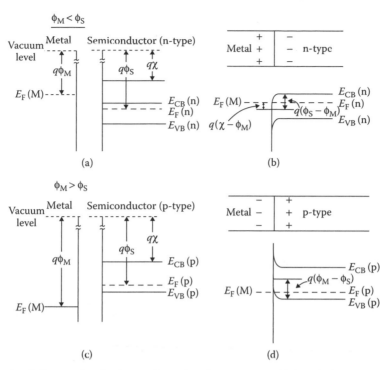

FIGURE 6.10   Band diagram for ohmic metal–semiconductor contact: (a) $\phi_M < \phi_S$ for an n-type semiconductor, (b) the corresponding equilibrium band diagram for the junction, (c) $\phi_M > \phi_S$ for a p-type semiconductor, and (d) the corresponding band diagram for the junction. (From Mahajan, S., and K. S. Sree Harsha. 1998. *Principles of Growth and Processing of Semiconductors*. New York: McGraw Hill. With permission.)

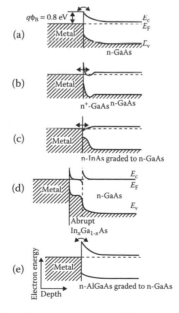

FIGURE 6.11   Strategies for creating ohmic contacts on n-GaAs. (a) Thermionic emission barrier; (b) field emission (tunneling) barrier; (c) graded composition low-resistance barrier; (d) heterojunction contact reduced barrier; (e) graded composition enhanced barrier. (Adapted from Mahajan, S., and K. S. Sree Harsha. 1998. *Principles of Growth and Processing of Semiconductors*. New York: McGraw Hill.)

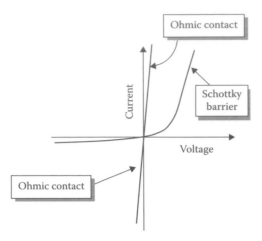

FIGURE 6.12 A comparison of the *I–V* curves for an ohmic and a Schottky contact. (From Singh, J. 2001. *Semiconductor Devices: Basic Principles*. New York: Wiley. With permission.)

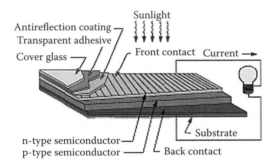

FIGURE 6.13 Schematic of the structure of a solar cell. (Courtesy of Solar Energy Technology Program, U.S. Department of Energy.)

The sun emits most of its energy in the wavelength of 2 to 4 μm. Solar cells are made using semiconductors that can absorb energy in this wavelength spectrum. The bandgap of the semiconductors used is $h\nu > E_g$, where ν is the frequency of light. *Absorptivity* is the ability of the semiconductor to absorb solar radiation. This is also important to solar cells. The semiconductors used in solar cells include crystalline silicon, amorphous silicon (a:Si:H), polycrystalline silicon, silicon ribbons, nanocrystalline silicon, and GaAs. Polycrystalline silicon is the most widely used because it costs less than single-crystal silicon, the second most widely used (Green 2003). Amorphous silicon is attractive because it can be used for deposition on large areas. Other compound semiconductors such as *copper indium diselenide* ($CuInSe_2$, CIS) and cadmium telluride (CdTe) can also be used, and provide a very high absorption of incident light.

A schematic of the structure of a solar cell is shown in Figure 6.13.

Usually, the top layer is an n-type material and the bottom layer is a p-type semiconductor. There are other elements in solar cell structure, such as an antireflective coating (Figure 6.13). The coating helps capture as much of the light energy incident on the solar cell as possible by minimizing reflection losses. Similarly, in addition to the p-n junction, electrical ohmic contacts are required for operating an electrical circuit. The bottom contact is made using metals such as aluminum or molybdenum. The top contact is in the form of metal grids or transparent conductive oxides such as indium tin oxide (ITO) so that light can still get through to the p-n junction. The symbol for a solar cell in an electrical circuit is shown in Figure 6.14.

6.4.1 Principles of Operation

A solar cell is basically an illuminated p-n junction that absorbs energy from light radiation (Figure 6.15). The absorption of light causes the creation of electron–hole pairs (EHPs). This effect is known as *photogeneration* or *optical generation* of carriers. The photogenerated carriers within the diffusion distances of the depletion layer (L_p or L_n, for holes or electrons, respectively) are swept up by the internal, built-in electric field. The photogenerated holes drift toward the p-side with the negatively charged space-charge region, in the direction of the built-in

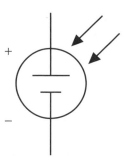

FIGURE 6.14 Symbol for a solar cell in an electrical circuit.

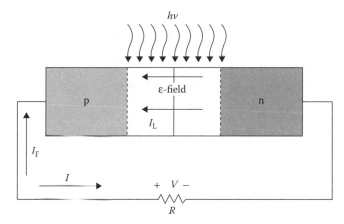

FIGURE 6.15 Schematic of an illuminated p-n junction that forms the basis of a solar cell. (From Neaman, D. 2006. *An Introduction to Semiconductor Devices*. New York: McGraw Hill. With permission.)

electric field (E). Photogenerated electrons drift toward the positively charged space-charge region on the n-side. Both of these motions result in a *photocurrent* (I_L) that is directed from the n-side to the p-side (Figure 6.15).

The photocurrent (I_L) is also known as the *short circuit current* (I_{sc}). This is the maximum current that flows through an external circuit with zero external resistance. The photocurrent magnitude depends upon the rate at which electron–hole pairs are created per unit volume (g_{op}). If A is the area of the p-n junction and L_p and L_n are the diffusion lengths for holes and electrons, respectively, then the photocurrent (I_L) is given by

$$I_L = q \times A \times (L_p + L_n) \times g_{op} \tag{6.10}$$

We assume in Equation 6.10 that only the carriers that are generated within the diffusion distances of the depletion layer will contribute to I_L. The rest of the carriers created in the semiconductor p- and n-regions will recombine, which will not result in the generation of a current. We also assume that the number of carriers generated in the depletion layer itself is small compared to those created in the neutral region n- and p-sides of the junction.

When an external resistance (R) is connected to a solar cell, the flow of the current creates a voltage drop (V) across the resistor. This voltage drop, in turn, creates a forward bias for the p-n junction and results in a forward current (I_F), which is directed in an opposite direction (compared to I_L). The appearance of this forward voltage across an illuminated p-n junction is known as the *photovoltaic effect*.

Thus, the net current (I) flowing through the external circuit is

$$I = (I_L - I_F) \tag{6.11}$$

Note that the photocurrent (I_L) and the net solar cell current (I) are always in the reverse-bias direction.

The maximum value of the solar cell current (I) is I_L; when there is no resistance ($R = 0$), it is also known as the short-circuit current (I_{sc}). The minimum value of the solar cell current (I) is zero when the resistance is infinite ($R = \infty$) and is known as the open-circuit condition.

We now derive expressions for the solar cell current (I) and *open-circuit voltage* (V_{oc}) starting with the diode equation

$$I_F = I_S \left[\exp\left(\frac{qV}{kT}\right) - 1 \right] \tag{6.12}$$

The saturation current (I_s) in Equation 6.12 can be written as

$$I_S = q \times A \times \left[\frac{D_p p_n}{L_p} + \frac{D_n n_p}{L_n} \right] \tag{6.13}$$

The product of the hole and electron concentrations is a constant for a given semiconductor at fixed temperature T. Therefore, $p_n \times n_n = n_i^2$ and $p_p \times n_p = n_i^2$ or $p_n \times N_d = n_i^2$ and $n_p \times N_a = n_i^2$. Recall that subscripts for the electron and hole concentrations refer to the side of the p-n junction; for example, p_n is the concentration of holes on the n-side. Thus, Equation 6.13 can be written as

$$I_S = q \times A \times n_i^2 \left[\frac{D_p}{L_p N_d} + \frac{D_n}{L_n N_a} \right] \tag{6.14}$$

where D_p and D_n are the diffusion coefficients for holes and electrons, respectively, and N_a and N_d are the dopant concentrations for the p- and n-sides, respectively.

Substituting the expression for I_S in Equation 6.12, we can write I_F as

$$I_F = q \times A \times n_i^2 \left[\frac{D_p}{L_p N_d} + \frac{D_n}{L_n N_a} \right] \left[\exp\left(\frac{qV}{k_B T}\right) - 1 \right] \tag{6.15}$$

The diffusion lengths for a carrier are related to the lifetime of the carrier (τ) by the following equations:

$$L_n = \sqrt{D_n \tau_n} \tag{6.16}$$

$$L_p = \sqrt{D_p \tau_p} \tag{6.17}$$

The longer the lifetime (τ) of a carrier, the higher the diffusion length (L). The longer the carrier can survive without recombining, the higher the probability of it contributing to the photocurrent.

Sometimes, we prefer to write I_F without directly involving diffusion coefficients. For example, using Equations 6.16 and 6.17 and substituting them into Equation 6.15 and rearranging, we get

$$I_F = q \times A \times \left[\frac{L_p}{\tau_p} p_n + \frac{L_n}{\tau_n} n_p \right]\left[\exp\left(\frac{qV}{k_B T} \right) - 1 \right] \tag{6.18}$$

Now, substituting Equation 6.10 for I_L and Equation 6.18 for I_F in the expression for total current I (Equation 6.11), we get

$$I = q \times A \times \left[\frac{L_p}{\tau_p} p_n + \frac{L_n}{\tau_n} n_p \right]\left[\exp\left(\frac{qV}{k_B T} \right) - 1 \right] - q \times A \times g_{op} \times (L_p + L_n) \tag{6.19}$$

For the limiting case of $I = 0$ (i.e., open circuit), when the photocurrent (I_L) is equal and opposite to the forward current (I_F), we get the open-circuit voltage (V_{oc}):

$$V_{oc} = (k_B T/q) \ln\left[1 + \left(\frac{I_L}{I_S} \right) \right] \tag{6.20}$$

V_{oc} is the maximum voltage that can be generated from the p-n junction forming a solar cell. We can see the appearance of this voltage on a band diagram upon the illumination of a p-n junction, called the photovoltaic effect (Figure 6.16).

We can rewrite Equation 6.20 as follows:

$$V_{oc} = \left(\frac{k_B T}{q} \right) \ln\left\{ 1 + \left[\left(\frac{(L_p + L_n)}{(L_p/\tau_p) p_n + (L_n/\tau_n) n_p} \right) \times g_{op} \right] \right\} \tag{6.21}$$

If the rate of the optical generation of carriers (g_{op}) is large compared to their rate of thermal generation (g_{th}) and the p-n junction is symmetric, that is, $p_n = n_p$ and $\tau_p = \tau_n$, then

$$V_{oc} = (k_B T/q) \ln\left(\frac{g_{op}}{g_{th}} \right) \tag{6.22}$$

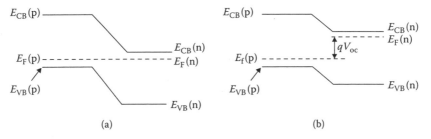

(a)   (b)

FIGURE 6.16 Band diagram for a p-n junction (a) under equilibrium, and (b) upon illumination. (From Mahajan, S., and K. S. Sree Harsha. 1998. *Principles of Growth and Processing of Semiconductors*. New York: McGraw Hill. With permission.)

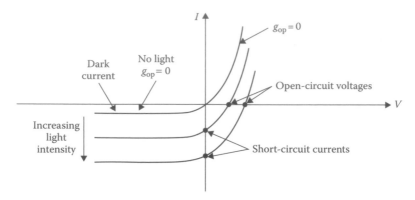

FIGURE 6.17 *I–V* curves for a solar cell with increasing illumination intensity. (From Kano, K. 1997. *Semiconductor Fundamentals.* Upper Saddle River, NJ: Prentice Hall. With permission.)

As the intensity of the illumination increases, the optical generation rate increases on both sides of the p-n junction. This leads to an increase in the open-circuit voltage (V_{oc}). We can see this in the *I–V* curves for a solar cell in Figure 6.17.

The open-circuit voltage (V_{oc}) cannot keep increasing indefinitely with the increasing g_{op}. As the optical rate of the carrier generation increases, the minority carrier concentration increases. This causes the lifetime (τ) of the carriers to become shorter. Thus, more carriers recombine, which prevents the voltage from exceeding the built-in potential (V_0).

Example 6.2 illustrates the calculation of the open-circuit voltage (V_{oc}) for a solar cell.

EXAMPLE 6.2: SOLAR CELL OPEN-CIRCUIT VOLTAGE

A solar cell is made using an Si p-n junction with $N_a = 10^{18}$ cm^{-3} and $N_d = 10^{16}$ cm^{-3}. The diffusion lengths for electrons and holes are 25 and 10 μm, respectively. Assume that $n_i = 1.5 \times 10^{10}$ cm^{-3} at $T = 300$ K, $D_p = 20$ and $D_n = 10$ cm^2/s, and the photocurrent density (J_L) is 10 mA/cm^2. (a) What is the open-circuit voltage (V_{oc})? (b) How does this compare with the contact potential (V_0) for this p-n junction?

SOLUTION

a. We rewrite Equation 6.20 current densities as

$$V_{oc} = (k_B T/q)\ln\left[1+\left(\frac{J_L}{J_s}\right)\right] \qquad (6.23)$$

At 300 K:

$$V_{oc} = (0.026\ \text{V})\ln\left[1+\left(\frac{J_L}{J_s}\right)\right] \qquad (6.24)$$

The value of J_L is given = 10 mA/cm^2. To get the value of saturation current density (J_s) for the solar cell diode, we rewrite Equation 6.14 as

$$J_s = q \times n_i^2\left[\frac{D_p}{L_p N_d} + \frac{D_n}{L_n N_a}\right]$$

Substituting the values for diffusion coefficients, diffusion lengths, and dopant concentrations:

$$J_S = (1.6 \times 10^{-19}) \times (1.5 \times 10^{10})^2 \left[\frac{20}{(10 \times 10^{-4}\,\text{cm})(10^{16})} + \frac{10}{(25 \times 10^{-4}\,\text{cm})(10^{18})} \right]$$

Therefore, J_s, saturation current density $= 7.21 \times 10^{-11}$ A/cm^2 or 7.21×10^{-8} mA/cm^2. We use both current densities in the same units to calculate V_{oc} from Equation 6.24:

$$V_{oc} = (0.026\text{ V})\ln\left[1 + \left(\frac{10\,\text{mA/cm}^2}{7.21 \times 10^{-8}\,\text{mA/cm}^2}\right)\right]$$

$$V_{oc} = 0.48\text{ V}$$

b. From Equation 5.21, the contact potential (V_0) is given by

$$V_0 = (k_B T/q)\ln\left(\frac{N_a N_d}{n_i^2}\right) \tag{6.25}$$

Therefore,

$$V_0 = (0.026\text{ V})\ln\left(\frac{10^{18}\,\text{atoms/cm}^3 \times 10^{16}\,\text{atoms/cm}^3}{(1.5 \times 10^{10}\,\text{electrons/cm}^3)^2}\right)$$

Therefore, $V_0 = 0.813$ V. The contact potential is the maximum forward bias that can appear across the p-n junction. Therefore, the value of the open-circuit voltage (V_{oc}) is always less than V_0.

6.4.2 CURRENT–VOLTAGE CHARACTERISTICS FOR A SOLAR CELL

The maximum voltage from the solar cell is V_{oc} for zero current in the external circuit. The minimum voltage is zero when the current is maximum under a short-circuit condition (I_{sc}). This gives us the diode current (I) as a function of the diode voltage generated as shown in Figure 6.18.

The power a solar cell delivers to a load is obtained by calculating the value of $I \times V$. Note that since the current generated is in the reverse-bias direction, the sign of I is negative, and since the

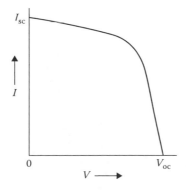

FIGURE 6.18 The I–V characteristics of a p-n junction solar cell. (From Neaman, D. 2006. *An Introduction to Semiconductor Devices*. New York: McGraw Hill. With permission.)

voltage is positive, the product $I \times V$ is negative. This means that the power is being generated but not consumed.

This can be written as

$$P = I \times V = (I_L \times V) - \left\{ I_S \left[\exp\left(\frac{qV}{k_B T} \right) - 1 \right] \times V \right\} \tag{6.26}$$

We can calculate the current (I_m) and voltage (V_m) that will result in maximum power (P_{max}) by equating $dP/dV = 0$. From Equation 6.26,

$$\frac{dP}{dV} = I_L - I_S \left[\exp\left(\frac{qV_m}{k_B T} \right) - 1 \right] - I_S V_m \left[\left(\frac{q}{k_B T} \right) \right] \exp\left(\frac{qV_m}{k_B T} \right) = 0$$

$$1 + \left(\frac{I_L}{I_S} \right) = \left[\exp\left(\frac{qV_m}{k_B T} \right) \times \left(1 + \frac{qV_m}{k_B T} \right) \right]$$

or

$$1 + \left(\frac{I_L}{I_S} \right) = \left[\exp\left(\frac{qV_m}{k_B T} \right) \times \left(1 + \frac{V_m}{V_t} \right) \right] \tag{6.27}$$

In Equation 6.27, $V_t = (k_B T/q) = 0.026$ V at $T = 300$ K. Since the right-hand side is known for a given p-n junction, we can solve for V_m by trial and error for a given I_L. The V_m and corresponding value of I_m are shown in Figure 6.19.

From Figure 6.19, we can see that the product $V_m \times I_m$ is less than $V_{oc} \times I_{oc}$. The ratio of these two products is called the *fill factor* (F_f) of a solar cell:

$$F_f = \frac{V_m \times I_m}{V_{oc} \times I_{oc}} \tag{6.28}$$

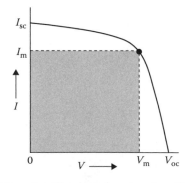

FIGURE 6.19 The *I–V* characteristics of an illuminated p-n junction. The shaded area defines the maximum power rectangle. (From Neaman, D. 2006. *An Introduction to Semiconductor Devices*. New York: McGraw Hill. With permission.)

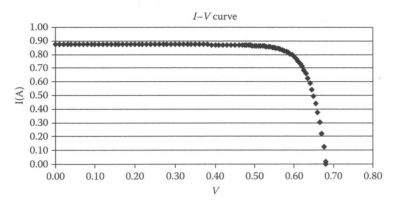

FIGURE 6.20 Experimental *I–V* curve for a silicon solar cell. (Courtesy of Professor Martin Green, University of New South Wales, Australia.)

Most solar cells have a fill factor of ~0.7. Please note that the subscript "m" in V_m and I_m does not stand for maximum voltage and maximum current. The subscripts instead represent the values of voltage and current, respectively, that lead to the maximum power.

Experimental data for a silicon solar cell *I–V* curve is shown in Figure 6.20. Note that the open-circuit voltage is slightly less than 0.7 V and the contact potential (V_0) for silicon is ~0.7 V.

6.4.3 SOLAR CELL EFFICIENCY AND MATERIALS

Solar cell conversion efficiency (η_{conv}) is defined as the ratio of maximum power delivered to the incident power (P_{in}):

$$\eta_{conv} = \frac{I_m \times V_m}{P_{in}} \tag{6.29}$$

If the incident photons have an energy ($h\nu$) less than the bandgap energy (E_g), then no electron–hole pairs are produced and the incident energy is wasted, that is, it is not used for conversion into electrical energy. Similarly, if the incident photon energy ($h\nu$) is too high compared to E_g, then electron–hole pairs are created, and the difference ($h\nu-E_g$) will appear as heat. This will also make the solar cell inefficient. Thus, the efficiency of the solar cell conversion is maximized by better matching the semiconductor bandgap with the solar spectrum. By this principle, GaAs is a better material; however, it is more expensive than silicon and is thus used only in specialized applications. Many polycrystalline silicon-based solar cells offer a conversion efficiency of ~15%.

Note that both direct and indirect bandgap semiconductors can be used in solar cells. Recall that in LEDs, electrons and holes recombine and cause the emission of light, so we must use direct bandgap semiconductors. In some ways, the solar cell and the LED make use of opposite effects. Currently, CuInSe$_2$ solar cells offer some of the highest conversion efficiencies, approaching ~15%. More complex solar cells based on multiple layers of different semiconductors offer higher efficiencies, up to 40%, because as the bandgap varies across the thickness of the solar cell, we can capture photons of different energies (Figure 6.20b). However, the cost of these solar cells is relatively high. There is also a growing interest in using the nanoparticles of semiconductors and organic materials for solar cell applications.

6.5 LIGHT-EMITTING DIODES

An LED normally uses a direct bandgap semiconductor in which the process of electron–hole recombination results in *spontaneous emission* of light. The symbol for an LED is shown in Figure 6.21.

The first practical LED was reported by Nick Holonyak, Jr., working at General Electric in 1962.

6.5.1 Operating Principle

This principle underlying the operation of an LED has already been mentioned in Chapter 5 in the discussion of indirect and direct bandgap semiconductors as well as the processes of radiative and nonradiative electron–hole recombination (Figure 6.22). The recombination dynamics, that is, how rapidly the electrons and holes can recombine and produce light, determine the speed at which an LED can be turned on and off. The speed with which an LED can be turned on and off is important in some fiber optics applications.

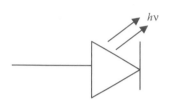

FIGURE 6.21   Symbol for an LED.

Most LEDs are made using direct bandgap materials. However, as has been noted in Chapter 5, it is possible to use indirect bandgap materials to make LEDs, provided a defect energy level can be introduced by doping (see Section 6.5.3). This defect level provides an intermediate state in which radiative recombination is possible.

LEDs are constructed as forward-biased p-n junctions made most often from direct bandgap semiconductors (Figure 6.23). When the p-n junction forming an LED is forward-biased, the electrons with a high energy can overcome the built-in potential barrier and arrive at the p-side of the junction. These minority carriers then combine with the holes (i.e., the majority carriers) on the p-side.

Similarly, some of the holes from the p-side of the junction also make it across the depletion layer onto the n-side and recombine with the electrons there. These recombination processes result in the emission of light in direct bandgap materials (Figure 6.23). For indirect bandgap materials such as silicon or germanium, the recombination process also leads to the generation of heat. Thus, no LEDs are made using silicon or germanium. Note that even in direct bandgap materials, some nonradiative recombination occurs and has a negative effect on the efficiency of an LED.

6.5.2 LED Materials

The light emitted from an LED may be in the visible, infrared, or ultraviolet ranges. LEDs that emit in the visible range are widely used in displays and many other consumer applications. Figure 6.24 shows some of the materials used for making LEDs.

The response of the eye to different wavelengths corresponding to various colors is also shown in this figure. The wavelength of light emitted (λ) is related to the bandgap of the semiconductor (E_g) by the following equation:

$$\lambda = \frac{h\nu}{E_g} = \frac{1.24}{E_g (\text{in eV})} \ \mu m \tag{6.30}$$

LEDs that emit in the infrared range are used in fiberoptic systems. Gallium nitride (GaN)–based LEDs are relatively new and emit in the near-ultraviolet and blue range. The bandgap of GaN is ~3.4 eV. GaN is alloyed with other nitrides (e.g., indium nitride and aluminum nitride) to obtain blue, amber, or green LEDs (Figure 6.25).

Thus, the bandgap of semiconductors used for making LEDs can be intentionally varied by forming solid solutions between multiple semiconductors. This bandgap engineering approach leads to variations in the color of the light emitted. For example, GaAs is a direct bandgap material with a bandgap of 1.43 eV and a corresponding wavelength of 870 nm in the infrared range. However, it can be alloyed with aluminum arsenide (AlAs) and other materials to form LEDs of different colors.

The quality of materials used in the manufacture of durable LEDs is extremely important from the viewpoint of the presence of atomic level defects such as vacancies and dislocations. For example, as pointed out by Nakamura et al. (1995), scientists knew for a long time that a blue or ultraviolet

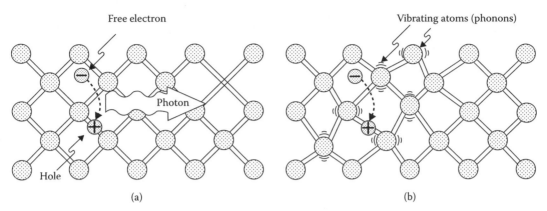

FIGURE 6.22 Illustration of radiative and nonradiative recombination. (a) Radiative recombination of an electron–hole pair accompanied by the emission of a photon with energy $h\nu = E_g$. (b) In nonradiative recombination events, the energy released during the electron–hole recombination is converted to phonons. (From Schubert, E. F. 2006. *Light-Emitting Diodes*. Cambridge, UK: Cambridge University Press. With permission.)

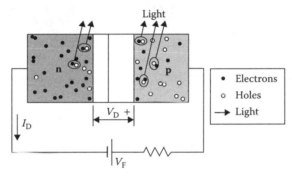

FIGURE 6.23 LED operating principle.

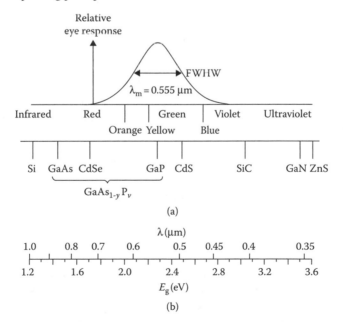

FIGURE 6.24 Semiconductors used in LEDs relative to (a) color wavelengths and relative eye response and (b) wavelength emitted. (From Sze, S. M. 2002. *Semiconductor Devices, Physics and Technology*, 2nd ed. New York: Wiley. With permission.)

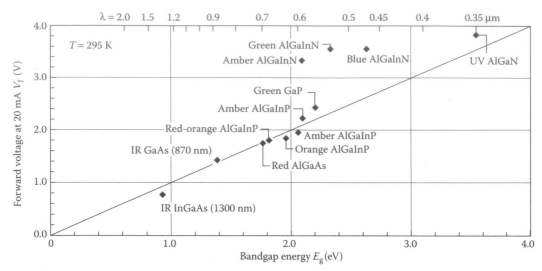

FIGURE 6.25 Bandgaps of LEDs emitting different visible colors of light. IR = infrared. (From Schubert, E. F. 2006. *Light-Emitting Diodes*. Cambridge, UK: Cambridge University Press. With permission.)

LED could be made using GaN-based materials. However, until recently, the available materials had too many defects, such as certain types of dislocations, which prevented the creation of a blue LED. Thus, defects such as dislocations should be minimized in LED device materials.

Organic light-emitting diodes (OLEDs) or *polymer light-emitting diodes* (PLEDs) have recently been developed in which there is an electron-injecting cathode instead of an n-layer. There is also a hole-injecting anode. This technology has considerable promise as a more energy-efficient alternative to lighting by conventional sources such as incandescent or fluorescent lights. Several companies have recently introduced high-quality displays based on OLEDs.

6.5.3 LEDs Based on Indirect Bandgap Materials

Gallium phosphide (GaP) has an indirect bandgap of 2.26 eV. When solid solutions of GaAs, a direct bandgap material, are formed with GaP, the bandgap of the solid solution $GaAs_{1-x}P_x$ is direct until $x = 0.45$ (Figure 6.26a). In fact, the most commonly used composition in this system is $GaAs_{0.6}P_{0.4}$, which emits in the red region (with $E_g \sim 1.9$ eV).

As the mole fraction x of phosphorus (P) in the $GaAs_{1-x}P_x$ increases, the materials become an indirect bandgap. These materials can be doped with a dopant such as nitrogen (N) to make LEDs that emit yellow to green wavelengths. When nitrogen, an isoelectronic dopant, is added to these materials, it substitutes for some of the phosphorus atoms at their locations. Although the valence of phosphorus and nitrogen is the same (+5), the positive nucleus of a nitrogen atom is less shielded than the positive nucleus of a phosphorus atom. Isoelectronic impurities such as nitrogen atoms tend to bind the conduction electrons more tightly, that is, the electron wave function is highly localized. According to the Heisenberg uncertainty principle, because the location of the electron is more fixed (i.e., Δx is small), its momentum is spread out (i.e., Δx is large). Hence, there are two possible vertical transitions via the isoelectronic nitrogen defect level. One of these is radiative transition, in which the nitrogen atoms introduce a defect energy level near the conduction band of GaP or $GaAs_{1-x}P_x$. The change in momentum that occurs when an electron makes the transition from the indirect conduction band (labeled X) to the valence band (labeled Γ) is absorbed by the impurity atom (Figure 6.26c). This enables the emission of light with an energy that is slightly less than that for the bandgap of GaP or $GaAs_{1-x}P_x$.

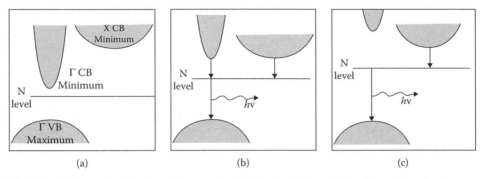

FIGURE 6.26 Schematic of the band diagram for (a) GaAs$_{1-x}$P$_x$ ($x < 0.45$) with a direct bandgap, and (b) nitrogen-doped indirect bandgap GaAs$_{1-x}$P$_x$ ($x > 0.45$). (c) Transitions in direct bandgap GaAs, crossover GaAsP, and indirect bandgap GaP. (From Schubert, E. F. 2006. *Light-Emitting Diodes*. Cambridge, UK: Cambridge University Press. With permission.)

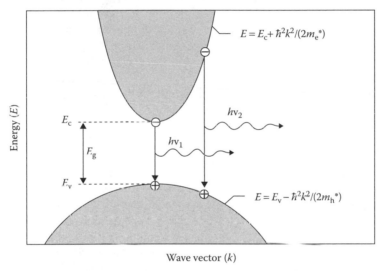

FIGURE 6.27 Vertical recombination of electrons and holes with energy $E = E_g$ and higher. (From Schubert, E. F. 2006. *Light-Emitting Diodes*. Cambridge, UK: Cambridge University Press. With permission.)

6.5.4 LED EMISSION SPECTRAL RANGES

We usually think of an LED as a device that emits the light of one wavelength. In practice, the light emitted from an LED is *not* all at one wavelength corresponding to the bandgap (E_g). The reason for this spectral width is that carriers in semiconductors have a range of energies. Thus, some electrons with an energy higher than E_c show a vertical recombination with holes in the valence band, without a change in carrier momentum. For these electrons, which recombine with holes at a higher energy, the frequency of light emitted is also higher (shown as $h\nu_2$; Figure 6.27).

The maximum emission intensity occurs at

$$E = E_g + (1/2)(k_B T) \tag{6.31}$$

At half mask, the full width of the luminescence intensity as a function of energy is 1.8 $k_B T$. Thus, the width of the energy is given by

$$\Delta E = 1.8 \ k_B T \tag{6.32}$$

The spectral width ($\Delta\lambda$) is given by

$$\Delta\lambda = \frac{1.8\,kT\lambda^2}{hc}$$

where λ is the wavelength of light emitted corresponding to the bandgap energy (E_g). For example, a GaAs LED emitting at 870 nm has a spectral width ($\Delta\lambda$) of ~28 nm, that is, the wavelengths range from 870 nm ($E = E_g = 1.43$ eV) to 870–28 nm = 842 nm. Note that higher electron energies mean that the wavelength of light emitted is smaller. Thus, an LED emits wavelengths in the range of λ to ($\lambda - \Delta\lambda$). Compared to the human eye's sensitivity, this spectral range is small, and hence for all practical purposes, the color of a given LED appears monochromatic.

Although the spectral purity is not a major concern in display applications, it is very important for some applications in fiber optics systems. In these applications, LEDs cannot be used to send signals over longer distances because the signal spreads; that is, different wavelengths start at the same point but arrive at the destination at different times. In such applications, devices known as *laser diodes* must be used. In a laser diode, photons of a given energy and wave vector are absorbed by the semiconductor, causing the generation of electrons and holes. The electrons and holes recombine and release *stimulated radiation*, an emission of coherent photons; that is, the photons emitted are in phase with the incident photons (i.e., with the same energy and wave vector). Blue laser diodes based on GaN offer a higher resolution for writing a digital video disk (DVD) and are used in the latest DVDs. This laser diode also forms the basis of the so-called Blu-ray disks and disk drives.

6.5.5 *I–V* Curve for LEDs

Since LEDs are usually operated as forward-biased p-n junctions, their *I–V* curves are similar to the curves we have seen for regular silicon-based p-n junctions. The *I–V* curves for different LED materials are compared to those for silicon in Figure 6.28.

As we can expect, the knee of these various curves appears at a voltage close to the built-in potential for these different materials. Thus, for germanium, the knee occurs around 0.4 V. For silicon, which cannot be used to make LEDs because of its indirect bandgap, the knee occurs at ~0.7 V. This is close to the built-in potential for a silicon p-n junction.

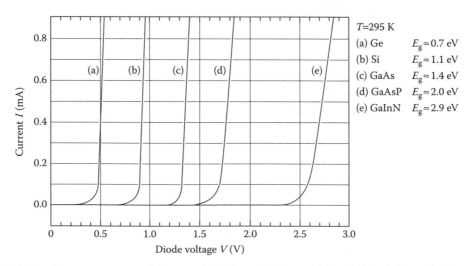

FIGURE 6.28 Room-temperature *I–V* curves for different LED materials and silicon. (From Schubert, E. F. 2006. *Light-Emitting Diodes*. Cambridge, UK: Cambridge University Press. With permission.)

The typical forward voltage required to drive an LED is about 1.2–3.2 V. This is much larger than that for a silicon-based diode. Similarly, the reverse-bias breakdown voltage of a typical LED is ~3 to 10 V. This is much smaller than that for a silicon diode.

A typical LED circuit uses a series resistance (R_s) to limit the forward current and prevent any damage to the LED because of excessive current and the associated heat. This series resistance (R_s) is separate from the internal resistance associated with the LED itself. The use of a series resistance is discussed in Example 6.3.

EXAMPLE 6.3: LED CIRCUITS WITH SERIES RESISTANCE

An LED is driven by a voltage supply of maximum 8 V. For the driving voltage across the LED of 1.8–2.0 V, what should be the series resistance (R_s)? Assume that the LED current is 16 mA.

SOLUTION

In this example, the maximum voltage for the circuit shown in Figure 6.29 is 8 V. This voltage appears between the resistor (R_s) and the LED. We assume that the resistance of the LED itself, that is, the smaller resistance of the neutral parts of the p-n junction, is small (~5 Ω) and can be ignored.

A maximum current results in the circuit when the voltage supply by the variable voltage supply is maximum and the voltage across the LED is minimum. This forward current (I_F) is given by

$$I_F = \frac{V_{Supply,max} - V_{F,min}}{R_S}$$

$$\therefore 16 \times 10^{-3}\,A = \frac{(8 - 1.8)\,V}{R_S}$$

$$\therefore R_S = \frac{7.2\,V}{16 \times 10^{-3}\,A} = 387.5\,\Omega$$

This can be rounded to 390 Ω. Therefore, we connect a 390-Ω resistor in the series to ensure that the current in this LED does not exceed 16 mA.

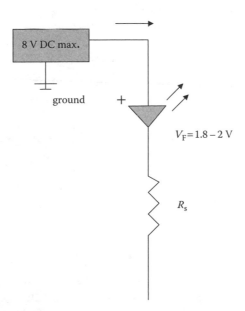

FIGURE 6.29   LED circuit showing a series resistance (R_S).

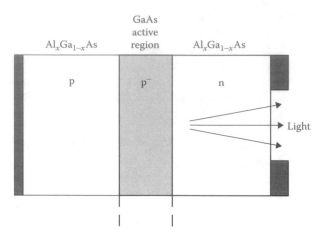

FIGURE 6.30 Schematic of a heterojunction LED. (From Singh, J. 1996. *Optoelectronics: An Introduction to Materials and Devices*. New York: McGraw Hill. With permission.)

6.5.6 LED EFFICIENCY

The efficiency of an LED depends upon what happens inside the active region in which the recombination of holes and electrons occurs, that is, the efficiency of the spontaneous emission process. The efficiency of an LED device also depends upon what happens to the light that is emitted.

The photons that are emitted can be lost due to reabsorption into the semiconductor, which leads to regeneration of electron–hole pairs. Some photons are reflected at the p-n junction–air interface and some are lost (i.e., they do not come out of the LED structure) because of total internal reflection. To minimize the losses by reflection at the semiconductor–air interface, the LED is encapsulated in a dielectric dome.

In advanced LED structures, the two sides of the junction are made from different materials (Figure 6.30). An LED with p- and n-sides made from different semiconductors is known as a *heterojunction* LED. In these LEDs, an active region is sandwiched between two layers of different semiconductors. The compositions are selected such that the emitted photons are not absorbed by the top or bottom layer. As a result, these heterojunction LEDs are more efficient than *homojunction* LEDs (Figure 6.31), in which both sides of the p-n junction are made from the same semiconductor.

As a light source, LEDs are more efficient than conventional light sources such as the incandescent lamp (Figure 6.32).

6.5.7 LED PACKAGING

In addition to the p-n junction, an LED used as a light source has other parts. Almost all LEDs are mounted into a package that has two electrical leads. The anode lead is usually longer. The assembly also has a reflector with a hemispherical epoxy encapsulant (Figure 6.33).

Some high-powered LEDs are mounted on a heat sink and use a silicone encapsulant and a plastic cover (Figure 6.34). Packaging of OLEDs used in flat-panel displays and as light panels is also a very important area of research and development.

6.6 BIPOLAR JUNCTION TRANSISTOR

The term "transistor" is an abbreviation of transfer resistor, a device whose electrical resistance can be changed by an external voltage. On December 16, 1947, Bardeen, Brattain, and Shockley

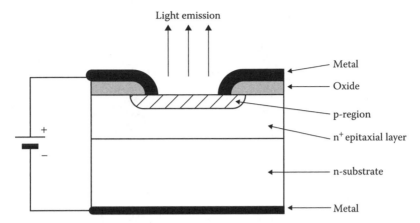

FIGURE 6.31 Schematic of a homojunction LED. (From Sparkes, J. J. 1994. *Semiconductor Devices.* London: Chapman and Hall. With permission.)

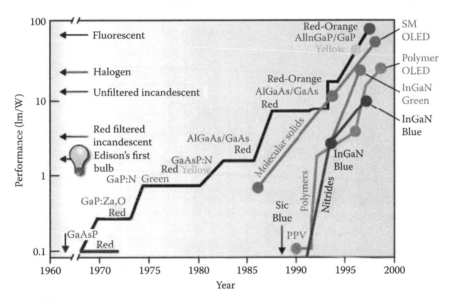

FIGURE 6.32 The performance of LEDs and OLEDs compared to incandescent and fluorescent lamps. (From Bergh, A., G. Craford, A. Duggal, and R. Haitz. 2001. *Physics Today* 54:12. With permission.)

reported a device known as the point contact transistor. They received the Nobel Prize in 1956 for the discovery of this important device.

Transistors are an indispensable component of modern-day integrated circuits (ICs). The two major types of transistors are the *bipolar junction transistor* (BJT) and the *field-effect transistor* (FET). A type of FET known as *metal oxide silicon field-effect transistor* (MOSFET) is the most widely used transistor in ICs (see Section 6.7). Millions of MOSFETs are integrated onto semiconductor chips. Transistors play a vital role in running most computers and other state-of-the-art electronic equipment. In this section, we discuss the BJT. We will discuss FET in Section 6.7.

The term "bipolar" is used to describe the BJT because this particular type of transistor uses both electrons and holes in its operation. In the BJT, the *transistor action* involves controlling the voltage at one terminal by applying voltages at two other terminals.

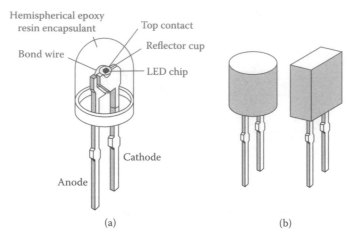

(a)   (b)

FIGURE 6.33 LED packaging. (a) LED with hemispherical encapsulant; (b) LEDs with cylindrical and rectangular encapsulant. (From Schubert, E. F. 2006. *Light-Emitting Diodes*. Cambridge, UK: Cambridge University Press. With permission.)

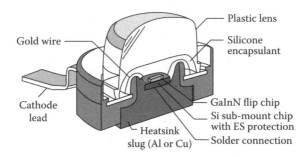

FIGURE 6.34 Packaging for high-powered LEDs showing the heat sink. (From Schubert, E. F. 2006. *Light-Emitting Diodes*. Cambridge, UK: Cambridge University Press. With permission.)

6.6.1 Principles of Operation of the Bipolar Junction Transistor

The BJT has two p-n junctions that are connected back-to-back (Figure 6.35). In the BJT, there are three regions: an *emitter*, a *base*, and a *collector* (Figure 6.35a). In the so-called *npn transistor* (Figure 6.35b), the emitter is the most heavily donor-doped n-type (n^{++}). The base is a p-type region that is relatively thin, moderately doped (p^+), and sandwiched between two n-regions. The collector is a lightly doped n-type region. A similar logic applies to how the different regions are arranged in a *pnp transistor* (Figure 6.35c; see Section 6.6.8). Doping levels are ~10^{19}, 10^{17}, and 10^{15} cm^{-3} for the emitter, base, and collector regions, respectively. The npn transistors are more popular than pnp transistors because electrons have a higher mobility.

As we will discuss in Section 6.6.2, the terms "emitter," "base," and "collector" originate from the transistor operation. The emitter region is the main source of injecting or emitting carriers, hence the name emitter. The carriers are emitted by the emitter and flow through the base and into the collector region. The original transistor reported in 1947 used a germanium (Ge) crystal as a mechanical base. This is how the term "base" came into being.

The symbols for npn and pnp transistors and their schematics showing different terminals and junctions are shown in Figure 6.36.

The BJT is a three-terminal, two-junction device. The direction of the arrow on the transistor symbols shows the direction of conventional current flow under *active mode* (see Section 6.6.6).

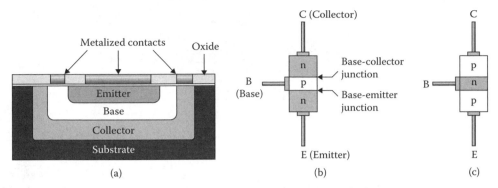

FIGURE 6.35 Schematic of bipolar junction transistors. (a) Basic epitaxial planar structure; (b) npn; (c) pnp. (From Floyd, T. L. 1998. *Electronics Fundamentals: Circuits, Devices, and Applications*, 4th ed. Boca Raton, FL: Chapman & Hall. With permission.)

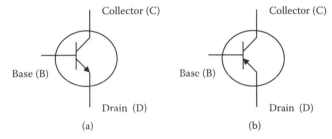

FIGURE 6.36 Symbols for bipolar junction transistors: (a) npn; (b) pnp. (From Floyd, T. L. 1998. *Electronics Fundamentals: Circuits, Devices, and Applications*, 4th ed. Boca Raton, FL: Chapman & Hall. With permission.)

Thus, for the npn transistor in active mode, the *conventional current* flows from the collector to the base and then to the emitter. For the npn transistor, a majority of the electrons injected from the emitter flow into the base and then into the collector.

Similarly, for a pnp transistor in active mode, the conventional current flows from the emitter to the base to the collector (Figure 6.36b). In this case, the positively charged holes are injected from the emitter into the base and then into the collector.

6.6.2 BIPOLAR JUNCTION TRANSISTOR ACTION

Consider the current flows for the npn transistor in the forward active mode Figure 6.37. In this mode, the base–emitter junction is forward-biased. The base–collector junction is reverse-biased. This is considered the *active mode* or *forward active mode* for the transistor. The band diagram for an npn transistor under these biases is shown in Figure 6.38. The band diagram for an npn transistor without bias is also shown for comparison.

The resultant current flows for an npn transistor in an active mode are shown in Figure 6.38.

In the *common base* (CB) *configuration*, the base electrode is common to the emitter and collector circuits. The transistor can also be connected in other configurations such as *common emitter* (CE) and *common collector* (CC; Figure 6.39).

6.6.3 CURRENT FLOWS IN AN NPN TRANSISTOR

To better understand the following discussion, the reader should review the discussion from Sections 5.10 and 5.11 concerning current flows in a forward-biased p-n junction.

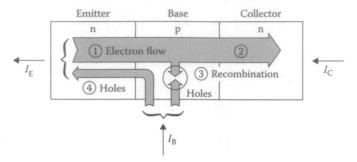

FIGURE 6.37 Common base configuration of current flows for an npn transistor in a forward active mode. (From Neaman, D. 2006. *An Introduction to Semiconductor Devices.* New York: McGraw Hill. With permission.)

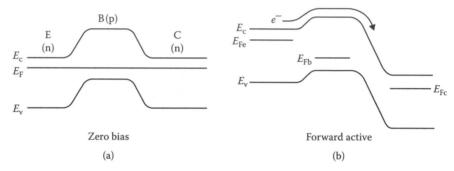

FIGURE 6.38 Band diagram for an npn transistor (a) without bias and (b) in active mode. (From Neaman, D. 2006. *An Introduction to Semiconductor Devices.* New York: McGraw Hill. With permission.)

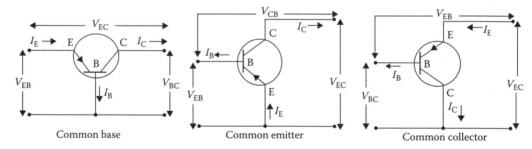

FIGURE 6.39 Common base (CB), common emitter (CE), and common collector (CC) configurations for an npn BJT. (From Singh, J. 2001. *Semiconductor Devices: Basic Principles.* New York: Wiley. With permission.)

Consider the current flows in the CB configuration and in active mode (Figure 6.37):

1. The base–emitter junction is forward-biased. Electrons from the emitter region are injected into the base region. This flow of negatively charged carriers sets up one part of the emitter current ($I_{E,1}$) from the base toward the emitter, because electrons have a negative charge.
2. The base–collector junction is under a reverse bias. Thus, at the collector–base interface, there is a very low concentration of electrons.
3. Therefore, on one side of the base, at the emitter–base junction, there is significant injection of electrons into the base. On the other side of the base, at the collector–base interface, there is a very low electron concentration. There is a significant concentration gradient of

electrons across the base, so that electrons diffuse across the base region into the collector region.

4. Once the electrons enter the collector region, they are driven by the internal electric field in the base–collector space-charge region. Note that this field is directed from the n-type collector toward the p$^+$ base region. This means that electrons are attracted toward the positively charged depletion layer on the n-side.

5. To summarize, for an npn transistor in an active mode, electrons are injected from the n^{++} region into the base region and continue on to the collector. Recombination of electrons occurs as they travel through the p-type base region.

6. One of the goals in the design of npn transistors is to transfer as much of the current from the emitter into the collector. This means that recombination occurring in the base region must be minimized. One way is to minimize the width of the base region (p-region in this case). A narrow base region is desirable because it minimizes the recombination of holes and electrons. The length of the neutral p-type material in the base must be much smaller than diffusion length of electrons (L_n).

7. The number of electrons that flow into the collector is controlled by the injection of electrons in the base. The injection of electrons depends upon the voltage applied to the base–emitter voltage (V_{BE}). This is the transistor action. In a BJT, the current at one terminal is controlled by the voltage at the other two terminals. In this case, the output, that is, the collector current (I_C), is controlled by the input voltage across the base and the emitter (V_{BE}).

8. Note that the transistor action is possible only if the two p-n junctions connected back to back are *interacting p-n junctions*. This means that junctions are designed so that the carriers injected from forward biasing of the emitter–base junction flow into the base–collector junction. For example, if the base region is very wide, then the injected carriers simply recombine and no transistor action results. This is the difference between a transistor and two diodes (p-n junctions) connected back-to-back.

9. In terms of magnitude, there are secondary current flows in addition to this flow of electrons from the emitter to the collector. They are very important in terms of the use of transistors as amplifiers.

10. In an npn transistor, some of the electrons are injected from the emitter into the p-type base region and recombine with the holes. The base contact connected to a power source supplies the replacement of these holes. This is one component of the base current ($I_{B,1}$).

11. For the forward-biased emitter–base junction, holes diffuse from the p-doped base region to the emitter n^{++} region. This is the second component of the base current ($I_{B,2}$) and also makes the second component of the emitter current ($I_{E,2}$). Since the base is lightly doped compared to the emitter, the current caused by the diffusion of holes is relatively small.

12. There is also a small reverse-bias current associated with the base–collector junction.

6.6.4   TRANSISTOR CURRENTS AND PARAMETERS

6.6.4.1   Collector Current

Due to the injection of electrons from the emitter into the base region and their journey into the collector region, the collector current (I_C) in an npn BJT is related to the base–emitter voltage (V_{BE}) by the following equation. This is also one part of the emitter current ($I_{E,1}$).

$$I_{E,1} = I_C = I_{S,1} \exp\left(\frac{V_{BE}}{V_t}\right) \qquad (6.33)$$

Equation 6.33 describes the transistor action. We control the current at the collector (I_C) by controlling the base–emitter voltage (V_{BE}).

6.6.4.2 Emitter Current

The total emitter current (I_E) in the npn transistor has two components. The first component ($I_{E,1}$) is due to the injection of electrons from the emitter into the base, which is equal to the collector current (I_C; Equation 6.33). The second component ($I_{E,2}$) is due to the diffusion of holes from the p-type base into the n-type emitter. Note that this current $I_{E,2}$ is part of the emitter current only. It does not become part of the collector current. The expression for this current ($I_{E,2}$) is

$$I_{E,2} = I_{S,2} \exp\left(\frac{V_{BE}}{V_t}\right) \tag{6.34}$$

where $I_{S,2}$ involves the minority-carrier (i.e., hole, in this case) parameters in the emitter.

Thus, the total emitter current (I_E) is given by

$$I_E = I_{E,1} + I_{E,2} = I_C + I_{E,2} = I_{SE} \exp\left(\frac{V_{BE}}{V_t}\right) \tag{6.35}$$

where I_{SE} represents the sum of $I_{S,1}$ (Equation 6.33) and $I_{S,2}$ (Equation 6.34).

The ratio of collector current (I_C) to the emitter current (I_E) is known as the *common base current gain* (α).

$$\alpha = \frac{I_C}{I_E} \tag{6.36}$$

Note that the $I_{E,2}$ component is not a part of the collector current. Therefore, I_C is always smaller than I_E. Thus, the common base current gain (α) is smaller than 1 and should be closer to 1.

6.6.4.3 Base Current

For the npn transistor in a forward active mode, some of the electrons injected from the emitter into the base recombine with the holes in the p-region. These holes must be replaced by a flow of positive charge into the base. This is the first component to the base current is ($I_{B,1}$). This part of the current is proportional to the rate at which the holes are recombining with the electrons being injected into the base region. This, in turn, is related to the V_{BE}, the magnitude of the forward bias. The other part of the base current ($I_{B,2}$ or $I_{E,2}$) in an npn transistor is due to the diffusion of holes from the p-type base into the n-type emitter.

There are two components of the base current (I_B), both of which are proportional to $\exp(V_{BE}/V_t)$. The ratio of the collector current (I_C) to the base current (I_B) is known as the base-to-collector *current amplification factor* (β).

$$\beta = \frac{I_C}{I_B} \tag{6.37}$$

This parameter is also known as *common emitter current gain* (β). The base current (I_B) is usually small, and the value of β is >100. The directions of the conventional current flow in npn and pnp transistors are summarized in Figure 6.40. The relationship between transistor parameters is shown in Example 6.4.

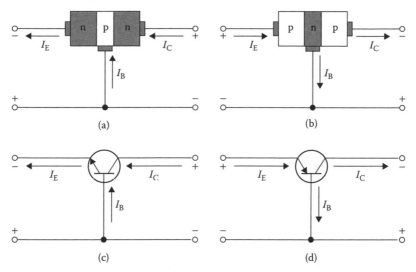

FIGURE 6.40 Directions of conventional current flows in bipolar junction transistor (a) npn, (b) pnp, (c) npn, (d) pnp. (From Floyd, T. L. 1998. *Electronics Fundamentals: Circuits, Devices, and Applications*, 4th ed. Boca Raton, FL: Chapman & Hall. With permission.)

EXAMPLE 6.4: THE RELATIONSHIP BETWEEN TRANSISTOR PARAMETERS α AND β

Show that the common base current gain (α) and common emitter current gain (β) are related by the following equation:

$$\beta = \frac{\alpha}{1-\alpha} \qquad (6.38)$$

SOLUTION

The emitter current (I_E) is equal to the sum of collector and base currents:

$$I_E = I_C + I_B \qquad (6.39)$$

Divide both sides of this equation by I_C:

$$I_E/I_C = 1 + (I_B/I_C)$$

The common base current gain $\alpha = I_C/I_E$ and the common emitter gain (β) is I_C/I_B.
 Therefore, we get

$$(1/\alpha) = 1 + (1/\beta)$$

or

$$\beta = \frac{\alpha}{1-\alpha}$$

Thus, the closer the value of α to 1 (i.e., the closer a transistor is to getting most of the emitter current to the collector), the higher the value of β.

6.6.5 ROLE OF BASE CURRENT

The base current (I_B) is small in magnitude, but is controllable and therefore plays a very important role in the functioning of a BJT. We have focused on showing that the collector current (I_C) can be controlled by controlling the emitter current (I_E) through V_{BE}. Small variations in the base current (I_B) can lead to large changes in I_C. This is how a BJT can be used as a *current amplifier*. A similar amplification effect is seen in situations involving alternating current (AC), as shown in Figure 6.41.

Consider an npn transistor in an active mode. Assume that the supply of holes available to compensate for the holes lost in recombination is now limited. If the electron injection from the emitter continues into the base region, the negative charge builds up in the base, which causes reduction of the forward bias for the emitter–base junction. This in turn creates a loss of electron injection and thus reduces the collector current (I_C). By controlling the base current (I_B), we can control the collector current (I_C). This situation is similar to the way the smallest step in a chemical reaction controls its overall kinetics, or how the weakest link in a structure determines its overall strength. Therefore, the base current (I_B) is sometimes known as the *controlling current*, while the collector current (I_C) or the emitter current (I_E) is known as the *controlled current*.

6.6.6 TRANSISTOR OPERATING MODES

The BJT is similar to a two-way valve that can be turned on in either direction and can deliver either a desired level of flow or no flow at all. We have considered a situation in which the emitter–base junction was forward-biased and the collector–base junction was reverse-biased. This combination of biases results in an active mode for transistor operation. In this active region, we use the base current to control the collector current. We can also use the emitter–base voltage to control the collector current.

If both the emitter–base and collector–base junctions are forward-biased, the transistor is said to be in *saturation mode*. A small biasing voltage results in a large current; the transistor is in the "on" state. If both junctions are reverse-biased, the transistor is in *cutoff mode*, in which no current flows. This is the "off" or "cutoff" state of the transistor. When the roles of the emitter and the collector are switched—that is, when the collector–base is forward-biased and the emitter–base is reverse-biased—the transistor is in *inverted* or *inverse active mode*. The polarities of the two junctions that describe the different modes of operation for a BJT are summarized in Figure 6.42.

6.6.7 CURRENT–VOLTAGE CHARACTERISTICS OF THE BIPOLAR JUNCTION TRANSISTOR

Consider the BJT in CB configuration (Figure 6.39). The *I–V* characteristics of a BJT in this configuration are shown by plotting the collector current (I_C) as a function of reverse-biased V_{BC} on the base–collector junction for various fixed values of I_E (Figure 6.43). Note the negative sign associated with V_{BC} on *x*-axis.

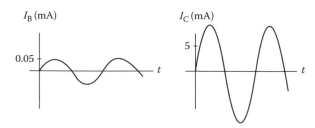

FIGURE 6.41 Schematic of the amplification achieved in a common emitter transistor circuit. Small changes in the base current (I_B) cause large changes in the collector current (I_C). (From Streetman, B. G., and S. Banerjee. 2000. *Solid State Electronic Devices*, 5th ed.Englewood Cliffs, NJ: Prentice-Hall. With permission.)

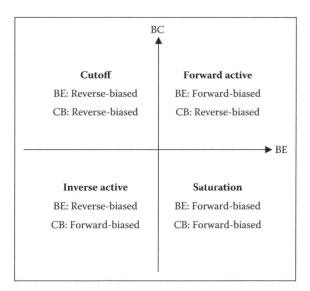

FIGURE 6.42 Different modes for operating a bipolar junction transistor.

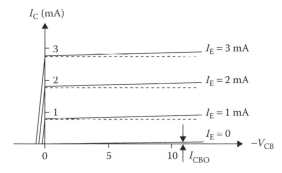

FIGURE 6.43 *I–V* characteristics of a bipolar junction transistor in common base configuration. (From Kasap, S. O. 2002. *Principles of Electronic Materials and Devices.* New York: McGraw Hill. With permission.)

For a BJT in CB configuration, when the transistor is in a cutoff mode, the emitter current (I_E) is nearly zero (a fraction of a microampere). This is shown as I_{CBO} in Figure 6.43. The emitter current increases exponentially with base–emitter voltage (V_{BE}; Equation 6.35). The collector current (I_C) increases with increasing emitter current (I_E). Once the emitter current (I_E) is finite, the collector current is not zero for $V_{BC} = 0$. The collector current (I_C) is independent of the base–collector voltage (V_{BC}).

In Figure 6.43, we have shown the I_C to be nearly constant with V_{BC}. The slight increase in the emitter current with a reverse-biased base–collector voltage (V_{BC}), shown by the solid lines, is called the *early effect*. As V_{BC} increases, the width of the depletion layer (W) associated with the base–collector junction increases. This means that the width of the neutral part of the base region through which the electrons must travel to get to the collector becomes shorter, and the number of electrons lost to the recombination with holes is reduced. This in turn causes a slight increase in I_C as a function of the increasing reverse bias (V_{BC}).

If the polarity of a base–collector junction is changed so that it is forward-biased, then the collector current is the difference between the forward currents associated with the two junctions, since the two forward currents subtract from each other.

The collector current (I_C) for an npn BJT in a CE configuration is shown as a function of the emitter–collector voltage (V_{EC}) in Figure 6.44.

In Example 6.5, we show the actual calculations of the transistor voltages and currents in a CE circuit.

6.6.8 CURRENT FLOWS IN A PNP TRANSISTOR

The current flows in a pnp transistor are similar in concept to those seen in Section 6.6.7 for the npn transistor (Figure 6.38). The doping levels follow a similar pattern; that is, the emitter has the highest level of doping (p⁺⁺), the n-type base is moderately doped, and p-type collector is lightly doped.

In this case, when the emitter–base junction is forward-biased, the emitter region injects holes into the base region. These holes flow through the relatively small, n-type base region and are then swept up by the internal electric field at the base–collector junction, ultimately arriving at the collector. Similar to an npn transistor, there is a current due to the diffusion of electrons from the base to the emitter region, shown with a filled arrow in Figure 6.45. The resultant current is directed from the emitter to the base because the electrons are negatively charged (Figure 6.45). There is also a recombination of injected holes and electrons in the base region. The electrons lost to the recombination are made up for by the base power supply. Similar to the npn transistor, there is a base current (I_B). A small reverse-biased current also exists at the base–collector junction. The particle motions and currents associated with this are not shown in Figure 6.45.

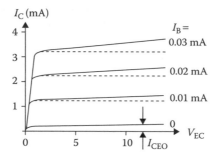

FIGURE 6.44 *I–V* characteristics of a bipolar junction transistor in a common emitter configuration. (From Kasap, S. O. 2002. *Principles of Electronic Materials and Devices*. New York: McGraw Hill. With permission.)

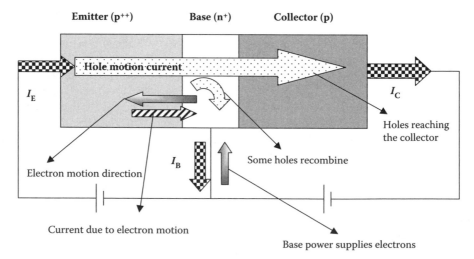

FIGURE 6.45 Summary of current and carrier flows in a pnp transistor in active mode.

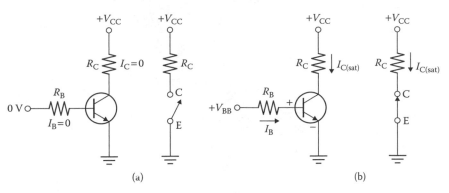

FIGURE 6.46 A transistor operating as a switch. (a) Cutoff, open switch; (b) saturation, closed switch. (From Floyd, T. L. 1998. *Electronics Fundamentals: Circuits, Devices, and Applications*, 4th ed. Boca Raton, FL: Chapman & Hall. With permission.)

The relationships among the base, emitter, and collector currents, as well as transistor parameters α and β, are the same as those defined in Equations 6.36 and 6.37 for the npn transistor.

6.6.9 APPLICATIONS OF BIPOLAR JUNCTION TRANSISTORS

The BJT can be used as an on-and-off switching device. In this application, the devices switch between the cutoff and saturation regions (Figure 6.46).

A transistor can be used as current amplifier (Figure 6.41). The current and voltage analysis for a BJT can be performed as shown in Example 6.5.

EXAMPLE 6.5: CURRENT AND VOLTAGE ANALYSIS FOR A BIPOLAR JUNCTION TRANSISTOR

Consider the transistor circuit shown in Figure 6.47. The value of resistance R_C connected between the collector and its power supply is 200 Ω. The resistance (R_B) connected between the base and the power supply is 15 kΩ. (a) Is this a pnp or npn transistor? (b) In what configuration is the transistor connected? (c) In what mode is the transistor? (d) What are the voltages across the different terminals (V_{CB}, V_{CE}, and V_{BE})? What are the transistor currents I_C, I_E, and I_B? Assume that $\beta = 200$, $V_{BB} = 5$ V, and $V_{CC} = 20$ V. Calculate the voltage drops across the resistors R_B and R_C.

SOLUTION

a. The arrow on the transistor symbol in Figure 6.47 indicates the direction of the conventional current flow. In this transistor, the conventional current flows from the collector → base → emitter, so the electrons flow from the emitter → base → collector. Therefore, this is an npn transistor (Figure 6.36a).

b. In the circuit diagram shown in Figure 6.47, we can see that this is a common emitter configuration, in which the emitter junction is common to the base and collector circuits.

c. The base–emitter junction is forward-biased because the p-type base is connected to the positive of the voltage supply (V_{BB}). The collector–emitter junction is reverse-biased because the n-type emitter is connected to the positive of the power supply (V_{CC}). Therefore, this transistor is in a forward active mode.

d. Since the base–emitter junction is forward-biased, the voltage drop at this junction is ~0.7 V, similar to most other forward-biased Si p-n junctions. Thus, $V_{BE} = 0.7$ V, and the voltage (V_{RB}) across the resistor (R_B) is

$$V_{RB} = V_{BB} - V_{BE}$$
$$\therefore V_{RB} = 5 - 0.7 = 4.3 \text{ V}$$

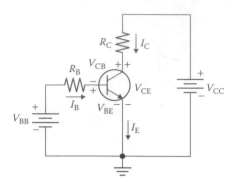

FIGURE 6.47 A transistor circuit for current and voltage analysis. (From Floyd, T. L. 2006. *Electric Circuit Fundamentals*, 7th ed. Boca Raton, FL: Chapman & Hall. With permission.)

By examining the base–emitter side of the circuit and applying Kirchhoff's law, which states that the algebraic sum of voltage drops around a closed loop is zero, the current I_B is given by

$$V_{BB} = (I_B \times R_B) + V_{BE}$$

Therefore,

$$I_B = \frac{V_{BB} - V_{BE}}{R_B}$$

$$\therefore I_B = \frac{(5 - 0.7)\,V}{15,000\,\Omega} = 287\,\mu A$$

$$I_B = 0.287\,mA$$

Since the current gain (β) is 200, from Equation 6.37 we get

$$\beta = \frac{I_C}{I_B}$$

$$\therefore I_C = 200 \times 0.287 = 57.3\,mA$$

Since the emitter current $I_E = I_C + I_B$, we get

$$I_E = 57.3\,mA + 0.287\,mA = 57.587\,mA$$

We will now look at the collector side of the circuit (Figure 6.47). The voltage drop (V_{RC}) across the resistor R_C is given by

$$V_{RC} = I_C \times R_C = 57.3 \times 10^{-3}\,A \times 200\,\Omega = 11.46\,V$$

Since the voltage V_{CC} is 20 V and the voltage across the resistor R_C is 11.46 V, the voltage V_{CE} is

$$V_{CE} = V_{CC} - V_{RC} = 20.0 - 11.46 = 8.54\,V$$

If we complete two electrical paths around the transistor, one from the collector to the base directly and the other from the collector to the emitter via the base, then the following relationship will be true:

The collector base voltage $V_{CB} = V_{CE} - V_{BE}$.

Therefore, $V_{CB} = 8.54 - 0.7 = 7.84$ V.

To summarize, the base current for this circuit $I_B = 0.287$ mA, and the collector current I_C is 200 times larger at 57.3 mA. The emitter current (I_E) is the sum of these two currents: $I_E = 57.587$ mA.

The base–emitter junction is forward-biased, so $V_{BE} = 0.7$ V (assumed for a typical p-n junction). The values of V_{CE} and V_{CB} are 8.54 and 7.84 V, respectively.

6.7   FIELD-EFFECT TRANSISTORS

The concept of the FET is relatively simple and was proposed in 1926 by Lilienfeld (Figure 6.48), before the discovery of the BJT. An FET is a device whose electrical resistance is controlled by the voltage.

The discovery of BJT was serendipitous, in that the research that originally aimed to develop the FET led to the discovery of the BJT. Although the concept of the FET existed for several years, the quality of materials and especially interfaces required to achieve the effect did not exist. Therefore, the initial FETs were based on the BJT. Today however, millions of types of FETs (known as MOSFETs) are routinely integrated onto computer chips (see Section 6.11).

In an FET, the *source* and *drain* regions are separated by the *channel* region. A *gate* is located between the source and the drain. The conductivity of the channel region is affected by the applied electric field at the gate. This is known as the *field effect*, hence the name "field-effect transistor." One of the main distinctions between the FET and the BJT is that in an FET, the current is mainly carried by one type of carrier (electrons or holes). Therefore, an FET is a *unipolar device*, whereas a BJT is a bipolar device.

The gate of an FET controls the flow of carriers between the source and the drain, and must be electrically isolated from the channel so that it can influence the flow of carriers without having to draw any significant amount of current flowing from the source to the drain through the channel region.

6.8   TYPES OF FIELD-EFFECT TRANSISTORS

The manner in which the gate isolation is achieved defines the different types of FETs (Figure 6.49). As shown in Figure 6.49, there are two main strategies for gate isolation. We have seen both of the basic effects associated with these strategies in the current flows at a p-n junction. The first strategy relies on changes in the width (W) of the depletion layer in the channel region as a function of

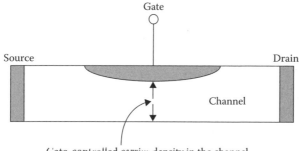

FIGURE 6.48   Illustration of the underlying principle for a field-effect transistor. (From Singh, J. 2001. *Semiconductor Devices: Basic Principles*. New York: John Wiley. With permission.)

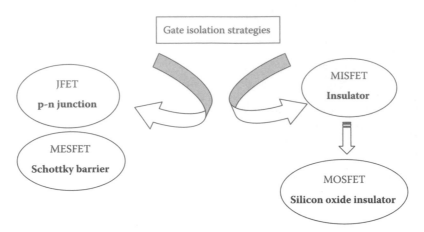

FIGURE 6.49 Different types of field-effect transistors based on different ways of isolating the gate region from the channel.

biasing voltage applied at the gate electrode. The second strategy relies on increasing or decreasing the conductivity of the channel region by creating a higher or lower carrier concentration.

In the first strategy, we create a reverse-biased p-n junction or a Schottky contact. If a reverse-biased p-n junction is used for gate isolation, we refer to the device as a *junction field-effect transistor* (JFET). If a Schottky contact is used for gate isolation, then we refer to the device as a *metal semiconductor field-effect transistor* (MESFET). These devices are used mainly with III–V semiconductors such as GaAs and indium phosphide (InP). In JFET and MESFET, we use a doped semiconductor to provide free carriers (electrons or holes). When we apply a bias to the gate, we change the depletion layer width in the channel region width (W). This in turn changes the flow of the current between the source and the drain for a given voltage bias applied between the source and the drain. We can then turn the JFET or MESFET on and off.

In the second strategy for gate isolation, we use an insulator deposited on the gate region. We deposit a metal contact onto the insulator. The device is known as a *metal insulator semiconductor field-effect transistor* (MISFET). The most widely used version of the MISFET is silicon oxide (SiO_2) as a *gate dielectric*. This device is known as the *metal oxide semiconductor field-effect transistor* (MOSFET).

There are two variations on MOSFET operation. In the first, the channel region is not conducting; that is, the transistor is in the off state. When we apply a voltage to the gate, we create a small region with a high carrier concentration underneath the gate insulator. After a voltage is applied between the source and drain regions, a current begins to flow through the channel, turning on the transistor. This mode is known as the *enhancement-mode MOSFET.*

In the second variation, the channel region already has a high conductivity; that is, the transistor is in the on state. When we apply a reverse bias to the gate, we deplete the carrier concentration in the channel region underneath the gate. If we now apply a voltage between the source and the drain, the transistor will not conduct because carriers are absent in the channel region, and thus the transistor is turned off. This way of operating a MOSFET is known as the *depletion-mode MOSFET.*

We will first discuss the operation of MESFET in Section 6.9, and then MISFET and MOSFET in Sections 6.10 and 6.11.

6.9 MESFET *I–V* CHARACTERISTICS

6.9.1 MESFET with No Bias

The MESFET uses a change in the width of the depletion layer (W) as a means of controlling the current flow from the source to the drain.

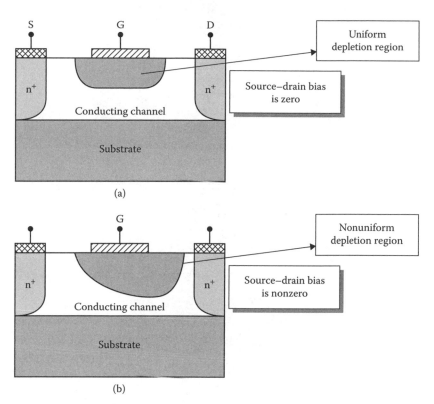

FIGURE 6.50   Changes in the width of the depletion layer with (a) zero bias and (b) non-zero bias. (From Singh, J. 2001. *Semiconductor Devices: Basic Principles*. New York: Wiley. With permission.)

For a MESFET with no bias applied to the gate (V_{GS}; Figure 6.50a), when a small bias is applied between the source and the drain (V_{DS}), a small amount of current flows to the drain (I_D). The magnitude of this current is given by (V_{DS}/R), where R is the resistance of the channel region. Thus, there is initially an ohmic region in that I_D increases with the increasing drain bias (V_{DS}). As the voltage V_{DS} increases further, the depletion layer width also increases (Figure 6.50). This causes the cross-sectional area of the channel current to decrease. This area is labeled as the conducting channel in Figure 6.50b. The channel resistance increases. The net result is that the channel current decreases. Therefore, with an additional drain bias (V_{DS}), the drain current (I_D) increases, but at a slower rate. A typical *I–V* characteristic curve for a MESFET, labeled $V_{GS} = 0$, shows this trend (Figure 6.51).

As the drain bias voltage increases even more, the depletion layer width also increases. This is shown as the shaded area under the gate region in Figure 6.50b. Since the increase in the voltage from 0 (at the source) to V_{DS} (at the drain) is not uniform across the channel, the width of the depletion layer is actually higher on the drain side. The increase of V_{DS} can be so high that it can actually pinch off the channel and saturate the drain current. If the applied drain bias (V_{DS}) increases further, the device will break down at $V = V_B$ (Figure 6.51).

6.9.2   MESFET WITH A GATE BIAS

When the gate bias (V_{GS}) is >0 (Figure 6.51), the depletion layer width of the channel region decreases, and overall the value of the drain current (I_D) increases for a given value of drain bias V_{DS}. If the gate bias is negative ($V_{GS} < 0$), the depletion layer width increases, and the overall I_D decreases for a given value of drain bias. The channel region pinch-off then occurs at lower voltages (Figure 6.51).

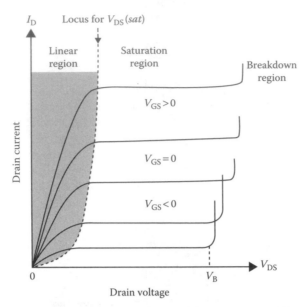

FIGURE 6.51 Typical *I–V* characteristics of a MESFET. (From Singh, J. 2001. *Semiconductor Devices: Basic Principles*. New York: Wiley. With permission.)

6.10 METAL INSULATOR FIELD-EFFECT TRANSISTORS

The second strategy for gate isolation involves the use of an electrical insulator, that is, a dielectric, nonconducting material, with a relatively large bandgap (Figure 6.49) known as the gate dielectric. To apply a voltage to the gate, a metal or a metal-like material is deposited onto the gate dielectric. This device, in which a metal is deposited onto an insulator to form an FET, is known as the MISFET. One of the best dielectrics for silicon is SiO_2. A MISFET with silica as the gate dielectric is known as MOSFET. A high conductivity polycrystalline silicon known as *polysilicon (poly-Si* or *polysil)*, is used as a gate electrode instead of a metal in the semiconductor processing of a MOSFET. The device is still traditionally referred to as MOSFET.

A thorough analysis and discussion of the operating principles for different FETs and the analysis of related circuits is beyond the scope of this book. Please consult introductory electrical engineering textbooks on microelectronics for a detailed discussion of these topics. In Section 6.11, we present an overview of the MOSFET, which is the most commonly used transistor.

6.11 METAL OXIDE SEMICONDUCTOR FIELD-EFFECT TRANSISTORS

6.11.1 MOSFET in Integrated Circuits

The MOSFET is the most important transistor device in ICs. Millions of transistors are integrated into a very small area of a silicon chip. The number of transistors on a computer chip has grown exponentially due to the significant advances in the science and technology of silicon microelectronic device processing (Figure 6.52).

The chart shown in Figure 6.52 is consistent with *Moore's law* (named for Roger Moore, cofounder of the Intel Corporation), which states that the number of active devices on a semiconductor chip doubles every 18 months.

6.11.2 Role of Materials in MOSFET

The structure of a conventional MOSFET is shown in Figure 6.53a. The current MOSFET design uses silica (SiO_2) as the gate dielectric and polysilicon as the conductor that forms the gate electrode.

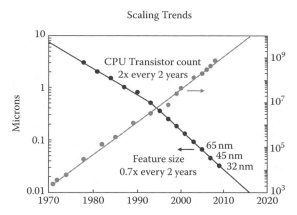

FIGURE 6.52 Transistor dimensions scale to improve performance, reduce power, and reduce cost per transistor. (Courtesy of Intel Corporation.)

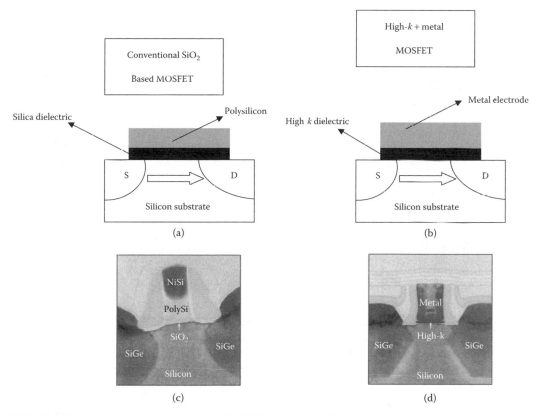

FIGURE 6.53 Structure of a conventional MOSFET based on (a) silicon oxide and polysilicon; (b) new high-k dielectrics and conductors; (c) silica gate dielectric with polysilicon gate electrode (65 nm transistor); and (d) hafnium-based dielectric with metal gate electrode (45 nm transistor). (Courtesy of Intel Corporation.)

Since the goal and trend in microelectronics has been to make transistors as small as possible (Figure 6.52), the thickness of the SiO_2 used as a gate dielectric has been getting smaller. The SiO_2 layer in state-of-the art transistors is currently only ~1.2 nm (12 Å; Figure 6.54). The thinness of the gate dielectric layer provides a serious challenge in scaling down the overall size of transistors in the future. Very thin oxide layers such as SiO_2 can electrically break down and start to "leak" current due to the fact that the oxide layer experiences a very high electric field.

SiO$_2$ Scaling

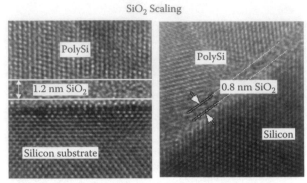

• 1.2 nm physical SiO$_2$ in production (90 nm logic node)
• 0.8 nm physical SiO$_2$ in research transistors

FIGURE 6.54 A transmission electron microscope image of a transistor cross-section showing the gate oxide layer (1.2 nm). (Courtesy of Intel Corporation.)

Researchers are therefore developing new gate dielectrics that insulate at even very small thicknesses. In 2007, the Intel Corporation reported a new hafnium oxide (HfO$_2$)–based insulator with a dielectric constant (k or ε_r) of ~25 (Figure 6.53b). The dielectric constant of SiO$_2$ is ~3.8. These higher dielectric constant materials, known as *high-k gate dielectrics*, allow the use of higher gate insulator thickness. The capacitance of a capacitor is directly proportional to its dielectric constant and inversely proportional to its thickness. Thus, higher-k gate dielectrics can be used with a higher insulator thickness, instead of using a lower-k materials with smaller thickness.

The higher thickness of the gate dielectric in turn helps reduce the current leakage, and thus, less energy is wasted as heat, resulting in a higher energy efficiency. Therefore, a computer can run for a longer time using the same battery power. Similarly, researchers in the semiconductor industry are also developing new alloys to replace polysilicon gate electrodes (Figure 6.53), because polysilicon is not compatible with HfO$_2$-based oxide.

6.11.3 NMOS, PMOS, AND CMOS DEVICES

The term "MOS" simply means a metal oxide semiconductor such as SiO$_2$ on silicon. MOSFETs can be designed so that the current from the source to drain is carried by electrons or holes; these devices are known as NMOS (or *n-channel*) and PMOS (or *p-channel*), respectively. Since the mobility of holes is lesser, the PMOS transistor occupies more area than a typical NMOS transistor.

To create an NMOS transistor, we start with a p-type substrate (e.g., silicon; Figure 6.55). We then create n-type source and drain regions, shown as n$^+$. The gate is isolated from the channel using SiO$_2$ and an n-type polysilicon is used to create the gate electrode. Note that the channel region is a p-type semiconductor.

When a positive voltage is applied to the gate, a channel of electrons is created between the source and the drain. This MOSFET is known as an n-channel MOSFET. The regions outside the transistor channel are heavily doped with acceptors, shown as p$^+$, to electrically isolate the drain and source regions from the substrate. This ability to electrically isolate devices is extremely important in integrating a large number of transistors into the small area of a silicon chip. When the gate voltage is sufficient with respect to the substrate or source (V_{GS}), there is a flowing drain current (I_D) shown in the *transfer characteristics* of an NMOS transistor (Figure 6.56a). These are different from the so-called *output characteristics*, which refer to the change in the drain current (I_D) as a function of source-to-drain voltage (V_{GS}) for a fixed-gate bias.

We can also create a complementary device to the NMOS transistor. In a PMOS transistor, there is an n-type substrate. The source and the drain are p$^+$ type. The gate region has an insulator with

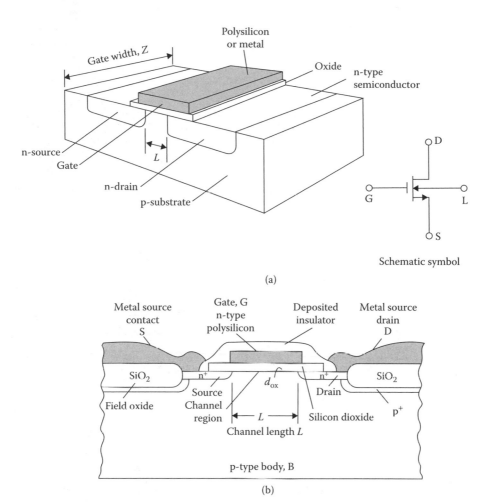

FIGURE 6.55 Schematic of an NMOS device: (a) structure and (b) cross-sectional view. (From Singh, J. 2001. *Semiconductor Devices: Basic Principles.* New York: Wiley. With permission.)

an electrode. When a *negative* voltage is applied to the gate, a region of n-type substrate under the gate develops a positive charge. The channel is then p-type, and the device is known as a p-channel or PMOS transistor. When a sufficiently large, negative gate voltage is applied, the PMOS transistor shows a drain current. This appears in the transfer characteristics of the PMOS transistor in enhancement mode (Figure 6.56b).

We can also operate the NMOS and PMOS devices in depletion mode (see Section 6.11.6). For an NMOS in depletion mode, we can turn off the channel conductivity by applying a negative gate voltage (Figure 6.56c). For a PMOS in depletion mode, we turn off the transistor by applying a positive voltage to the gate to deplete carriers in the channel (Figure 6.56d).

We can integrate the NMOS and PMOS devices onto the same substrate. This is known as a complementary MOSFET, or CMOS. CMOS devices consume relatively less power than NMOS and PMOS and are used in many applications of microelectronics (Figure 6.57).

6.11.4 ENHANCEMENT-MODE MOSFET

MOSFET can be operated in two modes (see Section 6.8). In enhancement-mode MOSFET, the conductivity of the channel region increases when a voltage is applied to the gate region and a current

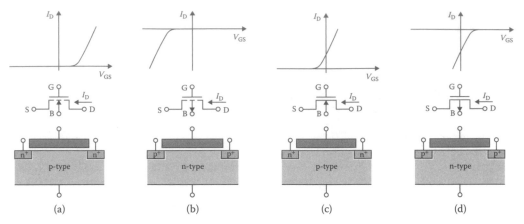

FIGURE 6.56 Transfer characteristics for the enhancement and depletion modes of NMOS (a and c) and PMOS (b and d). (From Dimitrijev, S. 2006. *Principles of Semiconductor Devices*. Oxford, UK: Oxford University Press. With permission.)

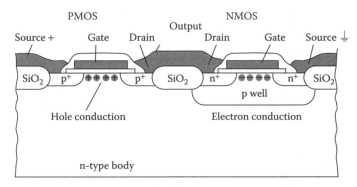

FIGURE 6.57 An illustration of a CMOS device. (From Singh, J. 2001. *Semiconductor Devices: Basic Principles*. New York: Wiley. With permission.)

begins to flow from the source to the drain. This is called the "on" state of the transistor. Enhancement-mode MOSFET can be NMOS (n-channel) or PMOS (p-channel; Figures 6.58 and 6.56a and b).

When no voltage is applied to the gate and the resistance of the channel region is too high, no current flows through from the source to the drain. This is called the "off" state of the transistor. Typical voltage–current characteristics are shown in Figure 6.59. This shows the increase in the drain current (I_D) as a function of drain bias with respect to the source (V_{DS}). We assume that the source and the substrate or the body are connected, but this is not always the case. If a voltage is applied between the body or the substrate and the source, this is known as the body effect and can be accounted for when estimating I_D.

In enhancement-mode MOSFET, applying a bias between the gate and the substrate (V_{GS}) increases the conductivity of the channel region. The mechanism through which this occurs is very different from that in MESFET, which makes use of changes in the width of the depletion layer (see Section 6.9).

6.11.5 Mechanism for Enhancement MOSFET

Consider an n-channel MOSFET, that is an NMOS made from a p-type semiconductor (Figure 6.55). The gate dielectric forms a parallel plate capacitor. The dielectric of a capacitor is normally sandwiched between two electrodes. However, in this case, one side of the capacitor is a conductor (polysilicon or some other metal), and there is a semiconductor underneath the gate dielectric. When a positive bias is applied to the gate electrode, negative charges are induced on the other side, that is, the dielectric–semiconductor interface. At a certain voltage called the *threshold voltage* (V_{TH}), a very thin layer of the original p-type semiconductor begins to behave like an n-type semiconductor. This is known as an *inversion layer*. When a small drain bias (V_{DS}) is applied, electrons can then flow from the source to the drain (Figure 6.59).

If the drain bias (V_{DS}) increases even more, then the voltage between the gate and the inversion layer near the drain decreases, as does the concentration of electrons near the drain region. This can continue with increasing V_{DS}. Eventually, this reaches a point where the voltage difference between the gate and channel becomes so small that an inversion layer cannot be maintained. This will pinch off the channel region, as there are usually very few electrons left near the drain region. This process causes the drain current (I_D) value to saturate (Figure 6.59).

If the gate bias with respect to the substrate (V_{GS}) increases, then the concentration of electrons present in the inversion layer increases. This causes an increase in the drain current when a small bias is applied to the drain (Figure 6.59).

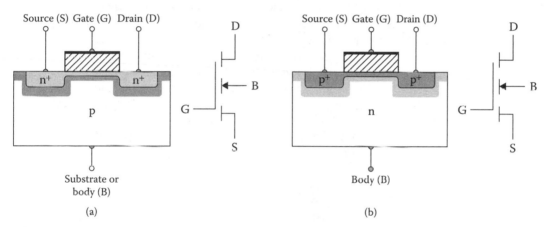

FIGURE 6.58 Schematic of (a) NMOS (n-channel) and (b) PMOS (p-channel) enhancement-mode MOSFETs. (From Neaman, D. 2006. *An Introduction to Semiconductor Devices*. New York: McGraw Hill. With permission.)

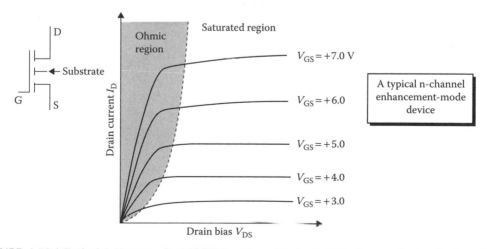

FIGURE 6.59 Typical *I–V* curves for MOSFETs. (From Singh, J. 2001. *Semiconductor Devices: Basic Principles*. New York: Wiley. With permission.)

The *I–V* characteristics for an enhancement-type MOSFET (Figure 6.59) and those for a MESFET look similar (Figure 6.51). However, the fundamental underlying phenomena responsible for their operations are very different. For MESFET, we change the depletion layer width (W) of the channel region as a function of the gate bias (see Section 6.9). For enhancement-type MOSFET (either NMOS or PMOS), we create an inversion layer in the channel region by inducing a charge at the insulator–semiconductor interface.

The most common MOSFET used is the n-channel or NMOS in enhancement mode.

6.11.6 DEPLETION-MODE MOSFET

In depletion-mode MOSFET, a MOSFET is created such that it is normally in an on state without any bias to the gate. This is a "ready-to-go" state achieved, for example, by creating a highly doped n-type region that forms a conductive channel between the source and the drain. In this case, the substrate is p-type. The free carriers in the n-type conductive channel are from donor doping. Similarly, we can create a p-channel MOSFET and operate it under the depletion mode. The structures of a depletion n-channel and p-channel MOSFET are shown in Figure 6.60.

When a negative voltage is applied to the gate of an NMOS, the conductivity of the channel region decreases because the carriers are depleted from the region. This switches the transistor into its off state, shown in the transfer characteristics (Figure 6.56c). When a positive gate voltage is applied to a PMOS in depletion mode, the transistor is switched off (Figure 6.56d).

The *I–V* curves for an NMOS depletion-layer MOSFET are shown in Figure 6.61.

Thus, if we consider the threshold voltage to be the gate voltage that is required to either just form the channel or just deplete it in the case of depletion-mode MOSFET, then the gate voltage is positive for the enhancement-mode NMOS and is negative for depletion-mode NMOS (Figures 6.56a and c). For PMOS, the threshold voltage is negative for the enhancement type and positive for the depletion type (Figures 6.56b and d).

6.12 PROBLEMS

6.1 Define the terms "work function" and "electron affinity." Why do we use the term work function for metals and electron affinity for semiconductors?

6.2 Sketch the band diagrams for a Schottky contact and an ohmic contact from an n-type semiconductor in contact with a metal.

6.3 What is the principle by which a solar cell operates?

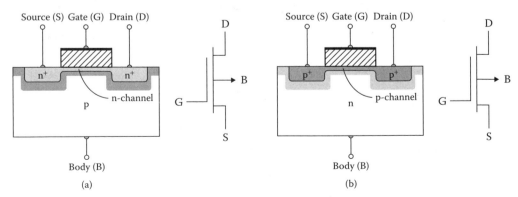

FIGURE 6.60 Schematic of depletion-mode (a) n-channel and (b) p-channel MOSFET. (From Neaman, D. 2006. *An Introduction to Semiconductor Devices.* New York: McGraw Hill. With permission.)

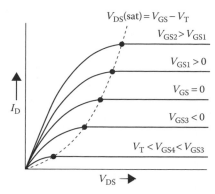

FIGURE 6.61 Characteristic I–V curves for an n-channel MOSFET in depletion mode. (From Neaman, D. 2006. *An Introduction to Semiconductor Devices*. New York: McGraw Hill. With permission.)

6.4 An Si p-n junction has an open-circuit voltage of 0.6 V. What value of J_L is required to produce this open-circuit voltage? Assume $T = 300$, $n_i = 1.5 \times 10^{10}$ cm^{-3}; $N_a = 10^{18}$ and $N_d = 10^{16}$ atoms/cm^3; $L_n = 10$ μm and $L_p = 25$ μm; and $D_p = 20$ and $D_n = 10$ cm^2/s.

6.5 Use the diffusion coefficient values in the previous example to calculate the carrier lifetimes.

6.6 Show that the maximum possible open-circuit voltage for a p-n junction in Si with $N_a = 10^{17}$ and $N_d = 3 \times 10^{16}$ atoms/cm^3 is 0.782 V.

6.7 If the photocurrent density J_L is 150 mA/cm^2, what is the open-circuit voltage for the p-n junction in Example 6.2?

6.8 Consider an Si p-n junction solar cell with electron and hole lifetimes of 3×10^{-7} s and 10^{-7} s, respectively. If the acceptor and donor doping levels are 3×10^{17} and 10^{16} atoms/cm^3, respectively, what is the open-circuit voltage (V_{oc})? Assume $D_p = 20$ and $D_n = 10$ cm^2/s and $J_L = 10$ mA/cm^2.

6.9 A solar cell with an area of 1 cm^2 has I_L of 15 mA. If the saturation current (I_S) is 3×10^{-11} A (a) calculate the open-circuit voltage (V_{oc}) and (b) calculate the short-circuit current (in milliamperes). (c) If the fill factor is 0.7, what is the power extracted for each solar cell (in milliwatts)?

6.10 Why is GaAs better suited than Si for making efficient solar cells? Why, then, is it not widely used?

6.11 Do solar cells require indirect or direct bandgap semiconductors? Explain.

6.12 In Figure 6.62, the I–V curve for a solar cell is shown. Calculate the fill factor for this solar cell. (Hint: Draw the largest possible rectangle that fits inside the given I–V curve.)

6.13 For the I–V curve shown in Figure 6.20, what is the open-circuit voltage (V_{oc}) for this solar cell? Show that the fill factor for this solar cell is about 0.8.

6.14 What is the basic principle of an LED? How is it different from that of a solar cell?

6.15 What is the difference between spontaneous and stimulated radiation?

6.16 Do LEDs always require a direct bandgap semiconductor?

6.17 What mechanisms cause LEDs' overall efficiency to be lower?

6.18 What techniques can one use to produce LEDs that emit a "white" light?

6.19 In Example 6.3, what is the current through the circuit with a resistance of 390 Ω that is used in a series? Assume that the voltage driving the LED changes to 2.0 V.

6.20 In Example 6.3, what is the current through the circuit with a resistance of 800 Ω that is used in a series? Assume that the voltage driving the LED changes is 1.8 V.

6.21 What is the minimum drive voltage for a blue LED that emits at 470 nm?

6.22 Show that the minimum drive voltage for an LED that emits at 1550 nm is 0.8 V.

6.23 Why is the BJT considered a bipolar device?

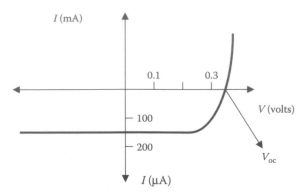

FIGURE 6.62 *I–V* curve for a solar cell. (From Edwards-Shea, L. 1996. *The Essence of Solid-State Electronics*. Englewood Cliffs, NJ: Prentice Hall. With permission.)

6.24 An engineer uses an Si crystal to create two p-n junctions connected back-to-back. However, he or she finds that this device does not function as a transistor. Explain why.

6.25 A transistor has a base current of 50 μA. If the collector current is 3.7 mA, what is the value of emitter current (I_E)? What are the values of transistor parameters α and β?

6.26 In Example 6.5, what is the value of α?

6.27 In the circuit shown in Example 6.5 (Figure 6.47), assume that the values of $R_B = 10$ kΩ, $R_C = 100$ Ω, $V_{BB} = 5$ V and $V_{CC} = 10$ V. Assume the current gain is 100. Calculate the transistor currents and voltages across different junctions.

6.28 In the circuit shown in Example 6.5 (Figure 6.47), assume that the values of $R_B = 50$ kΩ, $R_C = 500$ Ω, $V_{BB} = 5$ V, and $V_{CC} = 10$ V. Assume the current gain (β) is 90. Calculate the transistor currents and voltages across different junctions.

6.29 What does the term "field effect" mean?

6.30 Explain how the different gate-isolation strategies used for JFET and MESFET differ from those used for MOSFET.

6.31 What does enhancement-mode MOSFET mean? How is this different from depletion-mode MOSFET?

6.32 Explain the need for developing new high-*k* gate dielectrics for MOSFETs.

GLOSSARY

Active mode: An npn transistor in which the emitter–base junction is forward 0 biased and the base–collector junction is reverse 0 biased (Figure 6.37) is an example of a transistor in active mode.

Base: The moderately doped middle region sandwiched between the collector and emitter of a transistor.

Base current: A flow of positive charge into the base; controls the collector and emitter currents.

Base-to-collector current amplification factor (β): The ratio of the collector current to the base current $\beta = I_C/I_B$. Typical values are ~>100, also known as common emitter current gain.

Bipolar device: Any microelectronic device whose functioning depends upon the motions of both electrons and holes (e.g., BJT).

Bipolar junction transistor (BJT): A device with three terminals (emitter, base, and collector) based on two interacting p-n junctions connected back-to-back. The current at the collector can be controlled by regulating the voltage between the two other terminals, emitter and base. The most commonly used BJT is the npn type because of the higher mobility of the electrons.

CMOS: A MOSFET in which both NMOS and PMOS transistors are integrated. The acronym stands for complementary metal oxide silicon. The term FET is implied.

Collector: The lightly doped region in a transistor that ultimately receives the charge carriers emitted by the emitter.

Common base (CB) configuration: A BJT configuration in which the base electrode is common to the emitter and collector circuits. The transistor can be connected in other configurations as well (e.g., common emitter).

Common base current gain (α): The ratio of collector current to the emitter current: $\alpha = I_C/I_E$. Maximum value possible is 1.

Common emitter (CE) configuration: A BJT configuration in which the emitter electrode is common to the base and collector circuits.

Common emitter current gain (β): The ratio of the collector current to the base current $\beta = I_C/I_B$. Typical values are ~>100, also known as base-to-collector current amplification factor.

Controlled current: A collector or emitter current, which is controlled by the base current.

Controlling current: See **Base current**.

Copper indium diselenide (CuInSe$_2$): A compound semiconductor material for high-efficiency solar cells. Also known as CIS.

Current amplifier: The use of a BJT in which a small change in the base current (I_B) causes large changes in the collector current (I_C).

Cut-off mode: A mode of an npn operation in which both the emitter–base and base–collector junctions are reverse-biased.

Depletion mode: A MOSFET that is normally in the "on" state without bias. This is achieved by creating a highly doped region forming a conductive channel. When a negative bias is applied to the gate region, the conductivity of the channel region decreases and can turn "off" the device (Figure 6.60).

Early effect: In a BJT with common base configuration, when V_{BC} reverse bias increases, the width of the depletion layer (W) associated with the base–collector junction gets larger. This means the width of the base region through carrier diffusion gets a little shorter, causing a slight increase in I_C as a function of V_{BC}.

Effective Richardson constant: The term $\left(\dfrac{m^* q k_B^2}{2\pi^2 \hbar^3}\right)$, which is related to the reverse saturation current due to thermionic emissions in a Schottky contact.

Electron affinity of a semiconductor ($q\chi_S$): The energy required to remove an electron from the conduction band and set it free.

Emitter: The heavily doped region in a transistor that injects minority carriers into the base.

Enhancement mode: The mode in which a potential forward or reverse bias is applied to the gate region of a MOSFET that is normally off; the conductivity of the channel region increases and a current can begin to flow from the source to the drain (Figure 6.58).

Fill factor (F_f): The ratio of $V_m \times I_m$ to $V_{oc} \times I_{oc}$. Most solar cells have a fill factor of ~0.7.

Field effect: The effect of an applied electric field on the conductivity of the channel region.

Field-effect transistor (FET): A unipolar device based on the control of charge carrier flow between a source and a drain, using voltage applied to a region known as the gate. The applied electric field changes the conductivity of the channel region, hence the name field-effect transistor.

Forward active mode: See **Active mode**.

Gate: The region in an FET between the source and the drain.

Gate dielectric: A nonconducting, dielectric material that isolates the gate electrode from the channel region. Usually the gate dielectric is silicon oxide (SiO$_2$), although new gate dielectrics with higher dielectric constant are being developed.

Heterojunction: A p-n junction in which the n- and p-sides are made using two different semiconductors.

Homojunction: A p-n junction in which the n- and p-sides are made using the same semiconductor.

High-k gate dielectrics: Recently developed materials used as gate insulators (e.g., hafnium oxide with $k \sim 25$). These materials allow for a higher thickness of the dielectric compared to silica, which leads to smaller leakage of current and more efficient computer chips.

Interacting p-n junctions: Two p-n junctions connected back-to-back and designed such that transistor action is possible. This means that the carriers injected from forward biasing of the emitter–base junction flow into the base–collector junction. If the base region is too long, injected carriers simply recombine and there is no transistor action.

Inverse active or inverted mode: A mode of npn transistor operation in which the roles of emitter and collector are reversed—the emitter–base is reverse-biased and the collector–base is forward-biased.

Inversion layer: In a MOSEFT, a thin layer underneath the gate that exhibits semiconductivity opposite to that of the rest of the substrate when a voltage is applied. In a p-type substrate, it becomes enriched with electrons and behaves as n-type.

Junction field-effect transistor (JFET): An FET in which a reverse-biased p-n junction is used to isolate the gate region from the channel current flow. This is used mainly for III–V semiconductors such as GaAs.

Laser diode: In a laser diode, electrons and holes are generated when photons of a given energy and wave vector are absorbed by the semiconductor. The electrons and holes recombine and cause a stimulated emission of coherent photons, that is, the photons emitted are in phase with the incident photons, that is with the same energy and wave vector.

Majority device: A device such as a Schottky diode in which currents are generated primarily through the motions of majority carriers. The motion of minority carriers does not play a primary role.

Metal semiconductor field-effect transistor (MESFET): An FET in which a metal that forms a Schottky contact with the semiconductor is used to isolate the gate region from the channel current flow. This is used mainly for III–V semiconductors such as GaAs.

Metal insulator field-effect transistor (MISFET): An FET in which a dielectric material is used to isolate the gate region from the channel current flow. A metal electrode forming an ohmic contact with the dielectric allows application of the gate voltage.

Metal oxide field-effect transistor (MOSFET): An FET, usually based on silicon, in which SiO_2 is used as a dielectric for isolating the gate region from the channel current flow. A relatively high-conductivity polysilicon (in this context, a metal) forms the electrode that allows application of the gate voltage.

MOS: A metal oxide semiconductor, for example, silicon oxide on silicon. It is usually used to indicate a MOSFET whose operation can be explained using an MOS capacitor.

Moore's law: Named after Roger Moore, cofounder of Intel, this law states that the number of active devices on a semiconductor chip doubles every 18 months.

n-channel: The conduction from the source to the drain in the channel (which is p-type) occurs by electrons; hence the channel is known as an n-channel or NMOS device.

NMOS: A MOSFET in which electrons carry the channel current from source to drain.

npn transistor: A BJT with a p-type base sandwiched between the n-type emitter and source.

Ohmic contact: A contact between a metallic material and a semiconductor such that there is no barrier to block the flow of current at the junction. The contact resistance may or may not follow Ohm's law. This contact is seen if $\phi_M < \phi_S$ for an n-type semiconductor or if $\phi_M > \phi_S$ for a p-type semiconductor.

Open-circuit voltage (V_{oc}): The maximum possible voltage generated from a solar cell.

Optical generation: The generation of carriers (electrons and holes) as a result of the absorption of light energy in a semiconductor or a p-n junction. This process forms the basis for solar cell operation.

Organic light-emitting diodes (OLEDs): Diodes based on organic materials with an electron-injecting cathode instead of the n-layer and a hole-injecting anode. These devices have considerable promise for energy-efficient lighting and are also used in flat-screen displays.

Output characteristics: The plot of the drain current (I_D) as a function of the source to drain voltage (V_{DS}), for a given value of gate bias. These are different from the transfer characteristics.

p-channel: The conduction from the source to the drain in the channel (which is n-type) occurs by holes; the channel is known as p-channel or PMOS.

Photocurrent (I_L): A current directed from the n- to p-side of a p-n junction resulting from the process of photogeneration. Also known as the short-circuit current; flows in an external circuit with no resistance.

Photogeneration: See **Optical generation**.

Photovoltaics: A field of research and development involving converting light energy into electricity.

Photovoltaic effect: The appearance of forward voltage across an illuminated p-n junction.

PMOS: A MOSFET in which the current from source to drain is carried by holes.

pnp transistor: A BJT with an n-type base sandwiched between a p-type emitter and source.

Polymer organic light-emitting diodes (PLEDs): See **Organic light-emitting diodes**.

Polysilicon (poly-Si or polysil): A polycrystalline form of silicon. A relatively high-conductivity form of this material is used as a gate electrode for MOSFET and is referred to as a metal in this context.

Richardson constant (R^*): See **Effective Richardson constant**.

Saturation mode: A mode to operate an npn transistor in which both the emitter–base and base–collector junctions are forward-biased.

Schottky barrier (ϕ_B): The barrier that prevents injection of electrons from a metal into a semiconductor. Often, this barrier height is independent of the metal because the Fermi energy is pinned at the interface.

Schottky contact: The rectifying contact between a metal and a semiconductor, seen if $\phi_M < \phi_S$ for a p-type semiconductor or if $\phi_M > \phi_S$ for an n-type semiconductor.

Schottky diode: A diode based on a metal–semiconductor junction, requiring $\phi_M < \phi_S$ for a p-type semiconductor or $\phi_M > \phi_S$ for an n-type semiconductor.

Short-circuit current (I_{sc}): The same as a photocurrent (I_L), this refers to the maximum current flowing when a solar cell is connected to an external circuit with zero external resistance.

Silicides: Intermetallic compounds formed by reactions of elements with silicon.

Solar cell: A p-n junction–based device that generates electric voltage or current upon optical illumination.

Solar cell conversion efficiency (η_{conv}): The ratio of maximum power delivered to the power that is incident (P_{in}):

$$\eta_{conv} = \frac{I_m \times V_m}{P_{in}}$$

where I_m and V_m are the current and voltage, respectively, which lead to maximum power.

Spontaneous emission: Radiation of photons by electron–hole recombination such that the emitted photons have no particular phase relationship with each other and are thus incoherent.

Stimulated emission: Radiation of photons by electron–hole recombination such that the emitted photons are coherent, that is, they have the same energy and wave vector as the photons that cause the emission to occur.

Surface pinning: A constant energy level that does not depend upon doping. Defects at the surface or interface of a semiconductor cause the Fermi energy of the semiconductor to be pinned at a bandgap level. Surface pinning of the Fermi energy level causes the same Schottky barrier heights for different metals deposited on a semiconductor.

Thermionic emission: The process in which electrons on the semiconductor side, which have a high-enough energy, overcome the built-in potential (V_0) and flow onto the metal side in a Schottky contact. This creates a current known as the thermionic current.

Threshold voltage (V_{TH}): The gate voltage that is required to either just form the channel or just deplete it.

Transfer characteristics: The plot of drain current (I_D) as a function of the gate voltage for an NMOS or PMOS. These are different from the output characteristics that refer to the change in drain current (I_D) as a function of V_{DS} for a fixed value of gate bias.

Transistor: An abbreviation for a transfer resistor, a microelectronic device based on p-n junctions that is used as a tiny switch, as an amplifier, or for other functions.

Transistor action: In a transistor, the current at one terminal is controlled by the voltage at the other two terminals.

Unipolar device: A device in which the current is carried mainly by one type of carrier, electrons or holes. FET is a unipolar device. This is one of the main distinctions between an FET and a BJT.

Work function ($q\phi$): The energy required to remove an electron from its Fermi energy level and set it free.

REFERENCES

Baliga, B. J. 2005. *Silicon Carbide Power Devices*. Singapore: World Scientific.

Bergh, A., G. Craford, A. Duggal, and R. Haitz. 2001. The promise and challenge of solid-state lighting. *Physics Today* 54:12.

Dimitrijev, S. 2006. *Principles of Semiconductor Devices*. Oxford, UK: Oxford University Press.

Edwards-Shea, L. 1996. *The Essence of Solid State Electronics*. Englewood Cliffs, NJ: Prentice Hall.

Floyd, T. L. 1998. *Electronics Fundamentals: Circuits, Devices, and Applications*, 4th ed. Boca Raton, FL: Chapman & Hall.

Floyd, T. L. 2006. *Electric Circuit Fundamentals*, 7th ed. Boca Raton, FL: Chapman & Hall.

Green, M. A. 2003. *Solar Energy* 74:181–92.

Kano, K. 1997. *Semiconductor Fundamentals*. Upper Saddle River, NJ: Prentice Hall.

Kasap, S. O. 2002. *Principles of Electronic Materials and Devices*. New York: McGraw Hill.

Mahajan, S., and K. S. Sree Harsha. 1998. *Principles of Growth and Processing of Semiconductors*. New York: McGraw Hill.

Nakamura, S., M. Senoh, N. Iwasa, S. Nagahama, T. Yamaka, and T. Mukai. 1995. *Jpn J Appl Phys* 34:L1332.

Neamen, D. 2006. *An Introduction to Semiconductor Devices*. New York: McGraw Hill.

Schubert, E. F. 2006. *Light-Emitting Diodes*. Cambridge, UK: Cambridge University Press.

Singh, J. 1996. *Optoelectronics: An Introduction to Materials and Devices*. New York: McGraw Hill.

Singh, J. 2001. *Semiconductor Devices, Basic Principles*. New York: Wiley.

Solymar, L. and D. Walsh. 1998. *Electrical Properties of Materials*, 6th ed. Oxford, UK: Oxford University Press.

Sparkes, J. J. 1994. *Semiconductor Devices*: London: Chapman and Hall.

Streetman, B. G., and S. Banerjee. 2000. *Solid State Electronic Devices*. Englewood Cliffs, NJ: Prentice-Hall.

Sze, S. M. 2002. *Semiconductor Devices, Physics and Technology*, 2nd ed. New York: Wiley.

7 Linear Dielectric Materials

KEY TOPICS

- Dielectric materials
- Induction, free, and bound charges
- Dielectric constant and capacitor
- Polarization mechanisms in dielectrics
- Dielectric constant and refractive index
- Ideal versus real dielectrics
- Complex dielectric constant
- Dielectric losses
- Frequency and temperature dependence of dielectric properties
- Linear and nonlinear dielectrics

7.1 DIELECTRIC MATERIALS

A *dielectric material* is typically a large bandgap semiconductor ($E_g \sim$ >4 eV) that exhibits a high resistivity (ρ). The prefix *dia* means "through" in the Greek language. The word *dielectric* refers to a material that normally does not allow electricity (electrons, ions, and so on) to pass through it. There are special situations, for example, exposure to very high electric fields or changes in the composition or microstructure, which may lead to a dielectric material exhibiting semiconducting or metallic behavior. However, when the term "dielectric material" is used, it is generally understood that the material is essentially a nonconductor of electricity. An electrical *insulator* is a dielectric material that exhibits a high breakdown field.

7.1.1 ELECTROSTATIC INDUCTION

To better understand the behavior of nonconducting materials, let us first examine the concept of *electrostatic induction* and what is meant by the terms "free charge" and "bound charge." First, consider a dielectric such as a typical ceramic or a plastic that has a net positive charge on its surface. Now, assume that we bring a conductor B near this charged insulator A (Figure 7.1a). The electric field associated with the positively charged insulator A "pulls" the electrons toward it from conductor B. This is also described as the atoms in conductor B being "polarized," or affected by the presence of an electric field.

The process of the development of a negative charge on conductor B is known as electrostatic induction. In this case, the negative charge developed on conductor B is the "bound charge," because this charge is bound by the electric field caused by the presence of the charged insulator A next to it.

The creation of a bound negative charge on conductor B, in turn, creates a net positive charge on the other side of the conductor, because the conductor itself cannot have any net electric field within it. If we connect this conductor B with a grounded wire, then electrons will flow from the ground to this conductor and make up for the positive charge (Figure 7.1b). The flow of electrons from the ground to the conductor can also be described as the flow of positive charge from conductor B to the ground. Thus, the positive charge on conductor B is considered the "free charge." The word "free" in this context means that these charges are mobile, that is, they are free to move.

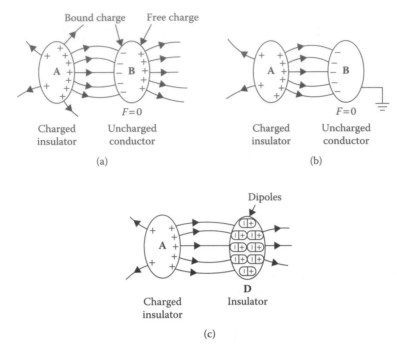

FIGURE 7.1 Illustration of electrostatic induction and bound and free charges, where A is the charged insulator, B is the conductor, and D is another dielectric. (a) Isolated induction, (b) grounded induction, and (c) induction with isolated insulator D. (From Kao, K. C. 2004. *Dielectric Phenomena in Solids*. London: Elsevier-Academic Press. With permission.)

If, instead of bringing in conductor B near the charged insulator A, we bring in another dielectric material D, then no induced charge is created on this dielectric material because no free carriers are available (Figure 7.1c). If we could look into this dielectric material D at an atomic scale, we would see that the electronic clouds are not perfectly symmetric around the nuclei of atoms. Instead, the electronic clouds are tilted toward the positive charges in insulator A. The presence of an electric field creates dipoles within the atoms of material D, which is polarized by the electric field emanating from dielectric A. If we now move dielectric D away from dielectric A, we find that after some time, the dipole moments induced in this material fade away because of fluctuations in thermal energy.

7.2 CAPACITANCE AND DIELECTRIC CONSTANT

7.2.1 PARALLEL-PLATE CAPACITOR FILLED WITH A VACUUM

A capacitor is a device that stores electrical charge. Consider two parallel conductive plates of area (A), separated by a distance d, and carrying charges of $+Q$ and $-Q$; assume that there is a vacuum between these plates. This is the basic structure of a parallel-plate capacitor.

The charge (Q) on the plates creates a potential difference V. We define the proportionality constant as the capacitance (C). Thus,

$$Q = C \times V \tag{7.1}$$

The SI unit of capacitance is a Farad (or coulombs per volt [C/V]). One Farad is a very large capacitance. Some *supercapacitors* have capacitances in this range. Most capacitors in microelectronics have a capacitance that is expressed in microfarads (10^{-6} F), nanofarads (10^{-9} F), picofarads (10^{-12} F), or femtofarads (10^{-15} F).

One of the laws of electrostatics is *Gauss's law*, which states that the area integral of the electric field (E) over any closed surface is equal to the net charge (Q) enclosed in the surface divided by the permittivity (ε).

$$\oint \vec{E} \times d\vec{A} = \frac{Q}{\varepsilon_0} \tag{7.2}$$

Gauss's law and the fundamentals of many electrical and magnetic properties and phenomena are derived from Maxwell's equations.

We define the surface charge density (σ or σ_s) as the charge per unit area.

$$\sigma_s = \frac{Q}{A} \tag{7.3}$$

The SI unit of charge density is coulombs per square meter (C/m^2). Note that the letter "C" represents the capacitance of a capacitor or charge in coulombs. The charge density is usually expressed as, for example, microcoulombs per square centimeter ($\mu C/cm^2$) or picofarads per square nanometer (pF/nm^2).

The generation of a voltage (V) between two plates separated by a distance d is also represented using the electric field (E) as follows:

$$E = \frac{V}{d} \tag{7.4}$$

We can rewrite Equation 7.1 for the capacitance (C) using Equations 7.3 and 7.4 as follows:

$$C = \frac{Q}{V} = \frac{\sigma_s \times A}{E \times d} \tag{7.5}$$

This equation tells us that the capacitance of a capacitor, that is, its ability to hold charge, depends on geometric factors, namely, the areas (A) of the plates, and the distance (d) between them.

We define the *dielectric permittivity* (ε) of the material between the plates as

$$\varepsilon = \frac{\sigma_s}{E} \tag{7.6}$$

Therefore, from Equations 7.5 and 7.6,

$$C = \varepsilon \frac{A}{d} \tag{7.7}$$

We use the special symbol ε_0 for permittivity of the vacuum or free space, which is equal to 8.85×10^{-12} F/m.

We can write an expression for the capacitance of a capacitor filled with vacuum (C_0) between the conductive plates as

$$C_0 = \frac{Q}{V_0} = \frac{\sigma_{s,0} A}{E_0 d} = \varepsilon_0 \frac{A}{d} \tag{7.8}$$

The subscript "0" has been added to indicate that this equation refers to a capacitor filled with a vacuum. Thus,

$$\varepsilon_0 = \frac{\sigma_{s,0}}{E_0} \tag{7.9}$$

We define the *dielectric flux density* (*D*) or *dielectric displacement* (*D*) as the total surface charge density.

$$D = \sigma_s \tag{7.10}$$

We will see in Section 7.3 that the dielectric flux density (*D*) can also be written as a sum of the free charge density and *bound charge density* (σ_b). The SI unit for dielectric displacement (*D*) is the same as that for charge density, that is, C/m^2.

7.2.2 Parallel-Plate Capacitors with an Ideal Dielectric Material

Consider a capacitor in which the space between the two plates is filled with an *ideal dielectric material* (Figure 7.2). The term "ideal dielectric" means that when charge is stored in a capacitor, no energy is lost in the processes that lead to storage of the electrical charge. In reality, this is not possible. In Section 7.13, we will define the term "dielectric loss," which represents the electrical energy that is wasted or used when charge-storage processes occur in a dielectric material. These processes are known as polarization mechanisms.

In one form of polarization, a tiny dipole is induced within an atom. As shown in Figure 7.2, an applied atom without an electric field has a symmetric electronic cloud around the nucleus. The centers of the positive and negative charges coincide, and there is no *dipole moment* (μ). However, when the same atom is exposed to an electric field, the electronic cloud becomes asymmetrical, that is, the electronic cloud of a polarized atom moves toward the positive end of the electric field. The nucleus remains essentially at the same location. Therefore, the centers of the positive and negative charges are now separated by a smaller distance (*x*). This process of creating or inducing a dipole in an atom is known as *electronic polarization* (Figure 7.3).

For a dipole with charges +*q* and –*q* separated by a distance *x*, the dipole moment is μ = *q* × *x*. The international system (SI) unit of dipole moment is C·m. We define one *Debye* (D) as being equal to 3.3356×10^{-30} C·m. Please note that the symbol D is also commonly used for dielectric displacement.

In a dielectric material exposed to an electric field, these induced atomic dipoles align in such a way that all the negative ends of the dipole line up near the positively charged plate (Figure 7.3b). This process means that some of the surface charges on the plates, which were originally free when there was a vacuum between the plates, will now become "bound" to the charges inside the dielectric material. When the capacitor is filled with a vacuum, all of the charges on the plates are free. The

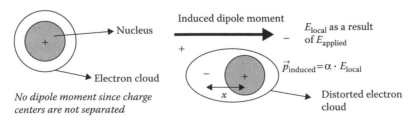

FIGURE 7.2 Illustration of electronic polarization of an atom.

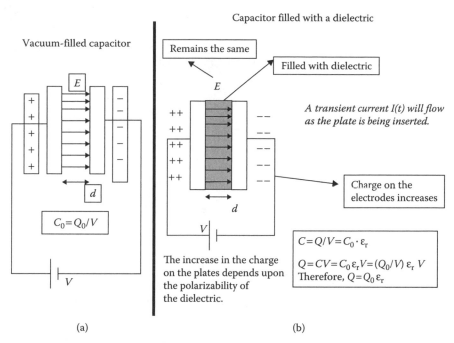

FIGURE 7.3 Structure of a parallel-plate capacitor filled with (a) a vacuum and (b) a real dielectric material.

original free surface charge density (σ_s) with a vacuum-filled capacitor is now reduced to ($\sigma_s - \sigma_b$), where σ_b is the bound charge density (see Section 7.3). This reduced charge density means that the voltage (V) across the plates is also reduced. Therefore, the new electric field (E) for a capacitor filled with a dielectric is smaller compared to E_0. A lower electric field is needed to maintain the same charge, because part of the surface charge is now held or bound by the dielectric material.

The ratio E_0/E—the electric fields existing in a capacitor filled with a vacuum and a dielectric material, respectively—is defined as the dielectric constant (k) or relative dielectric permittivity (ε_r).

$$\varepsilon_r = k = \frac{E_0}{E} \tag{7.11}$$

Now, because the total charge on the plates is the same,

$$Q = C \times V = C_0 \times V_0$$

The distance between the parallel plates (d) is the same.
Therefore, $Q = C \times E = C_0 \times E_0$ or $(E_0/E) = (C/C_0)$
We can write Equation 7.11 as

$$\varepsilon_r = k = \frac{C}{C_0} = \frac{\varepsilon}{\varepsilon_0} \tag{7.12}$$

Thus, the dielectric constant (k) is also defined as the ratio of the capacitance of a capacitor filled with a dielectric to that of an identical capacitor filled with a vacuum. One advantage of defining a dielectric constant (k) or relative dielectric permittivity (ε_r) as a dimensionless number is

TABLE 7.1

Approximate Room-Temperature Dielectric Constants (or Ranges) for Some Dielectric Materials or Classes of Materials (Frequency ~1 kHz–1 MHz)

Material	Dielectric Constant (k) (Approximate Range)
Vacuum	1 (by definition)
Oxygen (O) gas	1.000494
Argon (Ar) gas	1.000517
Mineral oil	2.25
Water (H_2O)	78
Polymers	~2–9
Teflon	2.1
Polyethylene	2.2
Silicon (Si)	11
Linear dielectric ceramics (SiO_2, TiO_2, Al_2O_3, and so on)	~4–200
Silica (SiO_2)	3.8–4.0
Alumina (Al_2O_3)	9.9
96% Al_2O_3 thick film	9.5
Magnesium oxide (MgO)	20
Nonlinear dielectrics or ferroelectrics	10–10,000
Polyvinylidene fluoride (a piezoelectric and ferroelectric polymer)	12–13
Barium titanate ($BaTiO_3$; a ferroelectric ceramic)	~2000–5000

that it becomes easy to compare the abilities of different materials to store charges. For example, the dielectric constant of the vacuum becomes 1. The dielectric constant of silicon (Si), alumina (Al_2O_3), and polyethylene are approximately 11, 9.9, and 2.2., respectively (Table 7.1).

Thus, the capacitance of a parallel capacitor containing a dielectric material with a dielectric constant (k) is given by modifying Equations 7.8 and 7.12 as follows:

$$C = \varepsilon_0 \times k \times \frac{A}{d} \tag{7.13}$$

If the dielectric material consists of atoms, ions, or molecules that are more able to be polarized, that is, if they are easily influenced by the applied electric field, then the dielectric constant (k) will be higher. In Section 7.12, we will examine how the dielectric constant (k) changes with electrical frequency (f), temperature, composition, and the microstructure of a material. Materials with a high dielectric constant are useful for making capacitors. Unlike transistors, diodes, solar cells, and so on, capacitors are considered "passive" components. One of the goals of capacitor manufacturers is to minimize the overall size of the capacitor while enhancing the total capacitance. This is usually achieved by arranging multiple, thin layers of dielectrics in parallel (Figure 7.4). This device is known as a *multilayer capacitor* (MLC).

The *volumetric efficiency* of a single-layer capacitor with area A and thickness d is given by:

$$\left(\frac{\varepsilon_0 \times k \times \dfrac{A}{d}}{A \times d} \right)$$

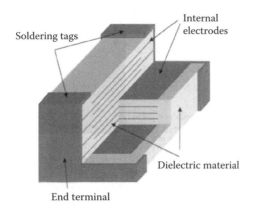

Soldering tags

Internal electrodes

Dielectric material

End terminal

FIGURE 7.4 Multilayer capacitors made using $BaTiO_3$-based formulations. (From Kishi, H., Y. Mizuno, and H. Chazono. 2003. *Jap J Appl Phys* 42:1–15. With permission.)

or

$$\text{volumetric efficiency of a single layer} = \left(\frac{\varepsilon_0 \times k}{d^2}\right) \tag{7.14}$$

Examples 7.1 and 7.2 illustrate how the volumetric efficiency is enhanced using multiple layers of a dielectric connected in parallel, instead of using a single, thick layer. An analogy to this is that we stay warmer during winter by wearing multiple layers of clothing, rather than wearing one very thick jacket.

EXAMPLE 7.1: CAPACITORS IN PARALLEL AND SERIES

a. Show that the capacitance of two capacitors with capacitances C_1 and C_2 connected in parallel is $C_1 + C_2$.

b. Show that the capacitance of the same capacitors connected in series is

$$\frac{1}{C} = \frac{1}{C_1} + \frac{1}{C_2} \tag{7.15}$$

or

$$C = \left(\frac{C_1 \times C_2}{C_1 \times C_2}\right) \tag{7.16}$$

SOLUTION

a. When capacitors are connected in parallel (Figure 7.5), the voltage across each capacitor is the same (V). The charge stored on capacitor 1 is Q_1 and that on capacitor 2 is Q_2. These are given by $C_1 = Q_1/V$ and $C_2 = Q_2/V$. The total charge stored on these capacitors connected in parallel is $Q = Q_1 + Q_2$. Thus, $Q = C_1V + C_2V$. If the total capacitance is C, then $C = Q/V$. Therefore,

$$C = \frac{(C_1V + C_2V)}{(V)}$$

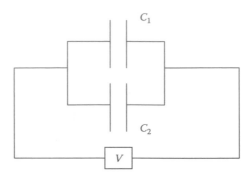

FIGURE 7.5 Capacitors connected in parallel.

or

$$C = C_1 + C_2 \text{ (for capacitors connected in parallel)} \qquad (7.17)$$

Thus, when capacitors are connected in parallel, the capacitances of the individual capacitors add up. This is why the layers in MLCs are arranged and electroded so that the layers are connected in parallel (Figure 7.4). Note that this is the opposite of what happens in resistors. Resistances add up when they are in a series. Capacitances add up when connected in parallel.

b. When connected in a series, the total charge on the plates of each capacitor is the same (Q). However, the voltage across each capacitor is different (V_1 and V_2). Thus, we can write $C_1 = Q / V_1$ and $C_2 = Q / V_2$. The total capacitance is given by $C = Q / V$.

$$C = \frac{Q}{\left(V_1 + V_2 \right)}$$

or

$$C = \frac{Q}{\left(\dfrac{Q}{C_1} \right) + \left(\dfrac{Q}{C_2} \right)} = \frac{1}{\left(\dfrac{1}{C_1} + \dfrac{1}{C_2} \right)}$$

or

$$\frac{1}{C} = \frac{1}{C_1} + \frac{1}{C_2} \text{ (for capacitors in series)}$$

or

$$C = \left(\frac{C_1 \times C_2}{C_1 + C_2} \right)$$

Note that when capacitors are connected in a series, the total capacitance (C) decreases (Figure 7.6).

EXAMPLE 7.2: MULTILAYER CAPACITOR DIELECTRICS

Billions of MLCs are made every year using barium titanate ($BaTiO_3$)-based dielectric materials formulated into temperature-stable dielectrics with high dielectric constants. Dielectric layers of

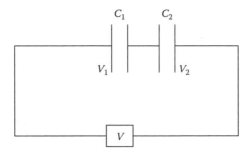

FIGURE 7.6 Capacitors connected in a series.

$BaTiO_3$ formulations are connected in parallel in an MLC because this provides the maximum volumetric efficiency, that is, the maximum capacitance per unit volume.

a. Calculate the capacitance of a parallel-plate *single-layer* capacitor made using a $BaTiO_3$ formulation of $k = 2000$. Assume $d = 3$ mm and $A = 5$ mm². What is the volumetric efficiency of this capacitor?
b. Calculate the capacitance of an MLC comprised of 60 dielectric layers connected in parallel using $N = 61$ electrodes. The thickness of each layer is 50 μm. The cross-sectional area of this capacitor is also 5 mm², and $k = 2000$. What is the volumetric efficiency of this capacitor?
c. What is the ratio of the volumetric efficiencies for multilayer and single-layer capacitors?

SOLUTION

a. From Equation 7.13,

$$C = \varepsilon_0 \times k \times \frac{A}{d}$$

$$C = 8.85 \times 10^{-12} \text{F/m} \times 2000 \times \frac{5 \times 10^{-6} \text{m}^2}{3 \times 10^{-3} \text{m}}$$

Thus, the capacitance of a single layer capacitor with a thickness of 3 mm is 2.95×10^{-11} F or 0.0295 nF.

The volumetric efficiency of this single-layer capacitor = (0.0295 nF)/(5 mm² × 3 mm)
$$= 1.96 \times 10^{-3} \text{ nF/mm}^3$$

b. In the MLC, there are 61 metal electrodes (i.e., $N = 61$) that connect $(N - 1) = 60$ dielectric layers.
   The capacitance of one layer of the dielectric in the MLC is given by

$$C_{layer} = 8.85 \times 10^{-12} \text{F/m} \times 2000 \times \frac{5 \times 10^{-6} \text{m}^2}{50 \times 10^{-6} \text{m}}$$

$$= 1.77 \times 10^{-9} \text{ F or 1.77 nF}$$

Note that the overall dimensions of the MLC and the single-layer capacitor are about the same (overall thickness is 3 mm, and the cross-sectional area is 5 mm²).
   The total capacitance = $(N - 1) \times C_{layer}$ = (60) × 1.77 nF = 106.2 nF.
   The volumetric efficiency of the MLC = (106.2)/(60 × 50 × 10⁻³ mm × 5 mm²) = 7.080 nF/mm³.

c. The ratio of the volumetric efficiencies (in nF/mm^3) is

$$= \frac{7.080}{1.96 \times 10^{-3}}$$
$$= 3600$$

If we have two capacitors that occupy the same volume, the MLC provides significantly more capacitance. This is very important because the goal is to minimize the size of the capacitors in integrated circuits (ICs) or printed circuit boards.

Example 7.3 illustrates some real-world situations in which there is a decrease in the capacitance because the device structure can transform into capacitors in a series.

EXAMPLE 7.3: HIGH-K GATE DIELECTRICS: HFO_2 ON SI

IC fabrication technology uses silica (SiO_2/SiO_x) ($k{\sim}4$) as a gate dielectric. The formula SiO_x indicates that the stoichiometry of the compound is not exactly known. The thickness of this SiO_2 gate dielectric in state-of-the-art transistors is ${\sim}2$ nm. (a) What is the capacitance per unit area of a 2-nm film made using SiO_2? Express your answer in $fF/\mu m^2$ (1 femtofarad [fF] = 10^{-15} F).

At such small thicknesses, the dielectric layer is prone to charge leakage. The use of materials with higher dielectric constants allows a relatively thicker gate dielectric (for the same capacitance), which will be less prone to electrical breakdown. Hafnium oxide (HfO_2; $k{\sim}25$) is one example of such a material. A 10-nm thin film of HfO_2 was deposited on Si using a low-temperature chemical vapor deposition (CVD) process. (b) What is the capacitance of this HfO_2 film per unit area? (c) When an engineer tested a HfO_2 capacitor structure, the thickness of the film formed on the Si was found to be 10 nm. However, the capacitance per unit area was smaller than expected. Explain why this is possible.

SOLUTION

a. The capacitance per unit area of a 2-nm film of SiO_2 is given by

$$\frac{C}{A} = \frac{\varepsilon_0 \times k}{d} \tag{7.18}$$
$$\frac{C}{A} = \frac{8.85 \times 10^{-12} \ F/m \times 4}{2 \times 10^{-9} \ m} = 1.77 \times 10^{-2} \ F/m^2$$

Because 1 F = 10^{15} fF and 1 m^2 = $10^{12} \ \mu m^2$, 1 F/m^2 = $10^3 \ fF/\mu m^2$.
Thus, the capacitance per unit area of a 4-nm SiO_2 film is 17.7 $fF/\mu m^2$.

b. For a 10-nm HfO_2 film on Si, the capacitance per unit area is

$$\frac{C}{A} = \frac{8.85 \times 10^{-12} \ F/m \times 25}{10 \times 10^{-9} \ m} = 2.212 \times 10^{-2} \ F/m^2 = 22.12 \ fF/\mu m^2$$

This is the capacitance per unit area if the film formed is from HfO_2 only. Thus, using a high-k material allows us to apply a greater thickness, which means better protection against dielectric breakdown and charge leakage, and still achieve comparable capacitance per unit area.

c. Because the measured capacitance per unit area for the HfO_2-on-Si structure is smaller, we expect that something must have caused the *effective* dielectric constant of the *device structure* to become smaller. One possibility is that the HfO_2 reacted with the Si and formed a material or phase that had an overall lower dielectric constant, that is, the film was not pure HfO_2. However, because a low-temperature CVD process was used, the formation of a different phase, although possible, is unlikely. A careful analysis of the microstructure using transmission electron microscopy (TEM) along with simultaneous nanoscale chemical

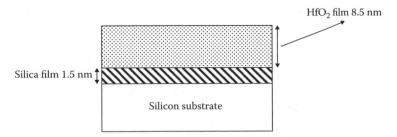

FIGURE 7.7   HfO$_2$ film on silicon, with a thin layer of SiO$_2$/SiO$_x$ formed at the interface, thus creating two capacitors in series and reducing the total capacitance per unit area.

analysis can resolve this dilemma. When this analysis was performed, it showed that Si had reacted with oxygen (O$_2$) during the CVD process and formed a thin layer (~2 nm) of SiO$_2$ at the interface between the Si and the HfO$_2$. We can assume the thickness of the HfO$_2$ to be $(10 - 1.5) = 8.5$ nm. We now have a structure that has two dielectric layers forming two capacitors connected in a series (Figure 7.7).

Similar to the previous calculation in part (a) of this example, the capacitance per unit area for a 1.5-nm SiO$_2$ film is 23.6 fF/µm^2. The capacitance of the 8.5-nm HfO$_2$ film is 26.03 fF/µm^2. Because these capacitors are in a series (Figure 7.7), the total capacitance per unit area for this structure is

$$\frac{1}{C_{total}} = \frac{1}{C_{SiO_2}} + \frac{1}{C_{HfO_2}} \tag{7.19}$$

Note that we are dealing with capacitances per unit area, and thus, each side of this equation is divided by area (A).

Therefore, we get

$$\frac{1}{\left(\dfrac{C_{total}}{A}\right)} = \frac{1}{\left(\dfrac{C_{SiO_2}}{A}\right)} + \frac{1}{\left(\dfrac{C_{HfO_2}}{A}\right)} = \frac{1}{23.6} + \frac{1}{26.03}$$

Thus, the new total capacitance per unit area $(C_{total}/A) = 12.37$ fF/µm^2.

We can see that this value is lower than the value 22.12 fF/µm^2, obtained when there is no interfacial layer of SiO$_2$ present. Note that in principle, it may seem that thinner layers of SiO$_2$/SiO$_x$ enhance the capacitance per unit area; however, such layers are prone to increased leakage of currents and electrical breakdown and are therefore not reliable. We can use strategies to prevent the lowering of capacitance by processing the alternative gate dielectric oxide films so that an SiO$_2$ layer does not form. The Intel Corporation has developed such dielectrics using HfO$_2$ and zirconium oxides. Another possibility not considered in this example is that the actual dielectric constant of HfO$_2$ may be lower than 25. The exact value of this apparent dielectric constant of HfO$_2$ can depend on the thin-film microstructure. Research has shown that HfO$_2$ films can have an apparent dielectric constant between 18 and 25.

7.3   DIELECTRIC POLARIZATION

We define *dielectric polarization* (P) as the magnitude of the bound charge density (σ_b). The application or presence of an electric field (the *cause*) leads to dielectric polarization (the *effect*). The situation is very similar to that encountered while discussing the mechanical properties of materials.

The application of a stress (the *cause*) leads to development of a strain (the *effect*). The stress and strain are related by Young's modulus. In this case, the electric field (E) applied and the polarization (P) created are related by the dielectric constant (k).

Assume that the dielectric polarization is caused by N number of small (atomic scale) dipoles, each comprised of two charges ($+q_d$ and $-q_d$) separated by a distance x (Figure 7.2). The dielectric polarization (P) is equal to the total dipole moment per unit volume of the material. This is another definition of dielectric polarization. The mechanisms by which such dipoles are created in a material are discussed in Section 7.5.

Assume that the concentration of atoms or molecules in a given dielectric material is N; if each atom or molecule is polarized, then the value of dielectric polarization is

$$P = \sigma_b = N \times < q_d \times x > \qquad (7.20)$$

If σ is the total charge density for a capacitor, then the other portion of the charge density, that is, $(\sigma - \sigma_b)$, remains the free charge. This creates a dielectric flux density (D_0), as in the case of a capacitor filled with a vacuum. From Gauss's law,

$$D_0 = \varepsilon_0 \times E \qquad (7.21)$$

Thus, the total dielectric flux density (D) for a capacitor filled with a dielectric material originates from two sources: the first source is the bound charge density (σ_b) associated with the polarization (P) in the dielectric material; and the other is the free charge density ($\sigma - \sigma_b$).

Therefore,

$$D = P + (\varepsilon_0 \times E) \qquad (7.22)$$

We can also rewrite the dielectric displacement as $D = \varepsilon \times E$. Therefore, we get

$$P = \sigma_b = (\varepsilon - \varepsilon_0) \times E \qquad (7.23)$$

If μ is the average dipole moment of the atomic dipoles created in a dielectric and N is the number of such dipoles per unit volume (i.e., the concentration), then we can also write the polarization as follows:

$$P = (N \times \mu) \qquad (7.24)$$

Polarizability describes the ability of an atom, ion, or molecule to create an induced dipole moment in response to the applied electric field. The average dipole moment (μ) of an atom can be written as the product of the polarizability of an atom (α) and the local electric field (E) that an atom within the material experiences.

$$\mu = (\alpha \times E) \qquad (7.25)$$

The SI units for dipole moment and electric field are $C \cdot m$ and V/m, respectively. Therefore, the unit of polarizability (α) is $C \cdot V^{-1} \cdot m^2$ or $F \cdot m^2$ (from Equation 7.1). The polarizability is often expressed as *volume polarizability* (α_{volume}), in the unit of cm^3 or $Å^3$:

$$\alpha \text{ volume} = \text{volume polarizability (cm}^3) = \frac{10^6}{4\pi\varepsilon_0} \times \alpha \ (C \cdot V^{-1} \cdot m^2 \text{ or } F \cdot m^2) \qquad (7.26)$$

If the volume polarizability is expressed in Å³ (as is often done in the case of atoms or ions), we use the following equation:

$$\alpha_{\text{volume}} \text{ in Å}^3 = \frac{10^{30}}{4\pi\varepsilon_0} \times \alpha \ \ (\text{C} \cdot \text{V}^{-1} \cdot \text{m}^2 \text{ or F} \cdot \text{m}^2) \tag{7.27}$$

The factors 10^6 and 10^{30} are used in these equations because $1 \text{ m}^3 = 10^6 \text{ cm}^3$ and $1 \text{ m}^3 = 10^{30} \text{ Å}^3$.

For now, let us assume that the electric field that an atom within a material experiences is the same as the applied field (E). From Equations 7.24 and 7.25, we get

$$P = (N \times \alpha \times E) \tag{7.28}$$

From Equations 7.28 and 7.23, we get

$$\alpha = \frac{(\varepsilon - \varepsilon_0)}{N} \tag{7.29}$$

We can rewrite this equation as shown below:

$$\varepsilon_r = \frac{\varepsilon}{\varepsilon_0} = \left[1 + \left(\frac{N\alpha}{\varepsilon_0} \right) \right] \tag{7.30}$$

Equation 7.30 is important because it links the dielectric constant (k or ε_r) of a material to the polarizability of the atoms (α) from which it is made and also to the concentration of dipoles (N).

We also use another parameter called the *dielectric susceptibility* (χ_e) to describe the relationship between polarization (the effect) and the electric field (the cause):

$$P = \chi_e \varepsilon_0 E \tag{7.31}$$

The subscript e in χ_e distinguishes the dielectric susceptibility from the magnetic susceptibility (χ_m), which is defined in Chapter 9. The dielectric susceptibility (χ_e) describes how susceptible or polarizable a material is, that is, how easily the atoms or molecules in the material are polarized by the presence of an electric field. Comparing Equations 7.23 and 7.31,

$$\chi_e = (k - 1) = (\varepsilon_r - 1) \tag{7.32}$$

Also, from Equations 7.31 and 7.32,

$$P = (\varepsilon_r - 1)\varepsilon_0 E \tag{7.33}$$

Another way to express the dielectric susceptibility is as follows:

$$\chi_e = (\varepsilon_r - 1) = \frac{P}{D_0} = \frac{\text{bound surface charge density}}{\text{free surface charge density}} \tag{7.34}$$

The dielectric susceptibility (χ_e), which is another way to express the dielectric constant (k), depends on the composition of the material. We can show from Equations 7.29 and 7.31 that

$$\chi_e = \frac{N\alpha}{\varepsilon_0} \tag{7.35}$$

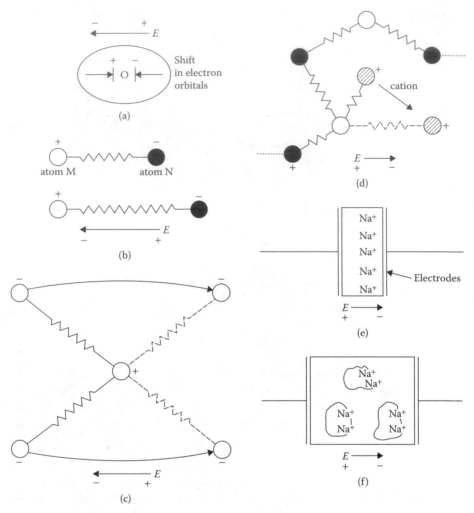

FIGURE 7.8 Schematic representation of polarization mechanisms. (a) Electronic, (b) atomic or ionic, (c) high-frequency oscillatory dipoles, (d) low-frequency cation dipole, (e) interfacial space-charge polarization, and (f) interfacial polarization. (From Hench, L. L. and J. K. West. 1990. *Principles of Electronic Ceramics*. New York: Wiley. With permission.)

From Equation 7.35, we can see that the dielectric susceptibility (χ_e) of the vacuum is zero or the dielectric constant (k) is 1. This is expected because there are no atoms or molecules in a vacuum, so $N = 0$. The more polarizable the atoms, ions, or molecules in the material, the higher is the bound charge density, and the higher is the dielectric susceptibility (χ_e) or dielectric constant (k). In Section 7.5, we will see that several different polarization mechanisms exist for a material (Figure 7.8).

The dielectric susceptibility values (χ_e) of silicon, Al_2O_3, and polyethylene are approximately 10, 8.9, and 1.2, respectively, because the dielectric constants are approximately 11, 9.9, and 2.2, respectively (Table 7.1).

7.4 LOCAL ELECTRIC FIELD (E_{LOCAL})

When we examine the effect of an externally applied electric field (E) on a dielectric material, we need to account for the electric field that exists inside a polarized material. For a solid or liquid exposed to an electric field (E), the actual field experienced by the atoms, molecules, or ions inside the material is different and is referred to as the *internal electric field* or *local electric field* (E_{local}).

In general, the greater the polar nature of the material, the higher the strength of the dipoles induced, and the larger the local electric field. Furthermore, the magnitude of the electric field experienced by the atoms, ions, or molecules inside a material depends on their arrangement.

For a cubic-structured isotropic material, a liquid, or an amorphous material, the local electric field is given by

$$E_{\text{local}} = E + \frac{1}{3\varepsilon_0} P \tag{7.36}$$

This expression is known as the *local field approximation*. Thus, we can rewrite Equation 7.28 by substituting E_{local} for E as

$$P = \left(N \times \alpha \times E_{\text{local}} \right)$$

or

$$N \times \alpha = \frac{P}{E_{\text{local}}} = \frac{P}{\left(E + \dfrac{P}{3\varepsilon_0} \right)} \tag{7.37}$$

Note that E is the applied electric field.

Therefore,

$$N \times \alpha = \frac{E(\varepsilon - \varepsilon_0)}{E + \dfrac{E(\varepsilon - \varepsilon_0)}{3\varepsilon_0}}$$

Eliminating E, dividing by ε_0, and rewriting $\varepsilon/\varepsilon_0 = \varepsilon_r$, we get

$$\frac{\varepsilon_r - 1}{\varepsilon_r + 2} = \frac{1}{3\varepsilon_0} (N \times \alpha) \tag{7.38}$$

Equation 7.38 is also known as the Clausius–Mossotti equation. It describes the relationship between the dielectric constant (k or ε_r), a *macroscopic* property, and the concentration of polarizable species (N) and their polarizability (α), which are *microscopic properties*. We have used the local field approximation (Equation 7.36), which is valid for either amorphous materials or cubic-structured materials. Strictly speaking, the Clausius–Mossotti equation should be used only for these types of materials. In some materials, molecules have a permanent dipole moment (such as water). Ferroelectrics develop spontaneous polarization because of the rearrangement of ions (Section 7.11). For such polar materials, the internal electric field is *not* given by Equation 7.37. As a result, the Clausius–Mossotti equation (Equation 7.38) *cannot* be used for such polar materials.

In Equation 7.38, N is the concentration of dipoles per unit volume (number of molecules/m³). If we assume that each molecule or atom becomes a dipole, then N is related to the N_{Avogadro} (6.023 × 10²³ molecules/mol), density (ρ in kg/m³), and molecular weight (M in kg/mol) as follows:

$$N = N_{\text{Avagadro}} \times \frac{\rho}{M} \tag{7.39}$$

Substituting for N in Equation 7.38, we get another form of the Clausius–Mossotti equation:

$$\left(\frac{\varepsilon_r - 1}{\varepsilon_r + 2} \right) \frac{M}{\rho} = \frac{N_{\text{Avagadro}} \times \alpha}{3\varepsilon_0} \tag{7.40}$$

We can verify that this equation is dimensionally balanced. The units are α in $F \cdot m^2$, ε_0 in F/m, and Avogadro's number in number per mole. Thus, the unit on the right-hand side of the above expression is m^3/mol. The unit on the left-hand side is also m^3/mol.

We can rewrite Equation 7.40 to get the polarization per mole or *molar polarization* (P_m), defined as follows:

$$\left(\frac{\varepsilon_r - 1}{\varepsilon_r + 2} \right) = \frac{P_m \times \rho}{M} \tag{7.41}$$

The units for molar polarization (P_m) are m^3/mol and cm^3/mol.

As we will see in Section 7.5 and Figure 7.8, there are five polarization mechanisms—electronic, ionic, dipolar, interfacial, and spontaneous (ferroelectric). The total polarizability due to ionic and electronic polarization is additive, because these polarizations occur throughout the volume of a material. Thus, we can write the total polarizability as

$$\alpha = \alpha_e + \alpha_{ionic} \tag{7.42}$$

where, α_e and α_{ionic} are the electronic and ionic polarizabilities of the atoms, respectively.

Electronic and ionic polarization are defined in Sections 7.6 and 7.7, respectively.

We can rewrite the Clausius–Mossotti equation (Equation 7.38) to separate out the ionic and electronic polarization effects as follows:

$$\frac{\varepsilon_r - 1}{\varepsilon_r + 2} = \frac{1}{3\varepsilon_0} \left[\left(N_e \times \alpha_e \right) + \left(N_i \times \alpha_i \right) \right] \tag{7.43}$$

In Equation 7.43, we *cannot* incorporate the effects of dipolar (Section 7.9), interfacial (Section 7.10), and spontaneous or ferroelectric polarization (Section 7.11), because the effects that these polarizations have on the local electric field (E_{local}) are complex. They cannot be described by Equation 7.36; thus, the Clausius–Mossotti equation *cannot* generally be used for polar materials, such as water, or ferroelectric compositions of materials, such as the tetragonal form of $BaTiO_3$. An exception to this is a situation where the Clausius–Mossotti equation may be used for polar materials if the electrical frequency (f) of the applied field is too high for these polarization mechanisms to exist.

7.5   POLARIZATION MECHANISMS—OVERVIEW

We will now examine the different ways in which atoms, ions, and molecules in a material can be polarized by an electric field (Table 7.2 and Figure 7.8).

These polarization mechanisms are shown schematically in Figure 7.8 and are discussed in the following sections.

7.6   ELECTRONIC OR OPTICAL POLARIZATION

7.6.1   Electronic Polarization of Atoms

All materials contain atoms (in the form of neutral atoms, ions, or molecules). When subjected to an electric field, each atom is polarized, in that the center of the electronic charge shows a slight shift toward the positively charged electrode.

TABLE 7.2
Summary of Polarization Mechanisms in Dielectrics

Polarization Mechanism	Causes of Net Dipole Moment	Approximate Frequency Range (Hz)	Temperature Dependence	Examples
Electronic or optical polarization (Section 7.6)	Displacement of electronic cloud with reference to nucleus	Up to 10^{14} Hz	Not strong	All materials
Ionic, atomic, or vibrational polarization (Section 7.7)	Displacement of ions with reference to each other	Up to ~10^{13} Hz	Not strong	Ionic solids such as oxide ceramics (e.g., Al_2O_3 and TiO_2)
Dipolar or orientational (Section 7.7)	Reorientation of permanent dipoles	Up to ~10^{12} Hz	Yes; decreases with increasing temperature	H_2O
Interfacial, space charge, Maxwell–Wagner, or Maxwell–Wagner–Sillars polarization (Section 7.10)	Movement of ions at an interface such as grain boundaries	Up to several MHz	Yes; involves the short-range movement of atoms or ions	Lithium tantalate ($LiTaO_3$), lithium niobate ($LiNbO_3$), inorganic glasses containing Li^+, Na^+, and so on
Spontaneous, ferroelectric polarization (Section 7.11)	Creation of spontaneous dipoles in the unit cells of a material via small displacements of ions; typically involves a phase transformation between polymorphs	Up to ~10^9–10^{10} Hz	Yes; decreases above a certain temperature, and eventually disappears at very high temperatures	$BaTiO_3$, PZT, and PVDF

This very slight elastic displacement of the electronic cloud (a few parts per million of the atomic radius), shown as δ or x, occurs very rapidly with reference to the nucleus (~10^{-14} seconds; Figure 7.9). This means that even if the electric field changes its polarity ~10^{14} times a second, the electronic polarization process can still follow this rapid change in the direction of the electric field. The electric field associated with visible light (which is an electromagnetic wave) oscillates with a frequency of ~10^{14} Hz. This field interacts with dielectric materials and causes electronic polarization. Thus, electronic polarization is related to the optical properties of materials, such as the refractive index. This is why electronic polarization is also known as *optical polarization*.

Because the electrons of the outermost shell are the ones most susceptible to the electric field applied, the electronic polarizability (α_e) of an atom or an ion depends primarily on its size and the number of electrons in its outermost shell. The larger the atom or ion, the farther the electrons are from the nucleus. Therefore, the electron clouds surrounding larger atoms or ions are more susceptible to electric fields and are more polarizable. Thus, larger atoms or ions have a higher electronic polarizability. The extent to which an atom or an ion is polarized via this mechanism is measured by the electronic polarizability (α_e).

Consider the nucleus of a monoatomic element (such as argon [Ar]) with radius R. When an electric field is applied, the electronic cloud is displaced by a distance δ with respect to the nucleus (Figure 7.9).

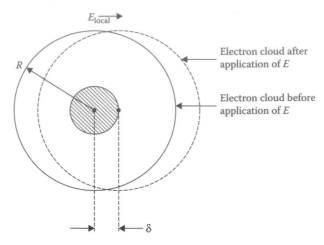

FIGURE 7.9 Displacement of the electronic cloud in an argon atom. (From Kao, K. C. 2004. *Dielectric Phenomena in Solids*. London: Elsevier-Academic Press. With permission.)

A dipole is created between the charge q_1 of the nucleus (Zq; where Z is the number of electrons surrounding the nucleus) and the charge q_2, which is the part of the charge in the electron cloud that no longer surrounds the nucleus because it is displaced. This charge q_2 is contained in a sphere of radius δ and is given by

$$q_2 = -\frac{Zq[(4/3)\pi\delta^3]}{[(4/3)\pi R^3]} = -Zq\frac{\delta^3}{R^3} \tag{7.44}$$

Thus, the Coulombic force of attraction between the nuclear charge q_1 and the negative charge q_2, not shielded by the nucleus and separated by distance δ, is given by Coulomb's law:

$$F = \frac{q_1 \times q_2}{4\pi\varepsilon_0\delta^2} \tag{7.45}$$

We can rewrite this as

$$F = \frac{(Zq)\times\left(-Zq\dfrac{\delta^3}{R^3}\right)}{4\pi\varepsilon_0\delta^2} = -\frac{(Zq)^2\delta}{4\pi\varepsilon_0 R^3} \tag{7.46}$$

The magnitude of this attractive Coulombic force is balanced by the force (F_d) that causes the displacement.

$$F_d = (Zq)\times E \tag{7.47}$$

Thus, equating the magnitude of the force that causes the displacement of the electronic cloud and the Coulombic restoring force, we get

$$\frac{(Zq)^2\delta}{4\pi\varepsilon_0 R^3} = (Zq)\times E \tag{7.48}$$

Solving for displacement (δ), we get

$$\delta = \frac{4\pi\varepsilon_0 R^3 E}{Zq} \tag{7.49}$$

The dipole moment (μ) caused by the electronic polarization is given by

$$\mu = (Zq) \times \delta \tag{7.50}$$

Substituting for δ from Equation 7.49,

$$\mu = 4\pi\varepsilon_0 R^3 E$$

Recalling Equation 7.25, we can write the dipole moment (μ) as

$$\mu = \alpha_e \times E \tag{7.51}$$

where α_e is the electronic polarizability of an atom or an ion. Note that the term "electronic polarizability" does *not* describe the polarizability of an electron. It describes the polarizability of an atom or an ion.

Comparing Equations 7.51 and 7.54, we get

$$\alpha_e = 4\pi\varepsilon_0 R^3 = 3\varepsilon_0 V_a \tag{7.52}$$

In Equation 7.52, V_a is the volume of the atom or ion being polarized. Because the unit of permittivity is F/m, and the unit of volume is m³, the unit of electronic polarizability (α_e) is F·m².

The electronic polarizability values for atoms of different elements are shown in Figure 7.10. These values are the volume polarizabilities. To convert them into SI units, they should be multiplied by $4\pi\varepsilon_0 \times 10^{-30}$ (Equation 7.27).

$$\alpha(\text{volume polarizability}) \text{ in } \mathring{A}^3 = \frac{10^{30}}{4\pi\varepsilon_0} \times \alpha \ \left(\text{in } C \cdot V^{-1} \cdot m^2 \text{ or } F \cdot m^2\right) \tag{7.53}$$

Example 7.4 will give us an idea of the magnitudes of the electronic polarizability of atoms and the electron-cloud displacement distances.

EXAMPLE 7.4: ELECTRONIC POLARIZABILITY OF THE Ar ATOM

The atomic radius of an Ar atom is 1.15 Å.

a. What is the electronic polarizability (α_e) of an Ar atom in units of F·m²?
b. What is the volume polarizability of an Ar atom in Å³?
c. What is the volume polarizability of an Ar atom in cm³?
d. If an Ar atom experiences a local electric field (E_{local}) of 10^6 V/m, what is the displacement (δ) when the atom experiences electronic polarization?
e. Calculate the ratio of the radius of atom (R) to the displacement (δ) caused by electronic polarization.

SOLUTION

a. We calculate the electronic polarizability (α_e) of Ar atoms from Equation 7.52 as follows:

$$\alpha_e = 4\pi\varepsilon_0 R^3 = \left(4 \times \pi \times 8.85 \times 10^{-12} \text{ F/m}\right)\left(1.15 \times 10^{-10} \text{ m}\right)^3 = 1.69 \times 10^{-40} \text{ F·m}^2$$

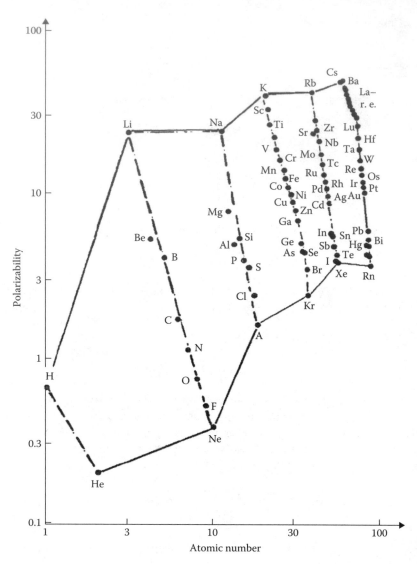

FIGURE 7.10 Volume electronic polarizability of atoms of different elements. Note: Multiply the y-axis value by $4\pi\varepsilon_0 \times 10^{-30}$ to get the polarizability in $F \cdot m^2$. (From Raju, G. G. 2003. *Dielectrics in Electric Fields*. Boca Raton, FL: CRC Press. With permission.)

b. We use Equation 7.27 to calculate the polarizability volume in $Å^3$

$$\alpha(\text{volume}) \text{ in } Å^3 = \frac{10^{30}}{4\pi(8.85 \times 10^{-12} \text{ F/m})} \times 1.69 \times 10^{-40} \text{ F} \cdot m^2 = 1.52 \text{ } Å^3$$

c. For calculating the volume polarizability, we use Equation 7.26

$$\alpha(\text{volume}) \text{ in } cm^3 = \frac{10^6}{4\pi\varepsilon_0} \times \alpha \text{ } (C \cdot V^{-1} \cdot m^2 \text{ or } F \cdot m^2)$$

$$\alpha(\text{volume}) \text{ in } cm^3 = \frac{10^6}{4\pi\varepsilon_0} \times (1.69 \times 10^{-40} \text{ F} \cdot m^2) = 1.52 \times 10^{-24} \text{ } cm^3$$

This value is similar to the value of 1.642×10^{-24} cm³, reported in literature (Vidal et al. 1984). We use Equation 7.49 to calculate the displacement (δ):

$$\delta = \frac{4\pi\varepsilon_0 R^3 E}{Zq} = \frac{4\pi\varepsilon_0 (1.15 \times 10^{-10} \text{ m})^3 (10^6 \text{ V/m})}{(8)(1.6 \times 10^{-19} \text{ C})}$$

$$= 1.32 \times 10^{-16} \text{ m} = 1.32 \times 10^{-6} \text{ Å}$$

d. The displacement (δ) is 1.32×10^{-6} Å. The radius $R = 1.15$ Å; and hence, the ratio of R/δ is ~870284. Thus, the displacement (δ) (Figure 7.9) is very small compared to the radius of the atoms.

We can use Bohr's model to calculate the electronic polarizability (α_e) of an atom. Using this approach, the electronic polarizability under a static electric field is given by the equation

$$\alpha_{e,static} = \frac{(Z \times q)^2}{m\omega_0^2} \tag{7.54}$$

where Z is the number of electrons orbiting the nucleus, q is the electronic charge, m is the electron mass, and ω_0 is the natural oscillation frequency of the center of the mass of the electron cloud around the nucleus. The *static electronic polarizability* represents the value of electronic polarizability when the electric field causing the polarization is not time-dependent.

The electronic polarizability (α_e) depends on the frequency of the electric field (ω) as follows:

$$\alpha_e = \frac{(Zq)^2}{m(\omega_0^2 - \omega^2) + j\beta\omega} \tag{7.55}$$

where j is the imaginary number and β is a constant that is related to the damping force attempting to pull the electron cloud back toward the nucleus. This equation is derived using a classical mechanics approach. A quantum mechanical-based approach yields a different equation, but the trend in the change in electronic polarizability as a function of frequency is similar. Example 7.5 illustrates a calculation of the value of static electronic polarizability.

EXAMPLE 7.5: STATIC ELECTRONIC POLARIZABILITY OF H ATOMS

If the natural frequency of oscillation (ω_0) of the electron mass around the nucleus in a hydrogen (H) atom is 4.5×10^{16} rad/s, what is the static electronic polarizability (α_e) of the H atom ($R = 1.2$ Å) calculated by Bohr's model and the classical approach? How do these values compare with the polarizability of the Ar atom calculated previously in Example 7.4?

SOLUTION

For Bohr's model, using Equation 7.54,

$$\alpha_e = \frac{(1 \times 1.6 \times 10^{-19} \text{ C})^2}{(9.11 \times 10^{-31} \text{ kg})(4.5 \times 10^{16} \text{ rad/s})^2} = 1.387 \times 10^{-41} \text{ F} \cdot \text{m}^2$$

The value can also be calculated using the classical approach (Equation 7.52), as shown below:

$$\alpha_e = 4\pi\varepsilon_0 R^3 = \left(4 \times \pi \times 8.85 \times 10^{-12} \text{ F/m}\right)\left(1.20 \times 10^{-10} \text{ m}\right)^3 = 1.92 \times 10^{-40} \text{ F} \cdot \text{m}^2$$

The value using the classical approach is larger than that predicted from Bohr's model.

7.6.2 ELECTRONIC POLARIZABILITY OF IONS AND MOLECULES

Ions and molecules have electronic polarizability similar to neutral atoms. The electronic polarizability of an ion is nearly equal to that of an atom, with the same number of electrons. Thus, the electronic polarizability of sodium ions (Na^+) (0.2×10^{-40} F · m^2) is comparable to the electronic polarizability of neon (Ne) atoms. From Equation 7.52, we can expect the larger atoms or ions to have a larger electronic polarizability. Because cations are typically smaller than anions, the electronic polarizability of anions is generally larger than that of cations. Similarly, larger ions such as lead (Pb^{2+}) have a larger electronic polarizability. Many real-world technologies make use of these effects, such as the development of lead crystal and optical fibers.

Electronic polarization is also linked to optical properties such as the index refractive (see Section 7.12). Molecules are comprised of several atoms and show electronic polarizability. In general, the polarizability of molecules is larger because they contain more electrons.

7.7 IONIC, ATOMIC, OR VIBRATIONAL POLARIZATION

Many ceramic dielectrics exhibit mixed ionic and covalent bonding. In ceramics with ionic bonds, each ion undergoes *electronic* polarization. In addition to this, ionic solids exhibit *ionic polarization*, also known as *atomic polarization* or *vibrational polarization*. In this mechanism of polarization, the ions themselves are displaced in response to the electric field experienced by the solid, creating a net dipole moment per ion (p_{av}). Consider pairs of ions in an ionic solid such as sodium chloride (NaCl; Figure 7.11a). Assume that these ions have an equilibrium separation distance of a. This is the average separation distance between an anion and a cation. Ions in any material are not stationary. They vibrate around their mean equilibrium positions; these vibrations of ions or atoms are known as phonons.

When an electric field (E) is applied (Figure 7.11b), the electronic polarization of both cations and anions is established almost instantaneously (in ~10^{-14} seconds). This effect, that is, distortion of the electronic clouds for both anions and cations, is *not* shown in Figure 7.11. Since anions are bigger than cations because of the extra electrons present, anions typically show higher electronic polarizability than cations.

In addition to this electronic polarization, a positively charged cation moves toward the negative end of the electric field. Similarly, anions move closer to the positive end of the electric field. Figure 7.11b shows the displacements of ions; cations 1 and 3 move to the right, that is, toward the negative end of the electric field. Anions 2 and 4 are displaced toward the positive end of the electric field. This means that the separation distance (Δx) between cation 1 and anion 2 is now reduced, compared to their separation (a) without the electric field. The separation Δx between cation 3 and anion 2 is now larger than their equilibrium separation distance (a). The extent to which these ions are displaced also depends on the magnitude of the restoring forces imposed by other neighboring ions. For example, as cation 3 moves toward anion 4 (Figure 7.11b), anion 2 tries to pull cation 3 back, and the cation to the right of anion 4 (not shown) repels it.

Such asymmetric displacements of ions in response to the presence of an electric field create a dipole moment; this effect is known as ionic polarization. The magnitude of ionic displacements encountered in ionic polarization is a fraction of an angstrom (Å). Because ions have a larger inertia than electrons, these movements are a bit sluggish, occurring in about 10^{-13} seconds. When the polarity of the electric field is reversed (Figure 7.11c), the directions of displacement are also reversed. If we have an alternating current (AC) electric field, then the ions move back and forth as long as the electric field does not switch too rapidly, as shown in Figures 7.11b and c. If the frequency (f) of the switching field is greater than ~10^{13} Hz, that is, if the field switches in less than 10^{-13} seconds, the ions cannot follow the changes in the electric field direction. In other words, the ionic polarization mechanism is seen in ionic solids for frequencies up to ~10^{13} Hz (Table 7.2). If the frequency is greater than ~10^{13} Hz, this polarization mechanism "drops out" and does *not* contribute to the total dielectric polarization (P) induced in the dielectric.

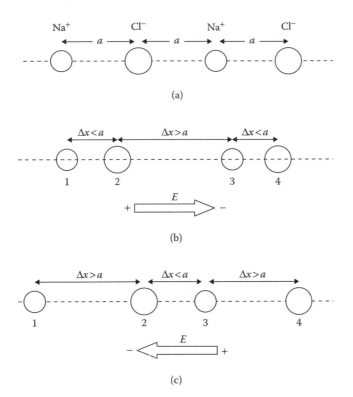

FIGURE 7.11 Illustration of ionic polarization: (a) no electric field; (b) electric field as shown; and (c) electric field direction reversed.

Under static electric fields, the magnitude of *ionic polarizability* (α_i) is given by

$$\alpha_i = \frac{(Z \times q)^2}{M_r \times \omega_0^2} \text{ (for static electric fields)} \tag{7.56}$$

where Z is the valence of the ion (not the number of electrons surrounding the nucleus, which Z represents in electronic polarization equations), q is the electronic charge, M_r is the reduced mass of the ion, and ω_0 is the natural frequency of oscillation for a given ion.

For AC fields with a frequency $\omega = 2\pi f$, where f is the electrical frequency in Hz, the magnitude of the ionic polarizability (α_i) is frequency-dependent and given by

$$\alpha_i = \frac{(Z \times q)^2}{[M_r(\omega_0^2 - \omega^2) + j\beta\omega]} \tag{7.57}$$

where ω is the frequency of the electric field (in rad/s), j is the imaginary number, and β is a coefficient related to the damping force that tries to bring the displaced ion back to its original position.

The details of the treatments for deriving these equations are beyond the scope of this book. However, it is important to recognize that the dielectric polarization mechanism is effective only up to a certain frequency ($\sim 10^{13}$ Hz).

In general, the magnitude of ionic polarizability (α_i) is about 10 or more times *larger* than the electronic polarizability (α_e). This is why most solids with considerable ionic bonding character exhibit much higher dielectric constants (Table 7.1). Also, note that in addition to ionic polarization, each ion undergoes *electronic* polarization. Thus, the dielectric constant (k) of ionic solids results from both electronic and ionic polarizations. However, the contributions from ionic polarization tend to be dominant, especially at lower frequencies.

If μ_{av} is the average dipole moment induced by the ionic polarization per ion, then we can write this dipole moment as follows:

$$\mu_i = \alpha_i \times E_{local}$$

where E_{local} is the electric field experienced by the ion. As mentioned in Section 7.4, this is the local electric field (E_{local}) and is different from the applied electric field (E). The polarization (P) induced in an ionic solid with N_i ions per unit volume is given by the following equation:

$$P = N_i \times \mu_{av} = N_i \times \alpha_i \times E_{local} \tag{7.58}$$

Recall from the Clausius–Mossotti equation (Equation 7.43) that the dielectric constant (k or ε_r) is linked to the different polarizabilities of ions. For an ionic solid with no permanent dipoles, we have contributions from both ionic and electronic polarizations. Note that this equation applies only to amorphous or cubic structures and nonpolar materials.

In Section 7.8, we will discuss an approach that is useful in predicting the dielectric constants of nonpolar materials from the values of the total (i.e., electronic and ionic) polarizability of ions.

7.8 SHANNON'S POLARIZABILITY APPROACH FOR PREDICTING DIELECTRIC CONSTANTS

7.8.1 OUTLINE OF THE APPROACH

Shannon measured the dielectric constants of several materials and back-calculated their ionic polarizabilities (Shannon 1993). The frequency range for the dielectric-constant measurements was 1 kHz–10 MHz. Thus, both the electronic and ionic polarization mechanisms contributed to the dielectric constant measured.

Shannon used experimentally determined values of dielectric constants to first estimate the polarizability of ions or simple compounds. Then, he used these values to estimate the *total dielectric polarizability* (α_D^T) of other compounds through the additive nature of ionic and electronic polarizabilities. This calculation requires knowledge of the molecular weight and unit cell volume (i.e., the theoretical densities) of the compound whose dielectric constant is to be estimated. For example, one can measure the dielectric constants of fully dense samples of magnesium oxide (MgO) and Al_2O_3 and estimate the polarizabilities of Al^{3+}, Mg^{2+}, and O^{2-} ions. We can use these polarizability values to calculate the total dielectric polarizability (α_D^T), and hence the dielectric constant of another compound such as magnesium aluminate ($MgAl_2O_4$).

The total dielectric polarizability of $MgAl_2O_4$ can be expressed as follows:

$$\alpha_{MgAl_2O_4} = \alpha_{Mg^{(2+)}} + 2\alpha_{Al^{(3+)}} + 4\alpha_{O^{(2-)}} \tag{7.59}$$

or

$$\alpha_{D,MgAl_2O_4}^T = \alpha_{D,MgO}^T + \alpha_{D,Al_2O_3}^T \tag{7.60}$$

Following this, we can use the value of total polarizability for $MgAl_2O_4$ to estimate its dielectric constant by using the following form of the Clausius–Mossotti equation:

$$\varepsilon_{r,cal} = \frac{3V_m + 8\pi\alpha_D^T}{3V_m - 4\pi\alpha_D^T} \tag{7.61}$$

where V_m is the molar volume of the compound whose dielectric constant or α_D^T is being estimated.

Shannon also used the following modified forms of the Clausius–Mossotti equation:

$$\alpha_D^T = \frac{3V_m}{4\pi}\left(\frac{\varepsilon_r - 1}{\varepsilon_r + 2}\right) \tag{7.62}$$

In Equations 7.60 through 7.62, α_D^T is the total dielectric polarizability of a material and is commonly expressed as Å^3.

7.8.2 LIMITATIONS OF SHANNON'S APPROACH

Shannon's approach for predicting the dielectric constant is useful, but it has some limitations. For many compounds, the calculated values of dielectric constants using Shannon's approach are very different from the measured values. The polarizability of the oxygen ion (O^{2-}), as estimated by Shannon (2.01 Å^3; Figure 7.12), is lower than that used by other researchers (2.37 Å^3). This leads to the prediction of lower dielectric constants for some oxides, for example, Al_2O_3. In many other materials exhibiting ferroelectric and piezoelectric behavior or for materials containing compressed or rattling ions, mobile ions, and impurities, the calculated and measured values of the dielectric constants do not match well. This can be due to the ionic or electronic conductivity of the material, the presence of interfacial polarization, the presence of polar molecules such as water (H_2O) or carbon dioxide (CO_2), and the presence of other dipolar impurities. Thus, Shannon's approach cannot be used to calculate the dielectric constants of ferroelectric materials (Section 7.11).

Despite these limitations, Shannon's approach serves as a powerful guide for the experimental development of new formulations of materials with high dielectric constants. The use of Shannon's approach is illustrated in Examples 7.6 and 7.7.

EXAMPLE 7.6: DIELECTRIC CONSTANT OF Li₂SiO₃ FROM ION POLARIZABILITIES

Use the ion polarizabilities in Figure 7.12 to estimate the dielectric constant of lithium silicate (Li_2SiO_3). The molar volume (V_m) of Li_2SiO_3 is 59.01 Å^3. The experimental value of the dielectric constant ($\varepsilon_{r,exp}$) for Li_2SiO_3 between 1 kHz and 10 MHz is 6.7. How does the calculated value of the dielectric constant compare with the value estimated using Shannon's approach?

Li 1.20	Be 0.19											B 0.05	C	N	O 2.01	F 1.62	Ne
Na 1.80	Mg 1.32											Al 0.79	Si 0.87	P 1.22	S	Cl	
K 3.83	Ca 3.16	Sc 2.81	Ti IV 2.93	V V 2.92	Cr III 1.45	Mn II 2.64	Fe II 2.23 III 2.29	Co II 1.65	Ni II 1.23	Cu II 2.11	Zn 2.04	Ga 1.50	Ge 1.63	As V 1.72	Se	Br	
Rb 5.29	Sr 4.24	Y 3.81	Zr 3.25	Nb 3.97	Mo	Tc				Ag	Cd 3.40	In 2.62	Sn 2.83	Sb III 4.27	Te IV 5.23	I	
Cs 7.43	Ba 6.40	La 6.07	Hf	Ta 4.73	W	Re				Au	Hg	Tl I 7.28	Pb II 6.58	Bi 6.12			

		Ce III 6.15 IV 3.94	Pr 5.32	Nd 5.01	Pm	Sm 4.74	Eu II 4.83 III 4.53	Gd 4.37	Tb 4.25	Dy 4.07	Ho 3.97	Er 3.81	Tm 3.82	Yb 3.58	Lu 3.64
		Th 4.92	Pa	U IV 4.45											

FIGURE 7.12 Polarizabilities of ions expressed in Å^3. Note: The frequency range of 1 kHz to 10 MHz means that the polarizability values include both electronic and ionic components. (From Shannon, R. D. 1993. *J Appl Phys* 73:348–66. With permission.)

SOLUTION

From Figure 7.12, the ionic polarizability of O^{2-} ions is 2.01 $Å^3$. The ionic polarizabilities of lithium (Li^+) and silicon (Si^{4+}) ions are 1.20 $Å^3$ and 0.87 $Å^3$, respectively. Thus, the dielectric polarizability (α_D^T) of Li_2SiO_3 is

$$\alpha_D^T \text{ Lithium silicate} = 2\alpha_{Li^{(+1)}} + \alpha_{Si^{(+4)}} + 3\alpha_{O^{(-2)}}$$

$$= 2(1.2) + 0.87 + 3(2.01) = 9.3 \text{ } Å^3$$

The molar volume of Li_2SiO_3 is $V_m = 59.01$ $Å^3$. Therefore, using Equation 7.61, the calculated dielectric constant ($\varepsilon_{r,cal}$) of Li_2SiO_3 is

$$\varepsilon_{r,cal} = \frac{3V_m + 8\pi\alpha_D^T}{3V_m - 4\pi\alpha_D^T}$$

$$\varepsilon_{r,cal} = \frac{(3 \times 59.01) + 8\pi(9.3)}{(3 \times 59.01) - 4\pi(9.3)} = 6.82$$

This compares well with the value $\varepsilon_{r,exp} = 6.70$. Note that for many compounds, the calculated and measured values do *not* match very well, for the reasons mentioned in Section 7.8.2. Furthermore, if the temperature is high and the frequency is low, interfacial polarization effects (see Section 7.10) may also develop in Li_2SiO_3. Under such conditions, the experimentally observed and calculated dielectric constants do not match well.

EXAMPLE 7.7: DIELECTRIC CONSTANT OF $MgAl_2O_4$ USING SHANNON'S APPROACH

The dielectric constant of MgO is 9.83, and its molar volume (V_m) is 18.69 $Å^3$. The dielectric constant of Al_2O_3 is estimated to be 10.126, and its molar volume (V_m) is 42.45 $Å^3$. What is the dielectric constant of $MgAl_2O_4$? The molar volume (V_m) of $MgAl_2O_4$ is 66.00 $Å^3$.

SOLUTION

We first calculate the total dielectric polarizability of MgO from its dielectric constant and volume using the modified form of the Clausius–Mossotti equation (Equation 7.62) that Shannon used.

$$\alpha_{D,MgO}^T = \frac{3(18.69)}{4\pi}\left(\frac{9.830 - 1}{9.830 + 2}\right) = 3.330$$

Similarly, we calculate the polarizability of Al_2O_3 from its dielectric constant and volume using Equation 7.62:

$$\alpha_{D,Al_2O_3}^T = \frac{3(42.45)}{4\pi}\left(\frac{10.126 - 1}{10.126 + 2}\right) = 7.663$$

From these values of polarizabilities, we estimate the polarizability of $MgAl_2O_4$ using the additivity rule developed by Shannon.

$$\alpha_{D,Al_2O_3}^T = 7.663 + 3.330 = 10.993$$

Now, we have estimated the total dielectric polarizability of $MgAl_2O_4$, and we know its molar volume ($V_m = 66$ $Å^3$); from these, we calculate the dielectric constant using Equation 7.61.

$$\varepsilon_{r,cal} = \frac{(3 \times 66) + 8\pi(10.993)}{(3 \times 66) - 4\pi(10.993)}$$

$$\varepsilon_{r,cal} = 7.923$$

Shannon carefully measured the dielectric constant of this material and reported the value to be 8.176. Thus, the value we estimated does not exactly match the experimentally determined value; however, it is relatively close (within a few percent).

7.9 DIPOLAR OR ORIENTATIONAL POLARIZATION

Molecules known as polar molecules have a permanent dipole moment. An example of a well-known polar material is H_2O (water). The polar molecules begin to experience torque when exposed to an external electric field and orient themselves along the electric field. This, in turn, causes an increase in the bound charge density (σ_b) and leads to an increase in polarization (P). The resulting polarization is known as *dipolar polarization* or *orientational polarization*. Orientational or dipolar polarization is the mechanism responsible for the relatively high dielectric constant of H_2O ($k{\sim}78$; Figure 7.13).

The main feature that distinguishes the orientational polarization mechanism from other mechanisms is the presence of permanent dipoles. This mechanism of dipolar polarization is seen only in materials that have molecules with a permanent dipole moment. Materials in which molecules develop a net polarization or have a permanent or built-in dipole moment are known as polar materials or *polar dielectrics*. Polar materials in which polarization appears spontaneously, even without an electric field, are known as *ferroelectrics* (Chapter 8). They do not contain molecules with a permanent dipole moment.

If each permanent dipole had a dipole moment of μ and if the concentration of such dipoles was N, then the maximum polarization that can be caused by this mechanism alone is $N \times \mu$. Not all dipoles can remain aligned with the applied electric field, because thermal energy tries to randomize their orientations. Thus, this polarization mechanism begins to fade away with increasing temperature. At substantially high temperatures, the orientations of dipoles with respect to the applied electric field become completely randomized. The net polarization begins to decrease, and this polarization mechanism stops. The *dipolar polarizability* (α_d) associated with this mechanism is given by

$$\alpha_d = \frac{1}{3}\frac{\mu^2}{K_B}$$

(7.63)

where μ is the permanent dipole moment of the molecules, K_B is the Boltzmann's constant, and T is the temperature.

Dipole moments associated with polar molecules are typically very large compared to those induced by the polarization of atoms or the displacements of ions. Dipolar or orientational polarizability (α_d) values and the resultant dielectric constants are therefore large for polar materials.

Dipolar polarization is typically seen in polar liquids, gases, or vapors (e.g., H_2O, alcohol, and hydrochloric acid [HCl]). Unlike in their vapor form, molecules with permanent dipoles are not free to rotate in polar solids (e.g., ice, instead of water vapor), even if they are present. This means that the effect of orientational polarization in polar solids is smaller compared to that in liquids and gases or vapors.

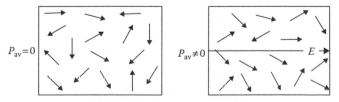

FIGURE 7.13 Illustration of dipolar or orientational polarization. (From Kasap, S. O. 2002. *Principles of Electronic Materials and Devices*. New York: McGraw Hill. With permission.)

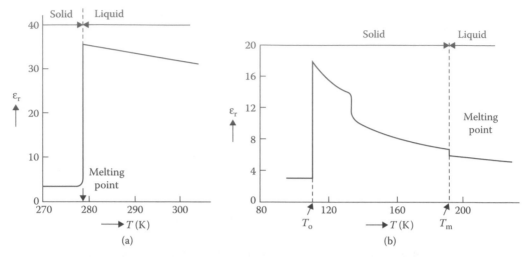

FIGURE 7.14 Dielectric constant of (a) nitrobenzene and (b) hydrogen sulfide. (From Kao, K. C. 2004. *Dielectric Phenomena in Solids*. London: Elsevier-Academic Press. With permission.)

For example, nitrobenzene ($C_6H_5CO_3$) is a polar liquid with $k{\sim}35$ (Figure 7.14). The dielectric constant of liquid $C_6H_5CO_3$ is expected to decrease with increasing temperatures (Equation 7.63). When liquid $C_6H_5CO_3$ freezes into a solid, the dipoles are present but are frozen and unable to rotate. This is why the dielectric constant of $C_6H_5CO_3$ decreases to about 3 (Figure 7.14a). This lower value for the dielectric constant reflects the smaller extent of electronic and ionic polarization, compared to the extent of orientational polarization and ignoring the effects of the differences in the molar volumes of the solid and liquid phases.

In some materials (such as hydrogen sulfide [H_2S]), the dielectric constant continues to increase with decreasing temperatures, even below the freezing or melting temperature (T_m). This continues up to the critical temperature T_0, below which the dipoles cannot rotate and the orientational polarization mechanism ceases (Figure 7.14b).

7.10 INTERFACIAL, SPACE CHARGE, OR MAXWELL–WAGNER POLARIZATION

Some dielectrics contain relatively mobile ions (e.g., H^+, Li^+, and K^+). At high temperatures, these ions can drift under the influence of an electric field. The movement of such charge carriers is eventually impeded by the existence of interfaces (such as grain boundaries) in a material or a device. This can create a buildup of double-layer-like capacitors at interfaces, such as grain boundaries, in a polycrystalline material or material–electrode interfaces. The increase in polarization due to such movements of mobile ions in a material and the creation of polarization at the interfaces is known as *space-charge polarization, interfacial polarization*, or Maxwell–Wagner (M–W) or Maxwell–Wagner–Sillars (M–W–S) polarization. Trapping electrically charged ions, electrons, and holes at the interfaces is the essential process behind interfacial polarization (Figure 7.15).

This polarization mechanism differs from other mechanisms we have discussed (Figure 7.8). First, the polarization mechanism is usually more prominent at higher temperatures. This is because the rate of the diffusion process, which often occurs through atoms or ions jumping or hopping from one location to another, increases exponentially with increasing temperature (Chapter 2). Second, the diffusion of atoms or ions is relatively slow compared to that of electrons and holes. Thus, this polarization mechanism is usually operated under static (DC) or AC fields, in which the electrical frequency (f) is small (a few mHz to several Hz). If the electrical frequency (f) is too high, the otherwise mobile ions cannot rapidly follow this frequency. The interfacial polarization mechanism then ceases to exist. Many silicate glasses and crystalline ceramic materials containing mobile ions

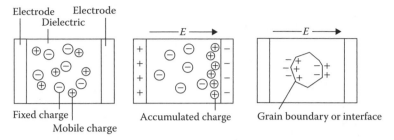

FIGURE 7.15 Illustration of interfacial polarization. (From Kasap, S. O. 2002. *Principles of Electronic Materials and Devices.* New York: McGraw Hill. With permission.)

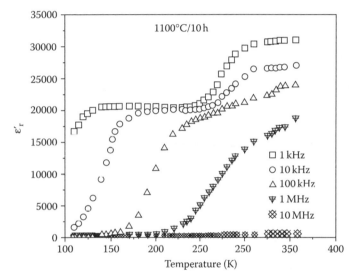

FIGURE 7.16 Apparent dielectric constant of $CaCu_3Ti_4O_{12}$ ceramics processed at 1100°C for 10 hours. The dielectric constant is measured at different temperatures up to 350 K. (From Prakash, B., and K. B. R. Varma. 2007. *J Phys Chem Solids* 68:490–502. With permission.)

(e.g., lithium niobate [$LiNbO_3$], lithium tantalate [$LiTaO_3$], and lithium cobalt oxide, [$LiCoO_2$]) exhibit this polarization mechanism. This polarization mechanism is also quite common at the liquid electrolyte–electrode interfaces because diffusion in liquids occurs rather readily. This polarization mechanism is commonly used in many electrochemical reactions encountered during the operation of *supercapacitors*, batteries, fuel cells, and similar devices.

The existence of interfacial polarization is often considered a strong possibility whenever unusually high dielectric constants are seen, especially at high temperatures and low frequencies. For example, the data for the dielectric constant for a calcium–copper–titanium oxide (CCTO) ceramic are shown in Figure 7.16.

7.11 SPONTANEOUS OR FERROELECTRIC POLARIZATION

A ferroelectric material is defined as a material that exhibits spontaneous and reversible polarization (Chapter 8). This polarization is typically very large in magnitude compared to other polarization mechanisms, such as electronic and ionic polarizations.

A prototypical example of a ferroelectric material is the *tetragonal* polymorph of $BaTiO_3$. The term "polymorph" means the particular crystal structure of a material. Consider the two polymorphs of $BaTiO_3$—one is cubic, and the other is tetragonal. The tetragonal form of $BaTiO_3$ is also

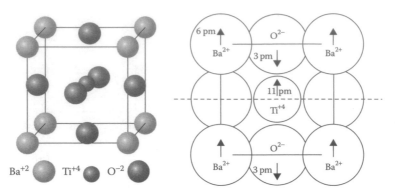

FIGURE 7.17 Small-scale displacements in a cubic structure lead to the tetragonal form of ferroelectric $BaTiO_3$. (Adapted from Moulson, A. J., and Herbert, J. M. 2003. *Electroceramics: Materials, Properties, Applications.* New York: Wiley; and Buchanan, R. C. 2004. *Ceramic materials for electronics.* New York: Marcel Dekker. With permission.)

known as the *pseudocubic form.* The cation/anion ratio for the tetragonal polymorph of $BaTiO_3$ at room temperature (300 K) is ~1.01. For the cubic polymorph, which is stable at higher temperatures, the c/a ratio is 1.00. The lattice constant is ~4 Å. Thus, the physical difference in the dimensions of the unit cells is actually very small. However, the differences in the electrical properties between the cubic and tetragonal forms of $BaTiO_3$ are significant.

For the typical single-crystal or polycrystalline $BaTiO_3$, the centrosymmetric cubic phase is stable at temperatures above ~130°C. The temperature at which a ferroelectric material transforms into a centrosymmetric paraelectric form is known as the *Curie temperature.* In cubic $BaTiO_3$, the titanium (Ti) ion "rattles" very rapidly around several equivalent but off-center positions present around the cube center. For each of these off-center positions of the titanium ion, the unit cell structure has a dipole moment. Thus, at any given time, the time-averaged position of the titanium ion *appears* to be exactly at the cube center. As a result, the cubic phase of $BaTiO_3$ has no net dipole moment from the viewpoint of electrical properties. From a structural viewpoint (e.g., while using x-ray diffraction) the crystal structure appears cubic and is centrosymmetric. In cubic $BaTiO_3$, all the dipole moments that are associated with the barium (Ba^{2+}), titanium (Ti^{4+}), and oxygen (O^{2-}) ions cancel one another out. This high-temperature phase, derived from an originally ferroelectric parent phase that now has no dipole moment per unit cell, is known as the *paraelectric phase.*

When the temperature approaches the Curie temperature (T_c~130°C for $BaTiO_3$), the titanium ions begin to undergo other very small displacements. At temperatures below T_c, barium ions are displaced by a distance of ~6 pm (1 pm = 10^{-12} m). Titanium ions are displaced in the same direction by ~11 pm. Oxygen ions (O^{2-}) are displaced by ~3 pm. After these displacements occur, the unit cell becomes tetragonal. The tetragonal structure is *not* centrosymmetric. It is a polar structure, that is, the tetragonal unit cell of $BaTiO_3$ has a net polarization (Figure 7.17).

7.12 DEPENDENCE OF THE DIELECTRIC CONSTANT ON FREQUENCY

The polarization mechanisms require displacements of ions, electronic clouds, dipoles, and so on (Table 7.2). These displacements are small; nevertheless, they require a small but finite amount of time. Consider a covalently bonded material such as silicon, where the electronic polarization is established quickly (~10^{-14} seconds). Now, consider changing the polarity of the applied electric field. The electronic clouds shift, and the induced dipoles realign with the new field direction within another ~10^{-14} seconds. The induced dipoles can align rapidly back and forth in an alternating electric field even if the electrical field switches at a frequency of, for example, 1 MHz. A frequency

(f) of 1 MHz means that the field switches back and forth 10^6 times per second, or in one microsecond (μs; 1 μs = 10^{-6} seconds). Under a static (i.e., $f = 0$) field or an electric field oscillating with a frequency of 10^6 Hz, we expect the electronic polarization process to contribute to the dielectric constant (k) of silicon. This continues to very high frequencies, ranging up to 10^{14} Hz, because electronic polarization is the only mechanism of polarization that survives up to such high frequencies. Thus, the dielectric constants of silicon and other covalently bonded solids (such as germanium [Ge] and diamond) are expected to remain constant with frequencies up to the range of ~10^{14} Hz.

Now, consider a material in which the bonding has some ionic character, such as SiO_2. In this material, we expect both the ionic and electronic polarization mechanisms to contribute to its dielectric constant (k). This will be true as long as the ionic and electronic polarization mechanisms can follow the alternating electric fields. The ionic polarization mechanism is slower compared to that of electronic polarization because it involves the displacement of ions. Thus, up to a frequency of ~10^{12}–10^{13} Hz, both electronic and ionic polarization mechanisms contribute to the dielectric constant of SiO_2. At higher frequencies, the ions cannot follow the back-and-forth switching of the polarity of the electric field. As a result, the ionic polarization mechanism stops, lowering the dielectric constant. The electronic polarization mechanism survives up to ~10^{14} Hz and continues to contribute to the dielectric constant. Thus, we expect the dielectric constant of SiO_2 to become smaller (or for it to relax) as we proceed from low to high frequencies. We can expect this trend for any material with more than one polarization mechanism (Figure 7.18).

The lowering of the dielectric constant with increasing frequency is known as *dielectric relaxation* or *frequency dispersion*. This *low-frequency dielectric constant* is also known as the *static dielectric constant* (k_s), although the value is not necessarily measured under DC fields. The *high-frequency dielectric constant* is designated as k_∞. This is also known as the *optical frequency dielectric constant*, and it is the value of dielectric constant when only the electronic (optical) polarization mechanism remains.

In practice, dielectric materials with multiple polarization mechanisms show a decrease in the dielectric constant (k). However, the decrease is rather steady and not as abrupt as that shown in Figure 7.18, because with multiple types of ions or atoms, the polarization mechanisms do not

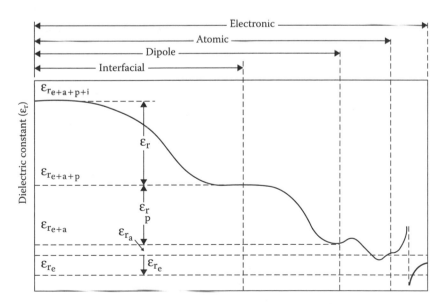

FIGURE 7.18 Variation of dielectric constant with frequency for a hypothetical dielectric with different polarization mechanisms (From Buchanan, R. C. 2004. *Ceramic Materials for Electronics*. New York: Marcel Dekker. With permission.).

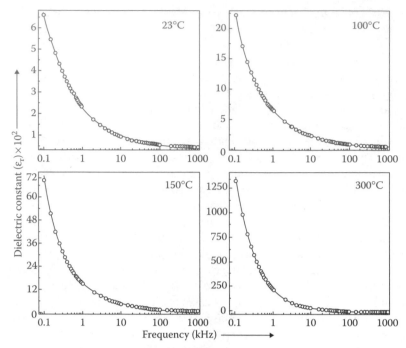

FIGURE 7.19 Relaxation in the dielectric constant of $LiFe_{1/2}Ni_{1/2}VO_4$ ceramics at different temperatures. (From Ram, M., and S. Chakravarty. 2008. *J Phys Chem Solids* 69(4):905–912. With permission.)

stop at a particular frequency but rather over a range of frequencies, even for the same type of polarization.

As an example, Figure 7.19 shows the lowering of the dielectric constant for lithium–iron–nickel–vanadium oxide ceramics. The data are shown for temperatures ranging from 23°C to 300°C.

At any given temperature, the dielectric constant decreases with increasing frequency. The lower-frequency dielectric constants increase with increasing temperature. This is very much an indication of interfacial polarization, or Maxwell–Wagner polarization. Its presence is not surprising because this material contains relatively mobile lithium ions.

7.12.1 Connection to the Optical Properties: Lorentz–Lorenz Equation

The only polarization mechanism that does not cease to exist at high frequencies is electronic polarization. The high frequencies at which only the electronic polarization mechanism survives correspond to a wavelength (λ) of light (Figure 7.20).

Therefore, electronic polarization is also known as optical polarization (see Section 7.6). We can show that the *high-frequency dielectric constant* (k_∞) of a material is equal to the square of its refractive index (*n*):

$$k_\infty = n^2 \tag{7.64}$$

Recall the Clausius–Mossotti equation (Equation 7.43) that correlates the dielectric constant with polarization.

Applying this equation for high-frequency conditions, that is, by replacing the term $\varepsilon_r = \varepsilon_\infty$ and removing the ionic polarization term, we get

$$\frac{\varepsilon_\infty - 1}{\varepsilon_\infty + 2} = \frac{1}{3\varepsilon_0}\left[\left(N_e \times \alpha_e\right)\right] \tag{7.65}$$

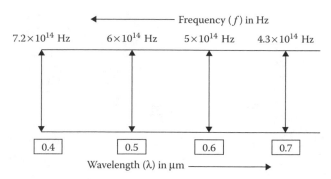

FIGURE 7.20 The relationship between frequency and wavelengths of light.

Combining Equations 7.64 and 7.65, we get the so-called Lorentz–Lorenz equation:

$$\frac{n^2 - 1}{n^2 + 2} = \frac{1}{3\varepsilon_0}\left[\left(N_e \times \alpha_e\right)\right] \tag{7.66}$$

This can be rewritten by replacing the concentration of atoms (N_e) as follows:

$$\left(\frac{n^2 - 1}{n^2 + 2}\right)\frac{M}{\rho} = \frac{N_{\text{Avogadro}} \times \alpha_e}{3\varepsilon_0} \tag{7.67}$$

A note of caution is in order regarding the use of the Lorentz–Lorenz equation. Recall that the Clausius–Mossotti equation was derived using the local internal electric field (E_{local}) calculation (Equation 7.36). Because the Lorentz–Lorenz equation is derived using the Clausius–Mossotti equation, it applies to *nonpolar* materials with a cubic symmetry or to amorphous materials. It *cannot* be applied to ferroelectric or other polar dielectric materials.

Materials that contain ions or atoms with a large electronic polarizability (α_e) have a higher refractive index (n). A common example of such a material is called the lead crystal (Figure 7.21). This is actually an *amorphous* silicate glass that contains substantial (up to 30–40 wt% PbO) concentrations of lead ions (Pb^{2+}). Because lead ions have a large electronic polarizability, the refractive index of the lead crystal is much higher (n up to 1.7) than that of a common soda-lime glass ($n \sim 1.5$). Note that the polarizability values shown in Figure 7.10 include both electronic and ionic polarizabilities.

Another important example of the use of higher electronic polarizability (α_e) to achieve a higher refractive index (n) is its application in optical fibers (Figure 7.22). In optical fibers, a small yet significant mismatch (~1%) is created between the refractive indices of the core and the cladding. By doping the core of the fibers with dopants such as germanium, the refractive index of the core region is maintained higher than that of the cladding. Optical fibers are usually made from ultra-high-purity SiO_2. The core is doped with germanium oxide (GeO_2), which enhances the refractive index of the core region. This increases the total internal reflection at the core-cladding interface, thereby restricting the light waves (i.e., information) to within the optical-fiber core. Doping the fibers with fluorine (F) causes the refractive index of SiO_2 to decrease.

The following example illustrates the extent of contribution of the electronic and ionic polarization mechanisms to the dielectric constant of ionic materials.

EXAMPLE 7.8: DIELECTRIC CONSTANT OF SILICATE GLASS

The refractive index (n) of a silicate glass is 1.5. What is the high-frequency dielectric constant of this glass? The dielectric constant of this glass at 1 kHz is $k = 7.6$. Based on the relaxation of the

FIGURE 7.21   Photograph of a lead crystal object.

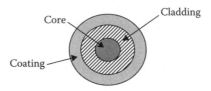

FIGURE 7.22   Core-clad structure of optical fibers.

dielectric constant, approximately what fraction of the low-frequency dielectric constant (k_s) can be attributed to ionic polarization?

SOLUTION

The refractive index (n) is 1.5. This means that the high-frequency dielectric constant (k_∞) is $= n^2 = 2.25$. The low-frequency dielectric constant is $k_s = 7.6$. Thus, the contribution of ionic polarization to the dielectric constant is $7.6 - 2.25 = 5.35$. The fraction of contribution due to ionic polarization is $= ((7.6 - 5.35)/7.6) = (5.35/7.6) \sim 0.7$. Thus, nearly 70% of the low-frequency dielectric constant (k_s) for this glass is due to ionic polarization, with the rest due to electronic polarization.

7.13   COMPLEX DIELECTRIC CONSTANT AND DIELECTRIC LOSSES

7.13.1   COMPLEX DIELECTRIC CONSTANT

Another feature associated with polarization mechanisms is the notion of *dielectric loss*. When ions, electron clouds, dipoles, and so on are displaced in response to the electric field (Figure 7.8), these displacements do not occur without resistance. This resistance is similar to the effect of friction on mechanical movement. The electrical energy lost during the displacements of ions, the electronic cloud, or any other entity that causes dielectric polarization, is known as the dielectric loss. One way to represent dielectric losses is to consider the dielectric constant as a complex number. Thus, we define the *complex dielectric constant* $\left(\varepsilon_r^*\right)$ as

$$\varepsilon_r^* = \varepsilon_r' - j\varepsilon_r'' \tag{7.68}$$

In Equation 7.68, j is the imaginary number $\sqrt{-1}$.

ε'_r, known as the *real part* of the dielectric constant, represents the charge-storage process, which is the same quantity that we have referred to so far as ε_r (or k). The *imaginary part* of the complex dielectric constant (ε''_r) is a measure of the dielectric losses that occur during the charge-storage process.

7.13.2 Real Dielectrics and Ideal Dielectrics

An *ideal dielectric* is a hypothetical material with zero dielectric losses (i.e., $\varepsilon''_r = 0$). This means that all the applied electrical energy is used to cause the polarization that leads to charge storage only. A *real dielectric* is a material that does have some dielectric losses. All dielectric materials have some level of dielectric loss because the displacements of ions, electron clouds, and so on, cannot occur without resistance from neighboring atoms or ions. The dielectric losses increase if the applied field switches in such a way that the polarization mechanisms can follow these changes in the applied electric field. It is usually desirable to minimize or lower the dielectric losses for microelectronic devices. However, dielectric losses can be useful for applications in which heat must be generated. A common example is that of a microwave oven. The water molecules in food, which are permanent dipoles, tumble around during the polarization caused by the microwave's electric field. The resultant dielectric losses cause the generation of heat (Vollmer 2004).

While developing materials for capacitors to store charge (Section 7.2.2), we prefer to use a *low-loss dielectric*; in other words, we want a small ε''_r. The need for increasing the dielectric constant (k, or now what we refer as ε'_r) while maintaining the dielectric losses at small levels poses a problem because polarization processes are required to achieve higher dielectric constants. When these polarization processes occur, they cause dielectric losses. Polarization and dielectric losses originate from the same basic processes (some type of displacement of ions and electron clouds, in addition to the reorientation or rotation of dipoles, etc; Figure 7.8).

7.13.3 Frequency Dependence of Dielectric Losses

Because the dielectric constant depends on frequency (f or ω), it is reasonable to assume that dielectric losses are also frequency-dependent.

Consider a hypothetical dielectric material with only the dipolar polarization mechanism. If the electrical frequency (f or ω) is too high, then the dipole cannot rotate or flip back and forth in response to the oscillating electric field. We may also change the magnitude of the electric field from some initial value E_0 to another value E without changing its direction. If this happens, the induced dipole moment (ionic or electronic polarization) adjusts from some starting value of μ_0 (at E_0) to another final value μ (at E). Such changes in the dipole moment require a small but finite time. The *relaxation time* (τ) is the time required for a polarization mechanism to revert and realign a dipole or change its value when the electrical field switches or changes in magnitude.

In a hypothetical dielectric material with a single polarization mechanism and a relaxation time τ, the dielectric losses are very small if the frequency of the electrical field switches too rapidly. When $\omega \gg 1/\tau$, the polarization mechanism is unable to follow the change in E. On the contrary, if $\omega \ll 1/\tau$, then the polarization process can follow the changes in the electric field. However, the induced dipole moment does not switch or readjust that often, so the dielectric losses are again small. When $\omega = 1/\tau$, similar to a resonance condition, the dielectric losses are maximized because when the polarization switches or adjusts itself in a synchronized fashion as the electric field switches again. Figure 7.23 shows this variation of dielectric losses with frequency for a hypothetical dielectric with a single polarization mechanism with relaxation time τ.

In any dielectric material, there are different relaxation times for different mechanisms of polarization. For example, the displacement of the electronic cloud around a nucleus occurs very rapidly,

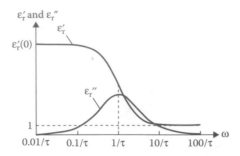

FIGURE 7.23 Dependence of ε_r' and ε_r'' on frequency for a material with a single polarization mechanism and relaxation time τ. (From Buchanan, R. C. 2004. *Ceramic Materials for Electronics*. New York: Marcel Dekker. With permission.)

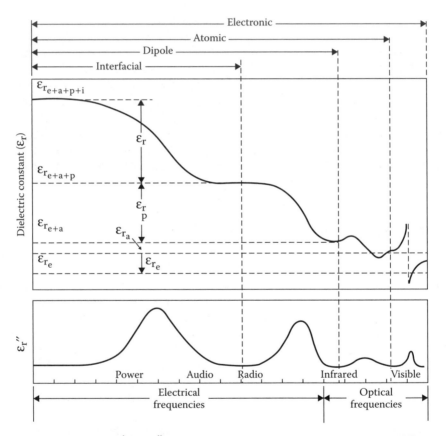

FIGURE 7.24 Dependence of ε_r' and ε_r'' on frequency for a hypothetical material with different polarization mechanisms and relaxation times. (From Buchanan, R. C. 2004. *Ceramic Materials for Electronics*. New York: Marcel Dekker. With permission.)

and the relaxation time τ is $\sim 10^{-14}$ seconds. On the contrary, for interfacial polarization, the relaxation time (τ) can be several seconds because this mechanism involves a relatively longer-range motion of ions. Thus, for materials with multiple polarization mechanisms, the real part of the dielectric constant $\left(\varepsilon_r'\right)$ shows changes similar to a set of cascades. The dielectric loss component $\left(\varepsilon_r''\right)$ shows a series of maxima that follow the different polarization mechanisms (Figure 7.24).

For a dielectric material with a single relaxation time (τ), the frequency dependence of ε_r' and ε_r'' can be written quantitatively as follows:

$$\varepsilon_r' = k' = k_\infty + \frac{k_s - k_\infty}{1 + \omega^2 \tau^2} \tag{7.69}$$

$$\varepsilon_r'' = k'' = (k_s - k_\infty)\left(\frac{\omega\tau}{1 + \omega^2\tau^2}\right) \tag{7.70}$$

Equations 7.69 and 7.70 describe changes in the real and imaginary parts of the dielectric constant and are also known as Debye equations.

We can see from Equation 7.70 that the maximum in k'', known as the Debye loss peak, occurs when $\omega = 1/\tau$ (see the problems at the end of this chapter and Example 7.12).

In many materials, there can be different relaxation times for the same polarization mechanism. This is because different atoms and ions may be involved. There is then a distribution of relaxation times due to the different polarization mechanisms and the distribution of relaxation times applicable for each mechanism for a given dielectric.

The following example illustrates an application of the Debye equations.

EXAMPLE 7.9: RELAXATION OF THE DIELECTRIC CONSTANT FOR H_2O

The static dielectric constant (ε_s) of H_2O is 78.4 at 298.15 K (Fernandez et al. 1995). It decreases to about $\varepsilon_r(\omega) = 20$ at a frequency of $f = 40$ GHz. If the high-frequency (optical) dielectric constant (ε_∞) is 5, what is the relaxation time τ for H_2O molecule dipoles? Why is it that ε_∞ is not equal to the square of the refractive index (n) of water?

SOLUTION

We first convert the frequency (f) into angular frequency (ω) by using

$$\omega = 2\pi f \tag{7.71}$$

Therefore, $\omega = 2\pi(40 \times 10^9 \text{ Hz}) = 2.51327 \times 10^{11}$ rad/s.

We assume that relaxation in the real part of the dielectric constant (k') of water follows the Debye equations, Equations 7.69 and 7.70.

For this problem, we substitute the following values in Equation 7.69: $\varepsilon_s = k_s = 78.4$, and $\varepsilon_\infty = k_\infty = 5$. We also know that $k' = 20$, when $f = 40$ GHz, that is, $\omega = 2\pi f = 2.51327 \times 10^{11}$ rad/s.

We rewrite the Debye equation as:

$$(1 + \omega^2\tau^2) = \frac{k_s - k_\infty}{k' - k_\infty}$$

or

$$\omega^2\tau^2 = \left(\frac{k_s - k_\infty}{k' - k_\infty}\right) - 1 \tag{7.72}$$

Substituting these values in Equation 7.72, we get

$$\omega^2\tau^2 = \left[\frac{78.4 - 5}{20 - 5}\right] - 1 = 3.893$$

Therefore,

$$\omega\tau = 1.973$$

or

$$\tau = \frac{1.973}{2.51327 \times 10^{11} \text{ rad/s}} = 7.85 \times 10^{-12} \text{ s}$$

One picosecond is equal to 10^{-12} seconds. Thus, the relaxation time (τ) for an H_2O molecule, which is a permanent dipole, is about 7.85 picoseconds, which is the average time that an H_2O molecule needs to "flip" and realign as the direction of the electrical field changes.

Note that because H_2O molecules are polar, neither the Clausius–Mossotti nor the Lorentz–Lorenz equations apply, because ε_∞ is not equal to n^2.

7.13.4 GIANT DIELECTRIC CONSTANT MATERIALS

The changes in the dielectric properties of dielectrics encountered in applications of microelectronics are more complex. For example, CCTO ($CaCu_3Ti_4O_{12}$) was reported as a giant dielectric constant material. The changes in the real and imaginary parts of the dielectric constant for CCTO are shown in Figure 7.25.

Note several details from the data in Figure 7.25:

1. First, the dielectric constant is very large. We should therefore immediately suspect that this material may be a ferroelectric or that a space-charge polarization (see Section 7.10) may be present. Crystal structure analysis and other measurements have shown that this material is not a ferroelectric. The measured dielectric constant must be an *apparent dielectric* constant.

   These experimental data show that the apparent or measured dielectric constant is a *microstructure-sensitive* property. If we were to predict the dielectric constant using the Clausius–Mossotti equation, then the dielectric constant, although expected to be a function of frequency, will *not* be expected to show microstructure-sensitive characteristics. This is because the Clausius–Mossotti equation does not account for interfacial or ferroelectric (spontaneous) polarization.

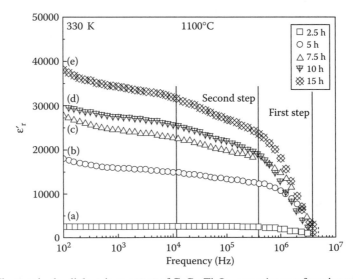

FIGURE 7.25 Changes in the dielectric constant of $CaCu_3Ti_4O_{12}$ ceramics as a function of the sintering time at 1100°C. Measurements of the dielectric constant (k') were carried out at 330 K. (From Prakash, B., and K. B. R. Varma. 2007. *J Phys Chem Solids* 68:490–502. With permission.)

2. Overall, as predicted by the Debye equations, the dielectric constant decreases with increasing frequency (Figure 7.25).

3. In this case, the increase in the dielectric constant with the sintering times may be because when the sintering times are low (e.g., 2.5 hours), the samples may not be dense. These materials are basically a composite of a dielectric and air. Because air has a low dielectric constant ($k' \sim 1$), the overall measured values of the dielectric constant are lower.

4. As the measured density increases with increasing sintering times, the overall dielectric constant also increases. The data in Figure 7.25 (marked as the first step) also show that the increase in the dielectric constant with sintering times is higher for the higher-frequency region ($\sim 5 \times 10^5$ to 3×10^6 Hz). In these materials, the interfacial or Maxwell–Wagner polarization at the grain boundaries plays an important role.

5. The relatively moderate increase in the dielectric constant in step 2 (frequency range $\sim 10^4$ to 5×10^5 Hz; Figure 7.25) is related to the formation of another phase, which involves a process known as liquid-phase sintering. In this process, a liquid phase is formed that can assist densification. However, it can also lead to the formation of grain boundaries or surface phases that have a chemical composition different from that of the original ceramic material.

6. The relaxation of the low-frequency dielectric constant (in the range of 10^2–10^4 Hz—the region to the left of that marked as the second step) may be due to interfacial polarization at the ceramic–electrode interface via a Schottky barrier effect. If this is the case, it may be possible to change the electrode materials to see if the apparent dielectric constant changes.

In summary, the interpretation of changes in the apparent or measured dielectric constant or losses with frequency requires a considerably detailed analysis. We can attempt to correlate these changes with the different polarization mechanisms and microstructures. To determine the controlling polarization mechanisms, the dielectric properties are measured as functions of the temperature and frequency while changing the processing conditions systematically. The change in the dielectric constant of CCTO ceramics as a function of the temperature is shown in Figure 7.16. The technique of measuring k' and k'' as a function of the frequency (f) is known as *impedance spectroscopy*. The measurements of k' and k'' can be accomplished using an instrument called an *impedance analyzer*. The variations of k' (x-axis) and k'' (y-axis) plotted as a function of the frequency appear as sets of arcs or semicircles and are known as Cole–Cole plots.

7.14 EQUIVALENT CIRCUIT OF A REAL DIELECTRIC

For a parallel capacitor, the capacitance is given by

$$C = \varepsilon_0 \times \varepsilon_r'(\omega) \times \frac{A}{d} \tag{7.73}$$

This is similar to Equation 7.13; the difference is that now we explicitly show the dependence of the dielectric constant on electrical frequency by incorporating the variation of the dielectric constant with frequency (Figures 7.18 and 7.23).

We can model a real dielectric material as a pure, lossless capacitor with capacitance C, connected in parallel to a resistor with a resistance (R_p). The subscript p tells us that in this equivalent circuit, the resistor is connected to the capacitor in parallel. This is an equivalent circuit of a real dielectric material (Figure 7.26). A circuit comprised of a capacitor and a resistor (in series or parallel) is also known as a resistor-capacitor (RC) circuit. For an ideal dielectric, there is no loss, that is, $R_p = 0$.

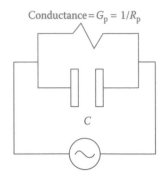

FIGURE 7.26 Equivalent circuit of a real dielectric material. (From Kasap, S. O. 2002. *Principles of Electronic Materials and Devices*. New York: McGraw Hill. With permission.)

The conductance ($G_p = 1/R_p$) of a dielectric is given by

$$G_p = \frac{1}{R_p} = \frac{\omega A \varepsilon_0 \varepsilon_r''(\omega)}{d} \tag{7.74}$$

Thus, for an ideal capacitor, if $R_p = 0$, then $G_p = \infty$. This means that as the capacitor charges and discharges, the current appears to flow "through" the capacitor. We have, of course, seen before that a dielectric material is a nonconductor, and very little, if any, current can actually flow through it. Equation 7.74 can be used to calculate the equivalent resistance or conductance of a dielectric material with specific capacitor geometry. The relatively simple appearance of Equations 7.73 and 7.74 is a bit misleading. If we want to calculate these values as a function of the frequency (ω), then we must keep in mind that both ε_r' and ε_r'' change with frequency.

7.15   IMPEDANCE (Z) AND ADMITTANCE (Y)

For a real dielectric, we define *impedance* (Z) as a measure of the resistance offered by a circuit under AC fields. The impedance of a real dielectric (Figure 7.26) can be written as follows:

$$Z = R_p + jX_C \tag{7.75}$$

where X_C is the *capacitive reactance*. It is the equivalent of a capacitor's resistance to the flow of current through it. The magnitude of the capacitive reactance (X_C) is given by

$$X_c = \frac{1}{\omega C} = \frac{1}{2\pi f C_p} \tag{7.76}$$

Thus,

$$Z = R_p + j\frac{1}{2\pi f C_p} \tag{7.77}$$

The subscript "p" is used to indicate a parallel arrangement of the resistor and capacitor (Figure 7.26).

The *admittance* (Y) is defined as the inverse of the impedance (Z). Note that the admittance (Y) and impedance (Z) are both complex numbers. The SI unit for admittance is Siemens, also often reported as mho, the inverse of Ohm. The admittance of a capacitor made using a real dielectric material can be written as a combination of a conductor, with conductance G_p, and a capacitor, with capacitance C_p (Figure 7.26). The total admittance is written as a complex number:

$$Y = G_p + j\omega C_p \tag{7.78}$$

or

$$Y = \frac{1}{R_p} + j\omega C_p$$

$$Y = \frac{1}{R_p} + j2\pi f C_p \tag{7.79}$$

The admittance (Y) can be rewritten as follows using Equations 7.73 and 7.74:

$$Y = \frac{\omega A \varepsilon_0 \varepsilon_r''(\omega)}{d} + j\frac{\omega A \varepsilon_0 \varepsilon_r'(\omega)}{d} \tag{7.80}$$

Examples 7.10 and 7.11 illustrate the calculations of capacitive reactance and its frequency dependence.

EXAMPLE 7.10: CAPACITIVE REACTANCE (X_C) OF A CAPACITOR

What is the magnitude of the capacitive reactance (X_C) of a 200-pF capacitor at a frequency of 10 MHz?

SOLUTION

Note that one picofarad (pF) is 10^{-12} F. The magnitude of the capacitive reactance (X_C) is given by

$$X_C = \frac{1}{\omega C} = \frac{1}{2\pi f C} = \frac{1}{2\pi(10 \times 10^6)(200 \times 10^{-12}\ \text{F})} = 79.5\ \Omega$$

Thus, this capacitor has a reactance of 79.5 Ω at a frequency of 10 MHz.

EXAMPLE 7.11: CHANGE OF CAPACITIVE REACTANCE (X_C) WITH FREQUENCY

A 10-V variable-frequency AC supply is connected to a 200-μF capacitor. What is the value of capacitive reactance for frequencies of (a) 0 Hz, (b) 60 Hz, and (c) 1 kHz? What is the value of the current (I) flowing through the circuit for each frequency?

SOLUTION

a. When the frequency (f) is zero, $X_C = \infty$, and the value of current flowing through the circuit is zero. In other words, when a DC voltage is applied to the capacitor, the dielectric material just blocks this voltage.
b. When $f = 60$ Hz, the value of the capacitive reactance (X_C) is given by

$$X_C = \frac{1}{2\pi f C} = \frac{1}{2 \times \pi \times 60\ \text{Hz} \times (200 \times 10^{-6})\ \text{F}} = 13.26\ \Omega$$

The current flowing through this circuit is given by the modified version of Ohm's law, in which we replace the resistance R with the capacitive reactance X_C. Thus, the current (I) is given by

$$I = \frac{V}{X_C} = \frac{10\ \text{V}}{13.26\ \Omega} = 0.754\ \text{A}$$

This current is not in phase with the voltage. It leads the voltage by 90° in the circuit of an ideal dielectric capacitor.

c. When $f = 1$ kHz, the value of capacitive reactance is given by

$$X_C = \frac{1}{2\pi f C} = \frac{1}{2 \times \pi \times (1 \times 10^3 \text{ Hz}) \times (200 \times 10^{-6}) \text{ F}} = 0.7958 \ \Omega$$

The current (I) is given by the equation

$$I = \frac{V}{X_C} = \frac{10 \text{ V}}{0.7958 \ \Omega} = 12.56 \text{ A}$$

Furthermore, the capacitive reactance (X_C) decreases with increasing frequency.

7.16   POWER LOSS IN A REAL DIELECTRIC MATERIAL

We know from Joule's law for a resistor that the power dissipated in a resistor is given as

$$\text{Power dissipated} = V \times I = \frac{V^2}{R} \tag{7.81}$$

The equivalent of this for a capacitor is

$$\text{Power input} = V \times I = V^2 \times Y \tag{7.82}$$

We substitute for Y from Equation 7.79 to get

$$\text{Power input} = j\omega C V^2 + \frac{V^2}{R_p} \tag{7.83}$$

The second term of this expression gives the power dissipated in a capacitor with equivalent resistance R_p (Figure 7.26).

Thus, the power lost in a real dielectric is given by

$$\text{Power loss in a real dielectric} = \frac{V^2}{R_p} \tag{7.84}$$

The dissipated power appears as heat in the dielectric material. It is sometimes important to note how much power is dissipated per unit volume. If we consider a capacitor (Figure 7.3) with volume $= A \times d$, the power loss per unit volume is given by substituting for $1/R_p$ from Equation 7.74 as shown below:

$$\text{Power loss in a real dielectric per unit volume} = \frac{V^2}{R_p \times A \times d} = \frac{V^2}{A \times d} \times \frac{\omega A \varepsilon_0 \varepsilon_r''(\omega)}{d}$$

or

$$\text{Power loss in a real dielectric per unit volume} = \omega \times E^2 \times \varepsilon_0 \varepsilon_r''(\omega) \tag{7.85}$$

7.16.1   CONCEPT OF TAN δ

Consider a resistor connected to voltage (V). Assure that the voltage changes with time in a sinusoidal fashion as follows:

$$V = V_0 \exp(j\omega t) \tag{7.86}$$

You may recall that the definition of the function exp (jx) is

$$\exp(jx) = \cos(x) + j\sin(x) \tag{7.87}$$

Thus,

$$\exp(j\omega t) = \cos(\omega t) + j\sin(\omega t) \tag{7.88}$$

In a pure resistor, the current (I) and the voltage are said to be "in phase." This means that as the voltage changes over time, the current instantly follows the change in voltage. This is shown in a *phasor diagram* in Figure 7.27. A *phasor* is a rotating vector that shows the phase angle for a particular type of current or voltage in a circuit component. A phasor diagram shows the relative positions of the phasors for currents and/or voltages corresponding to a circuit. The lengths of the phasor arrows are generally in scale with the magnitude of the voltage or current they represent. A phasor diagram is also known as a *vector diagram*.

At $\omega t = 0$, $V = V_0$; at $\omega t = \pi/2$ or 90°, $Vj = V_0$. As shown in Figure 7.27, as the voltage phasor rotates for a pure resistor from $\omega t = 0$ to $\omega t = \pi/2$, the phasor representing resistance also rotates from $\omega t = 0$ to $\omega t = \pi/2$.

When an AC voltage V (Equation 7.86) is applied to this capacitor, a charge Q appears on this capacitor and is given by

$$Q = C \times V = C \times V_0 \exp(jwt) \tag{7.89}$$

The buildup of charge on the capacitor can be described by defining a *charging current* (I_c). This current is given by

$$I_c = \frac{dQ}{dt} \tag{7.90}$$

Rewriting the expression for the charging current using Equation 7.89, we get

$$I_c = C\frac{dV}{dt} \tag{7.91}$$

Because $V = V_0 \exp(j\omega t)$, $dV/dt = V_0 j\omega \exp(j\omega t)$. Thus,

$$I_c = jCV_0\omega \exp(j\omega t) = j\omega CV \tag{7.92}$$

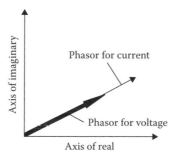

FIGURE 7.27 Phasor diagrams for a pure resistor.

Compare Equations 7.86 and 7.92 (for voltage V and I_c, respectively) when applied to an ideal capacitor. Note that the charging current I_c always leads the applied voltage V by exactly 90°. This is shown in the phasor diagram for an ideal capacitor, Figure 7.28a. The current and voltage waveforms for an ideal capacitor are shown in Figure 7.28b.

Referring to Figure 7.28, if the phase angle $\omega t = 0$, the charging current (I_c) will be along the axis of the imaginary and the voltage (V) will be along the axis of the real. As the phasor for I_c rotates through any angle, the phasor for the voltage also rotates by the same amount, thus maintaining a steady phase difference of 90° at all times (Figure 7.28b).

One way to represent a real dielectric is through a parallel combination of a resistor and a lossless capacitor (Figure 7.26). We then describe the total current (I_{total}) as the vector sum of a charging current (I_c) and a *loss current* (I_{loss}). The charging current associated with the ideal capacitor leads the applied voltage by 90°. The loss current originates from two sources. First, the dielectric material has a resistance under DC conditions that is usually very large. However, a very small amount of current still flows through a dielectric material. Second, for an AC voltage, as the polarization processes continue to occur, the back-and-forth displacements of electronic clouds, ions, molecular dipoles, and so on also cause a dissipation of energy and are thus part of the impedance. The loss current (I_{loss}) in a real dielectric due to both of these contributions is always in phase with the voltage, as these processes always follow the applied voltage (V).

The loss current (I_{loss}) in a dielectric can be written as follows:

$$I_{loss} = G \times V \tag{7.93}$$

where G is the conductance and V is the voltage. The conductance can be expressed as a sum of the DC and AC components.

The loss current can be written as

$$I_{loss} = \left(G_{dc} + G_{ac}\right) \times V \tag{7.94}$$

The total current (I) is given by the *vector sum* of these two currents, namely, the charging current (I_c) and the loss current (I_{loss}). This total current (I_{total}) always leads the voltage, but not by 90°. As shown in Figure 7.29, the total current (I_{total}) therefore leads the voltage at an angle $(90 - \delta)°$. The angle δ is known as the *loss angle*. Recall that the charging current (I_c) is $j\omega CV$ (from Equation 7.92). The total current (I_{total}) in a capacitor made using a real dielectric can be written as

$$I_{total} = I_c + I_{loss} = \left(j\omega C + G_{dc} + G_{ac}\right)V \tag{7.95}$$

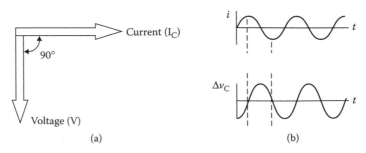

(a)   (b)

FIGURE 7.28   (a) Phasor diagram and (b) waveform relationships for an ideal capacitor.

From Figure 7.29, we can also note that

$$\tan \delta = \frac{I_{\mathrm{L}}}{I_{\mathrm{c}}} \qquad (7.96)$$

Thus, tan δ (pronounced tan delta) is also a measure of the dielectric losses. There are no dielectric losses in an ideal dielectric. This means that $I_{\mathrm{L}} = 0$, tan δ = 0, and the total current is equal to the charging current.

The concept of tan δ is very important for describing properties of dielectric materials. This parameter can be measured using an *impedance analyzer* similar to measuring the dielectric constant. We can compare how "lossy" different dielectric materials are relative

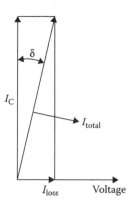

FIGURE 7.29 Corresponding waveforms for a real dielectric.

to each other. The real part of the complex dielectric constant is a measure of a material's charge-storing ability. The magnitude of tan δ indicates the inefficiency of the material in terms of the charge-storing ability.

In an ideal dielectric, which does not exist, polarization processes are assumed to occur without any electrical energy waste, with no energy wasted during charge storage, that is, tan δ = 0. In many real dielectrics, the tan δ values range from ~10^{-5} to 10^{-2}. Some special applications, such as ceramic materials used as dielectric resonators for microwave communications, require very low tan δ values. In this case, the values of a parameter defined as the *quality factor* (Q) are reported. The quality factor of a dielectric (Q_{d}) is defined as

$$Q_{\mathrm{d}} \approx \frac{1}{\tan \delta} \qquad (7.97)$$

We can also show that tan δ is the ratio of the imaginary and real parts of the complex dielectric constant (k^*).

The dielectric constant is defined as $\varepsilon_r^* = C/C_0$, and because $Q = CV$, $Q = \varepsilon_r^* C_0 V$.

The current $I_{\mathrm{total}} = dQ/dt = C(dV/dt)$, or

$$I_{\mathrm{total}} = \varepsilon_r^* C_0 \left(\frac{dV}{dt} \right) \qquad (7.98)$$

If $V = V_0 \exp(j\omega t)$, then $dV/dt = j\omega V_0 \exp(j\omega t) = j\omega V$; therefore, I_{total} is

$$I_{\mathrm{total}} = \varepsilon_r^* C_0 \left(\frac{dV}{dt} \right) = \left(\varepsilon_r' - j\varepsilon_r'' \right) C_0 j\omega V \qquad (7.99)$$

To derive Equation 7.99, we substituted for the complex dielectric constant in terms of its real and imaginary parts (see Equation 7.68).

$$I_{\mathrm{total}} = j\omega \varepsilon_r' C_0 V + \omega \varepsilon_r'' C_0 V \qquad (7.100)$$

Note that to derive this equation, we used the value $j^2 = -1$.

Now compare Equations 7.95 and 7.100. The first term of Equation 7.100 represents the charge-storage in the dielectric, that is, the charging current. The second term is the magnitude of the loss current (I_{loss}). Thus, the ratio of the *magnitude* of these two parts is tan δ:

$$\tan \delta = \frac{\omega \varepsilon_r'' C_0 V}{\omega \varepsilon_r' C_0 V} = \frac{\varepsilon_r''}{\varepsilon_r'}$$

Thus,

$$\tan \delta = \frac{\varepsilon_r''}{\varepsilon_r'} = \frac{k_r''}{k_r'} \tag{7.101}$$

Another interpretation for tan δ or loss tangent is that it is the ratio of the imaginary and real parts of the complex dielectric constant or the complex relative dielectric permittivity. One can think of tan δ as the ratio of the price we pay for storing the charge in a capacitor (in terms of the energy that is wasted and that appears as heat). Heating due to dielectric losses is useful in some applications. However, in many applications, the generation of heat changes the temperature of the dielectric material, and the dielectric properties may change with changes in temperature. In addition, heating a dielectric material can change its dimensions; the geometrical changes also lead to changes in capacitance. In general, we prefer to use temperature-stable and low-loss dielectric materials.

Because both the real and imaginary parts of the dielectric constant are frequency-dependent (Figure 7.24), tan δ also depends on the frequency.

Recall that for a hypothetical material with a single polarization mechanism with relaxation time (τ), the frequency dependences of the real and imaginary parts of the dielectric constants are given by the Debye equations (Equations 7.69 and 7.70). Therefore, the frequency dependence of tan δ for such a hypothetical material is given by

$$\tan \delta = \frac{\varepsilon_r''}{\varepsilon_r'} = \frac{(k_s - k_\infty) \omega \tau}{k_s + k_\infty \omega^2 \tau^2} \tag{7.102}$$

where k_s and k_∞ are the values of the low-frequency (static) and high-frequency dielectric constants.

While the value of ε_r'' reaches a maximum at $\omega t = 1/\tau$ (Figure 7.23), tan δ reaches a maximum at a slightly higher frequency, which is given by

$$\omega_{max} \text{(for } \tan \delta) = \frac{\left(\dfrac{k_s}{k_\infty} \right)^{1/2}}{\tau} \tag{7.103}$$

Recall that the power loss in a real dielectric per unit volume is given by Equation 7.85.

We can write this in terms of tan δ and ε_r' (dielectric constant) as follows:

$$\text{Power loss in a real dielectric per unit volume} = \omega \times E^2 \times \varepsilon_0 \times \varepsilon_r' \times \tan(\delta) \tag{7.104}$$

Example 7.12 shows how to calculate the frequency at which the value of tan δ reaches a maximum.

EXAMPLE 7.12: MAXIMUM FOR tan δ

Show that for a dielectric with only one polarization mechanism and a single relaxation time (τ), the maximum tan δ occurs at ω_{max} (for tan δ) (Equation 7.103).

Solution

We start with Equation 7.102, which describes the dependence of tan δ on frequency ω.

To locate the maximum, we take the derivative of tan δ with respect to ω and equate it to zero. Because the term $(k_s - k_\infty)$ does not depend on frequency, we can take the derivate of the term $\dfrac{\omega\tau}{(k_s + k_\infty\omega^2\tau^2)}$ with respect to ω and equate that term to zero. Recall the rule for the derivative of a function $h(x)$, which is a ratio of two functions $h(x) = f(x)/g(x)$; then

$$\frac{dh(x)}{dx} = \frac{\left[g(x)\cdot\dfrac{df}{dx} - f(x)\left(\dfrac{dg}{dx}\right)\right]}{g(x)^2} \tag{7.105}$$

$$\frac{d\left(\dfrac{\omega\tau}{k_s + k_\infty\omega^2\tau^2}\right)}{d\omega} = \frac{\left[\left(k_s + k_\infty\omega^2\tau^2\right)(\tau) - (\omega\tau)\left(\tau^2 k_\infty 2\omega\right)\right]}{\left(k_s + k_\infty\omega^2\tau^2\right)^2}$$

Equating this to zero gives us

$$\left(k_s + k_\infty\omega^2\tau^2\right)(\tau) - (\omega\tau)\left(k_\infty\tau^2 2\omega\right) = 0$$

This simplifies to

$$\left(k_s + k_\infty\omega^2\tau^2 - 2k_\infty\tau^2\omega^2\right) = 0$$

or

$$\omega = \frac{\left(\dfrac{k_s}{k_\infty}\right)^{1/2}}{\tau}$$

Tan δ will reach a maximum, i.e. ω_{max} (for tan δ) (Equation 7.103).

This maximum tan δ occurs at a frequency higher than $1/\tau$, which is the frequency at which k reaches a maximum.

7.17 EQUIVALENT SERIES RESISTANCE AND EQUIVALENT SERIES CAPACITANCE

A real dielectric can be described as a resistance and a capacitor connected in parallel (Figure 7.26). This description of a real dielectric is useful in understanding both the relationships of loss current with the charging current and the concept of tan δ. However, in many practical applications of capacitors, an important parameter that is often specified is the *equivalent series resistance* (ESR). This essentially entails describing the capacitor as a combination of an equivalent series capacitor (ESC) and an ESR, connected as shown in Figure 7.30.

Note that in both cases, the dielectric is modeled as either parallel or series connections of a capacitor and resistor, and there is no separate resistor connected to the dielectric for either model. We simply model the real dielectric as an arrangement of a capacitor and resistor.

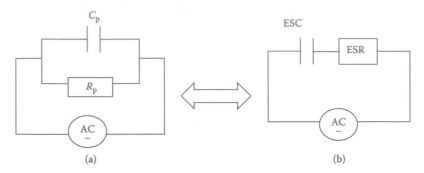

FIGURE 7.30 Circuit showing the model of a real dielectric as (a) parallel and (b) equivalent series resistance connected to an equivalent series capacitor.

We will show that the ESR and ESC are related to their parallel-circuit equivalents (i.e., R_p and C_p) using the following equations:

$$\text{ESR} = \frac{R_p}{\left(2\pi f C_p R_p\right)^2 + 1} \tag{7.106}$$

$$\text{ESC} = C_p\left[1 + \frac{1}{\left(2\pi f C_p R_p\right)^2}\right] \tag{7.107}$$

or

$$\text{ESC} = C_p + \frac{1}{\left(2\pi f R_p\right)^2 C_p} \tag{7.108}$$

In the parallel and series arrangements of the models of a real dielectric shown in Figure 7.30, the total impedance (Z) offered by the dielectric must be the same. The properties of the dielectric material do not depend on how we choose to describe it. Starting with the parallel arrangement, the total admittance of a real dielectric modeled as a parallel arrangement is written as Equation 7.79.

The total impedance of the parallel arrangement is

$$Z = \frac{1}{\dfrac{1}{R_p} + j\omega C_p} \tag{7.109}$$

In the series arrangement (Figure 7.30), we write the impedance as

$$Z = \text{ESR} + jX_{\text{ESC}}$$

or

$$Z = \text{ESR} + j\frac{1}{(\omega \times \text{ESC})} \tag{7.110}$$

Because the total impedance must be the same in both the series and parallel arrangements, the real part of this impedance (Z) in a parallel arrangement must be equal to the ESR. Similarly, the imaginary part of Z is equal to the equivalent series capacitive reactance (X_{ESC}). We start with Equation 7.109, or

$$Z = \frac{R_p}{1 + j\omega C_p R_p} \tag{7.111}$$

To separate the real and imaginary parts of this number, we multiply and divide by $(1 - j\omega C_p R_p)$, which is the complex conjugate of the term in the denominator.

$$Z = \frac{R_p}{1 + j\omega C_p R_p} \times \frac{\left(1 - j\omega C_p R_p\right)}{\left(1 - j\omega C_p R_p\right)}$$

or

$$Z = \frac{R_p}{\left(1 + \omega^2 C_p^2 R_p^2\right)} + j \frac{\left(\omega C_p R_p^2\right)}{\left(1 + \omega^2 C_p^2 R_p^2\right)} \tag{7.112}$$

We can compare Equation 7.112 to Equation 7.110, equate the real part of this derivation to ESR, and substitute $\omega = 2\pi f$ to get Equation 7.106.

We can equate the imaginary part of Z from Equation 7.112 to the equivalent capacitance for the series circuit (X_{ESC}; Equation 7.110) to get:

$$X_{ESC} = \frac{1}{\omega \times ESC} = \frac{\left(\omega C_p R_p^2\right)}{\left(1 + \omega^2 C_p^2 R_p^2\right)}$$

$$\omega \times ESC = \frac{\left(1 + \omega^2 C_p^2 R_p^2\right)}{\left(\omega C_p R_p^2\right)}$$

$$ESC = \frac{\left(1 + \omega^2 C_p^2 R_p^2\right)}{\left(\omega^2 C_p R_p^2\right)} = C_p \left[1 + \frac{1}{\omega^2 C_p^2 R_p^2}\right]$$

$$ESC = C_p \left[1 + \frac{1}{\omega^2 C_p^2 R_p^2}\right]$$

$$ESC = C_p \left[1 + \frac{1}{\left(2\pi f R_p\right)^2 C_p^2}\right]$$

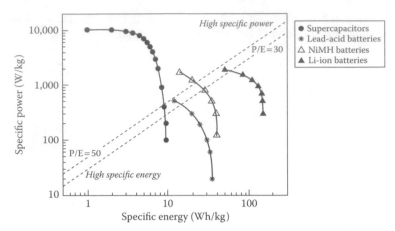

FIGURE 7.31 Ragone plot of electrochemical systems for hybrid electric vehicles. (From Mastragostino, M., and F. Soavi. 2007. *J Power Sources* 174:89–93. With permission.)

This equation is also rewritten as

$$\text{ESC} = C_p + \frac{1}{\left(2\pi f\, R_p\right)^2 C_p} \tag{7.108}$$

In applications involving supercapacitors, the emphasis is often on reducing the ESR, as opposed to increasing the volumetric efficiencies. Supercapacitors, which are based on electrochemical reactions that lead to a double-layer interfacial capacitance, have received considerable attention recently because of their potential for applications in electric hybrid vehicles. In these applications, supercapacitors can provide a high level of energy output in a short time (i.e., high specific power), compared to that provided by lithium ion or nickel hydride batteries and fuel cells, which can provide a high energy density (Mastragostino and Soavi 2007) but a lower energy output. Many supercapacitors have very high *specific capacitance*, that is, capacitance per unit mass, ranging from 700 to 1000 F/g (Yang et al. 2007).

A Ragone plot or Ragone chart is a diagram that compares the specific power density possible from a specific energy output for different power- or energy-generating devices. This plot is shown for some batteries and compared to supercapacitors in Figure 7.31.

7.18 PROBLEMS

7.1 A liquid-level sensor is to be designed for a car. Explain how such a sensor can be designed so that the change in capacitance can be used to measure the liquid level.

7.2 A tantalum oxide (Ta_2O_5) thin film (with a dielectric constant of ~25) was deposited on conductive polysilicon to form a capacitor. Another dielectric material under consideration for this application was SiO_2 (with a dielectric constant of ~3.9). If the thickness of the SiO_2 film required to achieve a certain value of capacitance is d_{silica}, what thickness of Ta_2O_5 (d_{Ta2O5}) will provide the same capacitance per unit area as that for the capacitor made using SiO_2?

7.3 A ferroelectric thin film of a particular composition of lead zirconium titanate (PZT), with an apparent dielectric constant of 400, was manufactured using a sol–gel process. This dielectric constant is lower than that seen for bulk or single crystals of PZT of the same composition. The film was 1 μm thick. What is the capacitance per unit area for this film? Express your answer in μF/cm². What is the total capacitance if the electrode area is 10 μm²?

7.4 Ferroelectric thin films (1000 nm thick) of apparently the same composition were made using a multitarget sputtering process. The dielectric constant of this PZT composition in a thin-film form should be 400. One set of samples was subjected to high temperatures to anneal the films. This caused some lead to evaporate and caused the formation of another layer (~100 nm thick), a phase known as the pyrochlore (assume a dielectric constant of ~50). The total film thickness of the PZT and pyrochlore layers was still 1000 nm. What is the capacitance of this composite film per unit area? Express your answer in $\mu F/cm^2$. What can be possibly done to prevent or minimize the formation of the pyrochlore phase having a low dielectric constant?

7.5 The dielectric constant of Si is 11.9. Because Si is covalently bonded, the only polarization mechanism present is electronic polarization. Use the Clausius–Mossotti equation to show that the electronic polarizability of Si atoms is 4.17×10^{-40} F $\cdot$ m^2.

7.6 The density of krypton (Kr) gas at 25°C and one atmosphere is reported to be 3.4322×10^{-3} g/cm^3. The polarizability volume is reported to be 2.479×10^{-24} cm^3 in literature (Vidal et al. 1984). Calculate the dielectric constant of Kr under these conditions, if its molecular weight is 83.8 g.

7.7 For a mixture of Ar (70% by weight) and Kr (30% by weight), show that the average molecular weight (M) is 53.1. Show that the average volume polarizability (using data from problem 7.5 and Example 7.4) is 1.80×10^{-24} cm^3. What is the dielectric constant of this mixture of gases at 25°C and one atmospheric pressure?

7.8 From the data in Example 7.4, calculate the dielectric constant of Ar gas.

7.9 Assuming that the dipole moment of H_2O molecules is 10^{-29} C $\cdot$ m and the electric field experienced by H_2O molecules is 10^5 V/m, calculate the dipolar polarizability of H_2O molecules in F $\cdot$ m^2. Note:

$$\alpha_d = \frac{1}{3} \frac{\mu^2}{K_B}$$

7.10 Calculate the heat generated from dielectric losses in a polymer cable subjected to an electric field of 80 kV/cm. Assume that the frequency is 60 Hz, the dielectric constant of the polymer is 2.1, and tan δ is 10^{-4}.

7.11 What are the limitations of using the Clausius–Mossotti equation? Explain.

7.12 The low-frequency dielectric constant of a silicate glass is 4.0. The refractive index of this glass is 1.47. What fraction of the dielectric constant can be attributed to ionic polarization?

7.13 The variation in the imaginary part of the dielectric constant (k'') with frequency is given by Equation 7.70. Prove that the maximum k'' occurs when $\omega = 1/\tau$.

7.14 Chalcogenide glasses are amorphous materials based on elements such as selenium (Se), Ge, and arsenic (As). These glasses have promise for many dielectric and optical applications. The dielectric constant of $Se_{55}Ge_{30}As_{15}$ was measured as a function of frequency ranging from 20 kHz to 4 MHz (Figure 7.32).

What is the trend in dielectric constant as the frequency increases at temperatures between 350 and 390 K? At temperatures greater than 400 K, the lower-frequency dielectric constant increases considerably. Suggest a possible reason for this.

7.15 The dependence of the imaginary part of the dielectric constant $\left(\varepsilon_r''(\omega)\right)$ for chalcogenide glasses of composition $Se_{55}Ge_{30}As_{15}$ is shown in Figure 7.33.

From the data shown in Figures 7.32 and 7.33, plot the real and imaginary parts of the complex dielectric constant as a function of ln(ω) for $T = 380$ K. Are the changes in the real and imaginary parts of the dielectric constant sharp, as in Figure 7.24? Explain. Show that at 300 K, the maximum $\varepsilon''(\omega)$ occurs at ~381 kHz. What is the value of ln(ω) corresponding

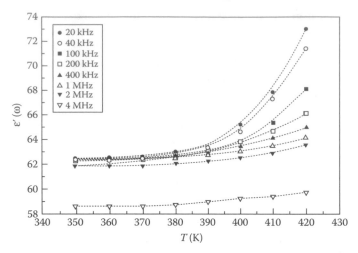

FIGURE 7.32 Dielectric constant of $Se_{55}Ge_{30}As_{15}$ glasses. (From El-Nahass, M. M. et al. 2007. *Physica B* 388:26–33. With permission.)

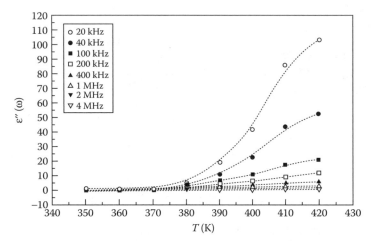

FIGURE 7.33 Dielectric losses ($\varepsilon_r''(\omega)$) of $Se_{55}Ge_{30}As_{15}$ glasses. (From El-Nahass, M. M. et al. 2007. *Physica B* 388:26–33. With permission.)

to this? Show that the most probable relaxation time for polarization in this material is ~4.11 $\times 10^{-6}$ s.

7.16 For the data on chalcogenide glasses shown in Figures 7.32 and 7.33, calculate the values of tan δ as a function of the frequency (τ) (ω). Plot these values on a graph. Assume $T = 380$ K.

7.17 Starting with the Debye equations (Equations 7.69 and 7.70) that describe the changes in the real and imaginary parts of the dielectric constant as a function of frequency, show that $(\varepsilon''/\omega) = (\tau) \times (\varepsilon' - \varepsilon_\infty)$, and therefore a plot of $\log(\varepsilon''/\omega)$ versus $\log(\varepsilon' - \varepsilon_\infty)$ results in a straight line. The y-axis intercept of this graph gives the value of $\log(\tau)$, from which the value of the relaxation time can be calculated.

7.18 A plot of $\log(\varepsilon''/\omega)$ versus $\log(\varepsilon' - \varepsilon_\infty)$ for a dielectric material of composition $Se_{70}Te_{30}$ is shown in Figure 7.34. The data were collected for temperatures ranging from 298 to 373 K.

   What are the $\log(\tau)$ values for $Se_{70}Te_{30}$ for these different temperatures? What happens to the values of the relaxation times as the temperature increases?

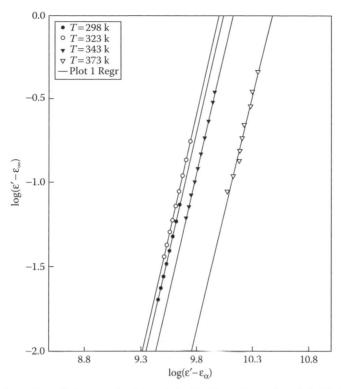

FIGURE 7.34   A plot of $\log(\varepsilon''/\omega)$ versus $\log(\varepsilon' - \varepsilon_\infty)$ for $Se_{70}Te_{30}$. (From Sayed, S. M. 2007. *Appl Surf Sci* 253:7089–93. With permission.)

7.19   The molar volume of a dielectric known as calcium molybdate ($CaMoO_4$) is 78.064 Å^3. Using Shannon's approach (see Section 7.8), write down the expression for calculating the molar polarizability of $CaMoO_4$. The measured dielectric constant of this material is 10.79 (Choi et al. 2007). What is the polarizability of the Mo^{6+} ion?

7.20   Many microwave ovens work at a frequency of 2.45 GHz. Examine how the real and imaginary parts of the dielectric constant of water change with the frequency for different temperatures (Figure 7.35). Show that the wavelength (λ) of the radiation associated with these waves is ~12.2 cm. The wavelength at which the dielectric losses (i.e., tan δ) are maximized (at room temperature) is 4 cm; what is the corresponding frequency in GHz? The speed of light can be assumed to be 3×10^8 m/s. Why is it that microwave ovens do not use the frequency at which the value of tan δ is maximized? Hint: Microwaves need to penetrate into the food. (See Figure 7.36.)

7.21   Many foods contain H_2O and salt, and the salt content of the food decreases the dielectric constant of H_2O. What is the effect of the salt content on the dielectric losses of H_2O?

7.22   The molar volume (V_m) for beryllium silicate (Be_2SiO_4) is 61.75 Å^3 (Shannon 1993). What is the expected dielectric constant of Be_2SiO_4 calculated using Shannon's approach? The experimental value of the dielectric constant for Be_2SiO_4 is 6.22.

7.23   What are some of the situations in which Shannon's approach for calculating the dielectric constant will probably not work?

7.24   The low-frequency dielectric constant of Ge is 16. What is its high-frequency dielectric constant? Why? What is the refractive index of Ge?

7.25   If the refractive index of NaCl is 1.5, what is the high-frequency dielectric constant of NaCl? If the low-frequency dielectric constant of NaCl is 5.6, approximately what fraction of the low-frequency dielectric constant can be attributed to ionic polarization?

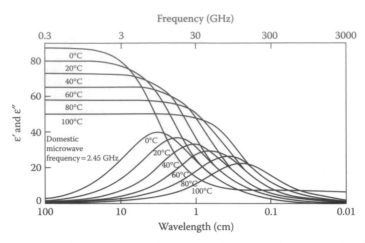

FIGURE 7.35 Dependence of the real and imaginary parts of the dielectric constant of H_2O on frequency for different temperatures. (From Vollmer, M. 2004. *Phys Educ* 39(1):74–81. With permission.)

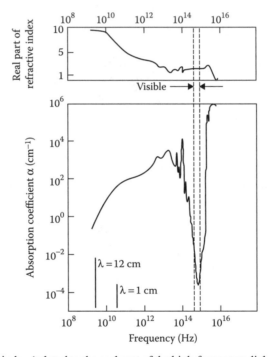

FIGURE 7.36 Refractive index (related to the real part of the high-frequency dielectric constant) and absorption coefficient (related to the imaginary part of the complex dielectric constant) of H_2O. (From Vollmer, M. 2004. *Phys Educ* 39(1):74–81. With permission.)

7.26 Compositions and properties of silicate glasses, as investigated by Wang et al. (2007), are shown in Tables 7.3 and 7.4.
  (a) At 10 MHz, what polarization mechanisms are expected to play a role in contributing to the dielectric constant for these glasses?
  (b) Why does the glass made using Li^+ (i.e., sample 11) exhibit the lowest dielectric constant?
  (c) Between samples 12 and 13 (i.e., samples containing sodium and potassium oxides), which sample has a higher total electronic polarizability and refractive index? Why?

TABLE 7.3
Compositions of Glasses (mol%)

Sample Number	SiO_2	B_2O_3	Al_2O_3	CaO	Li_2O	Na_2O	K_2O
11	62	25	1.2	5.4	6.4	–	–
12	62	25	1.2	5.4	–	6.4	–
13	62	25	1.2	5.4	–	–	6.4

Source: Wang, Z., Y. Hu, H. Lu, and F. Yu. 2008. *J Non-Cryst Solids* 354(12):1128–32.

TABLE 7.4
Dielectric Properties and Resistivities of Glasses

Sample Number	Dielectric Constant (at 10 MHz)	Dielectric Loss (tan δ) at 10 MHz	Resistivity ($\Omega \cdot cm$)
11	5.58	$4.97 \times 10{-2}$	0.73×1013
12	6.03	$4.5 \times 10{-2}$	0.66×1013
13	6.05	$4.41 \times 10{-2}$	1.17×1013

Source: Wang, Z., Y. Hu, H. Lu, and F. Yu. 2008. *J Non-Cryst Solids* 354(12):1128–32.

(d) Because the dielectric constant of samples 12 and 13 (at 10 MHz) is effectively the same, which sample has a higher ionic polarizability? Does this explain why the dielectric constant of samples 12 and 13 are approximately the same?

(e) The so-called modifier ions, that is, ions that cause disruption of the silicate tetrahedra in the local arrangements within glasses, play a key role in controlling the dielectric losses of these materials. Based on the values of tan δ, which modifier ion seems most mobile?

(f) The conductivity of these glasses depends on the type and concentration of modifier ions. For samples 11, 12, and 13, the concentration of modifier ions (Li^+, Na^+, and K^+) are the same. The conductivity then depends on the mobility of these ions (which, in turn, depends on the size of the ion) and the strength of the bond between the modifier and oxygen ions. Explain why sodium oxide–based glass has the lowest resistivity of these compositions.

GLOSSARY

Admittance (Y): The inverse of impedance (Z).

Atomic polarization: See **Ionic polarization.**

Bound charge: Charges on a capacitor plate, which are not free to move because they are bound by the dipoles, either induced or present, in the dielectric material.

Capacitive reactance (X_C): A capacitor's effective resistance (in Ohms) to the flow of electricity through it, which is given by $X_C = 1/\omega C = 1/2\pi C$. This component of the impedance does not cause any dissipation of electrical energy.

Capacitor: A device or structure filled with a dielectric layer that is separated by electrodes and is used for charge storage.

Charging current (I_c): The current that leads to the storage of charge in a dielectric (or capacitor) when a voltage is applied.

Clausius–Mossotti equation: The equation that relates the dielectric constant (a macroscopic property) to the polarizability of atoms and ions (a microscopic property). This equation *cannot* be used to account for interfacial (Maxwell–Wagner) or ferroelectric polarization.

$$\frac{\varepsilon_r - 1}{\varepsilon_r + 2} = \frac{1}{3\varepsilon_0}\left[\left(N_e \times \alpha_e\right) + \left(N_i \times \alpha_i\right)\right]$$

Coercive field (E_c): The electric field necessary to cause the domains in a ferroelectric with some remnant polarization to be randomized again in order to achieve a state of zero net polarization.

Cole–Cole plots: The sets of arcs or semicircles that appear when variations of k' (x-axis) and k'' (y-axis) are plotted as a function of the frequency.

Complex dielectric constant $\left(\varepsilon_r^*\right)$: A quantity that describes the ability of a real dielectric material to store electrical charge ε_r'; also a measure of the energy lost ε_r'' while the polarization processes that lead to charge storage occur. It is given by $\varepsilon_r^* = \varepsilon_r' - j\varepsilon_r''$.

Curie temperature: The temperature at which a ferroelectric material transforms into a centrosymmetric paraelectric material.

Debye: A unit of dipole moment, with one Debye (D) equal to 3.3356×10^{-30} C · m.

Debye equations: Equations describing both the relaxation in dielectric constant and the dependence of the imaginary part of the dielectric constant on frequency.

Debye loss peak: The frequency $\omega = 1/\tau$, at which the value of k'' reaches a maximum.

Dielectric constant (k or ε_r): A dimensionless parameter that expresses the ability of a dielectric to store charge, expressed as a ratio of the capacitance of a capacitor filled with a dielectric material to that filled with a vacuum (C/C_0).

Dielectric displacement (D): The total charge density, written as a total of the free and bound charge densities due to the polarization of a dielectric.

Dielectric flux density (D): See **Dielectric displacement**.

Dielectric material: A large-bandgap semiconductor ($E_g \sim >4$ eV) with high resistivity (ρ).

Dielectric permittivity (ε): A parameter that expresses the ability of a material to store a charge, expressed as $\varepsilon = \sigma_s/E$, where σ_s is the surface charge density and E is the electric field. The unit for permittivity is F/m, with the permittivity of free space being $\varepsilon_0 = 8.85 \times 10^{-12}$ F/m. We often use the dielectric constant $\varepsilon/\varepsilon_0$.

Dielectric polarization (P): The magnitude of the bound charge density in a dielectric material.

Dielectric relaxation: A reduction in the dielectric constant with an increasing frequency of the applied electric field, as some of the polarization mechanisms cannot keep up.

Dipolar polarization: Also known as orientational polarization, caused by the realignment of permanent dipoles present in certain materials.

Dipolar polarizability (α_d): Polarizability due to the presence of molecules with a dipole moment (μ); given by

$$\alpha_d = \frac{1}{3}\frac{\mu^2}{K_B}$$

Dipole moment (μ): For a dipole with charges $+q$ and $-q$ separated by a distance d, the dipole moment is $q \times d$, the SI unit of which is C · m.

Electrical insulators: Dielectric materials with high breakdown voltage, also known simply as insulators.

Electronic polarizability (α_e): A measure of the electronic polarization of an atom or an ion, with larger atoms and ions having higher electronic polarizabilities.

Electronic polarization: Creating or inducing a dipole in an atom or an ion by displacing the electronic cloud with respect to the nucleus.

Electrostatic induction: The development of a charge on a material when another material with a net charge is brought near it.

Equivalent series capacitance (ESC): The capacitance of a capacitor when a real dielectric material is modeled as a combination of a resistance and a capacitor connected in series. It is related to the equivalent elements of a parallel circuit (C_p and R_p) describing the same real dielectric by

$$\text{ESC} = C_p + \frac{1}{\left(2\pi f R_p\right)^2 C_p}$$

Equivalent series resistance (ESR): The resistance of a resistor when a real dielectric material is modeled as a combination of a resistance and a capacitor connected in series. It is related to the equivalent elements of a parallel circuit (C_p and R_p) describing the same real dielectric by

$$\text{ESR} = \frac{R_p}{\left(2\pi f C_p R_p\right)^2 + 1}$$

Ferroelectrics: Materials that show spontaneous and reversible polarization.

Free charge: A charge on a capacitor plate that is free to move and is not bound to the induced or permanent dipole in a dielectric material.

Frequency dispersion: See **Dielectric relaxation**.

Gauss's law: This law states that the area integral of the electric field (E) over any closed surface is equal to the net charge (Q) enclosed within the surface divided by the permittivity of space (ε_0) and is given by the following equation:

$$\oint \vec{E} \times d\vec{A} = \frac{Q}{\varepsilon_0}$$

High-frequency dielectric constant (k_∞): The dielectric constant measured at frequencies at which only the electronic polarization mechanism operates. It is also known as the optical frequency dielectric constant and is equal to the square of its refractive index (n).

Ideal dielectric: A dielectric material with no dielectric losses; such a material does not exist.

Imaginary part of the dielectric constant (k'' or ε_r''): The part of the complex dielectric constant that is associated with dielectric losses and the loss current for a hypothetical dielectric with only one relaxation time (τ). It is given by the following equation:

$$\varepsilon_r'' = k'' = \left(k_s - k_\infty\right)\left(\frac{\omega\tau}{1 + \omega^2\tau^2}\right)$$

Impedance (Z): The equivalent of resistance applicable to a circuit that can have a capacitor and an inductor. The units of impedance are ohms. Its inverse is admittance (Y).

Impedance analyzer: Equipment that can measure the inductance, capacitance, and resistance of a material or device, typically at different frequencies, which in turn helps to calculate the impedance (Z).

Impedance spectroscopy: The technique of measuring k' and k'' as a function of frequency.

Interfacial polarization: Polarization at the heterogeneities in a material or device, typically at the dielectric–electrode interfaces and at the grain boundaries of a polycrystalline material.

Internal electric field: See **Local electric field**.

Ionic polarization: A polarization mechanism in which the ions themselves are displaced in response to the electric field experienced by the solid, creating a net dipole moment per ion (p_{av}). This is also known as atomic or vibrational polarization.

Ionic polarizability (α_i): A parameter for describing the propensity of ions to undergo ionic polarization.

Local electric field (E_{local}): The actual field experienced by the atoms, molecules, or ions within a material, which is different from the applied field and is referred to as the internal or local electric field (E_{local}).

$$E_{local} = E + \frac{1}{3\varepsilon_o} P \text{ (for cubic structure or amorphous materials)}$$

This equation *cannot* be used for materials that either lack a cubic structure or are polar.

Local field approximation: See the above expression for local electric field.

Lorentz–Lorenz equation: An equation that relates the high-frequency dielectric constant to the refractive index through the electronic polarizability.

$$\frac{n^2 - 1}{n^2 + 2} = \frac{1}{3\varepsilon_0}\left[(N_e \times \alpha_e)\right]$$

Loss angle (δ): In an ideal dielectric, the charging current leads the voltage by 90°, whereas in a real dielectric, it leads by $(90-\delta)°$. The tangent of this angle represents the dielectric losses.

Loss current (I_{loss}): The dielectric losses that are encountered during the polarization processes in a real dielectric subjected to an electric field.

Low-frequency dielectric constant (k_s): See **Static dielectric constant**.

Low-loss dielectric: A material that has a low tan δ ($\sim < 10^{-3}$); typically, the value of the quality factor of the dielectric (Q_d) is reported as $Q_d = 1/\tan \delta$.

Loss tangent (tan δ): The ratio of the loss current to the charging current in a real dielectric. It is also the ratio of the imaginary to the real parts of the dielectric constant and is a measure of the quantity of the input electrical energy lost during the polarization processes.

Maxwell's equations: A set of four equations that describe the basis of many relationships pertaining to the electrical and magnetic properties of materials. These are the basis for many laws, such as Gauss's law.

Maxwell–Wagner polarization: Also known as Maxwell–Wagner–Sillars polarization. See **Interfacial polarization**.

Molar polarization (P_m): Another form of the Clausius–Mossotti equation, expressed such that we get the polarization per mole (P_m), with the unit of m^3/mol or cm^3/mol. This is given by the following equation:

$$\left(\frac{\varepsilon_r - 1}{\varepsilon_r + 2}\right) = \frac{P_m \times \rho}{M}$$

Multilayer capacitor (MLC): A capacitor comprised of alternating layers of dielectrics and electrodes. The layers are connected in parallel to maximize the volumetric efficiency of the capacitor.

Nonlinear dielectrics: Materials in which the developed polarization is not linearly related to the electric field; thus, the dielectric constant of these materials will be field-dependent. This includes ferroelectrics and other materials, such as water, in which molecules have a permanent dipole moment.

Optical-frequency dielectric constant: See **High-frequency dielectric constant**.

Optical polarization: See **Electronic polarization**.

Orientational polarization: See **Dipolar polarization**.

Paraelectric phase: The high-temperature phase derived from an originally ferroelectric parent phase that now has no dipole moment per unit cell.

Phasor: A rotating vector that shows the phase angle for a particular type of current or voltage in a circuit component.

Phasor diagram: A diagram showing the relative positions of phasors.

Piezoelectric: A material that develops electrical voltage or charge when subjected to stress. The material also develops a relatively large strain when subjected to an electric field.

Polar dielectrics: Materials that have a permanent dipole moment (e.g., water), or ferroelectric materials in which a dipole moment is spontaneously set up.

Polarization: The effect of an electric field on atoms, ions, and molecules in a material, resulting in the creation of induced dipoles or changes in the orientation of permanent dipoles.

Polarizability (α): The tendency of an atom or an ion to undergo polarization under the application of an electric field. The dipole moment created (μ) when an electric field (E) is applied is given by $\mu = (\alpha \times E)$.

Quality factor (Q): A measure of the dielectric losses; the higher the losses; the lower the Q. It is useful for comparing dielectric materials with very low dielectric losses; for these, Q is defined as the inverse of tan δ.

Real dielectric: A dielectric material that exhibits dielectric losses and an ability to store charge. An ideal dielectric, which does not exist, has no dielectric losses.

Real part of the dielectric constant (k'): The frequency-dependent part of the complex dielectric constant related to the polarization processes, which enables charge storage.

Relaxation time (τ): The time required for a polarization mechanism to revert and realign or change its value when the electrical field switches or changes in magnitude.

Ragone plot (Ragone chart): A diagram that compares a specific possible power density to a specific energy output for different power- or energy-generating devices.

Specific capacitance: Capacitance per unit mass.

Static electronic polarizability ($\alpha_{e,static}$): The value of electronic polarizability when the electric field causing the polarization is not time-dependent.

Static dielectric constant (k_s): The value of the dielectric constant under DC fields, sometimes also equated to the low-frequency dielectric constant.

Supercapacitors: Capacitors that make use of double-layer, interfacial polarization mechanisms to create a very high (~1 F) level of capacitance. The specific capacitance of these devices can be very high, at ~700–1000 F/g.

Tangent delta (tan δ): The ratio of the loss current to the charging current in a dielectric material; also the ratio of the magnitudes of the imaginary and real parts of the complex dielectric constant.

Tan delta: See **Tangent delta**.

Total dielectric polarizability $\left(\alpha_D^T\right)$: The dielectric polarizability of a compound calculated using Shannon's approach. This can be estimated from either the individual polarizabilities (both electronic and ionic) of the ions, the unit cell volume, and the molecular weight; or from the total dielectric polarizabilities of other compounds.

Vector diagram: See **Phasor diagram**.

Vibrational polarization: See **Ionic polarization**.

Volume polarizability (α_{volume}): Polarizability that is often expressed in the unit of cm^3 or Å^3:

$$\alpha_{volume} \text{ in cm}^3 = \frac{10^6}{4\pi\varepsilon_0} \times \alpha \ (C \cdot V^{-1} \cdot m^2)$$

$$\alpha_{volume} \text{ in Å}^3 = \frac{10^{30}}{4\pi\varepsilon_0} \times \alpha \ (C \cdot V^{-1} \cdot m^2)$$

Volumetric efficiency: The total capacitance per unit volume. This is maximized by using a large number of thinner dielectric layers connected in parallel and by using materials with higher dielectric constants.

REFERENCES

Buchanan, R. C. 2004. *Ceramic Materials for Electronics*. New York: Marcel Dekker.

Choi, G. K., J. R. Kim, S. H. Yoona, and K. S. Hong. 2007. *J Eur Ceram Soc* 27:3063–7.

El-Nahass, M. M., A. F. El-Deeb, H. E. A. El-Sayed, and A. M. Hassanien. 2007. *Physica* B 388:26–33.

Fernandez, D. P., Y. Mulev, et al. 1995. *J Phys Chem Ref Data* 24(1):34–44.

Hench, L. L. and J. K. West. 1990. *Principles of Electronic Ceramics*. New York: Wiley.

Jackson, J. D. 1975. *Classical Electrodynamics*. New York: Wiley.

Kao, K. C. 2004. *Dielectric Phenomena in Solids*. London: Elsevier-Academic Press.

Kasap, S. O. 2002. *Principles of Electronic Materials and Devices*. New York: McGraw Hill.

Kishi, H., Y. Mizuno, and H. Chazono. 2003. *Jap J Appl Phys* 42(July):1–15.

Mastragostino, M., and F. Soavi. 2007. Strategies for high-performance supercapacitors for HEV. *J Power Sources* 174:89–93.

Moulson, A. J., and J. M. Herbert. 2003. *Electroceramics: Materials, Properties, Applications*. New York: Wiley.

Prakash, B., and K. B. R. Varma. 2007. Influence of sintering conditions and doping on the dielectric relaxation originating from the surface layer effects in $CaCu_3Ti_4O_{12}$ ceramics. *J Phys Chem Solids* 68:490–502.

Raju, G. G. 2003. *Dielectrics in Electric Fields*. Boca Raton, FL: CRC Press.

Ram, M., and S. Chakravarty. 2008. Dielectric and modulus behavior of $LiFe_{1/2}Ni_{1/2}VO_4$ ceramics. *J Phys Chem Solids* 69(4):905–912.

Sayed, S. M. 2007. *Appl Surf Sci* 253:7089–93.

Shannon, R. D. 1993. Dielectric polarizabilities of ions in oxides and fluorides. *J Appl Phys* 73:348–66.

Vidal, D., L. Guengant, and J. Vermesse. 1984. Density and dielectric constant of dense argon and crypton mixtures. *Physica A* 127:574–86.

Vollmer, M. 2004. Physics of the microwave oven. *Phys Educ* 39(1):74–81.

Wang, Z., Y. Hu, H. Lu, and F. Yu. 2008. Dielectric properties and crystalline characteristics of borosilicate glasses. *J Non-Cryst Solids* 354(12):1128–32.

Yang, X., Y. Wang, H. Xionga, and Y. Xia. 2007. Interfacial synthesis of porous MnO_2 and its application in electrochemical capacitor. *Electrochim Acta* 53:752–7.

8 Ferroelectrics, Piezoelectrics, and Pyroelectrics

KEY TOPICS

- Ferroelectric materials
- Origin of ferroelectricity
- Ferroelectric hysteresis loop
- Properties and applications of ferroelectrics
- Electrostriction
- Piezoelectric materials
- Direct and converse piezoelectric effects and applications
- Properties and applications of soft and hard piezoelectrics
- Strain-tuned ferroelectrics
- Lead-free piezoelectrics
- Pyroelectric materials and applications

8.1 FERROELECTRIC MATERIALS

8.1.1 FERROELECTRICITY IN BARIUM TITANATE

Ferroelectrics are materials that possess a macroscopic spontaneous polarization that can be reoriented through the application of an external electric field (Schlom et al. 2007). Polarization in ferroelectric materials can exist in the absence of an electric field under certain ranges of temperature and pressure. Ferroelectric materials have crystal structures that lack inversion symmetry.

We can summarize the origin of ferroelectricity in tetragonal barium titanate ($BaTiO_3$), an archetypal ferroelectric, as follows.

The temperature at which the transformation occurs from a nonpolar, paraelectric phase to a polar, ferroelectric phase is known as the *Curie temperature* (T_c). At a given temperature higher than the Curie temperature, materials may exhibit a cubic structure. For $BaTiO_3$, this temperature is ~120°C. In this cubic phase, also called the *paraelectric phase*, the titanium (Ti^{4+}) ion appears to be exactly at the center of the cube. In reality, of course, ions are not stationary. The titanium ion, for example, vibrates very rapidly around several equivalent off-center positions. Each of these off-center configurations has a net dipole moment, which rotates very rapidly in space. The result is that the time-averaged position of the titanium ion in the cubic, paraelectric phase appears to be at the center, and this higher-temperature polymorph of $BaTiO_3$ has no net dipole moment. This cubic or paraelectric phase of the structure is described as centrosymmetric and nonpolar.

In the case of $BaTiO_3$, the paraelectric phase is cubic, and the ferroelectric phase is tetragonal (Figure 8.1). As the temperature is lowered, a phase transformation occurs at $T = T_c$, in which the nonpolar, paraelectric phase transforms into a polar, ferroelectric phase at the Curie temperature. The tetragonal structure develops when the titanium ions are locked into any of these six variants (Figure 8.2).

Other ions, such as Ba^{2+} and O^{2-}, are also displaced during the transformation from the cubic to the tetragonal phase in $BaTiO_3$. The actual displacement of ions is very small—only a few picometers, shown for $BaTiO_3$ in Figure 8.3. Recall that 1 pm = 10^{-12} m = 0.01 Å. As the titanium (Ti^{4+}) and

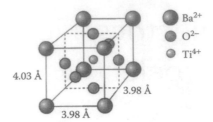

FIGURE 8.1 Schematic representation of the ionic arrangements in ferroelectric tetragonal BaTiO$_3$. (From Askeland, D., and P. Fulay. 2006. *The Science and Engineering of Materials*. Washington, DC: Thomson. With permission.)

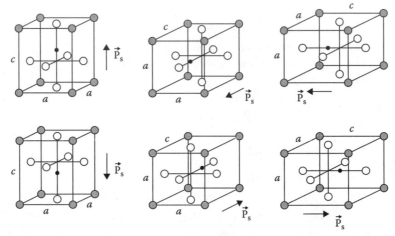

FIGURE 8.2 Six variants of the positions in the tetragonal phase of BaTiO$_3$ leading to six spontaneous polarization states and three spontaneous strain states. (From Mehling, V., C. Tsakmakis, and D. Gross. 2007. *J Mech Phys Solids* 55:2106–41. With permission.)

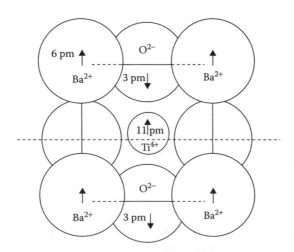

FIGURE 8.3 Actual picometer displacements of ions, leading to ferroelectric polarization in BaTiO$_3$. (From Moulson, A. J., and J. M. Herbert. 2003. *Electroceramics: Materials, Properties, and Applications*. New York: Wiley. With permission.)

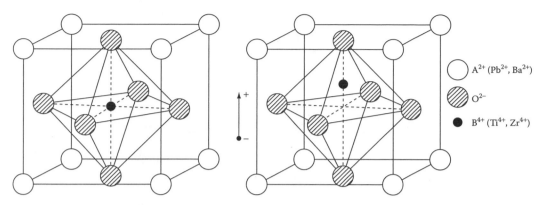

FIGURE 8.4 Ionic arrangements in the paraelectric and ferroelectric forms of PZT. (Adapted from Morgan Technical Ceramics, Guide to Piezoelectric and Dielectric Ceramics. http://www.morganelectroceramics .com/pzbook.html.)

barium (Ba^{2+}) ions in one unit cell move in one of the possible off-center directions (e.g., titanium ions moving up by 11 pm (picometers) and barium ions moving up by 6 pm, as shown in Figure 8.3), the titanium and barium ions in the unit cells above the unit cell shown in Figure 8.4 will also be pushed up.

These unit cells, with the asymmetric positions of titanium ions, create a net dipole moment in the tetragonal structure of $BaTiO_3$ (Mehling et al. 2007).

Other materials, such as lead zirconium titanate (PZT), also show a development of ferroelectric polarization. At higher temperatures, the dipole moments created in a ferroelectric structure become randomized, and the overall ferroelectric polarization begins to decrease. The ionic arrangements for the tetragonal ferroelectric and cubic paraelectric forms of PZT are shown in Figure 8.4.

Both $BaTiO_3$ and PZT have crystal structures other than the tetragonal form that exhibit ferroelectric behavior. Note that for these materials, the ferroelectric-to-paraelectric transition is not always from the tetragonal to the cubic structure.

8.1.2 Ferroelectric Domains

The cooperative displacement of the ions of several unit cells occurs during the formation of tetragonal $BaTiO_3$ from its cubic phase. This leads to the spontaneous formation of a region consisting of several unit cells, with all the unit cells' dipole moments lined up in the same direction. This region of a ferroelectric material, in which the polarization is in a given direction, is known as the *ferroelectric domain* or a *Weiss domain*.

Ferroelectrics are clearly different from other polar materials that have orientational or dipolar polarizations (e.g., water). In a ferroelectric material, there are no built-in or permanent dipole molecules to start with. Instead, a spontaneous polarization occurs in these materials as the atoms or ions self-assemble in an arrangement that causes the unit cells to develop a net dipole moment.

The process that causes the existence of a polar region or a ferroelectric domain in one part of a material cannot continue indefinitely because the process will gradually increase the free energy of the system. The growth of a ferroelectric domain will stop when it reaches a certain size. The total polarization in this domain will be compensated for by another domain next to it, in which the polarization is in the opposite direction. The polarizations in the neighboring domains may be at an angle of 90° or 180° in relation to one another. Thus, a ferroelectric material is comprised of many polar regions, in which each region has polarization in a certain direction. This is shown schematically in Figure 8.5. Domain walls in ferroelectrics are very thin boundaries (just a few angstroms) that separate domains polarized in different directions.

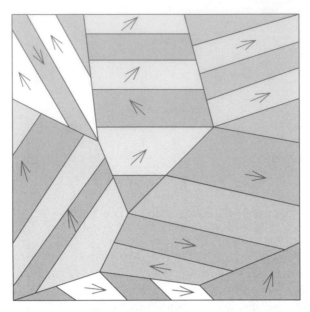

FIGURE 8.5 Schematic representation of the randomly arranged ferroelectric domains in an unpoled ferroelectric. (From Mehling, V., C. Tsakmakis, and D. Gross. 2007. *J Mech Phys Solids* 55:2106–41. With permission.)

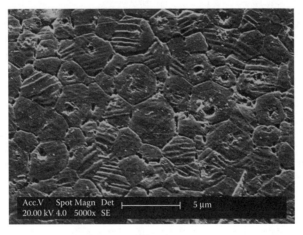

FIGURE 8.6 A scanning electron microscope (SEM) image of the microstructure of a lead zirconium titanate (PZT) ceramic modified using strontium (Sr). Sets of parallel lines within grains show the ferroelectric domain. (Courtesy of Dr. Raj Singh and I. Dutta, University of Cincinnati).

A scanning electron microscope (SEM) image of the ferroelectric domains in the tetragonal form of PZT is shown in Figure 8.6. The contrast between the domains has been obtained by carefully etching the sample with chemicals.

Unless we apply a substantial electric field, the polarizations associated with the different regions cancel each other out, and the ferroelectric material has no net polarization. Thus, when the domains are randomized and no substantial electric field has ever been applied to a ferroelectric material, the net dielectric polarization is zero. When an electric field is applied to a ferroelectric material, its domains begin to align with the electric field, and the material develops a net polarization. This is similar to the alignment of permanent dipoles in a polar material. This is shown in Figure 8.7, in which the electric field is relatively lower for the top part and higher for the bottom part.

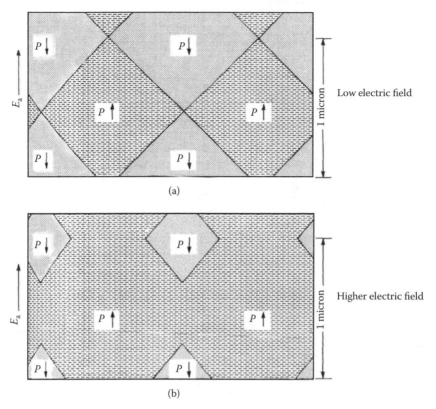

FIGURE 8.7 Schematic image of the domain growth in a ferroelectric material with (a) a low electric field and (b) a higher electric field. (From Hench, L. L., and J. K. West. 1990. *Principles of Electronic Ceramics.* New York: Wiley. With permission.)

The application of an electric field helps to align the domains or polar regions. This development of polarization in a ferroelectric material is nonlinear. This is a very important distinction between a linear dielectric material, such as silica (SiO_2), alumina (Al_2O_3), titanium oxide (TiO_2), or polyethylene, which shows only nonferroelectric polarization, and a nonlinear dielectric material such as the tetragonal form of $BaTiO_3$. Thus, Equation 8.1, which we have shown to be applicable to linear dielectrics, cannot be used to describe polarization as a function of the electric field in ferroelectric materials.

$$P = \chi \varepsilon_0 E \text{ (for linear dielectrics only, not for ferroelectrics)} \tag{8.1}$$

As will become clear in Section 8.1.3, the relationship between the polarization that develops in a ferroelectric material is not linear. Ferroelectric materials that have never been exposed to an electric field are said to be in a *virgin* or *unpoled state.*

8.1.3 DEPENDENCE OF THE DIELECTRIC CONSTANT OF FERROELECTRICS ON TEMPERATURE AND COMPOSITION

When an electric field is applied, domains begin to undergo alignment. This leads to the development of considerable ferroelectric polarization, which causes ferroelectrics such as $BaTiO_3$ to have a relatively large dielectric constant (Figure 8.8). As we continue to increase the magnitude of the electric field applied to a ferroelectric material, an increasing number of domains become aligned. However, similar to the law of diminishing returns, the additional increase in electric field does not produce much additional dielectric polarization. This means that at higher electric fields, the

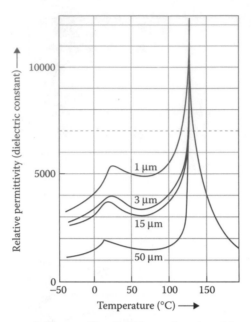

FIGURE 8.8   Dielectric constant of polycrystalline $BaTiO_3$ as a function of the temperature and average grain size. (From Moulson, A. J., and J. M. Herbert. 2003. *Electroceramics: Materials, Properties, and Applications.* New York: Wiley. With permission.)

dielectric constant (k) of ferroelectrics decreases. This is another important distinction between regular dielectric materials, also known as linear dielectrics, and nonlinear ferroelectric materials.

As we saw in Figure 8.8 and in Chapter 7, the measured or apparent dielectric constant of ferroelectrics and other dielectrics is a microstructure-sensitive property. The dielectric constant also depends strongly on the temperature. As the temperature increases beyond T_c, that is, if there is a switch from the ferroelectric phase to the paraelectric phase, the dielectric constant decreases. Above T_c, in the paraelectric state, the dielectric susceptibility ($1/\chi e = 1/(\varepsilon_r - 1)$) changes linearly with temperature. This behavior is known as the *Curie–Weiss law*. This variation in the dielectric constant of the paraelectric phase is written as follows:

$$\varepsilon = \varepsilon_0 + \frac{C}{(T - T_0)} \left(\text{applicable for } T > T_0\right)$$

where T_0 is equal to the Curie temperature (T_c) if the phase transformation from the ferroelectric to the paraelectric phase is of the second order. A second-order phase transformation means that the volume, energy, and structure change continuously as we go from the paraelectric phase to the ferroelectric phase. For $BaTiO_3$, the transformation is of the second order. If this phase transition is of the first order, then $T_0 < T_c$. The constant C is known as the Curie–Weiss coefficient or the Curie constant. The first term ε_0 is the temperature-independent part of ε. It is related to the polarization of ions.

Near T_0 or T_c, the second term dominates, and the first term can be ignored (Xu 1991).

Thus, the Curie–Weiss law is sometimes also written as

$$\varepsilon = \frac{C}{(T - T_0)} \left(\text{applicable for } T > T_0\right)$$

In Chapter 9, we will learn that essentially the same type of behavior is seen for magnetic materials.

An example of Curie–Weiss behavior in a material that is a solid solution of $BaTiO_3$ and strontium titanate ($SrTiO_3$) is shown in Figure 8.9. Note that the Curie temperature of this material is about ~+10°C. Although the data for dielectric constants are shown for temperatures below T_c, the Curie–Weiss law does *not* apply to the region below $T = T_c$. Note that even in a paraelectric state, the dielectric constant of materials is very high compared to other linear dielectrics such as Al_2O_3 and SiO_2. This property is very important for applications such as multilayer capacitors.

Lowering the Curie temperature of a ferroelectric material is useful for the applications of such dielectrics in capacitors. It is common to add *Curie-point shifters* to $BaTiO_3$. As we can see in Figure 8.9, the addition of $SrTiO_3$ to $BaTiO_3$ causes an overall lowering of the Curie temperature. Consequently, this dielectric formulation has a relatively high dielectric constant at and around room temperature (~27°C or 300 K), which offers high volumetric efficiency. At the same time, the material is paraelectric and thus is not piezoelectric (Section 8.11). This means that a capacitor made from such a material will not produce any spurious voltages. We can form solid solutions of $BaTiO_3$ with lead titanate (PT, $PbTiO_3$) and, in this case, the Curie temperature shifts to temperatures higher than T_c ~120°C (for $BaTiO_3$). Similarly, compounds such as barium zirconate ($BaZrO_3$) can form solid solutions with $BaTiO_3$, which can help suppress the Curie temperature. Such *Curie-point suppressors* can also be useful because they help to develop temperature-stable materials with a high dielectric constant. The dielectric constant of such materials will not vary significantly with temperature. This ability to form solid solutions allows the development of different formulations of capacitors.

The Electronic Industries Alliance in the Unites States has developed a scheme for the classification of different dielectrics based on their temperature stability. For example, an X7R dielectric means that in the temperature range of −55°C to +125°C the capacitance of the capacitor will be within ±15% relative to its value at 25°C. A Z5U dielectric formulation means that in the temperature range +10°C to +85°C, the capacitance can change between +22% and −56% relative to its value at 25°C.

For some compositions, we can create a composition and a nanostructure so that the resultant material shows a very high apparent dielectric constant, with a broad Curie peak. The dielectric constant of such a material depends significantly on the frequency of measurements. Such materials are

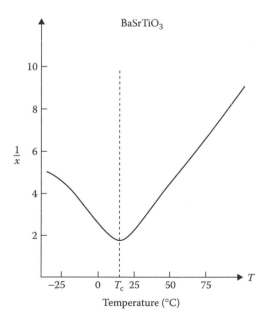

FIGURE 8.9 Curie–Weiss behavior in $BaTiO_3$–$SrTiO_3$ ceramic with $T_0 = T_c$ ~ +10°C. (From Hench, L. L., and J. K. West. 1990. *Principles of Electronic Ceramics*. New York: Wiley. With permission.)

known as *relaxor ferroelectrics*. The broadening of the Curie temperature occurring in these materials can be attributed to the disorder of the ions at different crystallographic sites in the crystal structures and to the nanoscale fluctuations in the compositions. These lead to nanoscale polar regions that have, in effect, a Curie temperature of their own. Collectively, the ferroelectric to paraelectric transition becomes diffused (Figures 8.10 and 8.11).

Relaxor ferroelectric materials include ceramics such as lead magnesium niobate (PMN), can be formulated in combination with other ferroelectrics such as $PbTiO_3$, and are useful in actuator applications because of their high electrostriction coefficients. The so-called coupling coefficient of these materials (see Section 8.16) is also better than PZT. Many novel relaxor ferroelectric compositions in the form of both single-crystal and polycrystalline materials have recently been made. The acoustic impedance of relaxor materials is better matched to human tissue than is PZT (Harvey et al. 2002). In principle, these materials are better suited for applications such as ultrasound imaging (see Section 8.11).

As an example, the dielectric constant (real part) and tan δ, a measure of the dielectric losses, are shown for strontium–bismuth–barium titanate (SBBT) ceramics in Figure 8.10.

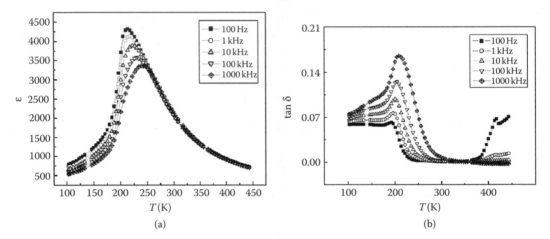

FIGURE 8.10   (a) Dielectric constant and (b) tan δ for strontium–bismuth–BaTiO$_3$ ceramics. (From Chen, W., X. Yao, and X. Wei. 2007. *Solid State Commun* 141:84–8. With permission.)

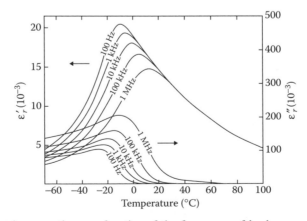

FIGURE 8.11   Dielectric properties as a function of the frequency of lead–magnesium niobate ceramics. (From Moulson, A. J., and J. M. Herbert. 2003. *Electroceramics: Materials, Properties, and Applications.* New York: Wiley. With permission.)

Because of compositional fluctuations occurring at the nanoscale, relaxor materials exhibit a diffuse phase transition and show a Curie range rather than a specific Curie temperature. A modified Curie–Weiss law expression is used to describe the variations in dielectric constants of these materials above the Curie temperature range.

The dielectric constant real part (ε_r') and imaginary part (ε_r'') for PMN ceramics is shown in Figure 8.11. Note the multipliers on the y-axis. The values of the dielectric constant range up to 20,000.

8.2 RELATIONSHIP OF FERROELECTRICS AND PIEZOELECTRICS TO CRYSTAL SYMMETRY

In a ferroelectric material, dielectric polarization appears spontaneously. The appearance of spontaneous dielectric polarization also causes a strain to develop in the ferroelectric material. For the six polarization-direction variations shown in Figure 8.2, there are three spontaneous strain states: one in the c direction and two in the a directions. A *piezoelectric* material develops a voltage when subjected to stress. The word "piezo" means "pressure."

All ferroelectric materials are also piezoelectric. However, not all piezoelectric materials are ferroelectric. For example, zinc oxide (ZnO), gallium arsenide (GaAs), and quartz (SiO_2) are piezoelectrics, but are not ferroelectrics.

The seven crystal systems are triclinic, monoclinic, orthorhombic, tetragonal, trigonal, hexagonal, and cubic, and lead to 32 crystallographic point groups (Gupta and Ballato 2007). Of these, 21 point groups have no center of symmetry (Figure 8.12). Of the 21 point groups without a center of symmetry, only 20 are piezoelectric. Among the 20 point groups that show piezoelectricity, 10 point groups are polar or *pyroelectric*. A pyroelectric is a material that shows a flow of charge to and from its surface consequent to a temperature change (see Section 8.20).

Thus, a ferroelectric material is also a pyroelectric; however, the converse is not true, that is, a pyroelectric material is not necessarily a ferroelectric (e.g., ZnO and cadmium selenide [CdSe]). Ferroelectrics are pyroelectric, are piezoelectric, and lack inversion symmetry.

Another definition of ferroelectrics is that they are pyroelectric materials whose direction of polarization can be reversed by a sufficiently strong electric field (Figure 8.12). Ferroelectric materials are also characterized by the presence of a Curie temperature (or a Curie temperature range for relaxor ferroelectrics), above which they become paraelectric.

We will discuss piezoelectric and pyroelectric materials in detail in Sections 8.12 and 8.20.

8.3 ELECTROSTRICTION

Every material, including liquids and amorphous materials such as glass, undergoes polarization when subjected to an electric field (E). This involves the movement of electron clouds, ions, and so on. Such movements associated with the polarization processes also cause the development of a small strain in the material. *Electrostriction* is a phenomenon that occurs in all materials in which the application of an electric field induces an elastic strain ε. This strain (ε) is proportional to the *square* of the electric field (E; Figure 8.13). The development of strain in piezoelectrics that are also ferroelectrics is quite different (Figure 8.26).

Electrostriction produces a strain that is independent of the sign of the applied field. The strain that develops due to ferroelectric polarization (Figure 8.2) is different from electrostriction in important ways. First, the strain produced because of ferroelectric polarization is proportional to the electric field (E). Second, it is typically much larger than that produced by electrostriction. Third, the strain produced by ferroelectric polarization is strongly temperature-dependent, especially as the material transitions from the paraelectric phase to the ferroelectric phase. In comparison, the strain produced by electrostriction does not show significant temperature dependence. Finally, because every material undergoes some sort of polarization, all materials show electrostriction.

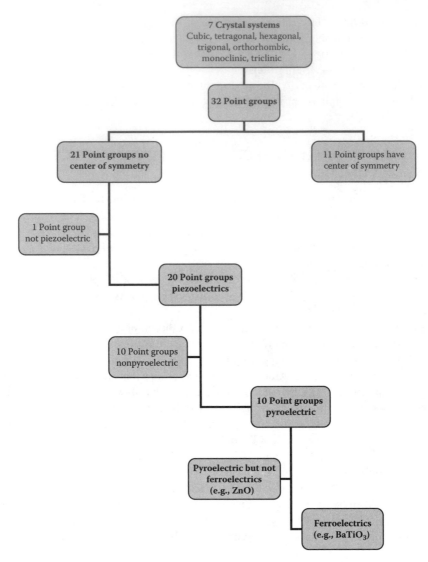

FIGURE 8.12   Crystal classes, point groups, and ferroelectrics.

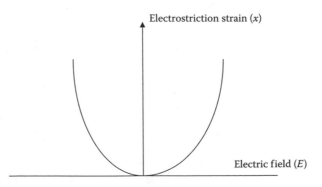

FIGURE 8.13   Electrostriction strain development with the application of an electric field.

In piezoelectric materials, the total strain developed will be the result of both the piezoelectric effects and electrostriction (Section 8.7). Electrostriction-induced strain contributes significantly to the high dielectric constant of ferroelectrics near their Curie temperature.

If Q_{ij} is the electrostriction coefficient for a material with remnant polarization P_r and dielectric constant ε_r, the piezoelectric coefficient d_{ij} is given by the expression

$$d_{ij} \sim 2 \times Q_{ij} \times \varepsilon_r \times \varepsilon_0 \times P_r \tag{8.2}$$

Consequently, relaxor ferroelectrics such as PMN–PT exhibit very high strains for a given level of applied electric field and are widely used in electromechanical actuators. Many relaxor ferroelectrics, such as PMN–PT, have a high electrostriction coefficient because of their crystal structures and chemical makeup. This is why they have very large piezoelectric coefficients. Even if we do not make use of the electrostriction effect as such, the consequence of this effect is embedded in the piezoelectric coefficients of many materials.

In some compositions of PZT, such as those near the morphotropic phase boundary (MPB; Section 8.14), the total piezoelectric effect strain development plays a major role. Electrostriction brings in a relatively smaller fraction of the total strain. Certain other materials, such as some lead lanthanum zirconium titanate compositions, show very high levels of polarization at high electric fields, and for these, the electrostriction strain is a major fraction of the total strain developed.

8.4   FERROELECTRIC HYSTERESIS LOOP

The dielectric constant (ε_r or k') of ferroelectric materials is dependent on the strength of the electric field (E) applied. When the electric fields applied are relatively high, the entire material can become a single-domain structure. If this stage is reached, the material cannot generate any more ferroelectric polarization; the polarization is said to be saturated, that is, either all domains are aligned or there is one large domain. This state of maximum possible ferroelectric polarization is known as the *saturation polarization* (P_s). The only small increase that is possible beyond P_s is due to the continued ionic and electronic polarization of the atoms or ions that make up a given ferroelectric material.

A ferroelectric material in which a substantial fraction of ferroelectric domains is aligned with the applied electric field due to exposure to a level of electric field is called a *poled ferroelectric*. The process of applying an electric field in order to align a substantial fraction of domains is known as *poling*. A schematic representation of the domain reorientation occurring during poling is shown in Figure 8.14. In Figure 8.14, for the sake of clarity, each grain is shown to have only one ferroelectric domain. In most cases, grains have multiple domains (Figure 8.16). Moreover, the material does not have to be polycrystalline.

The level of electrical field needed for poling a ferroelectric material depends upon the field that is needed to move the domain walls. In practice, the poling process is often conducted in an insulating oil bath maintained at a high temperature, but below T_c. The higher temperature makes it possible to carry out poling at lower electrical fields because of increased domain wall mobility. The insulating oil bath prevents any arcing that may occur between the electrodes applied on the material being poled.

Higher temperatures help in the alignment of the domains during poling. However, if an already-poled piezoelectric is again exposed to high temperatures, it can also "depole." *Depoling* means that the domains undergo randomization, which can cause the net piezoelectric effect to diminish or disappear. A general rule of thumb for using piezoelectrics is that they can be safely used up to a temperature of $\frac{1}{2}\,T_c$ without significant degradation of the piezoelectric activity. Another factor to consider is whether there are any potential changes in the crystal structure, even though the material will not become paraelectric (Figure 8.35a). Depoling is also caused by stress encountered during applications, because applied stress can cause domain switching.

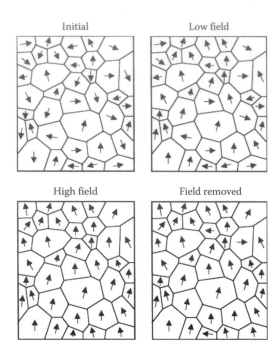

FIGURE 8.14 Schematic figure of the alignment of domains poling process for a polycrystalline material. (From Buchanan, R. C. 2004. *Ceramic Materials for Electronics*. New York: Marcel Dekker. With permission.)

Even at room temperature, domains in a freshly poled piezoelectric in time undergo some level of randomization after the poling electric field has been removed. This process is known as *aging*, and causes a decrease in the piezoelectric properties. The effects of aging on dielectric properties are usually logarithmic with respect to time and are accelerated at higher temperatures.

The dielectric constant (ε_r) of ferroelectrics will be very high at small electric fields and will continue to decrease as the electric field increases. If the applied field is too high, ferroelectric materials will exhibit an electrical breakdown. When the electric field is taken off, not all of the domains will return to their original random states of polarization (Figure 8.14). As a result, the ferroelectric material shows *remnant polarization* (P_r). We need to apply a magnitude equivalent in the opposite direction to what is known as the *coercive field* (E_c) in order to bring the polarization back to zero by again randomizing all the domains. This behavior, involving development of polarization in a ferroelectric material, is captured in what is described as a *ferroelectric hysteresis loop*. This trace of polarization (P) or displacement (D) as a function of the electric field is known as the polarization–electric field (P–E) or dielectric displacement–electric field (D–E) hysteresis loop (Figure 8.15).

There is a difference between the P–E and D–E loops. In a P–E loop, after all the domains are aligned, that is, when $P = P_s$, the trace describing P–E will become flat.

Recall that dielectric displacement (D) and polarization (P) have the following relationship:

$$D = \varepsilon_0 E + P \tag{8.3}$$

The ferroelectric polarization (P) that is typically induced in ferroelectric materials is significantly larger than $\varepsilon_0 E$, and therefore $D \cong P$. Thus, the difference between a D–E and P–E loop may not appear very large. However, it is important to note the fundamental difference between a P–E loop and a D–E loop. A P–E loop will show, leveling out of P, as electric field causes nearly complete domain alignment. In a D–E loop, the value of D will continue to increase with increasing E, even after domain alignment is complete.

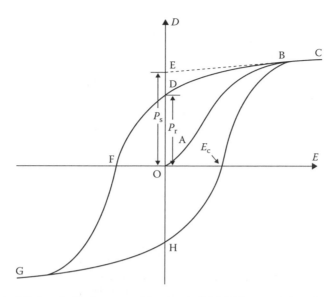

FIGURE 8.15 Typical dielectric displacement (D)–electric field (E) hysteresis loop for an unpoled ferroelectric material. (From Xu, Y. 1991. *Ferroelectric Materials and Their Applications*. Amsterdam: North Holland. With permission.)

8.4.1 Trace of the Hysteresis Loop

We will now follow the trace of the hysteresis loop shown in Figure 8.15. For an unpoled sample, the ferroelectric domains are at first randomly aligned. The starting point is the origin (O), where there is neither an applied electric field nor the development of net polarization or displacement. As an electric field is applied, a very small part of the hysteresis loop initially exhibits a polarization that is linear with the applied electric field. In this region, between points O and A, the ferroelectric material essentially behaves similarly to a linear dielectric because the applied electric field (E) is too small to cause any changes in the orientation of randomly arranged ferroelectric domains (Figure 8.14).

As we increase the applied field, the ferroelectric domains begin to realign. The realignment of ferroelectric domains between points A and B results in the development of relatively large polarization. Thus, the dielectric constant of ferroelectrics is relatively large in this region. As we can see from Figure 8.15, the development of this polarization is nonlinear.

As we reach point B, almost all the domains are aligned (Figure 8.14, high field). Any additional increase in the electric field will cause only a slight increase in polarization because of continued increases in the electronic and ionic polarization mechanisms. The polarization value corresponding to point B is known as the saturation polarization (P_s). This is the maximum polarization that can be expected from the alignment of domains in a ferroelectric material, ignoring the small levels of electronic and ionic polarizations that can continue to occur at high electric fields. The P_s value depends on the composition of the ferroelectric material; however, in contrast to the coercive field (E_c), P_s is not a microstructure-sensitive property.

Note that the slope of the region between points B and C is rather small. Thus, the dielectric constant of a ferroelectric at high electric fields is again rather small. Thus, another point that needs to be emphasized is that the dielectric constant (k or ε_r) of ferroelectrics is strongly dependent on the magnitude of the electric field applied. This is not true for so-called linear dielectrics, such as alumina, silica, and polyethylene. If we continue to increase the field indefinitely, the ferroelectric material will eventually experience dielectric breakdown.

Let us assume that we reverse the direction of the applied field after reaching point C. This state, in which all domains are forced into alignment with the field, is a high-energy state for the material. The natural tendency for the material will be to lower its internal energy by reverting to the random arrangement of domains. As the magnitude of the field decreases while going from point C toward point B and on toward point D, some domains are able to switch their polarization directions and others are not. Thus, as we decrease the magnitude of the applied field, the overall polarization (or dielectric displacement) decreases. However, a fraction of the domains continues to remain aligned (Figure 8.14) with the direction of the field applied (O → A → B → C). This causes the ferroelectric material to develop a remnant polarization (P_r), shown at point D in Figure 8.15. Please do not confuse this with the symbol used for dielectric displacement.

To remove this remnant polarization (P_r) from the ferroelectric material, we must apply an electric field in a direction opposite to the original direction. The region between points D and F (Figure 8.15) shows this effect. At point F, we reach a randomized domain configuration with no net polarization left in the ferroelectric material. The electric field at point F is known as the coercive field (E_c), which can be defined as the field required to coerce or force all the domains back to a random configuration, causing zero net polarization in a material that was subjected to an electric field.

As the magnitude of the field (applied in a reverse direction) increases from point F to point G, the domains realign with the new direction. At point G, all the domains are aligned with the new (i.e., reversed) field direction (similar to point C). As the magnitude of the field is decreased, some domains start to become randomized. We reach point H, where the applied field is reduced to zero. However, similar to point D, some domains remain aligned, causing a remnant polarization (P_r). The polarizations that remain at points D and H are essentially of the same magnitude but in opposite directions and are designated as $+P_r$ and $-P_r$, respectively. The polarization that remains at point H can be reduced to zero by again applying a field in the reverse direction until we reach a point that corresponds to the coercive field E_c.

One of the most widely-used piezoelectric polymers is polyvinylidene fluoride (PVDF). Figure 8.16 shows a *D–E* hysteresis loop for a poled PVDF film with a 25-μm thickness. In practice, such ferroelectric hysteresis loops are recorded using the Sawyer–Tower circuit. Automated instruments are currently available for the measurement of hysteresis loops. Note that because the PVDF sample has been poled, this loop does not have the "initial" region expected for an unpoled or virgin material (Figure 8.15). This means that even at zero electric field, the film already has a built-in dielectric polarization.

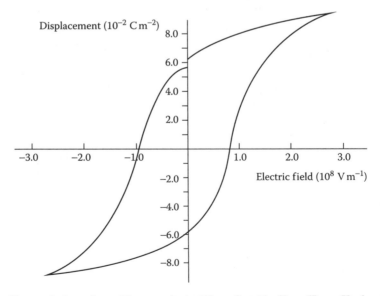

FIGURE 8.16 Hysteresis loop for a 25-μm polyvinylidene fluoride film. (From Kepler, R. G. 1995. In *Ferroelectric Polymers: Chemistry, Physics, and Applications*, ed. H. S. Nalwa. New York: Marcel Dekker Inc. With permission.)

Ferroelectric thin films of materials, such as PZT, barium strontium titanate, and PVDF, have been investigated for a number of applications, which include nonvolatile random-access memories, pyroelectric detector and imaging arrays, and microelectromechanical sensors and actuators. An example of a ferroelectric hysteresis loop for a PZT thin film doped with manganese (Mn) and antimony (Sb) is shown in Figure 8.17.

The remnant polarization for this manganese- and antimony-doped PZT film is about 15 μC/cm^2 (Figure 8.17). The saturation polarization (P_s) is about 34 μC/cm^2. Note that the x-axis data are shown in voltage and not as an electric field. The coercive voltage is ~3.0 V, and the coercive field based on the thickness of the thin film is 70 kV/cm (see Problem 8.1).

As mentioned before, the saturation polarization (P_s) is dependent on the composition of the ferroelectric material. For example, the polarization–voltage hysteresis loops for different compositions of Pb(Zr$_x$Ti$_{1-x}$)O$_3$ (PZT), with $x = 0.2$, 0.5, and 0.7, are shown in Figure 8.18.

As shown in Figure 8.19, the coercive field associated with 5-μm-thin films of PZT prepared using a sol–gel process changes considerably with changes in the Zr/(Zr + Ti) ratio. From these data, we can see that the coercive field of PZT decreases with increasing zirconium concentrations.

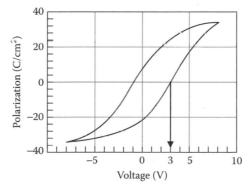

FIGURE 8.17 Hysteresis loop for manganese- and antimony-doped PZT thin films. (From Ignatiev, A. et al. 1998. *Materials Science and Engineering B* 56(2):191–4. With permission.)

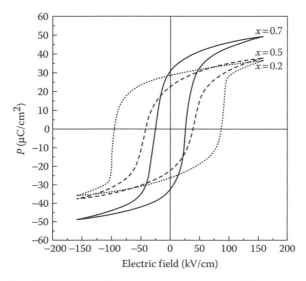

FIGURE 8.18 Composition dependence of ferroelectric properties on ferroelectric hysteresis loops. (From Osone, S., K. Brinkman, Y. Shimojo, and T. Iijima. 2008. *Thin Solid Films* 516:4325–9. With permission.)

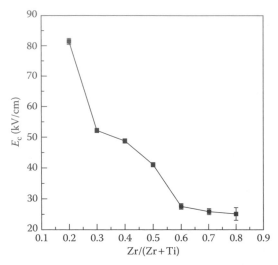

FIGURE 8.19 Coercive field of sol gel–derived PZT films as a function of the Zr/(Zr + Ti) ratio. (From Osone, S., K. Brinkman, Y. Shimojo, and T. Iijima. 2008. *Thin Solid Films* 516:4325–9. With permission.)

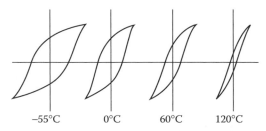

FIGURE 8.20 Qualitative appearance of *D*–*E* loops for BaTiO$_3$. (Adapted from Hench, L. L., and J. K. West. 1990. *Principles of Electronic Ceramics*. New York: Wiley.)

Even for the same composition of a material, the coercive field (E_c) can be made to change significantly with *microstructure* because different microstructural features, such as grain size and point defects, affect the mobility of domain walls. These composition- and microstructure-related effects are used to engineer the so-called soft and hard categories of piezoelectrics (Table 8.3).

In general, the saturation polarization (P_s), which is the maximum polarization that can be obtained from a given composition of a ferroelectric material, is *not* a strong microstructure-dependent quantity. However, the coercive field (E_c) is very much a microstructure-sensitive property. We will see similar trends in ferromagnetic and ferrimagnetic materials (Chapter 9).

The appearance of the ferroelectric loop and the associated values undergo changes with temperature and especially depend on the particular crystal structure of the phases. For example, for BaTiO$_3$, one of the phase transformations is that of a ferroelectric tetragonal structure changing to the paraelectric cubic structure at ~T_c = 120°C (Figure 8.20).

The corresponding changes in the polymorphic forms of BaTiO$_3$ are shown in Figure 8.21. BaTiO$_3$ also shows a hexagonal polymorph, which is not shown in Figure 8.21. At temperatures above T_c ~ 120°C, BaTiO$_3$ is cubic and undergoes a transformation to the tetragonal phase as the temperature decreases to less than 120°C. The crystal structure of BaTiO$_3$ then changes from tetragonal to orthorhombic at ~0°C. At an even lower temperature (−90°C), the crystal structure changes to a rhombohedral form (Figure 8.21).

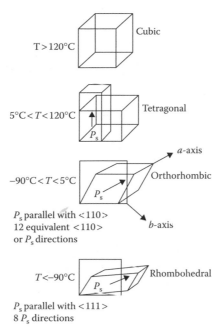

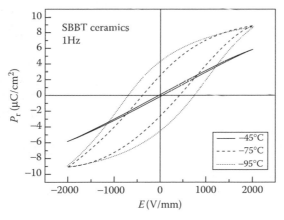

FIGURE 8.21 Polymorphs of $BaTiO_3$ as they change with temperature. (From Hench, L. L., and J. K. West. 1990. *Principles of Electronic Ceramics.* New York: Wiley. With permission.)

FIGURE 8.22 Changes in the hysteresis loops for strontium–barium–bismuth titanate (SBBT) ceramics. (From Chen, W., X. Yao, and X. Wei. 2007. *Solid State Commun* 141:84–8. With permission.)

We will also see similar changes in the ferroelectric hysteresis loop for relaxor ferroelectrics. For example, the changes in the hysteresis loops for SBBT ceramics at −45°C, −75°C, and −95°C are shown in Figure 8.22. As the temperature decreases, the nanopolar domains become stable, and the remnant polarization increases substantially.

8.5 PIEZOELECTRICITY

8.5.1 Origin of the Piezoelectric Effect in Ferroelectrics

The piezoelectric effect is present in all ferroelectrics; however, it is not limited only to ferroelectrics. Materials such as ZnO and SiO_2 also exhibit piezoelectricity (Figure 8.12). We begin by discussing how piezoelectricity occurs in ferroelectric materials.

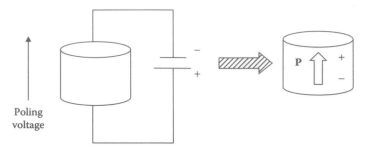

FIGURE 8.23 Poling and development of polarization in a piezoelectric material.

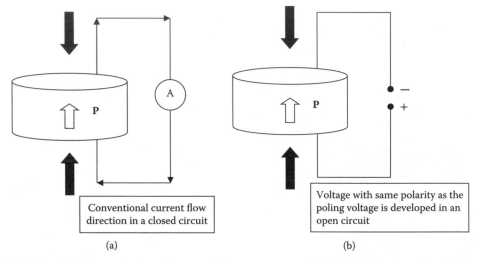

FIGURE 8.24 With the application of a compressive stress to a poled piezoelectric along the poling axis or a tensile stress in a direction perpendicular to the polar axis, (a) conventional current flows in a closed circuit as shown. (b) In an open circuit, a voltage with the same polarity as the poling voltage develops.

A poled piezoelectric ferroelectric material has a net polarization (P_r), which is caused by the alignment of the randomly aligned domains in the ferroelectric material by the poling process. When a poling voltage is applied to a piezoelectric material such as PZT, the positively charged titanium and zirconium ions in the asymmetric crystal structure are attracted toward the negative end of the poling field (Figure 8.23).

When the surfaces of a poled piezoelectric are connected using a wire, no current will flow because there is no free charge on the surfaces. However, when a poled piezoelectric material is subjected to a compressive stress, the dipole moment, and hence polarization, will change. Polarization is the bound charge density; as it decreases, free charge density increase. If excess free charge is generated on the surface, that charge will flow into an external circuit and create a transient current (Figure 8.24).

If we deliberately do not allow a current to flow by maintaining an open circuit, then a voltage with the same polarity as that used in the poling process is developed (Figure 8.24b). As soon as the wires are connected, a current flows as shown in Figure 8.24a.

Note that for PZT and other materials with the same basic crystal structure, essentially the same effect is obtained if a tensile stress is applied in a direction *perpendicular* to the poling direction.

Thus, the application of either a compressive stress along the polarization direction or a tensile stress perpendicular to the polarization direction generates a voltage with the same polarity as the poling voltage.

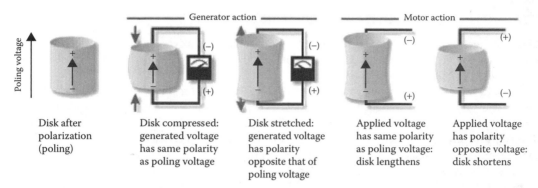

FIGURE 8.25 Schematic representation of the direct and converse piezoelectric effects. (From Moheimani, S. O. R., and A. J. Fleming. 2006. *Piezoelectric Transducers for Vibration Control and Damping.* Berlin: Springer. With permission.)

If a tensile stress is applied along the poling direction, these effects are reversed. This means that the current in the closed circuit flows in an opposite direction. If the circuit is open, then a voltage with polarity *opposite* to that of the poling voltage is set up across the piezoelectric. Thus, for materials such as $BaTiO_3$ and PZT, a tensile stress along or a compressive stress perpendicular to the poling direction will cause the development of a voltage with polarity opposite to that of the poling voltage. These effects are summarized in Figure 8.25.

If we do not connect the piezoelectric surfaces through an external conductor, the free charges generated on the surfaces of the poled piezoelectric will eventually "leak" inside the high (but finite)-resistivity material. If we bring the wires of the external conductor within 2 to 3 mm of each other but not close to the circuit, we will see an electrical arc! This is because the *voltage* created across the electrode gap is very high (a few thousand volts). This process builds an electric field that causes the electrical breakdown of air, thus creating an arc. This is the principle by which piezoelectric spark igniters work.

8.6 DIRECT AND CONVERSE PIEZOELECTRIC EFFECTS

In the *direct piezoelectric effect*, when a poled piezoelectric material is subjected to a compressive or tensile stress, a net charge develops on the surfaces of the material (Figure 8.25). This creates a voltage across the dielectric material. The direct piezoelectric effect is also known as the *generator effect* because its action generates a voltage or electrical charge.

The *converse piezoelectric effect* refers to the development of a strain when an electric field is applied to a poled piezoelectric (Figure 8.25). The converse piezoelectric effect is also known as the *motor effect* because the application of a voltage creates motion.

When we apply an electric field in the direction of poling, that is, the bottom surface connected to the positive terminal to a cylinder of a poled piezoelectric such as PZT, then the positively charged titanium and zirconium ions will move toward the top surface, which is connected to the negative terminal. This process will cause an extension of the cylinder. If we change the polarity, that is, apply a voltage opposite to the poling voltage and connect the positive electrode to the top surface, then the positively charged titanium and zirconium ions in the unit cells of poled PZT will move away from the top surface. This will cause the cylinder to become shorter in length (Figure 8.25).

8.7 PIEZOELECTRIC BEHAVIOR OF FERROELECTRICS

All ferroelectric materials also show *piezoelectric behavior*. However, not all piezoelectric materials are necessarily ferroelectrics. One common example of this is the quartz (SiO_2) crystal, which is piezoelectric but not ferroelectric.

Ferroelectric materials show a piezoelectric effect. However, for this effect to be measurable and useful, polycrystalline ceramics must be poled. In a polycrystalline ceramic, grains are randomly oriented and there are ferroelectric domains within each grain. If a ceramic with a randomized domain structure is subjected to stress, no measurable strain—or piezoelectric effect—will be developed. This is because the strains developed in each grain will cancel each other out. Therefore, for a net piezoelectric effect to be observed and to be useful, we must align a majority of the domains in a given direction. This is accomplished using the poling process. As discussed in Section 8.4, poling is usually carried out by heating the piezoelectric material to a higher temperature (not necessarily above the Curie temperature) in an oil bath and then applying an electric field for a few minutes. Poling is necessary in order for a piezoelectric ceramic to be useful for technological applications. Domain switching is a complicated process. It is similar to the manner in which nucleation and growth processes lead to phase transformations in materials.

A poled piezoelectric material that is either a single crystal or polycrystalline develops a voltage when subjected to stress. It also develops a strain when subjected to an electric field. The development of strain in a poled piezoelectric is described in what is known as a "butterfly loop" (Figure 8.26a). The corresponding *P–E* loop is shown in Figure 8.26b.

In Figure 8.18, the ferroelectric hysteresis loops are shown for $Pb(Zr_xTi_{1-x})O_3$ (PZT) with $x = 0.2, 0.5,$ and 0.7 (Osone et al. 2008). The corresponding butterfly loops showing the longitudinal strain that develops as a function of the electric field via the piezoelectric effect are shown in Figure 8.27.

8.8 PIEZOELECTRIC COEFFICIENTS

The direct and indirect effects in piezoelectric materials are commonly described using piezoelectric coefficients. These coefficients are also used to compare different piezoelectric materials with one another. The converse piezoelectric effect is commonly described by the *d* coefficient. The piezoelectric *d* coefficient is defined as the mechanical strain developed in a piezoelectric material per unit of electric field applied. Strain means mechanical displacement; hence, remember *d for*

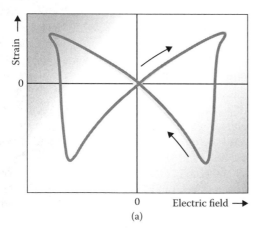
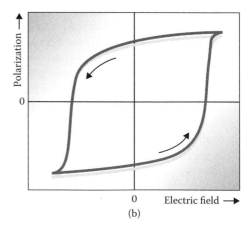

FIGURE 8.26 Relationship between (a) the butterfly loop and (b) the *P–E* loop showing the development of strain as a function of the electric field. (From Cross, E. 2004. *Nature* 432:24–5. With permission.)

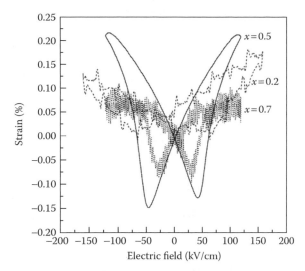

FIGURE 8.27 Longitudinal strain–electric field butterfly loops for Pb(Zr$_x$Ti$_{1-x}$)O$_3$ (PZT), with x + 0.2, 0.5, and 0.7. (From Osone, S., K. Brinkman, Y. Shimojo, and T. Iijima. 2008. *Thin Solid Films* 516:4325–9. With permission.)

displacement. If x is the strain developed in a piezoelectric material after the application of an electric field E, then the coefficient d is defined by the following equation:

$$d = \left(\frac{\partial x}{\partial \AA} \right)_{X,T} = \left(\frac{\partial D}{\partial X} \right)_{E,T} \tag{8.4}$$

The SI units of the coefficient d will be m/V if calculated as strain developed per unit electric field.

We can show that the coefficient d is equal to the displacement (D) induced per unit stress (X). It is also called the *piezoelectric charge constant*, and its unit is coulombs per newton (C/N).

The generator effect or direct piezoelectric effect is defined by the g coefficient. Remember, g stands for *generator*. The piezoelectric g coefficient is defined as the electric field (E) generated by the application of unit mechanical stress (X), which is given by the following equation:

$$g = -\left(\frac{\partial E}{\partial X} \right)_{D,T} = \left(\frac{\partial x}{\partial D} \right)_{X,T} \tag{8.5}$$

The piezoelectric g coefficient is also known as the *piezoelectric voltage constant* and is equal to the strain (x) induced per unit dielectric displacement (D) applied. The unit of g coefficient will be V · m/N if expressed as the electric field generated per unit stress, or m^2/C if described as the strain induced per dielectric displacement applied.

We have not derived Equations 8.4 and 8.5. However, the rationale for these equations is as follows.

We can induce the development of strain (x) in a dielectric material by using a mechanical stress (X), such as tensile or compressive stress. The strain that is produced depends on Young's modulus of the material. The stiffer the material, that is, the higher Young's modulus, the lesser the strain (x) produced for a given level of stress (X). When using a scalar quantity, the inverse of Young's modulus is compliance. Assume that the strain (x) produced in a material will be proportional to its compliance (S) and level of stress (X).

In a piezoelectric material, the strain (x) can also be produced by the application of an electric field (E) via the converse piezoelectric effect. Furthermore, the electrostriction effect also causes strain (Section 8.3). We assume that the strain produced by the converse piezoelectric effect is much greater than that produced by electrostriction.

Thus, we can consider the development of strain (x) as a function of the stress (X) and the electric field (E). If we want to consider only the effect of stress (X) on strain, then we must hold the electric field (E) constant. We show the compliance under a constant electric field as s^E. Similarly, if we want to express only the effect of electric field on the developing strain, then we must hold the stress X constant (this is the other variable that can also produce strain). This is why in the definition of piezoelectric coefficient d, we specified that X is constant (Equation 8.4).

Therefore, the development of strain x can be written as a function of the two causes: stress (X) and electric field (E):

$$x = s^E X + dE \tag{8.6}$$

Because ferroelectric materials are nonlinear dielectrics, and in some materials in which the stress–strain relationship may also be nonlinear, a more appropriate way to express the relationship may be to consider the change in strain (δx) as follows:

$$\delta x = s^E \delta X + d\delta E \tag{8.7}$$

Note that the piezoelectric d coefficient is also defined as the dielectric displacement (D) per unit stress (X; Equation 8.4). Dielectric displacement (D) is also induced when we apply an electric field (E). The extent of the dielectric displacement depends on the dielectric constant or permittivity (ε) of the material. Thus, we can write an equation that relates the dielectric displacement (D) as a function of the applied electric field and stress level (X). We must hold the stress constant when we consider the effect of electric field (E) on the dielectric displacement (D). Therefore, we use the permittivity value at constant stress (ε^X). In most measurements of the dielectric constant, the sample is not clamped or constrained. This allows the sample to expand or contract; thus, the strain varies but the stress (X) is constant.

As seen in the definition of d (Equation 8.4), the value of the electric field (E) is held constant. Thus, the *effect* of the developing dielectric displacement (D) due to two *causes* (1) electric field and (2) stress, can be written as:

$$D = dX + \varepsilon^X E \tag{8.8}$$

Considering that ferroelectrics are nonlinear dielectrics, we should write Equation 8.8 as

$$\delta D = d\delta X + \varepsilon^X \delta E \tag{8.9}$$

Using similar arguments, we can write equations for the direct effect, that is, application of stress, causing the generation of an electric field as follows:

$$E = -gX + D/\varepsilon^X \tag{8.10}$$

Considering that the g coefficient is also expressed as the strain produced per dielectric displacement applied (Equation 8.5), we get

$$x = s^D X + gD \tag{8.11}$$

We rewrite Equations 8.10 and 8.11, respectively, as follows:

$$\delta E = -g\delta X + \delta D/\varepsilon^X \tag{8.12}$$

$$\delta x = s^D \delta X + g\delta D \tag{8.13}$$

In addition to the definitions of the d and g coefficients, we also define another coefficient e as

$$e = \left(\frac{\partial D}{\partial x}\right)_{E,T} = -\left(\frac{\partial X}{\partial E}\right)_{x,T} \tag{8.14}$$

One way to understand the meaning of this e coefficient is that it describes the dielectric displacement caused by the creation of strain.

Similarly, the h coefficient is defined as

$$-h = \left(\frac{\partial E}{\partial x}\right)_{D,T} = \left(\frac{\partial X}{\partial D}\right)_{x,T} \tag{8.15}$$

Section 8.9 later describe the coefficients d and g under hydrostatic stress conditions, denoted as d_h and g_h. In these, the subscript h stands for "hydrostatic" and should not be confused with the coefficient h defined in Equation 8.15.

8.9   TENSOR NATURE OF PIEZOELECTRIC COEFFICIENTS

8.9.1   Conventions for Directions

The piezoelectric and pyroelectric coefficients (Section 8.20) are more accurately represented as tensors, as are the dielectric constant, compliance, and Young's modulus. However, in this book, we treat them as scalars. It is important to recognize that in many applications, the effects developed (e.g., development of strain) are in a different direction than the cause (e.g., the stress applied). For example, when we stretch a material along its length, its width also changes. The extent of this effect is given by Poisson's ratio (ν). Thus, although the stress is applied along one direction, its effect is felt in the directions perpendicular to it. Similar effects must also be accounted for in piezoelectrics. The notation shown in Figure 8.28 is generally used for directions of piezoelectrics.

According to this convention, the tensile or compressive forces are applied along directions 1, 2, and 3, also designated as the x-, y-, and z-axes. Planes designated as 4, 5, and 6 are for the shear stresses. Direction 3 is considered the poling direction. This means the electrodes used in the poling process have been applied to the piezoelectric on the faces perpendicular to direction 3.

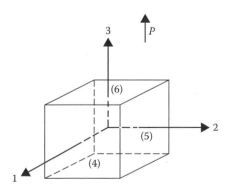

FIGURE 8.28   Notation for designating the directions for piezoelectrics; direction 3 (z-axis) is taken as the poling direction. (From Buchanan, R. C. 2004. *Ceramic Materials for Electronics*. New York: Marcel Dekker. With permission.)

8.9.2 General Notation for Piezoelectric Coefficients

If we apply a stress to the piezoelectric in a direction j, the expression for dielectric displacement (D) induced in the direction i would generally be written as follows:

$$D_i = d_{ij}X_j + \varepsilon_{ii}^X E_i \tag{8.16}$$

This is almost the same equation as Equation 8.8; the only difference is that the tensor nature of the piezoelectric coefficients and dielectric constant is now stated explicitly.

Similarly, we can write an expression for the strain developed in a piezoelectric as

$$x_j = s_{ji}^E X_j + d_{ij}E_i \tag{8.17}$$

This is essentially another form of Equation 8.6. The piezoelectric coefficient is d_{ij}. In this symbol, the first subscript (i) shows the applied electrical field that causes a mechanical strain in the direction j. Equation 8.16 also describes the effect of the stress X_j applied in direction j, causing a strain in the direction j. In Equation 8.17, s_{ij}^E is the compliance of the piezoelectric material under zero electric field. This is also known as compliance under a closed-circuit or short-circuit condition.

The value of compliance when the dielectric displacement (D) is constant is shown as s_{ij}^D and is known as compliance under the "open circuit" condition. Note that the values of compliance under short circuit and open circuit are very different. This leads us to the definition of a coefficient that can link the functioning of a piezoelectric as a *transducer*. A transducer is a device that can convert one form of energy into another. For example, a piezoelectric material can convert mechanical energy into electrical energy and vice versa.

Reduced notation is one in which both stress and strain are treated as first-rank tensors, instead of second-rank ones: the first subscript indicates the electrical direction and the second subscript indicates the mechanical direction.

If we apply an electric field to a piezoelectric material in direction 3 and measure the strain in direction 1, the corresponding converse piezoelectric coefficient is indicated as d_{31}. Similarly, the coefficient d_{33} will indicate the strain developed in direction 3 when an electric field is applied in that direction (i.e., along the poling axis; Figure 8.28). If we apply a field (E) in direction 1, that is, perpendicular to the poling direction, this will cause a shearing strain on the plane shown as 5 (Figure 8.28) and given by the d_{15} coefficient. To use piezoelectrics in this particular shear mode, the original electrodes used for poling in direction 3 must be removed, and new electrodes need to be applied in a direction perpendicular to direction 1.

We can subject piezoelectrics to a hydrostatic stress. In this case, we use a subscript h with the coefficients. The coefficient d_h is related to the d_{33} and d_{31} coefficients by the following equation:

$$d_h = d_{33} + 2d_{31} \tag{8.18}$$

8.9.3 Signs of Piezoelectric Coefficients

The d_{33} values for most commercially used PZT ceramics are in the range of 200 to 550 pm/V. The d_{31} values for PZT are negative and range from about −180 to −220 pm/V. The d_{31} values for PZT ceramics and $BaTiO_3$ and relaxor ferroelectrics, such as PMN–PT, are also negative. However, the d_{33} values for these materials are positive. This means that when an external voltage is applied to drive a piezoelectric cylinder, which has been poled previously in direction 3, in the direction of polarization, the cylinder expands in length, that is, d_{33} is positive (Figure 8.25). When the cylinder expands, the strain in the direction perpendicular to the longitudinal axis is negative, or d_{31} is negative.

In other piezoelectrics, such as PVDF, the mechanisms of piezoelectric polarizations are different. As a result, the d_{31} for PVDF is positive, and the d_{33} coefficients for PVDF are negative. This means that application of an electric field in the direction of poling (usually the thickness direction) will cause the length of a PVDF sample to increase.

8.10 RELATIONSHIP BETWEEN PIEZOELECTRIC COEFFICIENTS

We will now show that the piezoelectric coefficient d, which defines the strain induced per unit electric field applied or dielectric displacement per unit stress (Equation 8.4), and the g coefficient, which provides a measure of the electric field produced per unit stress (Equation 8.5), are related such that $d/g = \varepsilon^X$.

From Equation 8.4, the d coefficient is the strain developed by a unit change in the applied electric field. Both the stress (X) and temperature (T), which can also cause strain, are held constant. From Equation 8.4, we get

$$d = \left(\frac{\partial x}{\partial E}\right)_{X,T} \tag{8.19}$$

One definition of the g coefficient is that it describes the generator effect is that it is the electric field (E) developed per unit stress (X) or the strain (x) developed per unit of dielectric displacement (D) applied. Again, because strain can be developed in a material by the application of either stress (X) or a change in temperature (expansion or contraction), we hold these constant. Therefore, from Equation 8.5, the following equation is obtained:

$$g = \left(\frac{\partial x}{\partial D}\right)_{X,T} \tag{8.20}$$

Dividing Equation 8.19 by Equation 8.20, we get

$$d/g = \frac{\left(\dfrac{\partial x}{\partial E}\right)_{X,T}}{\left(\dfrac{\partial x}{\partial D}\right)_{X,T}} = \left(\frac{\partial D}{\partial E}\right)_{X,T} \tag{8.21}$$

Recall that the right-hand side of Equation 8.21 is the definition of dielectric permittivity (ε). Dielectric permittivity is the increase in the dielectric displacement (D) with an increase in the electric field (E) applied. Because the application of stress or a change in the temperature can also cause a change in the dipole moment and, consequently, a change in the dielectric displacement, these parameters—stress (X) and temperature (T)—must be held constant. This point was not emphasized previously in this chapter.

Therefore, from Equation 8.21, we get

$$d/g = \varepsilon^x \tag{8.22}$$

We could write a similar equation showing the tensor nature explicitly. For example, the ratio of d_{ij} and g_{ij} will be written as follows:

$$\frac{d_{ij}}{g_{ij}} = \frac{\left(\dfrac{\partial x_j}{\partial E_i}\right)_{X,T}}{\left(\dfrac{\partial x_j}{\partial D_i}\right)_{X,T}} \tag{8.23}$$

This can be rewritten as

$$\frac{d_{ij}}{g_{ij}} = \left(\frac{\partial D_i}{\partial E_i}\right)_{X,T} \tag{8.24}$$

or as

$$\frac{d_{ij}}{g_{ij}} = \varepsilon_{ii}^X \tag{8.25}$$

Applying this for d_{31} and g_{31}, we get

$$\frac{d_{31}}{g_{31}} = \varepsilon_{33}^X \tag{8.26}$$

or

$$g_{31} = \frac{d_{31}}{\varepsilon_{33}^X} \tag{8.27}$$

These are essentially the same as Equation 8.22.

Similarly, it can be shown that the following relationships exist between other piezoelectric coefficients.

$$g_{33} = \frac{d_{33}}{\varepsilon_{33}^X} \tag{8.28}$$

$$g_{15} = \frac{d_{15}}{\varepsilon_{11}^X} \tag{8.29}$$

In Table 8.1, a summary of the relationships between different piezoelectric and other coefficients is listed.

Recall that in the coefficients listed in Table 8.1, e relates the dielectric displacement D induced per unit strain or stress generated per unit electric field. The coefficient h represents the electric field generated per unit strain applied or stress generated per unit of dielectric displacement. The term h here should not be confused with the subscript h sometimes used to describe piezoelectric

---

TABLE 8.1

Relationships between Piezoelectric Coefficients

d Coefficients	e Coefficients	g Coefficients	h Coefficients
$d_{31} = e_{31}\left(s_{11}^E + s_{12}^E\right) + e_{33}s_{13}^E$	$e_{31} = \varepsilon_3^x h_{31} = d_{31}\left(c_{11}^E + c_{12}^E\right) + d_{33}c_{13}^E$	$g_{31} = \dfrac{d_{31}}{\varepsilon_3^X}$	$h_{31} = g_{31}\left(c_{11}^D + c_{12}^D\right) + g_{33}c_{13}^E$
$d_{33} = 2e_{31}s_{13}^E + e_{33}s_{33}^E$	$e_{33} = \varepsilon_3^x h_{33} = 2d_{31}c_{13}^E + d_{33}c_{33}^E$	$g_{33} = \dfrac{d_{33}}{\varepsilon_3^X}$	$h_{33} = 2g_{31}c_{13}^D + g_{33}c_{33}^E$
$d_{15} = e_{15}s_{44}^E$	$e_{15} = \varepsilon_1^x h_{15} = d_{15}c_{44}^E$	$g_{15} = \dfrac{d_{15}}{\varepsilon_1^X}$	$h_{15} = g_{15}c_{44}^D$

coefficients such as d or h under hydrostatic stress (Equation 8.18). Also note the difference between ε^x, the permittivity under constant stress, and ε^x, the permittivity under constant strain. To maintain constant strain, the sample must be clamped. Compliance is designated by s and stiffness by c. The terms E and D represent the electric field and the dielectric displacement, respectively. In addition, the permittivity is shown as a scalar quantity in these equations, that is, we have written ε_3 instead of ε_{33}. Finally, note the difference between the uses of permittivity (ε) and dielectric constant (ε_r). The following examples illustrate the use of piezoelectric coefficients.

EXAMPLE 8.1: THE MEANING OF THE PIEZOELECTRIC COEFFICIENTS D_{33} AND G_{15}

What is the meaning of (a) d_{33} and (b) g_{15}? What are their units?

SOLUTION

a. The coefficient d_{33} is one of the piezoelectric charge constants. It is defined as induced polarization (SI unit: C/m^2) in direction 3 per unit applied stress (SI unit: N/m^2), also applied in direction 3. This expresses the direct or the generator piezoelectric effect. (Figure 8.25)

$$d_{33} = \frac{\text{induced polarization in direction 3 } (C/m^2)}{\text{applied stress in direction 3 } (N/m^2)} \tag{8.30}$$

The unit for the d coefficient is coulomb per newton (typically written as pico-coulomb per newton [pC/N]).

Another way to express the piezoelectric d_{33} coefficient is as the strain induced in direction 3 (poling direction) by a field applied in direction 3.

$$d_{33} = \frac{\text{induced strain in direction 3}}{\text{applied electric field in direction 3 } (V/m)} \tag{8.31}$$

Because strain has no dimensions, another possible unit for the d coefficient is meters per volt, usually written as picometers per volt (pm/V).

b. Recall that the first subscript is the electrical direction, and the second subscript is the mechanical direction. Thus, g_{15} is defined as the induced electric field in direction 1 per unit shear stress on plane 5, that is, around direction 2 (Figure 8.28).

$$d_{15} = \frac{\text{induced electric field in direction 1 } (V/m)}{\text{applied stress on plane 5 } (N/m^2)} \tag{8.32}$$

The SI unit for g_{15} would then be Vm/N.
Alternatively, g_{15} is defined as the shear strain induced around direction 2 (i.e., plane 5) per unit dielectric displacement applied in direction 1 (Figure 8.28). The SI unit would be m^2/C.

$$g_{15} = \frac{\text{induced shear strain around direction 2 on plane 5}}{\text{applied dielectric displacement in direction 1 } (C/m^2)}$$

The SI unit will be m^2/C.

EXAMPLE 8.2: THE RELATIONSHIP BETWEEN PIEZOELECTRICS AND OTHER COEFFICIENTS

Show that $g_{33} = d_{33}/\varepsilon_3^X$, that is, prove Equation 8.28.

SOLUTION

To describe the relationship between d_{33} and g_{33}, we start with Equation 8.23.

Therefore,

$$\frac{d_{33}}{g_{33}} = \frac{\left(\dfrac{\partial x_3}{\partial E_3}\right)_{X,T}}{\left(\dfrac{\partial x_3}{\partial D_3}\right)_{X,T}} = \left(\frac{\partial D_3}{\partial E_3}\right)_{X,T}$$

The right-hand side of this equation is the dielectric displacement created in direction 3 per unit electric field applied in direction 3. This is ε_3, and we have assumed constant stress (X). Thus, Equation 8.28 is valid.

Note that lower dielectric constants lead to higher values of g. For example, the polymer PVDF has a relatively lower value of the d_{33} coefficient, ~ -30 pC/N (note the negative sign), compared to that of PZT, with $d_{31} \sim 200-400$ pC/N. As stated before, the coefficient d describes the converse piezoelectric effect, i.e. strain developed when a voltage is applied. However, the dielectric constant of PVDF is low ($\varepsilon_r \sim 10$) compared to that of PZT ($\varepsilon_r \sim 1000-2800$). Therefore, the piezoelectric voltage constant g of PVDF is still comparable to that of PZT. The coefficient d is important in applications where the strain developed in piezoelectrics causes a useful action or event.

When a sinusoidal voltage is applied to a piezoelectric disk, for example, the back-and-forth expansion and contraction of this disk (Figure 8.25) can lead to the generation of ultrasonic waves. For this application as an ultrasonic generator, we need materials with a high d value. We would prefer to have a material with a higher g coefficient in applications where we need to have a higher voltage generated when the material undergoes even very small levels of stress. For ultrasound detection, we would prefer to use materials with a higher g coefficient. In many real-life applications that involve piezoelectric materials, we need both ultrasound generation and detection capabilities. In such cases, we use the product d × g as a figure of merit. In some applications, such as underwater sound detectors (sonars or hydrophones), piezoelectric materials are exposed to a hydrostatic pressure, that is, they are subjected to stress from all directions. In this case, the *hydrostatic piezoelectric coefficients* are designated as d_h and g_h, and thus the product $d_h \times g_h$ is the important figure of merit.

Piezoelectric composites that make use of materials such as PZT and PVDF have been developed for many such applications (see Section 8.19). In other applications (e.g., pyroelectric detectors; Section 8.20), the piezoelectric voltage generation is actually undesirable. For these applications, it is possible to take advantage of the difference between the signs of the d coefficients and create composites in which the piezoelectric effect is substantially reduced. This causes the detector to respond more to the changes in temperature and less to vibration or shock.

EXAMPLE 8.3: CALCULATIONS FOR EVALUATING THE DIRECT AND CONVERSE EFFECTS IN LEAD ZIRCONIUM TITANATE

A poled PZT type I piezoelectric ceramic disk (length $l_0 = 2$ mm and poled in direction 3) of a certain composition has $d_{33} = 289$ pC/N, and its dielectric constant $\left(\varepsilon_{r33}^X\right)$ is 1300.

   a. What is the value of the piezoelectric voltage constant or the g_{33} coefficient?
   b. What is the voltage applied across the thickness of this material if the electric field (E) is 250 kV/m?
   c. How much strain (x) will this voltage generate along direction 3?
   d. What will be the increase in length, in micrometers?

SOLUTION

a. For the PZT ceramic sample here, the d_{33} coefficient is 289 pC/N, and the dielectric constant under constant stress $\left(\varepsilon_{r33}^{X}\right)$ is 1300. Recall that the dielectric constant is the ratio of the permittivity of a material to that of a vacuum, that is,

$$\varepsilon_r = \frac{\varepsilon}{\varepsilon_0} \qquad (8.33)$$

Therefore, in this case,

$$\varepsilon_{r33}^{X} = 1300 = \frac{\varepsilon}{8.85 \times 10^{-12} \text{F/m}} \quad \text{or} \quad \varepsilon_{r33}^{X} = 1.15 \times 10^{-8} \text{F/m}$$

Now,

$$g_{33} = \frac{d_{33}}{\varepsilon_{33}^{X}} \qquad (8.28)$$

Therefore, we get

$$g_{33} = \frac{\left(289 \times 10^{-12} \text{C/N}\right)}{\left(1.15 \times 10^{-8} \text{F/m}\right)} = 25.1 \times 10^{-3} \text{ V} \cdot \text{m/N}$$

b. Because the initial length of the material is 2 mm, the voltage applied to the material would be

$$250 \frac{\text{kV}}{\text{m}} \times 10^3 \frac{\text{V}}{\text{kV}} \times 2 \text{ mm} \times 10^{-2} \frac{\text{m}}{\text{mm}} = 5000 \text{ V}$$

c. The strain generated in direction 3 by applying an electric field of 250 kV/m in the same direction will be given by

$$x = d_{33} \times E_3 = (289 \times 10^{-12} \text{C/N})(250 \times 10^3 \text{V/m}) = 7.225 \times 10^{-5}$$

$$x = \frac{\Delta l}{l_0} = \frac{\Delta l}{2 \text{mm}} = 7.225 \times 10^{-5}$$

d. Now, from the definition of strain, that is, $x = \Delta l / l_0$, the change in the dimension will be given by

$$\Delta l = (2 \text{ mm} \times 1000 \text{ } \mu\text{m/mm})(7.225 \times 10^{-5}) = 0.145 \text{ } \mu\text{m}$$

Thus, applying a voltage of 5000 V to this cylinder of height 2 mm will cause an elongation of 0.145 µm.

8.11 APPLICATIONS OF PIEZOELECTRICS

Properties of some piezoelectric materials are summarized in Table 8.2. Various applications of piezoelectrics are summarized in Figure 8.29. We will now discuss some devices based on piezoelectrics.

8.12 DEVICES BASED ON PIEZOELECTRICS

8.12.1 EXPANDER PLATE

The following example helps to illustrate the meaning of some of the equations related to piezoelectric coefficients. Consider a piezoelectric plate of thickness (*t*) and area (*A*) made up of electrodes with length *l* and width *w* (Figure 8.30).

TABLE 8.2

Approximate Ranges for the Piezoelectric Coefficients and Other Relevant Properties of Some Single-Crystal Piezoelectric Materials

	$BaTiO_3$	α-Quartz (SiO_2)	$LiNbO_3$	PZN–PT
Piezoelectric				
d_{33} (pC/N or pm/N)	85.6	0	8.69	2140
d_{31} (pC/N or pm/N)	–34.5	0	–1.4	–980
d_{15} (pC/N or pm/N)		0	66.38	130
d_{11} (pC/N or pm/N)	0	2.331		
d_{14} (pC/N or pm/N)	0	–0.7763		
g_{33} (V · m/N)	57.5×10^{-3}	-50×10^{-3}	34.46×10^{-3}	
g_{31} (V · m/N)	-23×10^{-3}	0	-4.110×10^{-3}	
g_{14} (V · m/N)	0	0		
Coupling				
k_{33}	0.56	0.09		0.91
k_{31}	0.315	0		0.50
Dielectric				
"Free" dielectric constant at constant stress $\left(\varepsilon_{r33}^X\right)$	168	4.628	28.5	5242
"Clamped" dielectric constant at constant strain $\left(\varepsilon_{r33}^x\right)$	109	4.628	26.7	869
"Free" dielectric constant at constant stress $\left(\varepsilon_{r11}^X\right)$	2920	4.507	83.3	3099
"Clamped" dielectric constant at constant strain $\left(\varepsilon_{r11}^x\right)$	1970	4.420	44.9	2975
Mechanical				
Compliance $\left(s_{33}^E\right)$ at constant electric field (closed circuit) (m^2/N)	15.7×10^{-12}	9.7329×10^{-12}	5.058×10^{-12}	12×10^{-12}
Compliance at constant displacement (open circuit) $\left(s_{33}^D\right)$ (m^2/N)	10.8×10^{-12}	9.7329×10^{-12}	4.758×10^{-12}	
Compliance at constant electric field (closed circuit) $\left(s_{11}^E\right)$ m^2/N	8.05×10^{-12}	12.7791×10^{-12}	5.854×10^{-12}	8.3×10^{-12}
Compliance at constant displacement (open circuit) $\left(s_{11}^D\right)$ (m^2/N)	7.25×10^{-12}	12.6429×10^{-12}	5.291×10^{-12}	

Sources: Buchanan, R. C. 2004. *Ceramic Materials for Electronics.* New York: Marcel Dekker; Shrout, T. S., and S. J. Zhang. 2007. *J Electroceram* 19:111–24; and other sources.

When a voltage is applied along the thickness (t; also known as direction 3), a strain develops along the length (described as direction 1). The corresponding piezoelectric voltage constant (g), which can be described as the strain (x) induced per unit dielectric displacement (D) applied, will be written as

$$g_{31} = \frac{\text{strain developed in direction 1}}{\text{dielectric displacement applied in direction 3}} \tag{8.34}$$

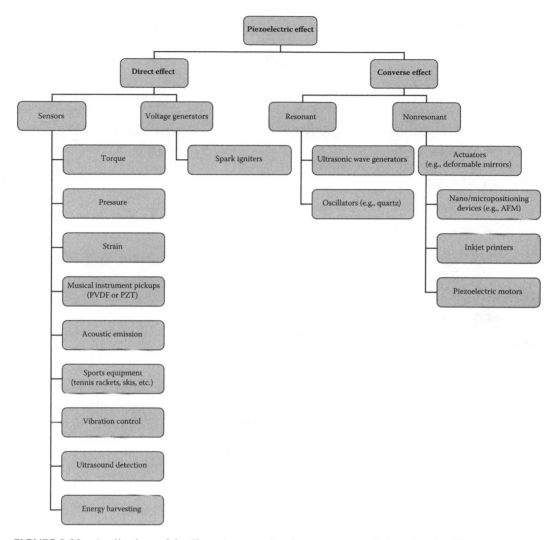

FIGURE 8.29 Applications of the direct (generator) and converse (motor) piezoelectric effects.

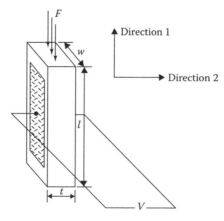

FIGURE 8.30 Piezoelectric plate geometry. The plate is poled in the thickness (t) direction. Stress is applied along the length direction. (Adapted from Morgan Technical Ceramics, Guide to Piezoelectric and Dielectric Ceramics. http://www.morganelectroceramics.com/pzbook.html.)

We can write Equation 8.34 as follows:

$$g_{31} = \frac{\left(\frac{\Delta l}{l}\right)}{\left(\frac{Q}{A}\right)} \tag{8.35}$$

In Equation 8.35, the numerator is simply the strain in direction 1, that is, the change in length (Δl) of the plate along direction 1 per unit length (l). The denominator is the dielectric displacement (D) applied, which is the charge Q divided by the area A. Considering that for this structure, $Q = V \times C$, and $A = l \times w$, we get

$$g_{31} = \frac{\left(\frac{\Delta l}{l}\right)}{\left(\frac{(V \times C)}{(l \times w)}\right)} = \frac{\Delta l \times w}{V \times C} \tag{8.36}$$

Recall that for a parallel-plate capacitor, the capacitance C is given by

$$C = \varepsilon_r \varepsilon_0 \frac{A}{t} = \varepsilon_r \varepsilon_0 \frac{l \times w}{t} \tag{8.37}$$

In this case, we assume that the capacitor structure, that is, the electroded piezoelectric, is free and not clamped. This means that the stress X to which the capacitor is subjected is constant. Therefore, the dielectric constant (ε_r) or dielectric permittivity (ε) we use is under constant stress X. Furthermore, the direction of the applied electric field and the poling direction are the same—direction 3. Thus, the correct value of permittivity to be used will be ε_{33}^X.

Substituting the expression for C from Equation 8.37 into Equation 8.36, we get

$$g_{31} = \frac{\Delta l \times w}{V \times C} = \frac{\Delta l \times w}{V \times \varepsilon_{r33}^X \varepsilon_0 \frac{l \times w}{t}}$$

or

$$g_{31} = \frac{\Delta l \times t}{V \times l \times \varepsilon_{r33}^X \times \varepsilon_0} = \Delta l \times \frac{t}{l} \times \frac{1}{V \varepsilon_{r33}^X \varepsilon_0}$$

Rewriting

$$g_{31} = \Delta l \times \frac{t}{l} \times \frac{1}{V \varepsilon_{r33}^X \varepsilon_0} \tag{8.38}$$

Note that in Equations 8.37 and 8.38, we denote the dielectric constant using the symbol ε_r and not k. This is because the symbol k^2 stands for the *coupling efficiency* of a piezoelectric. (Section 8.16).

We can write the strain developed in a piezoelectric along its thickness (direction 3) when a voltage is applied in direction 1 as follows:

$$\Delta l = g_{31} \times \frac{l}{t} \times V \times \varepsilon_{r33}^X \times \varepsilon_0 \tag{8.39}$$

We can similarly write an expression for the piezoelectric charge coefficient d_{31} as the ratio of strain developed in direction 1 (length) when a voltage is applied in direction 3 (thickness).

$$d_{31} = \frac{\text{strain developed in direction 1}}{\text{electric field applied in direction 3}} \tag{8.40}$$

$$d_{31} = \frac{\Delta l / l}{V / t} \tag{8.41}$$

or

$$\Delta l = d_{31} \times V \times \frac{l}{t} \tag{8.42}$$

This is the same as Equation 8.39.

A note of caution: These equations, for a change in length, are applicable only under nonresonant conditions. Every physical body or structure, such as the plate considered here, has natural mechanical resonant frequencies. If a piezoelectric material is excited at these resonant frequencies, the changes in dimensions we will get will be larger than those predicted by Equations 8.42 and 8.39. In other words, these equations are applicable to frequencies of excitation well below the resonant frequencies.

Both Equations 8.42 and 8.39 describe the change in the length (Δl) of a dielectric plate. By equating these, we get

$$\Delta l = g_{31} \times \frac{l}{t} \times V \varepsilon_r \varepsilon_0 = d_{31} \times V \times \frac{l}{t}$$

$$\Delta l = d_{31} \times V \times \frac{l}{t} \tag{8.43}$$

Therefore,

$$g_{31} = \frac{d_{31}}{\varepsilon_r \varepsilon_0} = \frac{d_{31}}{\varepsilon_{33}^X}$$

or

$$\frac{d_{31}}{g_{31}} = \varepsilon_{33}^X \tag{8.44}$$

This is the relationship between the g and d coefficients, previously shown in Equation 8.28.

In Example 8.4, we will show how to interpret the meaning of piezoelectric and mechanical coefficients and to calculate the magnitude of strain developed in a poled piezoelectric subjected to voltage.

EXAMPLE 8.4: PIEZOELECTRIC MICROPOSITIONERS

The small change in dimensions we can get by applying a voltage to a piezoelectric can be used to make devices known as micropositioners. A poled PZT ceramic plate, 50 mm long, 5 mm wide, and 2 mm thick, is used as a micropositioner device. Assume that the dielectric constant ε^X_{r33} is 1200 and $g_{31} = 10.5 \times 10^{-3}$ V·m/N.

a. What will be the value of d_{31}?
b. What will be the change in the *length* of this plate if a potential of 100 V is applied across its *thickness*?

Solution

a. We start with the relationship between d_{31} and g_{31} from Equation 8.28.

$$\frac{d_{31}}{g_{31}} = \varepsilon^X_{33} \tag{8.45}$$

Therefore,

$$d_{31} = g_{31} \times \varepsilon^X_{33} = 10.5 \times 10^{-3}\,\text{V·m/N} \times 1200 \times 8.85 \times 10^{-12}\,\text{F/m}$$

Note the conversion of the dielectric constant to dielectric permittivity using the permittivity of free space ($\varepsilon_0 = 8.85 \times 10^{-12}$ F/m).
   Thus, $d_{31} = 139.9 \times 10^{-12}$ m/V.

b. The change in the dimension of this plate along its length can be calculated by the expression

$$\Delta l = d_{31} \times V \times \frac{l}{t} \tag{8.46}$$

$$\Delta l = 139.9 \times 10^{-12}\,\text{m/V} \times 10\,\text{V} \times \frac{50 \times 10^{-2}\,\text{m}}{2 \times 10^{-2}\,\text{m}} = 0.348\ \mu\text{m}$$

Thus, by applying a voltage of 100 V, we get an increase of about 348 nm in the length of the plate. Such changes in the dimensions of piezoelectrics are used in micropositioning and nanopositioning devices. The relative strain developed in a piezoelectric PZT is shown in Figure 8.31 as a function of the electric field applied.

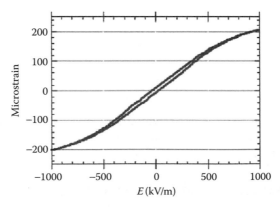

FIGURE 8.31   Typical converse piezoelectric effect in PZT ceramics. (From Pilgrim, S. *Piezoelectric Materials: A Unheralded Component, 11th ed.* FabTech, online at http://www.fabtech.org/white_papers/_a/ piezoelectric_materials_an_unheralded_component. With permission.)

8.13 TECHNOLOGICALLY IMPORTANT PIEZOELECTRICS

Properties of *single crystals* of some commonly encountered piezoelectric materials are listed in Tables 8.2 and 8.3. Although many piezoelectric materials have been investigated, only a handful of these are useful in commercial applications related to actuators, ultrasonic generators, ultrasound

TABLE 8.3

Piezoelectric Properties of Poled *Polycrystalline* Barium Titanate and Different Grades of PZT

	$BaTiO_3$	PZT–DOD-I/ PZT-4 (Hard PZT)	PZT–DOD-II/ or PZT5A (Soft PZT)	PZT–DOD-III (Hard PZT)	PZT-DOD-VI or PZT5H (Soft PZT)
Piezoelectric					
d_{33} (pC/N or pm/N)	191	289	374	218	593
d_{31} (pC/N or pm/N)	−79	−123	−171	−93	−274
g_{33} (V · m/N)	11.4×10^{-3}	25×10^{-3}	25×10^{-3}	25×10^{-3}	19.7×10^{-3}
g_{31} (V · m/N)	-4.7×10^{-3}				-9.1×10^{-3}
Coupling					
k_p	0.354	0.58	0.60	0.50	0.65
k_{33}	0.493	0.70	0.71	0.64	0.75
k_{31}	0.208	0.33	0.34	0.295	0.39
Dielectric					
"Free" dielectric constant at constant stress (ε_{r33}^X)	1900	1300	1700	1000	3400
"Clamped" dielectric constant at constant strain (ε_{r33}^x)	1420	635	830	600	1470
"Free" dielectric constant at constant stress (ε_{r11}^X)	1620	1475	1730		3130
"Clamped" dielectric constant at constant strain (ε_{r11}^x)	1260	730	916		1700
tan δ (dielectric loss)	0.01	0.001	0.02	0.004	0.02
Curie temperature (T_c) (°C)	~120	328	365	300	190
Coercive field (kV/cm)	~1–2	~18	~15	~22	~6–8
Mechanical					
Compliance (s_{33}^E) (constant electric field, i.e., closed circuit) (m²/N)	8.93×10^{-12}	15.5×10^{-12}	18.8×10^{-12}	13.9×10^{-12}	20.7×10^{-12}
Compliance (constant displacement, i.e., open circuit) (s_{33}^D) (m²/N)	6.76×10^{-12}	7.9×10^{-12}	9.46×10^{-12}	8.5×10^{-12}	8.99×10^{-12}
Compliance (constant electric field, i.e., closed circuit) (s_{11}^E) (m²/N)	8.55×10^{-12}	12.3×10^{-12}	16.4×10^{-12}	11.1×10^{-12}	16.5×10^{-12}
Compliance (constant displacement, i.e., open circuit) (s_{11}^D)(m²/N)	8.18×10^{-12}	10.9×10^{-12}	14.4×10^{-12}	10.1×10^{-12}	14.1

Data from Buchanan, R. C. 2004. *Ceramic Materials for Electronics*. New York: Marcel Dekker; Shrout, T. S., and S. J. Zhang. 2007. *J Electroceram* 19:111–24; and other sources.

imaging, and vibration control and dampening (Figure 8.29). We will discuss a few of these materials here, such as PZT, the most widely used piezoelectric ceramic.

Another technologically important material is PVDF, which is different because it is a polymer and is therefore quite flexible. Another class of technologically important materials is the group that contains relaxor ferroelectrics, such as lead magnesium niobate–lead titanate (PMN–PT). These materials, in both single-crystal and polycrystalline forms, have become particularly attractive because of their very high piezoelectric coefficients (d_{33} ~2000 pm/N) and electromechanical coupling coefficients (k_{33} ~0.9).

One of the limitations of piezoelectrics is that piezoelectrics can depole when they are exposed to temperatures approaching the Curie temperature. This means that the domains aligned during poling will become randomized, and the resulting material will have either a very weak or no piezoelectric response. In general, piezoelectrics cannot be used at temperatures above half the value of their T_c. This is the main reason why BaTiO$_3$, with a Curie temperature of 120°C, is not used widely as a piezoelectric material. Other factors that must be considered are changes in the crystal structure and the lowering of dielectric constants. The dielectric constants (ε_r) and the d_{33} coefficients at room temperature for different piezoelectric ceramics are shown in Figures 8.32 and 8.33.

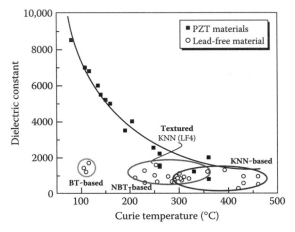

FIGURE 8.32   Room-temperature (300 K) values of dielectric constants as functions of the Curie temperature for PZT materials and novel lead-free piezoelectrics. (From Shrout, T. S., and S. J. Zhang. 2007. *J Electroceram* 19:111–24. With permission.)

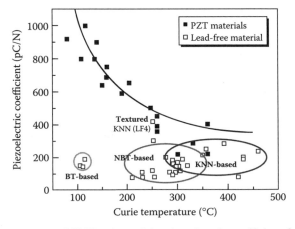

FIGURE 8.33   Room-temperature (300 K) values of the piezoelectric coefficient d_{33} (in pC/N) as a function of the Curie temperature for PZT materials and novel lead-free piezoelectrics. (From Shrout, T. S., and S. J. Zhang. 2007. *J Electroceram* 19:111–24. With permission.)

There is a need for piezoelectrics that can function at high temperatures. In Section 8.18, we will briefly discuss the latest developments related to strain-tuned piezoelectrics, which represent a new step in the development of piezoelectric devices that could function at higher temperatures. Another area of concern in the field of piezoelectrics is the presence of lead (Pb) in most commercially useful piezoelectrics. There is considerable interest in the development of lead-free piezoelectrics that are environmentally friendly (Figures 8.32 and 8.33).

8.14 LEAD ZIRCONIUM TITANATE

PZT and PZT-based ceramics are the most widely used piezoelectric materials. These materials are popular because of their relatively large piezoelectric, electromechanical coupling coefficients, and relatively high Curie temperatures (Table 8.3). Table 8.3 also contains data for $BaTiO_3$ for comparison purposes. Note that these are approximate ranges of the values of piezoelectric coefficients and other properties, and have been provided to provide a general idea. They should not be relied on for engineering design. The exact values will depend on many factors, including microstructure, processing, exact chemical composition, temperature, and the state and magnitude of any stresses present.

For PZT ceramics, the maximum strain develops at $x \sim 0.5$, which is a composition near the so-called morphotropic boundary (MPB; Figure 8.34). The MPB defines the change in the crystal structure of a ferroelectric material along with changes in its composition. Important piezoelectric properties of ferroelectric materials, such as the strain developed for a given level of electric field, are maximized when compositions near the MPB are used.

In PZT at room temperature, zirconium-rich compositions have rhombohedral crystal structure (Figure 8.35a). Titanium-rich compositions exhibit a tetragonal structure. At room temperature and at the mole fraction of zirconium 0.53 (i.e., titanium mole fraction of $x = 0.47$ in $PbZr_{1-x}Ti_xO_3$), the crystal structure changes from rhombohedral to tetragonal. The change in crystal structure has traditionally been used to define and describe the MPB (Figure 8.35).

It was previously believed that such MPB compositions led to maximum piezoelectric properties (e.g., piezoelectric coupling coefficients) and that the polarization vector could switch its orientation in the different variants of the tetragonal and rhombohedral phases. However, it has been suggested recently that the MPB is *not* a boundary but rather a phase with monoclinic symmetry (Ahart et al. 2008), and that there is actually a new monoclinic phase at or near the MPB (and *not* a mixture of nanotwin domains in the tetragonal and rhombohedral phases). This phase is intermediate between the tetragonal and rhombohedral PZT phases.

Research on essentially pure $PbTiO_3$ has shown changes in its crystal structure caused by increasing pressure (Figure 8.36). Because the original concept of the MPB was linked to changes in its chemical composition, we would have expected that an MPB could *not* exist in a pure compound like

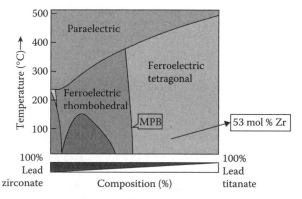

FIGURE 8.34 Phase diagram for the $PbZrO_3$–$PbTiO_3$ system showing the different ferroelectric phases and the morphotropic phase boundary. (From Cross, E. 2004. *Nature* 432: 24–5. With permission.)

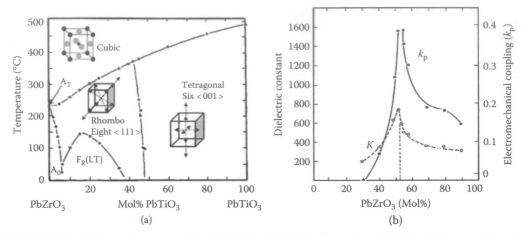

FIGURE 8.35 (a) Crystal structures for PZT, their dielectric constants, and (b) the changes in piezoelectric properties with composition. (Adapted from Moulson, A. J., and J. M. Herbert. 2003. *Electroceramics: Materials, Properties, and Applications.* New York: Wiley. With permission.)

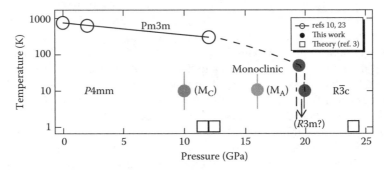

FIGURE 8.36 Changes in the crystal structure with increasing pressure in PT. (From Ahart, M., M. Somayazulu, R. E. Cohen, P. Ganesh, P. Dera, H. K. Mao, R. J. Hemley, et al. 2008. *Nature* 451:545–9. With permission.)

$PbTiO_3$. However, the tetragonal form of $PbTiO_3$ (shown as the space group P4mm) changes to the monoclinic phase (between 11 and 12 GPa; shown as M_c and M_A) with increasing pressure. Above 16 GPa, $PbTiO_3$ changes to the rhombohedral crystal structure (shown as the R3m space group).

Recent research suggests that the classic MPB seen in the PZT system is the result of "chemical pressure" that builds in $PbTiO_3$ as the lead ions are substituted with zirconium ions. This is similar to the development of strain-tuned ferroelectrics (see Section 8.18.1).

One of the concerns in the use of PZT materials is that they contain lead. A significant amount of recent research has been directed toward development of lead-free piezoelectrics (see Section 8.18.2).

8.14.1 PIEZOELECTRIC POLYMERS

Many polymers are piezoelectric (Table 8.4). The most widely used piezoelectric polymer is PVDF, which exhibits four crystalline structures. The beta (β) phase of PVDF exhibits a polar structure and is ferroelectric, pyroelectric, and piezoelectric. PVDF films are prepared by stretching or rolling the nonpolar alpha (α) phase sheets. During this process, the PVDF films undergo a phase transformation from the nonpolar α to the polar β phase. This phase transformation can also be achieved by annealing or by applying an electric field, also known as poling. Unlike PZT and $BaTiO_3$, the d_{31} coefficient for PVDF is positive, whereas the d_{33} coefficient is negative (Table 8.5). This means that

TABLE 8.4
Monomers or Repeat Units of Some Piezoelectric Polymers

Polymer	Repeat Unit	Polymer	Repeat Unit
Polyvinylidene fluoride	$[CH_2-CF_2]_n$	Nylon-7	$[-NH-(CH_2)_6-CO-]_n$
Polytrifluoro ethylene	$[CHF-CF_2]_n$	Nylon-9	$[-NH-(CH_2)_8-CO-]_n$
Polytetrafluoro ethylene, also known as Teflon	$[CF_2-CF_2]_n$	Nylon-11	$[-NH-(CH_2)_{11}-CO-]_n$
		Nylon-5,7	$[-NH-(CH_2)_5-NH-CO-(CH_2)_5-CO-]_n$

TABLE 8.5
Approximate Ranges of Values for the Piezoelectric, Dielectric, and Mechanical Properties of Copolymers in Comparison with PZT and Quartz

	PVDF	PVDF–55% TrFE	PVDF–75% TrFE	PZT–DOD-VI or PZT5H	Quartz
Piezoelectric					
d_{33} (pC/N or pm/N)	−35			+593	0
d_{31} (pC/N or pm/N)	+20	+25	+10	−274	0
d_h (pC/N or pm/N)	−3			45	
g_{31} (V · m/N)	174×10^{-3}	160×10^{-3}	110×10^{-3}	19.7×10^{-3}	0
h_{31} (V/m)	53×10^{-7}	19×10^{-7}	22×10^{-7}		
Coupling					
k_{33}				0.75	
k_{31}	0.10	0.07		0.39	0.09
Dielectric					
Dielectric constant	13	18	10	3400	4.5
tan δ (dielectric loss)	0.31		0.15		0.02
Curie temperature (T_c) (°C)				190	
Approximate operating temperature (°C)	80	70	100	110	573
Coercive field (kV/cm)					~6–8
Mechanical					
Compliance $\left(s_{33}^E\right)$ (constant electric field, i.e., closed circuit) (m²/N)	4.72×10^{-10}			20.7×10^{-12}	
Compliance $\left(s_{11}^E\right)$ (constant electric field, i.e., closed circuit) (m²/N)	3.65×10^{-10}			16.5×10^{-12}	
Density (kg/m³)	1.78×10^{-3}	1.9×10^{-3}	1.88×10^{-3}	7.45×10^{-3}	2.65×10^{-3}
Velocity of sound (m/s)	3×10^3		2×10^3	83.3	

Data from Nalwa, H. S. 1995. *Ferroelectric Polymers: Chemistry, Physics, and Applications*. New York: Marcel Dekker Inc; Buchanan, R. C. 2004. *Ceramic Materials for Electronics*. New York: Marcel Dekker; Moulson, A. J., and J. M. Herbert. 2003. *Electroceramics: Materials, Properties, and Applications*. New York: Wiley; and other sources.

TABLE 8.6

Piezoelectric and Other Properties of PVDF Films

Property	Coefficient	Biaxially Oriented Film	Uniaxially Oriented Film
Piezoelectric coefficients (pC/N)	d_{31}	4.34	21.4
	d_{32}	4.36	2.3
	d_{33}	−12.4	−31.5
	d_h	−4.8	−9.6
	d_{33} (calculated from d_h, d_{31}, and d_{32})	−13.5	−33.3
Pyroelectric coefficients (C/m^2K)	p_3	-1.25×10^{-5}	-2.74×10^{-5}
Thermal expansion coefficients (K^{-1})	α_1	1.24×10^{-4}	0.13×10^{-4}
	α_1	1.00×10^{-4}	1.45×10^{-4}
Mechanical properties	Compliance s_{11} (Pa^{-1})	4×10^{-10}	4×10^{-10}
	Compliance s_{12} (Pa^{-1})	-1.57×10^{-10}	-1.57×10^{-10}
	Poisson's ratio	0.392	0.392

Source: Nalwa, H. S. 1995. *Ferroelectric Polymers: Chemistry, Physics, and Applications.* New York: Marcel Dekker.

when a voltage is applied in the thickness direction (the poling direction, or direction 3), the PVDF film will *expand* in the thickness direction and shrink in the length direction. Also, although the piezoelectric coefficients for PVDF and related polymers (~10 pC/N) tend to be lower than those for ceramics (~10^2 pC/N; Table 8.3) by a factor of 10 to 50, the dielectric constants of PVDF and related materials are also lower than PZT and other ferroelectrics. This is why the voltage coefficients (i.e., the *g* values) of PVDF and related materials are similar to those for PZT and other ferroelectric ceramics (Table 8.5).

Blends of PVDF with trifluoroethylene (TrFE) and tetrafluoroethylene (TeFE), known as Teflon, are also piezoelectric. Similarly, odd-numbered and odd-odd nylons or polyamides (e.g., nylon 9 or nylon 5, 7) and polymers known as cyanopolymers (containing the C–CN group) are also piezoelectric. Ferroelectric nylons exhibit more useful piezoelectric properties (e.g., d_{31} ~15 pC/N) at higher temperatures (up to 200°C; Cheng and Zhang 2008). The monomers or repeat units of some of these are shown in Table 8.4.

In Table 8.5, the approximate ranges of selected piezoelectric properties of PVDF and some of its blends are shown. Note that these values are for illustration purposes only and should not be used for design. The properties are strongly dependent on the processing methods used, crystal structures, temperature, and frequency, and it is therefore important to get more accurate data from the suppliers. For example, the piezoelectric properties of a PVDF film depend strongly on the orientation of the films (Table 8.6). The following example illustrates the use of piezoelectric polymers.

EXAMPLE 8.5: POLYVINYLIDENE FLUORIDE FILM SENSOR

A commercially available poled PVDF film, electroded using silver metallic ink, is 110 μm thick. A stress of 20,000 Pa is applied to this film over a 1-in.^{2} area.

a. Assuming that the film has a rigid backing and deforms only in the thickness direction, what will be the open-circuit voltage generated?

b. If the film is now subjected to the same level of *force* as in part (a) but the film has a compliant backing, with a force acting on the cross section of the film such that the film stretches in the length direction, what will be the voltage generated across the thickness of the film? Assume that $g_{33} = 350 \times 10^{-3}$ and $g_{31} = 220 \times 10^{-3}$ V · m/N.

SOLUTION

a. The stress is acting on a 1-in.^{2} area, and because the film has a rigid backing, it can deform only along the thickness (direction 3). Thus, we must use the g_{33} coefficient.

From Equation 8.66, the electric field (E) generated $= -(350 \times 10^{-3}) \dfrac{V-m}{N} \times (20,000 \text{ Pa})$
$= -7000$ V/m.

The thickness of the film is 110 μm; therefore, the voltage generated across the film thickness will be

$$= (-7000 \text{ V/m}) \times (110 \times 10^{-6} \text{m}) = -0.77 \text{ V}$$

b. For this part, the film has a compliant, not rigid, backing and can change its length. We must use the g_{31} coefficient to calculate the electric field generated.

The example states that the *force* applied is the same as before. A stress of 20,000 Pa was previously applied over a 1-in.^{2} area. Let us first calculate the force that is applied.

The area was $= (1 \text{ in.}^2) \times (2.54 \text{ cm/in.})^2 \times (10^{-2} \text{ m/cm})^2 = 6.452 \times 10^{-4} \text{ m}^2$.

Thus, the force applied $= (20,000 \text{ N/m}^2) \times 6.452 \times 10^{-4} \text{ m}^2 = 12.903 \text{ N}$.

The same force is now applied over a cross section of the film that is 2.54 cm in length and 110 μm wide. The cross-sectional area is now $2.794 \times 10^{-6} \text{ m}^2$. This is a very small area. We now have a much higher level of stress for the same force as before:

$$= (12.903 \text{ N})/(2.794 \times 10^{-6} \text{ m}^2) = 4.62 \times 10^6 \text{ Pa}$$

The electric field generated across the thickness of the film now is

$$= -(220 \times 10^{-3} \text{ V} \cdot \text{m/N}) \times (4.62 \times 10^6 \text{ Pa}) = -1.02 \times 10^6 \text{ V/m}$$

Now note that this electric field still appears across the thickness of the film of 110×10^{-6} m. This is what the coefficient g_{31} represents.

Thus, the voltage generated will be $= (-1.02 \times 10^6 \text{ V/m}) \times (110 \times 10^{-6} \text{ m}) \sim -112$ V. This voltage is much higher because of the increased level of stress caused by the smaller cross-sectional area.

8.15 APPLICATIONS AND PROPERTIES OF HARD AND SOFT LEAD ZIRCONIUM TITANATE CERAMICS

Often, PZT ceramics are classified as soft PZT and hard PZT (Table 8.3). This terminology has its origins in the field of magnetism, wherein we refer to soft and hard magnets (Chapter 9). In general, soft PZTs have higher piezoelectric coefficient values. For example, DOD types II and VI are considered soft PZTs (Figure 8.37). These soft PZT compositions are donor-doped (e.g., Nb^{5+} substituting for Ti^{4+}; see Chapter 2) materials that have lower coercive fields (~6 – 15 kV/cm for PZT). We can move the domains in these materials relatively easily, so they are considered "soft" ferroelectrics or piezoelectrics. The donor-doping process creates positively charged point defects, on which the charge is balanced by cation vacancies, or defects with effective negative charge (Chapter 2). Typically, dielectric losses in these soft PZT materials are higher (tan δ ~ 0.02), since domain wall mobility is increased. Soft PZT materials are useful for low-power ultrasonic transducers, force and acoustic pickups, and so on.

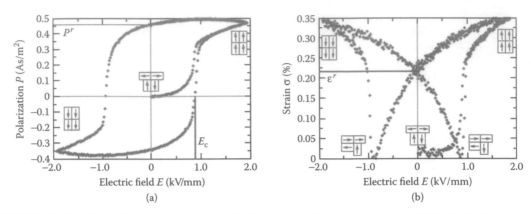

FIGURE 8.37 (a) *P–E* hysteresis loop and (b) the corresponding strain–electric field butterfly loop for a soft PZT ceramic DOD II or PIC 151. (From Schneider, G. A. 2007. *Annu Rev Mater Res* 37:491–538. With permission.)

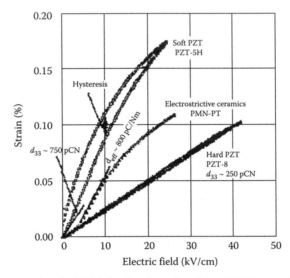

FIGURE 8.38 Strain versus electric field behavior for soft and hard PZT ceramics, compared to an electrostrictive PMN–PT ceramic. (From Park, S.-E., and T. R. Shrout. 1997. *J Appl Phys* 82(4):1804–11. With permission.)

Hard PZT compositions are acceptor-doped materials with relatively high coercive fields (~18 – 22 kV/cm). DOD types I and III are examples of hard PZT (Table 8.3). Their piezoelectric coefficients are smaller. Their coercivity is higher and their dielectric losses are smaller by an order of magnitude (tan δ ~ 0.001 – 0.004). Hard PZT materials are useful in applications such as resonant mode ultrasonic devices, piezomotors, and so on. Hard PZT materials are doped with acceptor dopants that create point defects with negative effective charge. This charge is balanced by the presence of oxygen ion vacancies, which are positively charged defects (Chapter 2).

The relationship of electric field versus strain for some soft and hard piezoelectrics are shown in Figure 8.38. This figure also shows data for some relaxor ferroelectrics, such as PMN–PT, that rely on electrostriction as a source for developing strain.

Soft piezoelectrics usually show higher piezoelectric coefficients. Their disadvantage is that their domain-wall motion is easy, so they tend to have higher dielectric losses. This means that when they are used as piezoelectric devices driven by an electric voltage, they will generate a considerable amount of heat. Another problem with soft piezoelectrics is that the electric-field-versus-strain relationship shows hysteresis. This means that when the material extends from one dimension to another and the field is removed, the material does not go back to its original dimension. This limits their applications as precise micropositioning devices.

Hard piezoelectrics show lower piezoelectric coefficients. Their advantage is that their dielectric losses are very small. These materials can be driven using higher electric fields and will not generate a lot of heat. They also do not exhibit much hysteresis in the electric-field-versus-strain relationship.

8.16 ELECTROMECHANICAL COUPLING COEFFICIENT

Consider a piezoelectric material that is subjected to a stress. An elastic strain will develop and the material will store some elastic energy. When the same piezoelectric material is subjected to an electrical field, it will develop a piezoelectric strain. This causes it to store additional energy. The *electromechanical coupling coefficient* (k) is defined as

$$k^2 = \frac{W_{12}}{W_1 W_2} \tag{8.47}$$

where W_{12}, W_1, and W_2 are the piezoelectric, mechanical, and electrical energy densities, respectively.

The strain development in a piezoelectric material as a function of stress (X) and electric field (E) was given by Equation 8.7.

For a stress δX, the total energy stored can be written as follows:

$$\frac{1}{2}\delta X \delta x = \frac{1}{2}s^E(\delta X)^2 + \frac{1}{2}d\delta X \delta E \tag{8.48}$$

$$\frac{1}{2}\delta X \delta x = W_1 + W_{12}$$

The first term, that is, $\frac{1}{2}s^E(\delta X)^2$, is the mechanical energy stored (W_1) and the second term, $\frac{1}{2}d\delta X\delta E$, is the piezoelectric energy stored (W_{12}).

The dielectric displacement D can be created or changed in a material by applying an electric field. The cause and effect are related by the dielectric constant. A change in the dielectric displacement can also occur by applying stress, which changes the distance between the entities that cause dipoles. Thus, the change in dielectric displacement is written as shown in Equation 8.9.

Now, if an electric field δE is applied, the total energy stored can be written as

$$\delta E \delta D = d\delta E \delta X + \varepsilon^X(\delta E)^2 \tag{8.49}$$

$$\frac{1}{2}\delta E \delta D = \frac{1}{2}d\delta E \delta X + \frac{1}{2}\varepsilon^X(\delta E)^2 \tag{8.50}$$

$$\delta E \delta D = W_{12} + W_2$$

From the definitions of W_1, W_2, and W_{12} in Equations 8.47, 8.49, and 8.50, we get

$$k^2 = \frac{d^2\delta E^2\delta X^2}{[s^E(\delta E)^2][\varepsilon^X(\delta E)^2]} \tag{8.51}$$

or

$$k = \frac{d}{\sqrt{s^E\varepsilon^X}} \tag{8.52}$$

Note that the compliance values for short-circuit (s^E) and open-circuit (s^D) conditions are very different.

We can now develop a relationship between these two as follows.

Now, $\delta x = s^E \delta X + d\delta E$, and also

$$\delta x = s^D \delta X + g\delta D \qquad (8.53)$$

Therefore, we get

$$s^D \delta X + g\delta D = s^E \delta X + d\delta E \qquad (8.54)$$

Substituting for δE (as given by Equation 8.12) and from Equations 8.12 and 8.54, we get

$$s^D \delta X + g\delta D = s^E \delta X + d(-g\delta X + \delta D/\varepsilon^X)$$
$$s^D \delta X + g\delta D = s^E \delta X - dg\delta X + (d/\varepsilon^X)\delta D \qquad (8.55)$$

Note that $g = d/\varepsilon^X$; therefore, simplifying the right-hand side of the second step in Equation 8.55, we get

$$s^D \delta X + g\delta D = s^E \delta X - \frac{d^2}{\varepsilon^X}\delta X + g\delta D$$

or

$$s^D = s^E - \frac{d^2}{\varepsilon^X} \qquad (8.56)$$

This can be rewritten as

$$s^D = s^E \left[1 - \frac{d^2}{s^E \varepsilon^X}\right] \qquad (8.57)$$

Recognizing that the last term in Equation 8.57 is k^2 (from Equation 8.52), we get

$$s^D = s^E [1 - k^2] \qquad (8.58)$$

Equation 8.58 is very important because it shows that the open- and closed-circuit compliances are related to each other via the electromechanical coupling coefficient. We can show that the stiffness (c) values under open- and closed-circuit conditions are also similarly related.

$$c^E = c^D [1 - k^2] \qquad (8.59)$$

Note that usually for stiffness, the left-hand side is the short-circuit value.

Instead of equating δx as we did to write Equation 8.54 and then deriving Equation 8.58, we can equate δE and show that

$$\varepsilon^x \approx \varepsilon^X (1 - k^2) \qquad (8.60)$$

This derivation involves an approximation that assumes that $k^4 \ll 1$.

We can rewrite Equation 8.60 as

$$k^2 = \frac{[\varepsilon^X - \varepsilon^x]}{\varepsilon^X} \tag{8.61}$$

Multiplying both numerator and denominator by $\frac{1}{2}E^2$, we get

$$k^2 = \frac{\left[\dfrac{1}{2}\varepsilon^X E^2 - \dfrac{1}{2}E^2\varepsilon^x\right]}{\dfrac{1}{2}\varepsilon^X E^2} \tag{8.62}$$

Equation 8.62 tells us that the electromechanical coupling coefficient can be written as the ratio of energy densities. The denominator is the total electrical energy that is stored in a piezoelectric body when a piezoelectric that is subjected to an electric field E is free to deform and is not constrained. Under these conditions, the stress X is maintained constant, and the strain x changes. The term $\dfrac{1}{2}E^2\varepsilon^x$ represents the electrical energy stored in a piezoelectric when the material is constrained such that the strain x is constant. Thus, the numerator term in Equation 8.62 represents the difference in electrical energy between a piezoelectric that is free to deform and a piezoelectric that is constrained. This is equal to the mechanical energy that results from the conversion of electrical energy and is stored in the piezoelectric body.

This stored mechanical energy can be recovered from the piezoelectric and used. The electrical energy stored in the piezoelectric can also be used. In this sense, the parameter k^2 should not be thought of as the efficiency in the same sense as we think of the efficiency of an engine. We define k^2 as the ratio of usefully converted energy to the input energy. Thus, for piezoelectric applications where the piezoelectric is used as a transducer, high coupling coefficients are desirable. If, for example, the value of k^2 is 0.7, this does not mean that the transducer is only 70% efficient.

If the piezoelectric material is clamped, it cannot develop any strain, that is, no mechanical deformation is allowed. Thus, no mechanical energy can be stored, so the only energy stored is electrical energy.

In Equation 8.62, the first term in the numerator is the total input electrical energy and the second term in the numerator is the electrical energy stored at constant strain (i.e., a clamped sample that cannot store any mechanical energy). The difference between these terms is in the conversion of the input electrical energy into mechanical energy. We define the ratio of the electrical energy converted into mechanical energy to the total input electrical energy, as the *effective coupling coefficient* (κ_{eff}), represented as follows:

$$\kappa_{eff}^2 = \frac{\text{input electrical energy converted into mechanical energy}}{\text{input electrical energy}} \tag{8.63}$$

Instead of starting with the relationship between compliances under open- and closed-circuit conditions (Equation 8.59), we can start with the stiffness under open- and closed-circuit conditions and show that this coefficient can also be defined as

$$\kappa_{eff}^2 = \frac{\text{input mechanical energy converted into electrical energy}}{\text{input mechanical energy}} \tag{8.64}$$

8.17   ILLUSTRATION OF AN APPLICATION: PIEZOELECTRIC SPARK IGNITER

Piezoelectrics are used in a variety of applications (Figure 8.29). Most applications involve a relatively sophisticated system that takes advantage of the direct or converse piezoelectric effect.

We have chosen a relatively simple application in the piezoelectric spark igniter to illustrate the applications of some of the equations derived so far. We will show that for a piezoelectric spark igniter made from a poled piezoelectric cylinder of diameter (d) and length (or height) l, the voltage generated (V) under open-circuit conditions is given by

$$V = g_{33} \frac{F \times l}{\pi d^2} \tag{8.65}$$

When the piezoelectric igniter is subjected to a compressive force F across a cross-sectional area,

$$A = \pi d^2$$

The definition of the piezoelectric voltage constant (g) is

$$g = -\left(\frac{\partial E}{\partial X}\right)_{D,T} = \left(\frac{\partial x}{\partial D}\right)_{X,T}$$

The electric field generated and the stress applied are related by the following expression:

$$E = -g \times X \tag{8.66}$$

Another relationship we have seen in Table 8.1 is

$$g_{33} = \frac{d_{33}}{\varepsilon_{33}^X} \tag{8.67}$$

Note that in this case, we are dealing with the stress (X) applied in direction 3, and the electric field is generated across the length of the cylinder; hence, we relate these properties by g_{33}.

Therefore,

$$E = -g_{33} \times \frac{F}{\pi d^2} \tag{8.68}$$

Representing the electric field generated across the length (i.e., height) of the cylinder as $E = V/l$, we get

$$V = -g_{33} \times \frac{F \times l}{\pi \times d^2} \tag{8.69}$$

This is written as follows:

$$V = \frac{d_{33}}{\varepsilon_{33}^X} \times \frac{F \times l}{\pi \times d^2} \tag{8.70}$$

When the piezoelectric igniter is first subjected to a compressive stress by using a force F on area A, the strain (x) developed under open-circuit condition (i.e., constant dielectric displacement D) is given by

$$x = -\frac{\delta l^D}{l} = -s_{33}^D \frac{F}{A} \tag{8.71}$$

The first part of this equation is the definition of strain. The second part relates the strain (x) to the stress (F/A) via compliance s_{33}^D under open-circuit conditions.

The mechanical work (w_m) done by the force F in creating a change in length (δl^D) is given by the equation

$$w_m = \frac{1}{2} F \delta l^D = \frac{1}{2} s_{33}^D \frac{F^2 l}{A} \tag{8.72}$$

From the definition of the electromechanical coupling coefficient (k^2), the result of this mechanical work done and the electrical energy available will be

$$w_{el} = \frac{1}{2} k_{33}^2 s_{33}^D \frac{F^2 l}{A} \tag{8.73}$$

We can also view the compressed piezoelectric cylinder as a capacitor with generated voltage V; for this capacitor with capacitance C and voltage V, the electrical energy stored in this capacitor is

$$w_{el} = \frac{1}{2} C V^2 \tag{8.74}$$

Equating these expressions for w_{el} and substituting for C in terms of dielectric permittivity and geometrical parameters, we get

$$w_{el} = \frac{1}{2} k_{33}^2 s_{33}^D \frac{F^2 l}{A} = \frac{1}{2} C V^2 = \frac{1}{2} \varepsilon_{33}^x \frac{A}{l} V^2$$

Solving for V,

$$k_{33}^2 s_{33}^D F^2 = \varepsilon_{33}^x \frac{A^2}{l^2} V^2$$

or

$$V^2 = \frac{k_{33}^2 s_{33}^D F^2 l^2}{\varepsilon_{33}^x A^2} = \frac{g_{33}^2 s_{33}^D \varepsilon_{33}^X F^2 l^2}{s_{33}^E \varepsilon_{33}^x A^2} \tag{8.75}$$

Substituting $\varepsilon_{33}^x \sim \varepsilon_{33}^X (1 - k_{33}^2)$ and $s_{33}^D = s_{33}^E (1 - k_{33}^2)$, we get the following expression:

$$V^2 = \frac{g_{33}^2 s_{33}^E (1 - k_{33}^2) \varepsilon_{33}^X F^2 l^2}{s_{33}^E \varepsilon_{33}^X (1 - k_{33}^2) A^2} \tag{8.76}$$

This simplifies to the following equation that we derived previously:

$$V = g_{33} \frac{Fl}{A} = g_{33} \frac{Fl}{\pi d^2} \tag{8.77}$$

If this voltage V generated is high enough to overcome the gap between the two electrodes connected to the piezoelectric, a spark will be created between the ends of the wires connected to the circular electroded faces of the piezoelectric. If this happens, we have a change from an open-circuit condition to a closed-circuit condition. The compliance of the material undergoes a change to its short-circuit value $\left(s_{33}^D\right)$ from its open-circuit value $\left(s_{33}^D\right)$. Because the short-circuit compliance is smaller, this transition allows for the development of an additional strain. Let us assume that if we were to apply a force F on area A under a closed circuit, then the total change in length would be δl^E. What we considered before was the strain x developed under open-circuit conditions, that is, the constant dielectric displacement. Let us refer to this change in length under open-circuit conditions as δl^D.

When we apply a force F on an area A under open-circuit conditions, the displacement will be δl^D. If a spark is generated such that a closed-circuit condition is created, an additional change in length will occur, with a magnitude of $(\delta l^E - \delta l^D)$.

This additional change in length resulting from the transition from closed- to open-circuit conditions will be given by

$$\delta l^E - \delta l^D = \left(s_{33}^E - s_{33}^D \right)\frac{F \times l}{A} = k_{33}^2 s_{33}^E \frac{Fl}{A} \tag{8.78}$$

The corresponding additional mechanical energy stored with this extra deformation, resulting from the formation of spark causing the closed-circuit condition, will be given by the following equation:

$$w_{\text{mechanical,extra}} = \frac{1}{2} F(dl^E - dl^D) = \frac{1}{2} F\left(k_{33}^2 s_{33}^E \frac{Fl}{A} \right) \tag{8.79}$$

This corresponds to the extra electrical energy (using the definition of coupling coefficient) as shown below:

$$w_{\text{electrical,extra}} = k_{33}^2 \frac{1}{2} F\left(k_{33}^2 s_{33}^E \frac{Fl}{A} \right) = \frac{1}{2} k_{33}^4 s_{33}^E \frac{F^2 l}{A} \tag{8.80}$$

Thus, the total energy stored in a piezoelectric before and after the short-circuit condition will be (from Equations 8.73 and 8.80):

$$w_{\text{total}} = \frac{1}{2} k_{33}^2 s_{33}^D \frac{F^2 l}{A} + \frac{1}{2} k_{33}^4 s_{33}^E \frac{F^2 l}{A} \tag{8.81}$$

or

$$w_{\text{total}} = \frac{1}{2} k_{33}^2 \left(s_{33}^D + k_{33}^2 s_{33}^E \right)\frac{F^2 l}{A} \tag{8.82}$$

In Equation 8.82, if we substitute $s_{33}^D = s_{33}^E[1 - k_{33}^2]$ from Equation 8.58, we get

$$w_{\text{total}} = \frac{1}{2} k_{33}^2 \left(s_{33}^E - s_{33}^E k_{33}^2 + k_{33}^2 s_{33}^E \right)\frac{F^2 l}{A}$$

$$= \frac{1}{2} k_{33}^2 s_{33}^E \frac{F^2 l}{A}$$

Substituting for $k_{33}^2 = \dfrac{d_{33}^2}{s_{33}^E \varepsilon_{33}^X}$, we get

$$w_{\text{total}} = \frac{1}{2}\left(\frac{d_{33}^2}{s_{33}^E \varepsilon_{33}^X} \right) s_{33}^E \frac{F^2 l^2}{A} = \frac{1}{2}\frac{d_{33}^2}{\varepsilon_{33}^X}\frac{F^2 l^2}{A} = \frac{1}{2}\frac{d_{33}^2}{\varepsilon_{33}^X}\frac{F^2 l}{A} = \frac{1}{2} d_{33} \frac{g_{33}\varepsilon_{33}^X}{\varepsilon_{33}^X}\frac{F^2 l}{A}$$

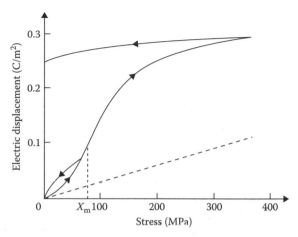

FIGURE 8.39 Increase in the dielectric displacement in certain PZT compositions plotted as a function of the compressive stress. (From Moulson, A. J., and J. M. Herbert. 2003. *Electroceramics: Materials, Properties, and Applications.* New York: Wiley. With permission.)

Therefore,

$$w_{total} = \frac{1}{2} d_{33} g_{33} \frac{F^2 l}{A} \tag{8.83}$$

In reality, the electrical energy stored in a piezoelectric igniter is even higher than this, because compressive stress causes an additional increase in the dielectric displacement (D). This occurs because some of the so-called 90° ferroelectric domains switch their orientation under the influence of the compressive stress. This increase in dielectric displacement as a function of the compressive stress is shown in Figure 8.39.

EXAMPLE 8.6: CALCULATION OF THE VOLTAGE OF A PIEZOELECTRIC IGNITER

A poled PZT cylinder with diameter 2 mm and thickness 5 mm is to be used as a spark igniter. What voltage is generated by applying a compressive force of 100 N to the circular cross section of this disk? Assume that the d_{33} for this PZT is 289 pC/N and the dielectric constant in a free, that is, the unclamped, state (ε_{r33}^X) is 1300.

SOLUTION

The electric field generated and the stress applied are related by Equations 8.66 and 8.67 Therefore, for this ceramic, the g_{33} will be given by

$$g_{33} = \frac{289 \times 10^{-12}\,C/N}{1300 \times 8.85 \times 10^{-12}\,F/m} = 0.0255\ V/N = 25.5\ mV/N \tag{8.84}$$

The area of the circular cross section is given by $\pi d^2/4 = 3.1416 \times 10^{-6}$ m².

The stress applied is the force divided by the area of the circular cross section and works out to 31,830,988.6 Pa or 31.83 MPa.

The electric field generated = −(31,830,988.6 Pa) × (0.0255 V/m) = 799,578.941 V/m or 0.799 MV/m.

This field runs across the height of the piezoelectric spark igniter; thus, the voltage this field generates is = (799,578.941 V/m) × (5 × 10⁻³ m) = 3997 V. If the applied stress is compressive, the voltage generated is in the same direction as the poling voltage (Figure 8.25).

Thus, a relatively small force (~100 N) that can be generated by a human hand leads to the generation of 4000 V. If the ends of this cylinder are connected by wires and if the two wires are brought within a few millimeters of each other but are not allowed to touch, a spark jumps between the tips of the wire. Although this voltage is high, the amount of charge accumulated because of the direct piezoelectric effect and the resultant current is actually quite small and is not likely to be very dangerous.

8.18   RECENT DEVELOPMENTS

8.18.1   Strain-Tuned Ferroelectrics

One of the limitations of materials developed to date is that most of them are not useful in high-temperature applications due to depoling. *Strain-tuned ferroelectrics* are ferroelectric materials or devices whose dielectric properties have been altered due to a built-in strain (Schlom et al. 2007). This approach could lead to a better high-temperature performance of current ferroelectrics.

Strain is used as a method for controlling the electrical properties of materials or devices in several semiconductors (e.g., silicon–germanium) and has recently been applied to ferroelectrics. One of the limitations of many well-known ferroelectrics is that their Curie temperatures (T_c) are low. For example, the T_c in $BaTiO_3$ is only ~120°C. Thus, $BaTiO_3$ loses its ferroelectric and piezoelectric behavior at temperatures greater than 120°C. $BaTiO_3$ epitaxial thin films were recently grown on $ReScO_3$ substrates (Re indicates a rare-earth element). The lattice-constant mismatch (Figure 8.40) leads to the development of a biaxial compressive strain of ~1.7% in these films. This in turn causes the remnant polarization to increase. More importantly, the T_c of such films increases by almost 500°C. Thus, strain-tuned ferroelectrics offer ways to use existing ferroelectric and piezoelectric materials at higher temperatures. In strain-tuned $BaTiO_3$, the increase in T_c also offers the hope of using this environmentally-friendly material in applications instead of lead-containing piezoelectrics such as PZT.

8.18.2   Lead-Free Piezoelectrics

Considerable research has been directed toward the development of lead-free piezoelectrics. Many countries around the world have directives requiring strict control of hazardous materials arising from all sources, including electronics, automotives, and consumer appliances. This is especially important for materials that contain toxic elements such as lead. The two major classes of materials that have been developed are potassium sodium niobate ($K_{0.5}Na_{0.5}NbO_3$, KNN) and sodium

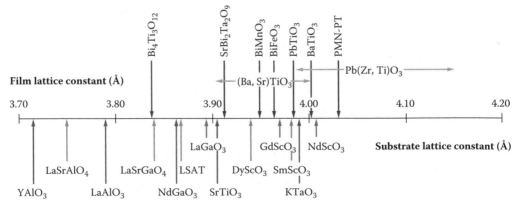

FIGURE 8.40   Lattice constants of several pseudocubic ferroelectrics (above the horizontal line) compared to the lattice constants of several substrate materials. (From Schlom, D. G. et al. 2007. *Annu Rev Mater Res* 37:589–626. With permission.)

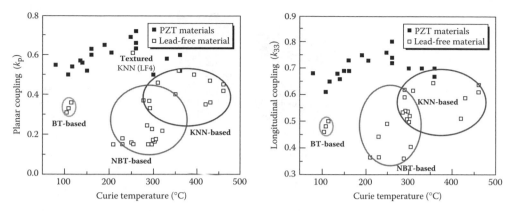

FIGURE 8.41 Room-temperature values of planar (k_p) and longitudinal (k_{33}) coupling coefficients. (From Shrout, T. S., and S. J. Zhang. 2007. *J Electroceram* 19:111–24. With permission.)

bismuth titanate ($Na_{0.5}Bi_{0.5}TiO_3$, NBT). The dielectric constants and the d_{33} coefficients for some of the lead-free piezoelectrics were shown in Figures 8.32 and 8.33, respectively. In these diagrams, BT is $BaTiO_3$, and LF refers to liquid flux–grown crystals, which refers to a method for growing single crystals.

The planar coupling coefficients (k_p) and k_{33} values for these materials are shown in Figure 8.41.

8.19 PIEZOELECTRIC COMPOSITES

Of all of the piezoelectric materials that have been developed, there is no single "perfect" or "ideal" material. Piezoelectric composites have been developed in order to try to optimize the properties of different piezoelectric and nonpiezoelectric materials.

One major advantage of polymer–ceramic piezoelectric composites is that they are quite flexible. The so-called macro fiber composites (MFCs; Figure 8.42) of ceramic ferroelectrics such as PZT, a brittle material, are arranged in a polymer matrix (the polymer is ferroelectric or linear dielectric; Sodano et al. 2004). Such MFCs have applications in the control of vibrations in defense-related applications like aircrafts and machine guns, and in civil structures like buildings and cable-stayed bridges (Song et al. 2006). One limitation in the structural applications of MFCs is that piezoelectric actuators require a source of power, which may be disrupted during certain events, such as earthquakes, when control of vibration is needed most.

MFC structures have been used for vibration control in "smart" tennis rackets, "smart" skis, and other applications. Unlike applications involving vibration control in large structures, the main advantage composites offer in such applications is that they function without the need for an external power source.

Since the dielectric constant of polymers is relatively low, ceramics such as PZT are embedded in the polymer mix. The piezoelectric voltage constant or the g coefficient of such composites is higher (Equation 8.67). This is similar to a point that has already been made: that PVDF has a lower d coefficient compared to PZT. However, its g coefficient is comparable to that of PZT because PVDF has a lower dielectric constant compared to PZT.

Another advantage of composites is that polymer composites can lower overall acoustic impedance. The acoustic impedance of water is ~1.5 MRayl. Many polymers have an acoustic impedance of ~3.5 MRayl. Most muscle, fat, tissue, and so on have an acoustic impedance of ~1.3–1.7 MRayl. Bones have an acoustic impedance of ~3.8–7.4 MRayl (Gururaja 1996). PZT has a high acoustic impedance (~35 Mrayl). If PZT is used by itself for ultrasound detection or imaging, there is a tremendous acoustic impedance mismatch at the PZT–biological tissue or PZT–water interfaces

FIGURE 8.42 Photograph of a PZT-based flexible macrofiber composites. (Courtesy of Smart Material Corporation, Sarasota, Florida.)

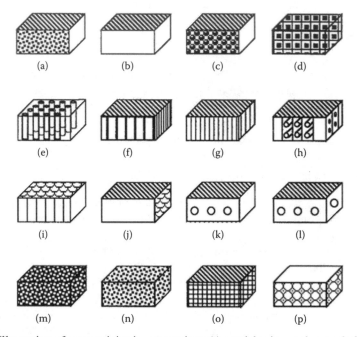

FIGURE 8.43 Illustration of connectivity in composites: (a) particles in a polymer: 0–3; (b) PVDF composite: 0–3; (c) PZT spheres in a polymer: 1–3; (d) diced composite: 1–3; (e) PZT rods in a polymer: 1–3; (f) sandwich composite: 1–3; (g) glass–ceramic composite: 1–3; (h) transverse reinforced composite: 1–2–0; (i) honeycomb composite: 3–1P; (j) honeycomb composite: 3–1S; (k) perforated composite: 1–3; (l) performed composite: 3–2; (m) replamine composite: 3–3; (n) burps composite: 3–3; (o) sandwich composite: 3–3; (p) ladder-structured composite: 3–3. (From Neelkanta, P. 1995. *Handbook of Electromagnetic Materials*. Boca Raton, FL: CRC Press. With permission.)

(Harvey et al. 2002). This impedance mismatch causes high reflection or "ringing" of the signal. To better match (i.e., to lower) the acoustic impedance, composites of PZT fibers with epoxy are used. These can be prepared by arranging the fiber bundles and then filling them with epoxy, called "arrange and fill." This is followed by a dicing step in which cubic blocks are obtained by cutting the cured material perpendicular to the fiber length direction. This provides the so-called 1–3 *connectivity* (Figure 8.43). The effect of the arrangement of different phases is known as connectivity, and certain properties in composites of all types depend not only on the relative volume fractions but also on the geometry that is used to connect the different phases.

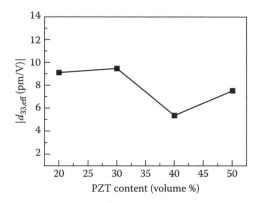

FIGURE 8.44 Effective piezoelectric coefficients of PVDF–PZT composites. Note the minimum in piezoelectric $d_{33,eff}$ coefficient at about 40% (vol/vol) PZT. (From Dietze, M., and M. Es-Souni. 2008. *Sens Actuators* A143:329–34. With permission.)

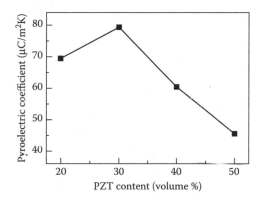

FIGURE 8.45 Effective pyroelectric coefficient (p_{eff}) of PVDF–PZT composites. (From Dietze, M., and M. Es-Souni. 2008. *Sens Actuators* A143:329–34. With permission.)

We can develop composites using different types of piezoelectric materials. As mentioned in Section 8.10, PZT–PVDF composites are formed to decrease the piezoelectric effect in PVDF-based infrared (IR) detectors or imaging elements (Section 8.20). The mechanism by which piezoelectricity occurs in PVDF is different from that in PZT, hence the piezoelectric coefficients of PVDF and PZT are $d_{33} \sim -30$ pC/N (note the negative sign) and $\sim +200$ to $+400$ pC/N, respectively. Thus, we can create composites of PZT particles/crystals dispersed in a PVDF matrix (the 0–3 composite; Figure 8.43a), in which the overall piezoelectric effect is substantially reduced (Dietze and Es-Souni 2008). This helps make a better pyroelectric detector that responds more to temperature change than to mechanical shock or vibration.

The effective piezoelectric coefficients of such PZT–PVDF–TrFE composites are shown in Figure 8.44. The corresponding changes in the effective pyroelectric coefficients of such composites are shown in Figure 8.45.

8.20 PYROELECTRIC MATERIALS AND DEVICES

A pyroelectric material is one that shows temperature-dependent spontaneous polarization (Lang 2005). To understand the origin of the pyroelectric effect, consider a disk of poled single-crystal or polycrystalline $BaTiO_3$ such that the polarization axis is perpendicular to the electrodes.

In the poled material or a single crystal, we have dipoles throughout the volume of the material. These dipoles lead to a certain level of bound charge density on the surface, attracting nearby charges, such as ions or electrons, toward the faces of the material. This is how the pyroelectric effect was discovered in the aluminum borosilicate mineral tourmaline (then known as *lyngourion* in Greek or *lyncurium* in Latin) by the Greek philosopher Theophrastus almost 2400 years ago! These external charges are attracted to the pyroelectric material and neutralize the bound charges on the surfaces of the pyroelectric crystal. If the temperature is constant ($dT/dt = 0$) and the electrodes on this pyroelectric $BaTiO_3$ sample are connected to an ammeter, no current will flow through the sample because no free charges are left behind on the surfaces (Figure 8.46).

If the temperature is increased ($dT/dt > 0$), the dipole moment for the ferroelectric material decreases. This means that the total polarization (P) decreases. Consequently, the total bound charge density on the surfaces of the ferroelectric material also decreases (Figure 8.46). This means that there will now be free charge on the surfaces of the pyroelectric material. As shown in Figure 8.46, the negative charges (i.e., electrons) flow from the top surface toward the ammeter, while a conventional current would flow in the opposite direction (Figure 8.46c). The current that flows when the temperature of a pyroelectric material is changing is known as the *pyroelectric current*.

If we further cool the sample (i.e., $dT/dt < 0$), the ferroelectric polarization will increase (Figure 8.46b). This will cause an increase in the bound charge density, and electrons will flow from the wire toward the pyroelectric material surface in order to compensate for the increased bound charge density. Thus, we will again see a pyroelectric current to flow, but its direction will be opposite to that when the material was being heated ($dT/dt > 0$). This is analogous to the development of a transient current due to the development of piezoelectric voltage (Figure 8.24).

If we do not connect the surfaces of the pyroelectric material to an ammeter or some other resistor using conductive wires, then there will be no electrical current. Instead, we can measure a pyroelectric voltage. Again, the sign of this voltage will change as we heat or cool the material.

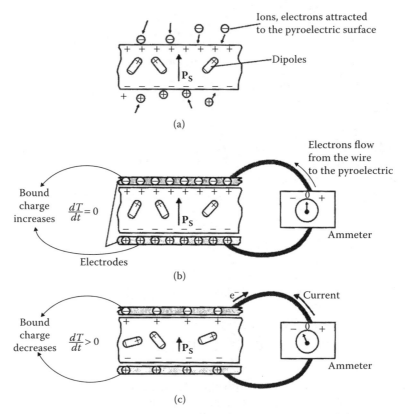

FIGURE 8.46 Origin of the pyroelectric effect. (From Lang, S. B. 2005. *Phys Today* (August): 31–5. With permission.)

If the pyroelectric material is perfectly insulated from its surroundings so that the charges cannot flow in an external circuit, then the charges built on the surface will eventually be neutralized by the intrinsic conductivity of the pyroelectric material. This means the pyroelectric charge developed will leak away inside the material.

The *pyroelectric coefficient* (*p*) of a material is defined as the change in the dielectric displacement (*D*) caused by a unit change in temperature (*T*).

$$dD = pdT \tag{8.85}$$

The change in dielectric displacement (*D*) or saturation polarization (P_s) as a result of the temperature change is known as the *primary pyroelectric effect*.

Because the units of *D* are in C/m^2, the units of pyroelectric coefficient (*p*) will be C m^{-2}K^{-1}. If we apply an electric field (*E*) or a stress (*X*) to a material, the dipole moment will also change.

The dielectric displacement (*D*) for a material is given by

$$D = \varepsilon_0 E + P_{total} \tag{8.86}$$

The total polarization originates from the induced polarization ($P_{induced}$) and ferroelectric polarization (P_s). Thus, we can write

$$D = \varepsilon_0 E + P_{induced} + P_s \tag{8.87}$$

Now, $(\varepsilon_0 E + P_{induced}) = \varepsilon E$, where ε is the permittivity of the pyroelectric material; we can rewrite Equation 8.87 as follows:

$$D = P_s + \varepsilon E \tag{8.88}$$

Taking the derivative of both sides with respect to temperature (*T*), we get

$$\left(\frac{\partial D}{\partial T}\right) = \left(\frac{\partial P_s}{\partial T}\right) + E\left(\frac{\partial \varepsilon}{\partial T}\right) \tag{8.89}$$

The true pyroelectric coefficient (*p*) is defined as

$$p = \left(\frac{\partial P_s}{\partial T}\right)_{X,E} \tag{8.90}$$

In this definition, the constant stress (*X*) means that the material is not clamped and is free to expand or contract. Thus, this definition of the true pyroelectric coefficient includes both the so-called primary pyroelectric effect and the *secondary pyroelectric effect*. In the secondary pyroelectric effect, the changes in temperature or electric field cause a change in the dimension of the material (i.e., contraction or expansion due to thermal expansion or via the direct piezoelectric effect). The strain (*x*) is then not constant. The change in dimensions, from either thermal expansion/contraction or the piezoelectric effect of the applied electric field, causes changes in both the dipole moment and the bound charge density, and causes the development of a pyroelectric current or pyroelectric voltage.

Secondary pyroelectric effect should *not* be confused with a correction to the true pyroelectric coefficient (*p*) because of the temperature dependence of permittivity (i.e., $(\partial \varepsilon/\partial T)$; Equation 8.91). We can write a *generalized pyroelectric coefficient* (p_g) as

$$p_g = p + E\left(\frac{\partial \varepsilon}{\partial T}\right) \tag{8.91}$$

The effect of the change in dielectric permittivity with temperature $(\partial\varepsilon/\partial T)$ also results in the development of a change in the bound charge density, which can induce the development of a current. This is *not* considered the true pyroelectric effect because this change in the dielectric constant with temperature occurs in all dielectrics and not just in polar dielectrics. In ferroelectrics, the term $(\partial\varepsilon/\partial T)$ can be quite high; this effect is comparable to the true pyroelectric effect.

To accurately measure the primary pyroelectric effect, we must consider the effect of a temperature change on spontaneous polarization (P_s) while maintaining a constant strain (x), that is, the material is not allowed to change its dimension and it is rigidly clamped. The primary pyroelectric effect is defined as the change in saturation polarization (P_s) with temperature (T) when we keep the sample dimensions constant. This means we do not include any effects of the change in dielectric displacement (D) due to a change in the dielectric permittivity (ε) with temperature (T) or electric field (E).

It is usually very difficult to isolate and measure only the primary pyroelectric effect. The secondary pyroelectric effect can be calculated using the thermal expansion coefficient, Young's modulus, and the piezoelectric coefficient. The total pyroelectric effect (due to both the primary and secondary effects) is typically measured. The values of the pyroelectric coefficients for some materials are shown in Table 8.7; the total pyroelectric coefficients represent the algebraic sum of the primary and secondary coefficients.

TABLE 8.7
Pyroelectric Coefficients of Some Materials

Material	Primary Pyroelectric Coefficient ($\mu Cm^{-2} \cdot K^{-1}$)	Secondary Pyroelectric Coefficient ($\mu Cm^{-2} \cdot K^{-1}$)	Total Pyroelectric Coefficient ($\mu Cm^{-2} \cdot K^{-1}$)
Ferroelectrics			
Poled $BaTiO_3$	−260	+60	−200
Poled $PbZr_{0.95}Ti_{0.05}O_3$	−305.7	+37.7	−268
Poled $PbTiO_3$			−180
Single crystal			
$LiNbO_3$	−95.8	+12.8	−83
$LiTaO_3$	−175	−1	−176
$PbGe_3O_{11}$	−110.5	+15.5	−95
$Ba_2NaNb_5O_{11}$	−141.7	+41.7	−100
$Sr_{0.5}Ba_{0.5}Nb_2O_6$	−502	−48	−550
$(CH_2CF_2)n$	−14	−13	−27
Triglycine sulfate	+60	−330	−270
Not ferroelectrics			
Single crystal			
CdSe	−2.94	−0.56	−3.5
CdS	−3.0	−1.0	−4.0
ZnO	−6.9	−2.5	−9.4
Tourlamine	−0.48	−3.52	−4.0
$Li_2SO_4.2H_2O$	+60.2	+26.1	+86.3

Sources: Lang, S. B. 2005. *Phys Today* (August): 31–5; Gupta, M. C., and J. Ballato. 2007. *The Handbook of Photonics*. Boca Raton, FL: CRC Press; and Herbert, J. M. 1982. *Ferroelectric Transducers and Sensors*. New York: Gordon and Breach Science Publishers.

For a pyroelectric crystal or poled material with a metalized electrode area A (perpendicular to the poling axis), the pyroelectric current is given by

$$I = A\left(\frac{dP_s}{dt}\right) \tag{8.92}$$

This can be rewritten as follows:

$$I = A\left(\frac{dP_s}{dT}\right)\left(\frac{dT}{dt}\right) \tag{8.93}$$

From this equation and Equation 8.90, we get

$$I = A \times p \times \left(\frac{dT}{dt}\right) \tag{8.94}$$

If the temperature change (ΔT) is small, then the pyroelectric coefficient (p) can be assumed to be constant over a small temperature range. Under these conditions, the pyroelectric current will be proportional to the rate of change of temperature (heating or cooling, as in Equation 8.94).

8.20.1 PYROELECTRIC DETECTORS FOR INFRARED DETECTION AND IMAGING

An important application of pyroelectric materials is to detect IR radiation. There are two "windows" in the atmosphere in which IR radiation can travel through air without significant absorption (primarily by water vapor) and without being scattered by dust particles. One is in the range of 3–5 μm, and the other between 8 and 14 μm. At a temperature of ~300 K, objects and warm-blooded animals emit heat in the form of IR radiation of a wavelength (λ) ~10 μm. Pyroelectrics can be used to make IR detectors. We can use IR radiation detectors to get a thermal image of an intruder hiding behind a wall or tree, as long as there is a temperature difference between the body temperature of the intruder and his or her surroundings. This is one approach to making pyroelectric IR detection and imaging systems. Such pyroelectric material–based IR detectors are known as *thermal detectors*.

Another approach to making IR detectors is to use narrow-bandgap semiconductors (such as mercury cadmium telluride or HCT). These detectors, also known as *quantum detectors* or *photon detectors*, operate on the principle that an electron from the valence band is promoted to the conduction band by absorbing a photon (IR radiation, in this case). Such detectors are extremely sensitive in the far-IR range. However, detectors based on narrow bandgap semiconductors require cooling to liquid nitrogen temperatures (~77 K). If we do not cool the narrow-bandgap semiconductor, there is enough thermal energy at room temperature to flood the conduction band simply by the thermal excitation of electrons from the valence band into the conduction band.

8.21 PROBLEMS

8.1 In Figure 8.17, the coercive voltage is ~3.0 V. What will the thickness of the film be if the corresponding coercive field is 70 kV/cm?

8.2 For ferroelectric materials, is the saturation polarization (P_s) a microstructure-sensitive property? Explain.

8.3 For ferroelectric materials, is the coercive field a microstructure-sensitive property? Explain.

8.4 What will happen if a piezoelectric ferroelectric material used in a device, such as an actuator, gets exposed to temperatures that are close to the Curie temperature or Curie-temperature range?

8.5 Explain how the piezoelectric and electrostriction effects can be used to create mirrors, whose surface can be adjusted using an electric voltage (known as adaptive optics). What materials will you use for this application, and why?

8.6 PZN and PZN-8%–PT (PZN–8PT) single crystals of relaxor ferroelectric materials oriented in the (001) direction were investigated by Park et al. (1996) for their potential as electromechanical actuators. The percent strain developed in direction 3 is shown as a function of the electric field, also in direction 3 (Figure 8.47). From these data, show that the d_{33} coefficients are approximately 1700 and 1150 pC/N for PZN–8PT and PZN samples, respectively, as marked on Figure 8.47. What would the g_{33} values be if the dielectric constants for PZN–8PT and PZN materials are 4200 and 3600, respectively?

8.7 The electric field–strain response for the PZN and PZN–8PT samples is linear. The dielectric loss (tan δ) values for these samples are ~0.008 and 0.012 and are considered rather low. How is the linear relationship between the electric field–strain response connected to the dielectric loss? Explain.

8.8 Ultrahigh strain actuators based on relaxor ferroelectrics were investigated by Park and Shrout (1997). The data on PMN–PT and soft and hard PZT samples are shown in Figure 8.48.

    (a) Show that for soft PZT samples, the d_{33} coefficient is ~700 pC/N. Use the initial linear portion of the data shown in Figure 8.48.

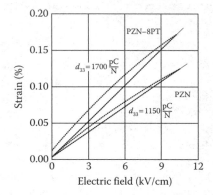

FIGURE 8.47 Strain–electric field data for PZN and PZN (001)-oriented single crystals. (From Park, S.-E. et al. 1996. *IEEE Int Symp Appl Ferroelectr* 1:79–82. With permission.)

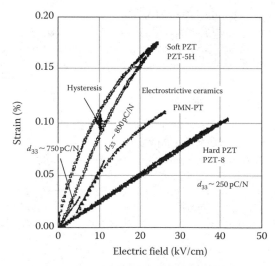

FIGURE 8.48 Strain–electric field relationship for PMN-PT and soft and hard PZT. (From Park, S.-E., and T. R. Shrout. 1997. *J Appl Phys* 82(4): 1804–11. With permission.)

(b) What does the nonlinear nature of the electric field–strain curve tell us about the domain motion in these materials?

(c) As we can see from the data, there is a significant hysteresis in the strain–electric field behavior. What does this say about the suitability of this material for making an accurate positioning device?

(d) Would the dielectric losses in soft PZT–SH be relatively low or high? What is the problem if the dielectric losses are high?

8.9 From the data shown in Figure 8.48, show that the d_{33} coefficient for the hard PZT sample will be ~250 pC/N. What is the advantage of this material if it is used as an actuator or positioning device, compared to the soft PZT material?

8.10 From the data shown in Figure 8.48, show that the d_{33} coefficient for the electrostrictive PMN–PT ceramic sample is ~800 pC/N.

8.11 There is virtually no hysteresis in PMN–PT. Thus, the dielectric losses will be expected to be smaller than those for the soft PZT sample, and the micropositioning that can be achieved using these materials will be quite accurate. What are the limiting factors for using this material for micropositioning applications? What will happen if we attempt to induce larger strains using higher electric fields (e.g., to go up to 0.15% strain)?

8.12 A poled piezoelectric cylinder made using PZT–DOD I material, of 5-mm diameter and 12-mm height, is subjected to a force of 700 N. What is the voltage generated across the height of the cylinder? What is the electrical energy (in millijoules) stored when the cylinder is under open-circuit conditions? What is the extra electrical energy stored when the cylinder undergoes additional compression under closed-circuit conditions? Use the properties of PZT–DOD I type included in Table 8.3.

8.13 A piezoelectric PVDF thin film is 4 cm in length, 2 cm in width, and 10 μm in thickness. Assume that the film is poled in the thickness direction. What will be the change in the dimension of the film along its length if a voltage of 150 V is applied in the thickness direction? What will be the change in the thickness of the film under the same conditions? Assume that $d_{31} = 25$ pC/N and $d_{33} = -35$ pC/N.

8.14 A piezoelectric plate is made from poled polycrystalline DOD type II ceramic (Table 8.3). The original dimensions of the plate are $9 \times 4 \times 1$ mm, and those of the electrodes are 9×4 mm. If a voltage of 50 V is applied along the thickness direction, what are the new dimensions of this plate?

8.15 Piezoelectrics are used in dental and bone surgery. This involves using ultrasonic vibrations to cut tissue. Which piezoelectric effect, direct or converse, is used for this application? A frequency of 25 to 29 kHz is used to make micromovements of about 60–210 μm. These movements cut only mineralized tissue. In this frequency range, the neurovascular and other soft tissues are not cut (Labanca 2008).

8.16 Vibrations in civil structures, such as buildings and bridges, can be controlled using piezoelectric materials formed into multilayer actuators or amplified piezoelectric actuators (Song et al. 2006). How do you think such actuators can be used for the control of vibrations in these structures?

8.17 $BaTiO_3$, one of the first ferroelectric ceramic materials developed, is widely used to make multilayer capacitors. However, this material has no widespread applications as a piezoelectric. Explain.

8.18 Piezoelectric fiber macrocomposites are used in some tennis rackets. Explain how such piezoelectrics work to reduce the transmission of vibrations from the racket to the player's elbow.

8.19 Can we use a lead-free PZT material that has been optimized for outstanding piezoelectric properties as a pyroelectric detector? Explain.

8.20 Find out how the focusing problems associated with the Hubble space telescopes were corrected using piezoelectrix actuators.

GLOSSARY

Aging: The small level of randomization undergone with time by domains in a freshly poled piezo-electric after the poling electric field has been removed. This causes a decrease in the piezoelectric properties with time.

Butterfly loop: A diagram showing the development of strain in a ferroelectric material (Figure 8.26b).

Closed-circuit compliance (s^E): The compliance of a material when the electrodes providing the field for the piezoelectric effect are short-circuited (i.e., $E = 0$). This is the same as short-circuit compliance, and is related to open-circuit compliance (s^D) by the following equation

$$s^D = s^E[1 - k^2]$$

Coercive field (e_c): The electric field necessary to cause domains in a ferroelectric with some remnant polarization to be randomized again in order to get a state of zero net polarization (Figure 8.15).

Connectivity: A particular geometrical arrangement of the different phases in a composite; affects many properties (but not all) of a composite (Figure 8.43).

Converse piezoelectric effect: Generation of a mechanical strain by applying an electric field to a poled piezoelectric material; also known as the motor effect (Figure 8.25), it results in mechanical strain and is characterized by the d coefficient. The d coefficient, also known as the piezoelectric charge constant, is equal to the polarization induced per unit stress.

Curie-point shifters: Compounds added to a ferroelectric to lower its Curie-temperature.

Curie-point suppressors: Compounds added to a ferroelectric to broaden its Curie transition so that the dielectric constant is relatively stable with temperature.

Curie temperature: The temperature at which a ferroelectric material transforms into a centro-symmetric paraelectric form. Relaxor ferroelectrics have a Curie-temperature range.

Curie–Weiss law: The variation in the dielectric constant of the paraelectric phase at temperatures above T_0 (or T_c), which is written as

$$\varepsilon = \varepsilon_0 + \frac{C}{(T - T_0)} \quad \text{(applicable for } T > T_0)$$

or

$$\varepsilon = \frac{C}{(T - T_0)} \quad \text{(applicable for } T > T_0)$$

Also stated as "the inverse of dielectric susceptibility (χ_e) varies linearly with temperatures above T_c or T_0."

Depoling: The randomization of previously aligned domains by the application of higher tempera-tures or stress. Piezoelectric activity can cease to exist, and it limits the applications of piezoelectrics to relatively low temperatures.

Direct piezoelectric effect: Generation of a voltage or charge by applying a stress to a poled piezoelectric, also known as the generator effect, and characterized by the g coefficients (Figure 8.25). It is also known as the piezoelectric voltage constant. It is also equal to the strain induced per unit dielectric displacement (D) applied.

Electromechanical coupling coefficient (k): The ratio of piezoelectric energy density (W_{12}) stored in a material to the product of the electrical- (W_1) and mechanical-energy (W_2) densities stored; is given by

$$k^2 = \frac{W_{12}}{W_1 W_2}$$

Electrostriction: The elastic strain (x) induced by the application of an electric field, which is proportional to the *square* of the electric field (E). This effect is seen in all materials and is embedded in the strain developed in ferroelectric piezoelectric materials.

Ferroelectric domains: A region of a ferroelectric material in which the polarization of all cells is in the same direction (Figure 8.6).

Ferroelectrics: Materials that show a spontaneous and reversible polarization.

Generalized pyroelectric coefficient (p_g): A pyroelectric coefficient that includes the primary and secondary coefficients, in addition to a correction needed because of the dependence of permittivity on temperature. It is given by the following equation:

$$p_g = p + E\left(\frac{\partial \varepsilon}{\partial T}\right) \tag{8.95}$$

Generator effect: See **Direct piezoelectric effect**.

Hard piezoelectrics: Compositions of lead zirconium titanate (PZT; often acceptor-doped) or other materials that offer lower piezoelectric coefficients, low dielectric losses, a higher coercive field, and little or no hysteresis in the strain–electric field relationship. These materials are well-suited for micropositioning and other applications that may require higher electric fields.

Hydrostatic piezoelectric coefficients: Piezoelectric coefficients under hydrostatic conditions that are designated with a subscript (e.g., d_h or g_h); the product $d_h \times g_h$ is the figure of merit that is important for applications including underwater sonar probes, hydrophones, and imaging applications.

Hysteresis loop: The trace of change in polarization or dielectric displacement for a ferroelectric material as a function of the electric field (Figure 8.15).

Morphotropic phase boundary (MPB): The boundary on a phase diagram across which there is a change in the crystal structure of a piezoelectric material; for compositions of ceramics, such as PZT, which are at or near the MPB, the dielectric properties are maximized. Recent research has raised questions about the existance of such a boundary in the PZT system.

Nonlinear dielectrics: Materials in which the polarization developed is not linearly related to the electric field. The dielectric constant of these materials is field-dependent; these include ferroelectrics and other materials, such as water, in which molecules have a permanent dipole moment.

Open-circuit compliance (s^D): The compliance of a material under open-circuit conditions, related to the short-circuit compliance by the following equation

$$s^D = s^E[1 - k^2]$$

Paraelectric phase: The high-temperature phase derived from an originally ferroelectric material's parent phase, which now has no dipole moment per unit cell. This phase behaves as a linear dielectric.

Photon detectors: IR detectors based on the creation of an electron–hole pair in narrow-bandgap semiconductors by absorption of IR radiation. They are also known as quantum detectors and are fundamentally different from thermal detectors.

Piezoelectric: A material that develops an electrical voltage or charge when subjected to stress and a relatively large strain when subjected to an electric field.

Piezoelectric charge constant: The d coefficient that describes the converse piezoelectric effect. This is the strain generated per unit electric field applied or the induced polarization per unit stress; the units are m/V or C/N.

Piezoelectric voltage constant: The g coefficient that describes the direct piezoelectric effect. This is the induced electric field per unit stress or strain induced per unit dielectric displacement (D) applied.

Poled ferroelectric: A ferroelectric material in which the domains are aligned in a particular direction; only a poled ferroelectric that shows a measurable piezoelectric effect.

Poling: The process of applying an electric field to a ferroelectric or piezoelectric material in order to cause the alignment of domains. This process is typically carried out at higher temperatures, often using a heated oil bath.

Primary pyroelectric effect: The change in dielectric displacement (D) or saturation polarization (P_s) as a result of the temperature change while maintaining a constant strain (x), that is, with the material clamped.

Pseudocubic: The cubic crystal structure polymorph of materials such as $BaTiO_3$.

Pyroelectric current: The current that flows when the temperature of a pyroelectric material is changing.

Pyroelectric coefficient (p): The change in dielectric displacement (D) caused by a unit change in temperature (θ).

Quantum detectors: Pyroelectric infrared detectors. See **Thermal detectors**.

Relaxor ferroelectrics: Ceramic materials with high dielectric constants, high electrostriction coefficients, high-frequency dispersion, and a broad Curie transition, such as lead magnesium niobate (PMN). They are useful in actuator applications.

Remnant polarization (P_r): The polarization that remains after removing the electric field from a ferroelectric material that has been subjected to sufficiently high fields to cause saturation.

Saturation polarization (P_s): The maximum possible ferroelectric polarization that can be obtained for a given ferroelectric material.

Secondary pyroelectric effect: The change in the spontaneous polarization of a material with a change in temperature (which causes thermal expansion or contraction) or applied electric field (which causes a change in dimension because of the piezoelectric effect); here, the strain is not constant.

Short-circuit compliance: See **Closed-circuit compliance**.

Soft piezoelectrics: Materials with smaller coercive fields and higher d_{33} values that exhibit easy domain switching, causing dielectric losses and hysteresis in the electric field–strain relationships. They are typically donor-doped, such as PZT.

Strain-tuned ferroelectrics: Ferroelectric materials or devices whose properties are altered because of strain created during their processing.

Thermal detectors: IR detectors based on the pyroelectric effect. These are different from the quantum or photon detectors, which are based on generation of an electron–hole pair in narrow-bandgap semiconductors.

Transducer: A device that can convert one form of energy into another.

Unpoled state: A ferroelectric or piezoelectric material that has not been subjected to the poling process.

Virgin state: A ferroelectric or piezoelectric material that has not been subjected to any electric field and therefore has a random configuration of domains and no net polarization.

Weiss domains: See **Ferroelectric domains**.

REFERENCES

Ahart, M., M. Somayazulu, R. E. Cohen, P. Ganesh, P. Dera, H. K. Mao, R. J. Hemley, et al. 2008. Origin of morphotropic phase boundaries in ferroelectrics. *Nature* 451:545–9.

Askeland, D., and P. Fulay. 2006. *The Science and Engineering of Materials*. Washington, DC: Thomson.

Berlincourt, D. 1971. In *Ultrasonic Transducer Materials: Piezoelectric Crystals and Ceramics*, ed. O. E. Mattiat, Ch. 2. London: Plenum Press.

Buchanan, R. C. 2004. *Ceramic Materials for Electronics*. New York: Marcel Dekker.

Chen, W., X. Yao, and X. Wei. 2007. Relaxor behavior of (Sr, Ba, Bi)TiO3 ferroelectric ceramic. *Solid State Commun* 141:84–8.

Cheng, Z., and Q. Zhang. 2008. Field-activated electroactive polymers. *MRS Bull* 33:183–7.

Cross, E. 2004. Lead-free at last. *Nature* 432:24–5.

Dietze, M., and M. Es-Souni. 2008. Structural and functional properties of screen-printed PZT–PVDF–TrFE composites. *Sens Actuators* A143:329–34.

Gupta, M. C., and J. Ballato. 2007. *The Handbook of Photonics*. Boca Raton, FL: CRC Press.

Gururaja, T. R. 1996. Piezoelectric transducers for ultrasonic imaging. *IEEE Trans*.

Harvey, C. J., J. M. Pilcher, R. J. Eckersley, M. J. K. Blomley, and D. O. Cosgrove. 2002. Advances in ultrasound. *Clin Radiol* 57:157–77.

Hench, L. L., and J. K. West. 1990. *Principles of Electronic Ceramics*. New York: Wiley.

Herbert, J. M. 1982. *Ferroelectric Transducers and Sensors*. Gordon and Breach Science Publishers.

Ignatiev, A., Y. Q. Xu, N. J. Wu, D. Liu. 1998. Pyroelectric, ferroelectric, and dielectric properties of Mn- and Sb-doped PZT thin films for uncooled IR detectors. *Materials Science and Engineering B* 56(2):191-4.

Jaffe, B., W. Cook, and H. Jaffe. 1971. *Piezoelectric Ceramics*. New York: Academic Press.

Kepler, R. G. 1995. In *Ferroelectric Polymers: Chemistry, Physics, and Applications*, ed. H. S. Nalwa. New York: Marcel Dekker.

Labanca, M., F. Azzola, R. Vinci, L. F. Rodella. 2008. Piezoelectric surgery: Twenty years of use. *British Journal of Oral and Maxillofacial Surgery* 46(4): 265–9.

Lang, S. B. 2005. Pyroelectricity: From ancient curiosity to modern imaging tool. *Phys Today* (August): 31–5.

Mehling, V., C. Tsakmakis, and D. Gross. 2007. Phenomenological model for the macroscopical material behavior of ferroelectric ceramics. *J Mech Phys Solids* 55:2106–41.

Moheimani, S. O. R., and A. J. Fleming. 2006. *Piezoelectric Transducers for Vibration Control and Damping*. Berlin: Springer.

Morgan Technical Ceramics, Guide to Piezoelectric and Dielectric Ceramics. http://www.morganelectroceramics.com/pzbook.html

Moulson, A. J., and J. M. Herbert. 2003. *Electroceramics: Materials, Properties, Applications*. New York: Wiley.

Nalwa, H. S. 1995. *Ferroelectric Polymers: Chemistry, Physics, and Applications*. New York: Marcel Dekker.

Neelkanta, P. 1995. *Handbook of Electromagnetic Materials*. Boca Raton, FL: CRC Press.

Osone, S., K. Brinkman, Y. Shimojo, and T. Iijima. 2008. Ferroelectric and piezoelectric properties of $Pb(Zr_xTi_{1-x})O_3$ thick films prepared by chemical solution deposition process. *Thin Solid Films* 516:4325–4329.

Park, S.-E., and T. R. Shrout. 1997. Ultrahigh strain and piezoelectric behavior in relaxor based ferroelectric single crystals. *J Appl Phys* 82(4):1804–11.

Park, S.-E., M. L. Mulvihill, P. D. Lopath, M. Zipparo, and T. R. Shrout. 1996. Crystal growth and ferroelectric related properties of $(1-x)Pb(Zn_{1/3}Nb_{2/3})O_3-x$ $PbTiO_3$ (A _ Zn^{2-}, Mg^{2-}). *IEEE Int Symp Appl Ferroelectr* 1:79–82.

Pilgrim, S. *Piezoelectric Materials: An Unheralded Component*, 11th ed. FabTech. Online at http://www.fabtech.org/white_papers/_a/piezoelectric_materials_an_unheralded_component.

Schlom, D. G., L.-Q. Chen, C.-B. Eom, K. M. Rabe, S. K. Streiffer, and J.-M. Triscone. 2007. Strain tuning of ferroelectric thin films. *Annu Rev Mater Res* 37:589–626.

Schneider, G. A. 2007. Influence of electric field and mechanical stresses on the fracture of ferroelectrics. *Annu Rev Mater Res* 37:491–538.

Shrout, T. S., and S. J. Zhang. 2007. Lead-free piezoelectric ceramics: alternatives for PZT? *J Electroceram* 19:111–24.

Sodano, H. A., G. Park, and D. J. Inman. 2004. An investigation into the performance of macro fiber composites for sensing and structural vibration applications. *Mech Syst Signal Process* 18:683–97.

Song, G., V. Sethi, and H.-N. Li. 2006. Vibration control of civil structures using piezoceramic smart materials: a review. *Eng Struct* 28:1513–24.

Xu, Y. 1991. *Ferroelectric Materials and Their Applications*. Amsterdam: North Holland.

9 Magnetic Materials

KEY TOPICS

- Origin of magnetism
- Diamagnetic, paramagnetic, and superparamagnetic materials
- Ferromagnetic, ferrimagnetic, and antiferromagnetic materials
- Ferromagnetic hysteresis loop
- Soft and hard magnetic materials
- Technologies and devices based on magnetic materials

9.1 INTRODUCTION

The word "magnet" has its origin in a magnetic material known as magnetite, which is a form of iron oxide or lodestone used as a magnetic compass. Lodestone was mined in the province of Magnesia. It is believed that among the minerals found in Magnesia (a part of Macedonia), magnesium carbonate was white, manganese dioxide was brown, and the third—magnetite—was black iron oxide. The latter was probably the first material known to be magnetic.

As we will see, in reality, *all* materials in this world are magnetic, that is, they respond to magnetic fields in some fashion. One objective of this chapter is to introduce the fundamental concepts related to magnetic materials. In this regard, we will explore the origin of magnetism in materials. We will define different types of magnetism in materials that include ferromagnetic and ferrimagnetic materials. The second objective is to explore different technologies based on the use of magnetic materials, including magnetic materials used in information storage (e.g., magnetic hard disks). We will also briefly mention materials called *multiferroics*. These materials simultaneously exhibit two or more switchable properties, such as ferroelectric and ferromagnetic behaviors.

Most concepts regarding magnetic materials and technologies would be better followed if you have already learned the basics of linear dielectric materials and ferroelectric materials from Chapter 8. You would start to recognize that many of the equations we would deal with here for magnetic materials are very similar to those used for ferroelectrics. Thus, the so-called *phenomenology* underlying the group of dielectrics and ferroelectrics and that of ferromagnetic and ferrimagnetic materials is similar. However, it is important to keep in mind that although many of the phenomena and equations appear very similar, the *physical origins of magnetic and dielectric behavior are quite different.*

9.2 ORIGIN OF MAGNETISM

All materials are magnetic. This means that every material responds to an externally applied magnetic field in a specific manner. The origin of magnetism lies in a very basic principle, that is, a moving charge (namely, current) produces a magnetic field. This connection between an electrical current and magnetism was probably first noted by Hans Oersted, who discovered that a compass needle placed next to a current-carrying wire was deflected in a perpendicular direction. Thus, it was found that a current-carrying coil behaved similar to a bar magnet and produced a magnetic moment (μ), which was found to be equal to the product of the current (I) and the area (A) of the loop (Figure 9.1). The unit of magnetic moment is ampere·square meter. Please note the symbol A is used for both area and ampere.

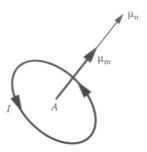

FIGURE 9.1   Magnetic moment (μ) induced by a current loop. (From Kasap, S. O. 2002. *Principles of Electronic Materials and Devices.* New York: McGraw Hill. With permission.)

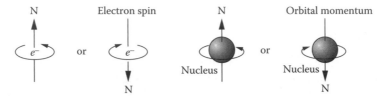

FIGURE 9.2   Orbital and spin motions of an electron. (From Askeland, D., and P. Fulay. 2006. *The Science and Engineering of Materials.* Washington, DC: Thomson. With permission.)

The origin of magnetic behavior can be developed from two types of electron motions, namely spin and orbital (Figure 9.2). These motions can be compared to the motion of the earth that spins around its own axis while simultaneously having an orbit around the sun.

These motions of the electrons create a spin magnetic moment and an orbital magnetic moment. Niels Bohr first considered the magnetic behavior of an atom based on the planetary or orbital motion of electrons that resulted in the generation of a magnetic field, and proposed the so-called Bohr's theory of magnetism. Later on, it was shown that the spin magnetic moment plays a critical role in determining the magnetic properties of materials.

A charged particle rotating in an orbit creates a magnetic moment (μ), which is given by

$$\mu = \frac{q \times L}{2 \times m} \tag{9.1}$$

In this equation, q is the charge, L is the angular momentum, and m is the mass of the charged particle. Bohr showed that the angular momentum of an electron is quantized.

Another parameter known as the gyromagnetic ratio (γ) is defined as

$$\gamma = \frac{q}{2m} \tag{9.2}$$

The basic unit of charge we often use is electronic charge ($q_e = 1.6 \times 10^{-19}$ C). Similarly, we define a basic unit of magnetic moment. This is known as a *Bohr magneton* (μ_B):

$$\mu_B = \frac{h}{2\pi} \frac{q_e}{2m_e} = \hbar \frac{q_e}{2m_e} = 9.274 \times 10^{-24} \text{ A} \cdot \text{m}^2 \tag{9.3}$$

In this equation, $\hbar = h/2\pi$, where h is the Planck's constant, q_e is the electronic charge, and m_e is the mass of the electron.

Defining the Bohr magneton as a unit for the magnetic moment allows us to express the magnetic moment of an ion, atom, or a molecule. Now, let us see how we can calculate the total magnetic moment of a material. The total magnetic dipole moment per unit volume of a material is known as the *magnetization* (*M*). We can calculate the magnetization of a material by calculating the vector sum of different magnetizations of electrons and nuclei that make up the atom. The magnetization that develops in a material is similar to the development of polarization (*P*) in dielectrics, because the unit of the magnetic dipole moment is ampere square meter, and the unit of magnetization will be ampere square meter/cubic meter or ampere/meter.

The nucleus does *not* have a planetary motion. However, it does have a spin. There is a magnetic moment associated with the spin of the nucleus. However, the magnetic moment due to nuclear spin is very small ($\sim 10^{-3}$ μ_B) and can be ignored while calculating the total magnetic moment of an *atom* or an *ion*. Important applications where we make use of the nuclear magnetic moment are the techniques of nuclear magnetic resonance (NMR) and magnetic resonance imaging (MRI).

Thus, to calculate the total magnetic moment of an atom, we need to add up the magnetic moments of all the electrons. In doing so, we need to account for the magnetic moments due to both the angular motion and the spin of the electrons (Figure 9.2). The expressions for the magnetic moments involve concepts of quantum mechanics. Additional information regarding these concepts can be found in physics textbooks.

What is important is to recognize that most "free" atoms possess a net magnetic moment. The word "free" here means that atoms are not bonded to other atoms, that is, they are isolated and do not form part of any solid or liquid. Although most *free* atoms have a magnetic moment, most often (but not always), when atoms bond to other atoms to form a solid or liquid, the electrons of different atoms interact, and the resultant material has no net magnetic moment. This is why, although all atoms contain electrons and each electron can be viewed as a magnet, most materials behave essentially as nonmagnetic materials.

Nevertheless, in a few materials, such as transition elements (with incomplete 3d, 4d, or 5d orbitals), lanthanides (partially filled 4f orbital), or actinides (partially filled 5f orbital), some of the electron shells are incomplete. In some of these materials, even after the atoms form a solid, the atoms in the resultant structure have a net effective magnetic moment. Similarly, ions of many elements may also have a net effective magnetic moment even when they are part of a material.

Thus, in an element such as manganese (Mn, atomic number $Z = 25$), with an electronic configuration of $1s^2 2s^2 2p^6 3s^2 3p^6 3d^5 4s^2$, up to and including the level 3p, the shells are completely filled. The rules of quantum mechanics tell us that before filling up the 3d sublevel, the 4s level must be filled. After filling up the 3p orbital, we first add two electrons to the 4s level, and these are paired. This means that the only difference between the quantum numbers of the electrons in the 4s level is that one electron has an upward spin ($\uparrow$) whereas the other has a downward spin ($\downarrow$). Five more electrons remain to be filled in this case. From quantum mechanics, the rule for filling the 3d subshell is that all electrons must remain unpaired until the subshell is half-filled; five electrons can thus have the same spin. Therefore, in a manganese atom, there are five unpaired electrons. The spin of each unpaired electron produces a magnetic moment of one Bohr magneton. Therefore, a "free" Mn atom has a net magnetic moment of $5\,\mu_B$ (Table 9.1).

However, when Mn atoms form an Mn crystal, the magnetic moments of the spin and the orbital motions cancel each other out. Thus, the atoms in a manganese *crystal* do *not* have a net magnetic moment. Similarly, in other compounds of Mn, bonding will have to be accounted for to find whether the resultant material will have any net magnetic moment. Most often, the spin and orbital magnetic moments are canceled out or "quenched" when elements form bonds with other atoms, resulting in zero net magnetic moment for most materials. From a practical viewpoint, these materials (e.g., copper, aluminum, silica, and polyethylene) are considered nonmagnetic. When we say that a material is not magnetic or that it is nonmagnetic, what is meant is that the material has no net magnetic moment. Most materials have a relatively weak (but nonzero) response to an external magnetic field.

We now consider another example, namely, iron (Fe). The electronic structure of an iron atom (atomic number $Z = 26$) is *expected* to be $1s^2 2s^2 2p^6 3s^2 3p^6 3d^8$. However, the actual electronic structure

TABLE 9.1

Pairing of 3d and 4s Electrons in Different Transition Elements

Element	Atomic Number (Z)	3d-Electron Spin Pairing					4s-Electron Spin Pairing
Sc	21	↑					↑↓
Ti	22	↑	↑				↑↓
V	23	↑	↑	↑			↑↓
Cr	24	↑	↑	↑	↑	↑	↑
Mn	25	↑	↑	↑	↑	↑	↑↓
Fe	26	↑↓	↑	↑	↑	↑	↑↓
Co	27	↑↓	↑↓	↑	↑	↑	↑↓
Ni	28	↑↓	↑↓	↑	↑	↑	↑↓
Cu	29	↑↓	↑↓	↑↓	↑↓	↑↓	↑

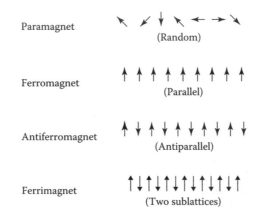

FIGURE 9.3 Schematic representation of the alignment of magnetic moments in different materials. (From Jiles, D. C. 2001. *Introduction to the Electronic Properties of Materials*. Cheltenham, UK: Nelson Thornes. With permission.)

of iron is $1s^2 2s^2 2p^6 3s^2 3p^6 3d^6 4s^2$. This is because the 4s orbital (sublevel) is filled first. Thus, in this electronic configuration, the 3d orbital (capable of accepting 10 electrons) again remains incomplete. Out of these six electrons, the first five electrons in the 3d level can remain unpaired. However, one more electron is left over to be filled. Because the 3d orbital is now half-full, the sixth electron pairs up with one of the five electrons in the 3d level. Thus, in an iron (Fe) atom, four electrons remain unpaired (Table 9.1).

Each unpaired electron produces a magnetic dipole moment equal to one Bohr magneton (μ_B). Consequently, an iron atom has a net magnetic moment of 4 μ_B. In iron, when the atoms form a phase with a particular crystal structure, some, but not all, of these magnetic moments are quenched, that is, canceled. Therefore, a material such as iron has a net magnetic moment. In iron, all the magnetic moments of the atoms are in the same direction. A material in which all the magnetic moments of atoms or ions are aligned in the same direction is known as a *ferromagnetic material* (Figure 9.3).

In some materials (e.g., iron oxide [Fe_3O_4]), the magnetic moments of some of the ions are in opposite directions or are antiparallel. However, they are not completely canceled out. A material in which the magnetic moments of atoms or ions are antiparallel but are not canceled out is known as

TABLE 9.2
Magnetic Moments of Some Atoms and Ions

	Magnetic Moment (μ_B)
Atom	
Iron	4
Cobalt	3
Nickel	2
Ion	
Fe^{2+} (ferrous)	4
Fe^{3+} (ferric)	5
Co^{2+}	3
Ni^{2+}	2
Mn^{2+}	5
Mg^{2+}, Zn^{2+}, and Li^+	0

Source: Goldman, A. 1999. *Handbook of Modern Ferromagnetic Materials*. Boston: Kluwer. With permission.

a *ferrimagnetic material*. The alignment of electron spins for atoms in different types of magnetic materials is shown schematically in Figure 9.3.

Note that ferromagnetic or ferrimagnetic materials do *not* have to contain iron or other ferromagnetic elements (e.g., Ni, Co, Gd). For example, Cu_2MnAl, ZrZn, and InSb are ferromagnetic, even though the latter two are ferromagnetic only at very low temperatures (O'Handley 1999).

If the spin magnetic moments due to different ions or atoms are completely canceled out, the material is known as *antiferromagnetic* (e.g., Cr, α-Mn, and MnO). Magnetic moments associated with some atoms and ions are shown in Table 9.2.

The following examples illustrate how to calculate the magnetic moments of ions.

EXAMPLE 9.1: MAGNETIC MOMENTS OF FERROUS (Fe^{2+}) AND FERRIC (Fe^{3+}) IONS

The atomic number of iron (Fe) is 26. From Table 9.1, what are the magnetic moments of (a) ferrous (Fe^{2+}) and (b) ferric (Fe^{3+}) ions?

SOLUTION

a. As seen before, the electronic structure of an iron atom is $1s^2 2s^2 2p^6 3s^2 3p^6 3d^6 4s^2$, with four unpaired electrons in the 3d level. When a ferrous ion (Fe^{2+}) is formed, two electrons are removed from an iron atom. This occurs by removing the two electrons that belong to the 4s energy level. Thus, the electronic configuration of the divalent ferrous (Fe^{2+}) ion is $1s^2 2s^2 2p^6 3s^2 3p^6 3d^6$. The first five electrons in the 3d level remain unpaired. The sixth electron is paired with one of the five. This again leaves us the Fe^{2+} ion with four unpaired electrons. Consequently, the magnetic moment of a free ferrous (Fe^{2+}) ion is 4 μ_B.

b. For the ferric (Fe^{3+}) ion, we can think of starting with an Fe^{2+} ion and removing one more electron (or, starting with a neutral iron atom and taking out a total of three electrons). This third electron comes from one of the paired electrons in Fe^{2+} to give us a total of five unpaired electrons. Thus, the magnetic moment of a trivalent or ferric ion (Fe^{3+}) is 5 μ_B. In Example 9.2, we will see how the magnetic moments of the Fe^{2+} and Fe^{3+} ions can be used to calculate the total magnetization per unit volume in Fe_3O_4.

EXAMPLE 9.2: FERRIMAGNETISM IN IRON OXIDE

From the magnetic moments of the ferrous and ferric ions, calculate the net magnetic moment per formula in ferrimagnetic iron oxide (Fe_3O_4).

SOLUTION

Another way to write the formula for this material is $FeO:Fe_2O_3$. This representation distinguishes between the divalent and trivalent forms of iron. For one formula unit of this material, the number of unpaired electrons will be 4 (from Fe^{2+}) and 10 (five each from a total of two Fe^{3+} ions). Thus, the total number of unpaired electrons is $4 + 10 = 14$ μ_B. However, this material is ferrimagnetic.

In Fe_3O_4, one-half of the ferric (Fe^{3+}) ions occupy the so-called tetrahedral sites (Chapter 2). The other half of the Fe^{3+} ions occupy the so-called octahedral sites. The spin magnetic moments of the ferric ions located in these different crystallographic sites are antiparallel, and they cancel each other out. The only magnetic moments remaining in Fe_3O_4 are due to the divalent ferrous ions that occupy the octahedral sites (same type of site that was occupied by one-half of the Fe^{3+} ions). Thus, in Fe_3O_4, the actual magnetic moment per unit formula will be only 4 μ_B since there is only one Fe^{2+} ion per formula unit.

9.3 MAGNETIZATION (*M*), FLUX DENSITY (*B*), MAGNETIC SUSCEPTIBILITY (χ_m), PERMEABILITY (μ), AND RELATIVE MAGNETIC PERMEABILITY (μ_r)

We now examine some of the relationships between the applied magnetic field (the cause) and the magnetization created (the effect). The equations will be similar to those we saw in Chapter 8 for dielectrics. However, there are fundamental and important differences between the origins of these phenomena. For example, ferromagnetic behavior is seen in amorphous materials, because the magnetic response is linked to the motion of electrons. Ferroelectricity, on the contrary, is seen only in crystalline materials or crystalline regions in materials such as polyvinylidene fluoride (PVDF). Ferromagnetism and ferrimagnetism are seen in materials that are either conductors or insulators, whereas ferroelectric behavior is seen only in insulating materials. Some of these similarities and differences will become clearer as we discuss these materials in further detail.

When a magnetic field (*H*) is applied to a material, it creates magnetization (*M*). This is equivalent to the application of an electric field (*E*), which creates a polarization (*P*). When we use the term "magnetic field" (*H*), we are referring to an *externally generated magnetic field*, such as that created using either a current-carrying coil or a permanent magnet.

The applied magnetic field (*H*) results in the creation of a *magnetic flux density* (*B*) or *magnetic induction* inside the material. When a magnetic flux (ϕ) is created by the applied field (*H*), there would be a certain number of magnetic flux lines per unit area. The number of flux lines per unit area is known as the magnetic flux density (*B*) or magnetic induction. Consider the flux density as the intensity of magnetic field at any given location in a magnetic material. The unit for flux density is Weber/square meter, written as Wb/m^2, also equivalent to Tesla.

The equivalent of the magnetic flux density generated due to the magnetic field is the dielectric displacement (*D*) created by the application of an electric field. This equivalence between different electric and magnetic quantities is shown in Table 9.3a.

9.3.1 MAGNETIC FIELD (*H*), MAGNETIZATION (*M*), AND FLUX DENSITY (*B*)

A magnetic field is often created by passing electrical current through a wire. Uniform magnetic fields can be created using a toroidal solenoid (Figure 9.4a) or a solenoid made using an insulated conductive wire wrapped around a cylinder (Figure 9.4b). Consider a coil of length l with n turns, carrying a current I; then, the magnetic field created is given by

$$H = \frac{nI}{l} \tag{9.4}$$

$$If\ N = \frac{n}{l},\ then \tag{9.5}$$
$$H = N \times I$$

TABLE 9.3A

Equivalence between Properties and Phenomena of Dielectric and Magnetic Materials

Electric	Magnetic
Electric field (E)	Magnetic field (H)
Polarization (P)	Magnetization (M)
Dielectric displacement (D)	Magnetic induction or flux density (B)
Dielectric susceptibility (χ_e)	Magnetic susceptibility (χ_m)
Dielectric constant (ε_r)	Relative magnetic permeability (μ_r)
Ferroelectricity	Ferromagnetism, Ferrimagnetism
Linear dielectric behavior	Paramagnetism, Diamagnetism
—	Superparamagnetism
Electrostriction	Magnetostriction
Piezoelectricity	Piezomagnetism

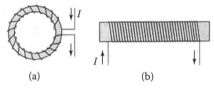

(a)   (b)

FIGURE 9.4 Illustration of (a) a toroidal solenoid and (b) a solenoid made using a current-carrying coil. (From du Trémolet de Lacheisserie, E., D. Gignoux, and M. Schlenker. 2002. *Magnetism: Fundamentals*. Berlin: Springer. With permission.)

Recall the relationship between the dielectric displacement (D), polarization (P), and electric field (E), which is

$$D = \varepsilon_0 E + P \tag{9.6}$$

In magnetic materials, the magnetic field (H) is the *cause*, and the *effect* is magnetization (M). The amount of magnetization (M) created within a material will depend on both the composition of the material, that is, how susceptible the material is to a magnetic field, and the strength of the applied field.

We can show that the magnetic flux density created within the material (B_{int}) is related to the magnetization M and magnetic field H by the following equation:

$$B_{int} = \mu_0 H + \mu_0 M \tag{9.7}$$

In this equation, the magnetic permeability of free space (μ_0) is given by

$$\mu_0 = 4\pi \times 10^{-7}\,\text{Wb/A} \cdot \text{m} \ \ \text{or} \ \ \text{H/m}$$

Because the unit of magnetic induction (B) is Weber/square meter (Wb/A $\cdot$ m), and the unit of magnetic field (H) is ampere/meter, the unit for magnetic permeability (μ) is Weber/ampere $\cdot$ meter or Henry/meter (H/m). Also, note that earlier in Section 9.2, the symbol μ has been used for the magnetic moment.

The first term on the right-hand side of Equation 9.7, that is, $\mu_0 H$, can be described as the external magnetic induction B_0.

Thus, the total magnetic induction created inside the material (B_{int}) is a sum of two factors, namely, the externally applied induction and the magnetization created inside the material, that is, Equation 9.7.

In vacuum, there is no material, and hence, magnetization created inside a current-carrying coil is zero ($\mu_0 M = 0$). Then, the magnetic flux density (B_{int}) and the applied magnetic field (H) are related by the following equation:

$$B = \mu_0 H \text{ (for vacuum)} \tag{9.8}$$

We neglected the subscript B_{int} because there is no material in this case.

Now, consider a given level of applied field (H), which creates a given value of external magnetic induction B_0.

Now, consider placing different types of magnetic materials in this field of external magnetic induction (B_0). For example, we can place a piece of copper or aluminum, silver, superconductor, or iron in this field of magnetic induction B_0. Also, imagine that we have the magical ability to look into this material and observe the magnetic flux lines. We will observe that these different materials "react" to the external magnetic induction in different ways.

For example, in a superconductor, the flux density or induction inside the material is actually zero, that is, $B_{int} = 0$. This means that the entire magnetic flux is excluded from the inside of a superconductor in a superconducting state. For silver, the B_{int} or the flux density will be slightly less than B_0, that is, some of the flux is excluded from the inside of the material. Such materials in which the B_{int} is less than B_0 are known as *diamagnetic materials*. Another way to state this is to say that the flux lines will be diluted inside a diamagnetic material such as silver and completely excluded from the inside of a superconductor.

For materials such as copper and aluminum, we will see that the B_{int} will be slightly *greater* than B_0. These materials are known as *paramagnetic* materials. In these materials, the flux lines will be slightly concentrated inside the material.

For materials such as iron (or any other ferromagnetic or ferrimagnetic material), we will find that the flux lines are heavily concentrated, that is, the flux density inside the material (B_{int}) will be significantly higher than the external magnetic induction (B_0).

For the same value of an externally applied magnetic field (H), the magnetic flux density (B_{int}) inside a ferromagnetic or ferrimagnetic material will be considerably higher than that for either vacuum or a diamagnetic or paramagnetic material (Figure 9.5 and Example 9.3). This is similar to the

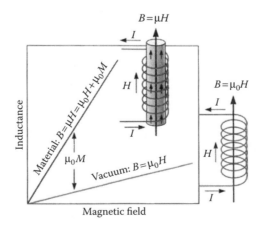

FIGURE 9.5 A current-carrying coil generates higher flux density (B) inside a magnetic material. (From Askeland, D., and P. Fulay. 2006. *The Science and Engineering of Materials*. Washington, DC: Thomson. With permission.)

dielectric displacement (D) inside a parallel-plate capacitor being higher for a capacitor filled with a material having a high dielectric constant in comparison to the value for one filled with vacuum. In ferromagnetic and ferrimagnetic materials, the flux density caused by induced magnetization dominates (i.e., $\mu_0 M \gg \mu_0 H$).

We will discuss these different classes of materials in detail. For now, let us revert to see how the flux density created inside the material (B_{int}) relates to the external induction (B_0).

9.3.2 Magnetic Susceptibility (χ_m) and Magnetic Permeability (μ)

Similar to dielectrics, we define *magnetic susceptibility* (χ_m) as a property that relates magnetization M (the effect) to the applied field H (the cause). Thus, magnetic susceptibility is defined as

$$\chi_m = \frac{M}{H} \tag{9.9}$$

Susceptibility values represent the tendency of a magnetic material to "fall for" the presence of a magnetic field.

For diamagnetic and paramagnetic materials, the internal flux density (B_{int}) created can be shown to be directly proportional to the applied magnetic field (H) according to the equation

$$B_{int} = \mu H \tag{9.10}$$

In this equation, the symbol μ is the magnetic permeability of a material. We note that similar to the dielectric constant (ε_r), magnetic permeability, and susceptibility are tensors that relate two vectors. We will treat them here as scalar quantities. Equation 9.10 indicates that the magnetic induction created within a diamagnetic or paramagnetic material depends on the strength of the external magnetic field applied and the magnetic permeability of the material.

It is easier to express the magnetic permeability of a material relative to the magnetic permeability of free space. We define the *relative magnetic permeability* (μ_r) as the ratio μ/μ_0. We rewrite Equation 9.10 as follows:

$$B_{int} = \mu_0 \mu_r H \tag{9.11}$$

We can now establish a relationship between the magnetic susceptibility (χ_m) and the relative magnetic permeability (μ_r).

We start with Equation 9.7:

$$B_{int} = B_0 + \mu_0 M = \mu_0 H + \mu_0 M$$

For diamagnetic and paramagnetic materials, the magnetic field created *inside the material* (H_{int}) is approximately equal to the externally applied magnetic field (H).

From Equations 9.7 and 9.11, we get

$$B_{int} = \mu_0 \mu_r H = (\mu_0 H + \mu_0 M)$$

or, solving for M, we get

$$M = (\mu_r - 1)H \tag{9.12a}$$

From Equations 9.9 and 9.12a, we get a relationship between relative magnetic permeability (μ_r) and magnetic susceptibility (χ_m):

$$\chi_m = \mu_r - 1 \tag{9.12b}$$

This is similar to dielectrics and can be written as follows:

$$\mu_r = 1 + \chi_m \tag{9.13}$$

9.3.3 DEMAGNETIZING FIELDS

Note that in deriving Equation 9.12, we assumed that for diamagnetic and paramagnetic materials, the *magnetic field* created inside the material (H_{int}) is approximately the same as the externally applied field (H). This is *not* true for ferromagnetic and ferrimagnetic materials—commonly known as magnetic materials. This is because inside these magnetic materials (when they are in their ferromagnetic or ferrimagnetic state), strong demagnetizing fields are present. These demagnetizing fields cause the *magnetic flux lines* (showing B) *inside* the magnetic material to move in a direction *opposite* to that of the *magnetic field lines* (showing H) inside the magnetic material. This can be shown using Maxwell's equations. Please pay careful attention to the arrows showing induction (B) in Figure 9.5.

Note that the direction of both the magnetic field lines (depicting H) and flux lines (showing B) is the *same* on the outside of the ferromagnetic or ferrimagnetic material.

Thus, for a ferromagnetic or ferrimagnetic material, the magnetic field inside the material is not equal to the applied field H; it is actually reduced by a certain factor known as the *demagnetizing factor* (N_d). Instead of writing $B_{int} = \mu_0 H + \mu_0 M$ (Equation 9.7), as we did for diamagnetic and paramagnetic materials, we write the internal flux density for a ferromagnetic or ferrimagnetic material as

$$B_{int} = \mu_0 (1 - N_d) M \tag{9.14}$$

The demagnetizing factor (N_d) is geometry-dependent and ranges from 0 to 1 (Table 9.3b).

As can be seen from these values of N_d, in many sample calculations, we often make use of toroid or long cylinders for which the N_d is zero or small so that the effect of the demagnetizing fields does not have to be considered. The demagnetizing factors indicate the ease with which magnetization within a material can be switched to different directions.

TABLE 9.3B
Demagnetizing Factors (N_d) for Different Shapes

Geometry	l/d	N_d
Toroid		0
Long cylinder		0
Cylinder	20	0.006
Cylinder	10	0.017
Cylinder	5	0.040
Cylinder	1	0.27
Sphere		0.333

Source: Buschow, K. H., and F. R. De Boer. 2003. *Physics of Magnetism and Magnetic Materials.* Boston: Kluwer. With permission.

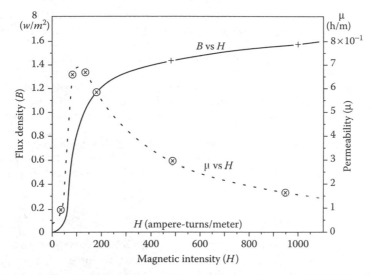

FIGURE 9.6 Dependence of magnetic permeability (μ) of iron on the applied field (H) (dashed curve). The development of initial magnetization is also shown (thick line) as the flux density. (From Jiles, D. C. 1991. *Introduction to Magnetism and Magnetic Materials*. London: Chapman and Hall. With permission.)

9.3.4 FLUX DENSITY IN FERROMAGNETIC AND FERRIMAGNETIC MATERIALS

Thus, the magnetic flux density (B) induced in a magnetic material is related to the applied magnetic field (H) by its magnetic permeability (μ) (Figure 9.5). Note that for convenience, in this diagram for ferromagnetic or ferrimagnetic materials, only the initial portion of the B–H relationship is shown. In reality, these materials are magnetically nonlinear. The more permeable the material, the higher will be the flux density inside the material. This is what the term magnetic permeability means. In some ways, it is useful to think of magnetic permeability (μ) as being similar to electrical conductivity—the higher the magnetic permeability (μ), the higher the ability of a material to carry magnetic field.

The relative magnetic permeability of ferromagnetic and ferrimagnetic materials is large (~100–600,000) and field-dependent. This is why, similar to ferroelectrics, ferromagnetic and ferrimagnetic materials are considered *nonlinear magnetic materials.*

The field-dependence of the permeability (μ) of a magnetic field (Henry/meter) for a magnet of annealed iron (Fe) is shown in Figure 9.6.

The following example illustrates the effect of inserting a magnetic material on the flux density created within a solenoid.

EXAMPLE 9.3: FLUX DENSITY FOR AIR AND IRON CORE

The applied magnetic field (H) for a toroidal solenoid (Figure 9.4a) is given by $N \times I$, where N is the number of turns per meter and I is the current in amperes (Equation 9.5). (a) What is the flux density (B) created for a current of 0.5 A if the toroidal solenoid has an air core? (b) What is the flux density (B) if the core is filled with iron? Assume that the relative magnetic permeability of iron is 900 for the field generated using 0.5 A. The average circumference of the solenoid is 75 cm, and there are 1000 turns.

SOLUTION

a. The average circumference of the solenoid winding is $I = 75$ cm, and there are 1000 turns. Therefore, the value of N (turns per meter) = 1000 turns/(0.75 m) = 1333.33. The magnetic field created is $H = N \times I = (1333.33) \times (0.5$ A$) = 666.66$ A/m.

The flux density generated from this field in air will be $(B_{air}) = (\mu_0 H) = (4\pi \times 10^{-7} \text{ H/m})$ $(666.66 \text{ A/m}) = 8.38 \times 10^{-4}$ T. We assumed that the magnetic permeability of air (μ_{air}) is the same as that of vacuum (μ_0). A very small level of flux density (B) is generated inside the toroidal solenoid filled with air. For comparison, this is the type of flux density created by the earth's natural magnetic field.

b. Now the solenoid gap is filled with iron. The relative magnetic permeability (μ_r) of iron is 1000 for the level of magnetic field created inside the solenoid. In magnetic circuit design, such values of μ_r are usually obtained from data similar to those shown in Figure 9.6.

Therefore, for this scenario, the flux density created inside the solenoid will be (from Equation 9.12) $B = \mu_0 \mu_r H = (4\pi \times 10^{-7} \text{ H/m}) (1000) (666.66 \text{ A/m}) = 0.838$ T. This is a relatively high flux density created inside the solenoid using iron, which is a ferromagnetic material.

9.4 CLASSIFICATION OF MAGNETIC MATERIALS

There are different ways to classify magnetic materials. One of these is based on the value of *coercivity* (H_c), that is, the level of magnetic field necessary to force a magnetization direction in a material. This leads to the definition of soft and hard magnetic materials (see Section 9.8). Another way of classification described here is based on the relative values and signs of the magnetic susceptibility.

9.4.1 DIAMAGNETIC MATERIALS

In diamagnetic materials, when an external magnetic field is applied, the material tries to exclude the magnetic flux. Many materials (e.g., most metals, inert gases, and organic compounds) often classified as "nonmagnetic" actually exhibit a weak diamagnetic response. Diamagnetism is an inherent property of the orbital motion of individual electrons in a field (Goldman 1999). It is basically a Lenz's law–like effect, occurring at an atomic scale. The relative magnetic permeability (μ_r) of diamagnetic materials is < 1. This means that the magnetic susceptibility (χ_m) is negative (Equation 9.12b).

Diamagnetic materials actually tend to expel the magnetic flux lines out from their interior. Certain materials known as superconductors are perfect diamagnetic materials. This means that in superconductors, the magnetic field can be completely expelled. This occurs when the temperature T of a superconductor is less than the so-called critical temperature (T_c), at which the superconducting state exists. This critical temperature (T_c) below which a material exhibits superconductivity is not related to the *Curie temperature* (T_c) of ferromagnetic and ferrimagnetic materials (see Section 9.6.1). The magnetic susceptibility of superconductors is $\chi_m = -1$. This important difference between a conductor and a superconductor is shown in Figure 9.7.

The expulsion of magnetic flux lines due to the diamagnetic behavior of a superconductor is known as the *Meissner effect*, which is the basis of magnetic levitation using superconductors. When a permanent magnet is brought near a superconductor (when it is in this superconducting state), it tries to expel the magnetic field. The result is that the magnet can float or levitate over the superconductor (Figure 9.8).

9.4.2 PARAMAGNETIC MATERIALS

When a paramagnetic material is placed in a field of uniform induction (B_0), the flux density inside (B_{int}) is slightly larger than B_0. This effect originates from the atoms or ions that have unpaired electrons, and this leads to a net magnetic moment for the free atom or an ion. These are known as *paramagnetic atoms* or *paramagnetic ions*. Examples include materials such as aluminum and copper. The atoms in these materials have a permanent magnetic moment. When a magnetic field is applied, these

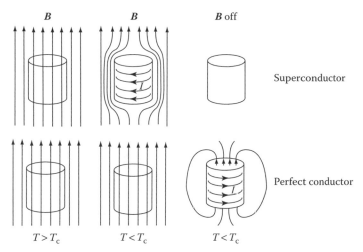

FIGURE 9.7 Expulsion of magnetic flux lines in a superconductor at $T < T_c$. In a conductor, the magnetic flux lines continue to penetrate. (From Kasap, S. O. 2002. *Principles of Electronic Materials and Devices.* New York: McGraw Hill. With permission.)

FIGURE 9.8 Illustration of Meissner effect in superconductors. (From http://www.wondermagnet.com/ images/super5.jpg. With permission.)

magnetic moments get oriented and create a magnetization in the same direction as the applied field. Thus, in paramagnetic materials, the susceptibility is very small ($\chi_m \sim + 10^{-6} - 10^{-3}$) but positive.

This can be shown by a demonstration in which a stream of liquid oxygen is deflected by a strong permanent magnet held next to it (Featonby 2005). As a side note, the unpaired electrons that create a magnetic moment in oxygen are also related to the absorption of red light (~630-nm wavelength) and impart a blue color to liquid oxygen! Paramagnetic materials show a slight attraction in the presence of a permanent magnet. However, the paramagnetic response is so small that, for all practical purposes, these materials are also considered nonmagnetic. Another important point is that at the Curie temperature, ferromagnetic and ferrimagnetic materials transform into paramagnetic materials.

The relative value and sign of magnetic susceptibility (χ_m) or magnetic permeability (μ) are often used to classify magnetic materials. The relationships between magnetic field and magnetization for different materials are shown in Figure 9.9. Note that this diagram is not to scale. Also, it does *not* show the nonlinear nature of ferromagnetic and ferrimagnetic materials.

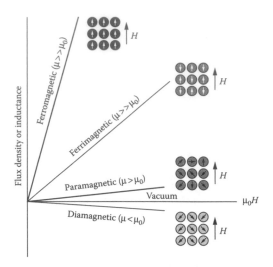

FIGURE 9.9 Magnetic permeability for different types of magnetic materials. Values are not to scale. (From Askeland, D., and P. Fulay. 2006. *The Science and Engineering of Materials*. Washington, DC: Thomson. With permission.)

TABLE 9.4
Magnetic Susceptibilities of Some Elements

Element	Magnetic Susceptibility (χ_m)
Antimony (Sb)	-7.0×10^{-5}
Bismuth (Bi)	-1.7×10^{-4}
Copper	-0.94×10^{-5}
Lead	-1.7×10^{-5}
Silver	-2.6×10^{-5}
Aluminum	$+0.21 \times 10^{-4}$
Neodymium	3.0×10^{-3}
Palladium	8.2×10^{-4}
Platinum	2.9×10^{-5}
Oxygen (under NTP conditions)	17.9×10^{-7}

Data from: Scott, W. T. 1959. *Physics of Electricity and Magnetism*. New York: Wiley; and Goldman, A. 1999. *Handbook of Modern Ferromagnetic Materials*. Boston: Kluwer. With permission.

The magnetic susceptibility values for some diamagnetic and paramagnetic elements are presented in Table 9.4.

9.4.3 SUPERPARAMAGNETIC MATERIALS

Another technologically important class of materials is known as *superparamagnetic materials*. This behavior is seen in ferromagnetic and ferrimagnetic materials fabricated in the form of nanoparticles or nanoscale structures, that is, the grain size of a polycrystalline material becomes of the order of a few nanometers. For example, in the "bulk" form, body-centered cubic (BCC; Chapter 2) iron (Fe) is ferromagnetic. As we decrease the size of iron particles or grains from the bulk to a few micrometers and then down to a few nanometers, the thermal energy at room temperature ($\sim k_B T$) becomes comparable to the magnetic energy. Such nanoparticles or nanoscale grains eventually become single domain. Thus, a nanoparticle or a nanosized grain (~5–10 nm) of iron or similar ferromagnetic or ferrimagnetic materials has a net dipole moment. However, if the energy associated with the magnetic moment of this single-domain nanoparticle or grain is comparable to the thermal energy, the magnetic dipole moment rotates rapidly at room temperature. Single-domain nanoparticles and nanograined materials of ferromagnetic and ferrimagnetic materials consequently behave as if they have no net magnetic moment. This behavior, in which a material that is originally either ferromagnetic or ferrimagnetic in its bulk form behaves similar to a paramagnetic material at a nanoscale level, is known as *superparamagnetism*. The materials in this nanoscale form that exhibit such behavior are known as superparamagnetic materials.

The existence of this effect imposes a limit on how small the grain size of magnetic nanostructures that are used for magnetic data storage can be.

The exact dimension or size at which the thermal energy can exceed the magnetic energy and render the material superparamagnetic depends on the composition, coercivity, and magnetic anisotropy of the material. However, in general terms, the superparamagnetic effects become dominant once the grain or particle size becomes of the order of less than 100 nm and more typically ~10 nm.

Magnetic materials form the basis of different classes of what can be described as passively *smart materials*. A smart material is one whose whose properties are controllable using an external stimulus (temperature, stress, electric field, magnetic field, etc.)

Materials known as *ferrofluids* are based on the use of superparamagnetic materials (Figure 9.10a). A ferrofluid is a stable dispersion of superparamagnetic particles (Figure 9.10b). Ferrofluids behave as *liquid magnets*, that is, when a permanent magnet is placed next to them, the entire material shows a body motion. The carrier fluid is typically an organic substance, but it could be water. Because the particles are colloidal in nature and surface-active agents (surfactants) are used, the particles of magnetic materials used to make ferrofluids do not settle out easily or form agglomerates. Commercially available ferrofluids are made using iron, iron oxide, and other materials. Ferrofluids are used commercially as cooling media. The advantage in using them as such, as for example for cooling of high-wattage speakers, is that this heat-transfer fluid can be held in place by the magnetic fields created by the permanent magnets used to make the audio speakers (Rosensweig et al. 2008).

The formation of "spikes" in a ferrofluid caused by its movement in response to a nonuniform magnetic field (created by a permanent magnet held below, as shown in Figure 9.10a) makes an interesting demonstration of the behavior of these materials.

Superparamagnetic nanoparticles of materials such as Fe_3O_4 have also been functionalized through surface modification so that biological species, molecules, cells, and so on can attach to them. These biological species then can be separated from the rest of the material using a permanent magnet.

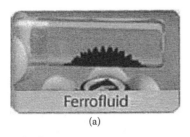
(a)

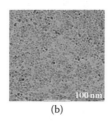
(b)

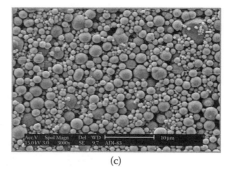
(c)

FIGURE 9.10 (a) Formation of spikes in a ferrofluid. (b) Superparamagnetic nanoparticles of iron oxide (Fe_3O_4). (Courtesy of Ferrotec Corporation and Dr. Kuldip Raj, Ferrotec Corporation.) (c) Soft magnetic, multidomain iron (Fe) particles (~2 μm) used in magnetorheological fluids. (Courtesy of Pradeep Fulay, University of Pittsburgh.)

Ferrofluids are distinctly *different* from the so-called *magnetorheological* (MR) *fluids* that are based on dispersions of larger (not superparamagnetic) and magnetically multidomain ferromagnetic or ferrimagnetic particles with very low coercivity. Applications of MR fluids include vibration control and have been commercialized (Figure 9.10c).

9.4.4 ANTIFERROMAGNETIC MATERIALS

In some materials, the individual atoms or ions have a net spin magnetic moment. However, in other materials, these spin magnetic moments cancel out completely, that is, the magnetic moments have an antiferromagnetic coupling because of the way that the ions or atoms are arranged in a given crystal structure. The result is that these materials have no net magnetization. Chromium (Cr) and α-manganese (α-Mn) are examples of antiferromagnetic *elements*. An example of an antiferromagnetic *compound* is manganese oxide (MnO), in which the spin magnetic moments of Mn^{2+} ions located at different crystallographic sites cancel each other out (Figure 9.11). Note that in this compound, oxygen ions do not have a net magnetic moment. However, through a quantum mechanical interaction, they mediate the magnetic moments of Mn^{2+} ions.

If we heat an antiferromagnetic material to a high temperature, eventually all the antiferromagnetic coupling will be destroyed and the material will become paramagnetic. When we cool an antiferromagnetic material from a high temperature, known as the *Néel temperature* (T_N or θ_N), at which it is in a paramagnetic state, the antiferromagnetic coupling will set in on the sublattice. This will be discussed later in Section 9.5 (Figure 9.12c).

9.5 FERROMAGNETIC AND FERRIMAGNETIC MATERIALS

These materials have the strongest response to an externally applied magnetic field. In other words, these materials are very susceptible to magnetic fields. In general, when we refer to a "magnetic material," it is understood that this refers to a ferromagnetic or ferrimagnetic material (Figure 9.12a). The term "nonmagnetic" material is commonly used for diamagnetic or paramagnetic materials.

9.5.1 HYSTERESIS LOOP

Ferromagnetic and ferrimagnetic materials are characterized by a *magnetic hysteresis loop*, similar to the polarization–electric field loops exhibited by ferroelectric materials. The terminology used for a magnetic hysteresis loop is very similar to that used for ferroelectrics and is shown in Figure 9.13a. Note that in Figure 9.9, we have shown only the initial portion of the hysteresis loop. In Figure 9.13a, the nonlinear nature of these materials is quite clear. Furthermore, in general,

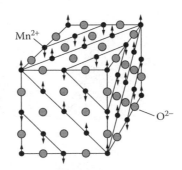

FIGURE 9.11 Antiferromagnetic coupling of magnetic moments of Mn^{2+} ions in MnO. (From Askeland, D., and P. Fulay. 2006. *The Science and Engineering of Materials*. Washington, DC: Thomson. With permission.)

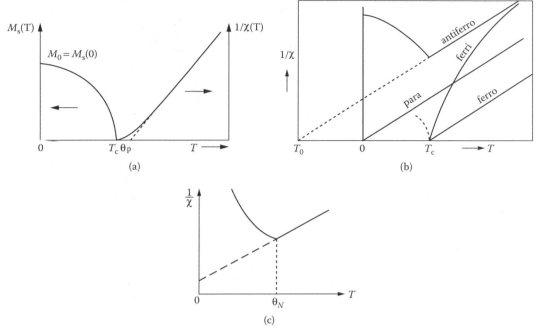

FIGURE 9.12 (a) Curie–Weiss law behavior for ferromagnetic materials. Note that the Curie–Weiss law is the variation of $1/\chi$ at $T > T_c$. (From du Trémolet de Lacheisserie, E., D. Gignoux, and M. Schlenker. 2002. *Magnetism: Fundamentals*. Berlin: Springer. With permission.) (b) Comparison of $1/\chi_m$ versus T for ferromagnetic, ferrimagnetic, antiferromagnetic, and paramagnetic materials. (From Goldman, A. 1999. *Handbook of Modern Ferromagnetic Materials*. Boston: Kluwer. With permission.) (c) Néel temperature (T_N or θ_N) for antiferromagnetic materials. (From Goldman, A. 1999. *Handbook of Modern Ferromagnetic Materials*. Boston: Kluwer. With permission.)

the magnetic susceptibility values of ferromagnetic and ferrimagnetic materials will be very high compared to those of diamagnetic and paramagnetic materials.

The hysteresis loop is the trace of magnetization (M) developed in these materials as a function of an applied magnetic field (H). It is typically not possible to differentiate between a ferromagnetic and a ferrimagnetic material simply by examining the shape of the hysteresis loop.

It is a common practice to show the values of $\mu_0 M$ (Tesla) instead of the value of magnetization M (ampere/meter). The quantity $\mu_0 M$ is known as the *magnetic polarization* (J). Often, the hysteresis loop is also shown as the magnetic induction (B) as a function of the applied magnetic field (H). Again similar to linear dielectrics, for linear magnetic materials (i.e., diamagnetic and paramagnetic), by definition, the magnetic susceptibility (χ_m) is independent of the applied field (H). However, for ferromagnetic and ferrimagnetic materials, the magnetic susceptibility or permeability is indeed dependent on the applied field (Figure 9.13a).

Similar to ferroelectrics (Chapter 8), ferromagnetic and ferrimagnetic materials exhibit the presence of *magnetic domains*. A magnetic domain is a region of a material in which the magnetic moments have a coupling that results in net magnetization. A ferromagnetic or ferrimagnetic material typically consists of many magnetic domains. If the material is polycrystalline, typically, there will be many magnetic domains within each grain. As mentioned before in Section 9.4.3, if a particle or grain becomes too small, it becomes a single-domain body and eventually becomes a superparamagnetic material (Figure 9.13b). In addition, notice the concomitant changes in coercivity of these materials.

The maximum possible magnetization we can get in a ferromagnetic or ferrimagnetic material is known as the *saturation magnetization* (M_s). This is also sometimes shown with the symbol

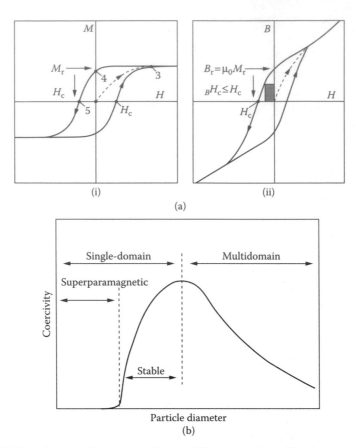

FIGURE 9.13 (a) Typical magnetic hysteresis loops of ferromagnetic and ferrimagnetic materials shown here as (i) *M–H* loop and (ii) *B–H* loop, respectively. (From Skomski, R., and J. M. D. Coey, eds. 1999. *Permanent Magnetism*. Bristol, UK: Institute of Physics. With permission.) (b) Schematic illustration showing transition of multidomain particles or grains to single domain and superparamagnetic particles or grains. (From Hadjipanayis, G. C. 1999. *J Magn Magn Mater* 200:373. With permission.)

$J_s = \mu_0 M_s$. It is defined as the maximum possible magnetic dipole moment per unit volume of a given material. Very often, the saturation magnetization is expressed as the flux density (B) that it creates in a material. Because $\mu_0 M \gg \mu_0 H$ for ferromagnetic and ferrimagnetic materials, the saturation magnetization is often expressed as $B_{sat} \sim \mu_0 M_s$ (Figure 9.13a).

Note that in literature, the coercivity is sometimes written as H_{cB} or $_BH_c$. The subscript or prefix B shows that this is the coercivity for the induction–magnetic field ($B–H$) loop. If the coercivity symbols appear as H_{cM} or $_MH_c$, it indicates the coercivity value for the magnetization–magnetic field ($M–H$) loop. For materials with $\mu_0 H_{cM} > B_r$, the magnitude of H_{cM} is larger than H_{cB} (du Trémolet de Lacheisserie, Gignoux, and Schlenker 2002).

Sometimes, especially in the older literature on magnetic materials, you will encounter cgs and emu units for magnetic properties. For example, magnetization is expressed in Gauss. One Tesla is 10,000 Gauss. Conversions of different units for the various magnetic properties are listed in Table 9.5. The conversion factors provided are for conversion of a Gaussian quantity to the corresponding unit in the SI system.

The following examples illustrate how the magnetic moment of "free" atoms or ions can be used to calculate the saturation magnetization in ferromagnetic and ferrimagnetic materials, and how the results compare to the measured values.

TABLE 9.5
Units for the Various Magnetic Properties

Quantity	Symbol	Gaussian System	Conversion Factor	SI System
Magnetic flux	ϕ	Mx, Gem^2	10^{-8}	Wb · Vs
Magnetic flux density, magnetic induction	B	G	10^{-4}	T, Wb · m^{-2}
Magnetic potential difference, magnetomotoric force	U, F	Gb (gilbert)	$10/4\pi$	A
Magnetic field strength (also unit of coercivity H_c)	H	Oe	$1000/4\pi$	A · m^{-1}
Volume magnetization	$4\pi M$	G	$1000/4\pi$	A · m^{-1}
Volume magnetization	M	emu cm^{-3}, G	1000	A · m^{-1}
Magnetic polarization	J	emu cm^{-3}, G	$4\pi \times 10^{-4}$	T, Wb · m^{-2}
Mass magnetization	M, σ	emu g^{-1}, G cm^{3}g^{-1}	1	A · m^2 · kg^{-1}, J · T^{-1} · kg^{-1}
Magnetic moment	m	emu, erg G^{-1}	1/1000	A · m^2, J · T^{-1}
Magnetic dipole moment	j	emu, erg G^{-1}	$4\pi \times 10^{-10}$	Wb · m, Vs · m
Volume susceptibility	χ, κ	Dimensionless, emu cm^{-3}	4π	Dimensionless
Mass susceptibility	χ, κ	emu g^{-1}, cm^{-3} g^{-1}	$4\pi \times 1000$	m^3 · kg^{-1}
Molar susceptibility	χ_{mol}	emu mol^{-1}, cm^3 mol^{-1}	$4\pi \times 10^{-6}$	m^3 · mol^{-1}
Permeability, $\mu = \mu_0\mu_r$	μ	$\mu^* = \mu_r$	$4\pi \times 10^{-7}$	H · m^{-1}, Vs · A^{-1} · m^{-1}
Relative permeability, μ/μ_0	μ_r	Dimensionless	1	Dimensionless
Energy density, energy product	w	erg cm^{-3}	1/10	J · m^{-3}
Demagnetization factor	N_d	Dimensionless	$1/4\pi$	Dimensionless

Source: Bloor, D. et al., eds. 1994. *Encyclopedia of Advanced Materials*, vol. 4. Oxford, UK: Pergamon Press. With permission.

EXAMPLE 9.4: SATURATION MAGNETIZATION (M_S) AND MAGNETIC POLARIZATION (J_S) OF BCC IRON

The BCC form of iron (Fe) is ferromagnetic.

a. If the room-temperature unit-cell lattice constant of this structure is 2.866 Å, what will be the saturation magnetic polarization of BCC iron?
b. Compare this with the experimental value of 2.1 T, which is the room-temperature saturation magnetization of this form of iron.
c. What is the "effective" magnetic moment of iron atoms in a BCC iron crystal?

SOLUTION

a. As shown in Figure 9.14, in BCC iron, we have eight iron atoms at the corners and one atom at the center of the unit cell. Each of the eight corner atoms counts only as one-eighth because it is shared among a total of eight unit cells (Chapter 2). The atom at the cube center counts as one. Thus, the actual number of atoms per unit cell of BCC iron is two.

Each iron atom has a magnetic moment of 4 μ_B because there are four unpaired electrons per atom (Table 9.1).

Thus, the total dipole moment per unit cell of BCC iron will be = (2 atoms/unit cell) × 4 μ_B = $8 \times 9.274 \times 10^{-24}$ A · m^2/unit cell.

The volume of each unit cell in BCC iron = $(2.866 \times 10^{-10}$ m$)^3 = 23.544 \times 10^{-30}$ m^3. Now, BCC iron is ferromagnetic. This means that all the magnetic dipole moments associated with all the atoms point in the same direction.

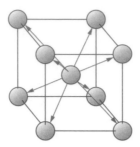

FIGURE 9.14   Crystal structure of body-centered cubic iron.

Thus, the magnetization (M_s) will be equal to the total magnetization per unit volume.

$$M_s = \frac{8 \times 9.274 \times 10^{-24} \text{ A} \cdot \text{m}^2}{23.544 \times 10^{-30} \text{ m}^3} = 3.15 \times 10^6 \text{ A/m}$$

Magnetization is often expressed as $\mu_0 M_s$, rather than as M_s, because then, it has units of Tesla and it is easier to compare $\mu_0 M$ with $\mu_0 H$ (see Example 9.6).

   b. The M_s value of 3.15×10^6 A/m is equivalent to $J_s = \mu_0 M_s = 3.95$ T. The measured value of saturation magnetization is 2.1 T. This difference is due to cancelation of some of the magnetic spin moments when the "free" iron atoms get together and form a BCC crystal.

   c. Because the actual value of magnetization is 2.1 T, we can back-calculate the "effective" magnetic moment of an iron atom in BCC iron. Let the effective magnetic moment be x Bohr magnetons per iron atom in BCC iron. Then, the total saturation magnetization will be

$$M_s = \frac{x \dfrac{\text{magnetons}}{\text{atom}} \times \dfrac{\text{atoms}}{\text{unit cell}} \times 9.27 \times 10^{-24} \dfrac{\text{A/m}}{\text{Bohr magneton}}}{(23.544 \times 10^{-30}) \dfrac{\text{m}^3}{\text{unit cell}}}$$

Therefore, $M_s = 7.87 \times x \times 10^5$ A/m. This corresponds to the magnetic polarization $J_s = \mu_0 M_s = (7.87 \times x \times 10^5 \text{ A/m}) (4\pi \times 10^{-7} \text{ Wb/A} \cdot \text{m}) = (0.9895 \times x)$ T.

   This is given as 2.1 T, which means $x = 2.122$.

   Thus, although a free iron atom has a magnetic moment of 4 μ_B, an iron atom in a BCC iron crystal behaves as if its magnetic moment is only 2.122 μ_B.

   Similar calculations can be done for other materials. The properties of some magnetic materials, including the effective magnetic moments of atoms, expressed as the number of Bohr magnetons per atom or per formula unit (n_B/formula unit), are shown in Table 9.6a. Compare these values with those listed in Table 9.2.

EXAMPLE 9.5: SATURATION MAGNETIZATION IN CERAMIC FERRITES

Ceramic ferrites are useful magnetic materials. Iron oxide (Fe_3O_4) is one example of a ceramic ferrite (Figure 9.15a). This material shows ferrimagnetic behavior, characterized by the antiferromagnetic coupling of the magnetic moments of Fe^{3+} ions on different crystallographic sites (Example 9.2). These crystallographic arrangements are shown in a subcell of Fe_3O_4 (Figure 9.15b). Oxygen ions mediating these magnetic interactions are not shown for the sake of clarity. (a) Calculate the saturation magnetization ($\mu_0 M_s$) for Fe_3O_4 if the lattice constant (a_0) of the larger unit cell, which consists of eight smaller unit cells (Figure 9.15a), is 8.37 Å. (b) What will be the saturation flux density (B_s) for this material?

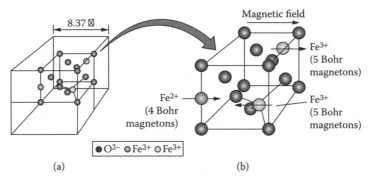

FIGURE 9.15 (a) Crystal structure of iron oxide and (b) arrangements of ferrous (Fe^{2+}) and ferric (Fe^{3+}) ions in a subcell of Fe_3O_4. (From Askeland, D., and P. Fulay. 2006. *The Science and Engineering of Materials*. Washington, DC: Thomson. With permission.)

SOLUTION

a. In Fe_3O_4, the magnetic moments of ferric ions (Fe^{3+}) (each Fe^{3+} has a magnetic moment of 5 μ_B) are canceled out because of the antiferromagnetic coupling of an equal number of Fe^{3+} ions located at both tetrahedral and octahedral sites. Thus, only the ferrous ions (Fe^{2+}) contribute to the net magnetic moment.

In each subcell, the single Fe^{2+} contributes to the total magnetic moment. We have a total of eight subcells that make the larger unit cell (Figure 9.15a). The magnetic moment for each cell being 4 μ_B for eight unit cells, the total magnetic moment will be 32 μ_B. Under saturation conditions, all these magnetic moments will be aligned with the field due to the divalent ferrous ions and hence the total magnetization will be

$$M_s = (32 \times \mu_B) = \left(32 \, \text{Bohr magnetons} \times 9.27 \times 10^{-24} \, \frac{A \cdot m^2}{\text{Bohr magneton}} \right)$$

The volume of the unit cell is $(8.37 \times 10^{-10} \, m)^3 = 5.86 \times 10^{-30} \, m^3$, and the saturation magnetization per unit volume is

$$M_s = \frac{2.96 \times 10^{-22} \, A \cdot m^2}{5.86 \times 10^{-30} \, m^3} = 5.1 \times 10^5 \, A/m$$

This can be expressed in Tesla as the magnetic polarization $J_s = \mu_0 M_s = (4\pi \times 10^{-7} \, Wb/A \cdot m)$ $(5.1 \times 10^5 \, A/m) = 0.64 \, T$.

This is close to the experimentally measured value of ~0.6 T at 290 K (Table 9.6a). For ferrites, the differences in the measured and calculated values of magnetization may be due to the incomplete quenching of the orbital magnetic moments, changes in the relative ratios of divalent and trivalent ions (i.e., changes in Fe^{2+}/Fe^{3+} ratio), and changes in the distribution of these ions among different tetrahedral and octahedral sites.

b. Now we can calculate the magnetic flux density under saturation (B_{sat}). The relationship between B and M is given by

$$B = \mu_0 H + \mu_0 M \tag{9.8}$$

For ferromagnetic and ferrimagnetic materials, the second term dominates, that is, $\mu_0 M \gg \mu_0 H$, especially under high fields. Thus, ignoring the first term, $B_{sat} \sim \mu_0 M_s = 0.64 \, T$.

In Table 9.6b, the values of magnetic moments and the coupling of different moments for some ceramic ferrites are shown.

When the divalent and trivalent ions of a ferrite with a general formula AB_2O_4 are positioned in the tetrahedral (A sites) and octahedral (B sites), the structure is known as a normal spinel structure.

TABLE 9.6A

Magnetic Properties of Some Magnetic Materials

Material	Crystal Structure	$\mu_0 M_s$ (290 K) Tesla	M_s (290 K) emu/cm³	$\mu_0 M_s$ (0 K) Tesla	M_s (0 K) emu/cm³	n_B/formula unit (T = 0 K)	Curie Temperature (K)
Fe	BCC	2.1	1707	2.2	1707	2.22	1043
Co	HCP, FCC	1.8	1440	1.82	1446	1.72	1388
Ni	FCC	0.61	485	0.64	510	0.606	627
$Ni_{80}Fe_{20}$	FCC	1.0	800	1.17	930	1.0	–
Gd	HCP			2.6	2060	7.63	292
Dy	HCP			3.67	2920	10.2	88
CrO_2		0.65	515			2.03	386
$MnOFe_2O_3$	Spinel	0.51	410			5.0	573
$FeOFe_2O_3$	Spinel	0.6	480			4.1	858
$Y_3Fe_5O_{12}$	Garnet (YIG)	0.16	130	0.25	200	5.0	560
$Nd_2Fe_{14}B$	Tetragonal	1.6	1280				585
Amorphous $Fe_{80}B_{20}$	Amorphous	1.6	1260	1.9	1480	2.0	650

Source: O'Handley, R. C. 1999. *Modern Magnetic Materials, Principles, and Applications.* New York: Wiley. With permission.

TABLE 9.6B

Magnetic Moments for Different Ferrites

Ferrite	Tetrahedral Sites A	Octahedral Sites E	Bohr Magnetons per Formula Unit
$NiFe_2O_4$	$Fe^{3+} \downarrow 5 \, \mu_B$	$Ni^{2+} \uparrow 2 \, \mu_B$ $Fe^{3+} \uparrow 5 \, \mu_B$	$2 \, \mu_B$
$MnFe_2O_4$	$Mn^{2+} \downarrow 0.8 \times 5 \, \mu_B$ $Fe^{3+} \downarrow 0.2 \times 5 \, \mu_B$	$Mn^{2+} \uparrow 0.2 \times 5 \, \mu_B$ $Fe^{3+} \uparrow 0.8 \times 5 \, \mu_B$ $Fe^{3+} \uparrow 5 \, \mu_B$	$5 \, \mu_B$
$ZnFe_2O_4$	$Zn^{2+} \downarrow 0 \, \mu_B$	$Fe^{3+} \uparrow 5 \, \mu_B$ $Fe^{3+} \uparrow 5 \, \mu_B$	$0 \, \mu_B$
$Zn_xMn_{(1-x)} Fe_2O_4$	$Zn^{2+} \downarrow 0 \, \mu_B$ $Mn^{2+} \downarrow (1-x) \times 5 \, \mu_B$	$Fe^{3+} \uparrow 5 \, \mu_B$ $Fe^{3+} \uparrow 5 \, \mu_B$	$(1+x) \times 5 \, \mu_B$
$Zn_xNi_{(1-x)} Fe_2O_4$	$Zn^{2+} \downarrow 0 \, \mu_B$ $Fe^{3+} \downarrow (1-x) \times 5 \, \mu_B$	$Ni^{2+} \uparrow (1-x) \times 2 \, \mu_B$ $Fe^{3+} \uparrow 5 \, \mu_B$ $Fe^{3+} \uparrow x \times 5 \, \mu_B$	$(1+4x) \times 2 \, \mu_B$

Source: Fiorillo, F. 2004. *Measurement and Characterization of Magnetic Materials.* Oxford: Elsevier. With permission.

For example, zinc ferrite ($ZnFe_2O_4$) has a normal structure. However, as seen in the case of Fe_3O_4 (written as $FeFe_2O_4$), the trivalent ions (i.e., what are supposed to be the B-site cations) are evenly distributed across both tetrahedral and octahedral sites. This structure is known as the inverse spinel structure. Examples of inverse spinel ferrites include $FeFe_2O_4$, $NiFe_2O_4$, and $CoFe_2O_4$. Some spinels, such as $MnFe_2O_4$, may exhibit partly normal and inverse spinel structures.

Note that addition of zinc ferrite (with zero net magnetic moment) to nickel ferrite or $MnFe_2O_4$ *increases* the net magnetic moment of these materials. This is counterintuitive and can be explained as follows. When zinc ions substitute on some of the tetrahedral sites in the crystal structure of either nickel or manganese ferrite, they *reduce* the antiparallel coupling between Fe^{3+} ions distributed on the tetrahedral sites in the inverse spinel structure. This, in turn, causes an *increase* in the net magnetization.

9.6 OTHER PROPERTIES OF MAGNETIC MATERIALS

9.6.1 CURIE TEMPERATURE (T_c)

Ferromagnetic and ferrimagnetic materials undergo a transformation at high temperatures, by which they become paramagnetic. The temperature at which the spontaneous magnetization in a ferromagnetic or ferrimagnetic material vanishes is known as the Curie temperature. Similar to ferroelectrics (Chapter 8), the Curie–Weiss law describes the variation of magnetic susceptibility with temperature for levels above the Curie temperature, written as follows:

$$\frac{1}{\chi_m} = \frac{T - \theta_p}{C}$$
(9.15)

In this equation, χ_m is the magnetic susceptibility, T is the temperature, θ_p is the paramagnetic Curie temperature (often, slightly greater than T_c), and C is the Curie constant.

The variation in $1/\chi_m$ as a function of temperature *for $T > T_c$* is what the Curie–Weiss law represents. Note that this law does *not* apply to the variation of magnetization at temperatures below T_c. The variation of susceptibility above T_c (Curie–Weiss law) and the changes in magnetization *below T_c* are shown in Figure 9.12. Because the material behaves as a paramagnetic material above T_c, the Curie–Weiss law describes the variation in permeability and susceptibility of the paramagnetic phase. Curie temperature (Table 9.6a) is very important from a technological perspective because, often, this parameter limits the highest temperature that can be used for a magnetic material.

For ferrimagnetic materials, the dependence of $1/\chi_m$ with temperature departs from linearity as we approach T_c. This is shown in Figure 9.12b.

In antiferromagnetic materials, there is no net magnetic moment because the magnetic moments cancel out. However, as we increase the temperature, the extent of *antiparallel coupling is reduced*. This causes the magnetic susceptibility to *increase* (i.e., $1/\chi_m$ decreases) with increase in temperature. This occurs until a temperature known as the Néel temperature (T_N or θ_N), at which an antiferromagnetic material becomes paramagnetic (Figure 9.12c).

9.6.2 MAGNETIC PERMEABILITY (μ)

In addition to saturation magnetization, many other properties of ferromagnetic and ferrimagnetic materials, important for different applications, are often specified. These include the initial (low-field) and high-field permeabilities.

As can be seen from Figures 9.6 and 9.16, the magnetic permeability (μ) of ferromagnetic and ferrimagnetic materials depends strongly on the strength of the applied magnetic field (H).

At very low fields, the magnetic permeability is low because the applied field is not sufficient to cause any kind of domain growth. This is often referred to as the initial permeability (μ_i). As the field increases, the domains are aligned along the direction of the magnetic field. This is similar to the orientation of domains in ferroelectric materials. As the domains become oriented, a relatively sharp increase in magnetization occurs. This is why, in this region, the magnetic permeability increases significantly to μ_{max} (Figure 9.16). As the field approaches relatively high values,

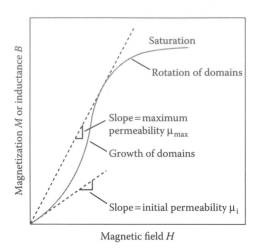

FIGURE 9.16 Field dependence of permeability for ferromagnetic and ferrimagnetic materials. (From Askeland, D., and P. Fulay. 2006. *The Science and Engineering of Materials.* Washington, DC: Thomson. With permission.)

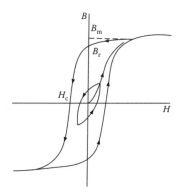

FIGURE 9.17 Minor hysteresis loops. (From Goldman, A. 1999. *Handbook of Modern Ferromagnetic Materials.* Boston: Kluwer. With permission.)

in which essentially all domains are aligned, the permeability begins to decrease again (Figures 9.6 and 9.16). This is the saturation region. What this means is that the material cannot accept any more flux, that is, it essentially becomes impermeable to magnetic field. In the designing of magnetic circuits, often, care has to be taken not to create such "bottlenecks" that effectively will not allow any more magnetic flux to pass through. As shown in Figure 9.17, if the applied field is not enough to cause saturation, we get what is called a *minor loop*.

9.6.3 Coercive Field (H_c)

Similar to ferroelectrics, application of a certain level of *coercive field* ($\mu_0 H_c$ or H_c) is needed to remove any remnant magnetization in the material. This field is known as the *coercivity* (H_c) or coercive field of a magnetic material. The relative value of coercivity defines soft and hard magnetic materials. These materials are further discussed in Section 9.8.

The coercivity values define one way to classify ferromagnetic and ferrimagnetic materials as *hard* (see Section 9.9) or *soft magnetic materials* (see Section 9.11). A hard material is easily distinguished from a soft magnet by comparing the values of coercivity relative to the saturation magnetization (Figure 9.18).

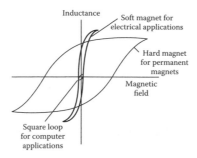

FIGURE 9.18 Typical shapes of hysteresis loops for soft and hard magnetic materials, diagram not to scale. (From Askeland, D., and P. Fulay. 2006. *The Science and Engineering of Materials*. Washington, DC: Thomson. With permission.)

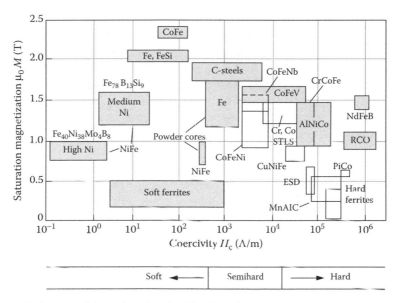

FIGURE 9.19 Relative coercivity values for classification of magnetic materials. (From Chin, G. Y. et al. 1994. In *Encyclopedia of Advanced Materials*, vol. 1, eds. R. Bloor, M. Flemings, and S. Mahajan, 1424. Oxford: Pergamon Press. With permission.)

Ferromagnetic and ferrimagnetic materials are used for flux enhancement or as materials that can store energy or information (data). The first application, that is, flux enhancement, requires high permeability and low coercivity (see Section 9.11). In general, materials that have a coercivity of less than ~5000 A/m are considered magnetically soft materials (see Section 9.11), whereas those with coercivity larger than approximately 10^4 A/m are known as hard magnetic materials (Figure 9.19). In the so-called cgs (emu) system of units, coercivity is expressed in Oersteds (Oe). One ampere/meter is equal to $4\pi \times 10^{-3}$ Oe.

Hard magnetic materials are also known as *permanent magnets* (see Section 9.9). Materials whose coercivity values range between a few hundred and ~10^4 A/m are known as *semihard magnetic materials*. These semihard materials are useful for magnetic storage of data (e.g., audio and video cassettes or magnetic hard disks used in computers and other electronic gadgets; see Section 9.12).

Similar to ferroelectrics, the coercivity (H_c) of magnetic materials is a microstructure-sensitive property. For essentially the same material composition, coercivity can be changed by orders of magnitude. Of course, we can change the composition and design an appropriate microstructure to develop a material with the desired magnetic properties. Currently, through such careful design, we have obtained extremely low coercivity magnetic materials, such as $Fe_{84}Zr_7B_9$ that has a coercivity of $\mu_0 H_c \sim 10^{-7}$ T. Note the negative sign in the exponent. On the contrary, we have a material such as $Fe_{84}Nd_7B_9$ with very high coercivity of $\mu_0 H_c \sim 1$ T. We have a large range of coercivities in engineered magnetic materials. In real-world applications, the cost-to-performance ratio and many other factors, such as durability, weight, and mechanical properties, are also very important.

9.6.4   NUCLEATION AND PINNING CONTROL OF COERCIVITY

When a magnetic field is applied, the magnetic domains in a material are forced or coerced to orient in the direction of the field. The coercive field and the initial part of the $B–H$ ($M–H$) loop depend on the mechanism by which the domains nucleate and grow. For example, consider a hypothetical hard magnetic material that has been subjected to a sufficiently high magnetic field to make it essentially a single domain. Now, if we start applying a field in the reverse direction, the field will try to align this single domain along the new direction. This will require nucleation of other domain walls and domains. This situation is referred to as *nucleation-controlled coercivity*. A very good hard magnet, on application of a field in the reverse direction, will resist nucleation of domains as much as possible. If the coercivity is controlled by domain nucleation, then the process starting from the origin of the hysteresis loop (i.e., demagnetized state) to saturation (see the dotted curve in Figure 9.20a) is achieved using a much weaker field, compared to reversing the magnetic field from $-M_r$ to M_s where the coercivity is nucleation-controlled. Most state-of-the art materials show nucleation-controlled coercivity (Figure 9.20b).

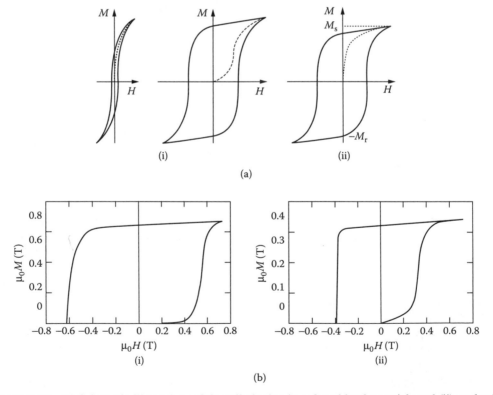

FIGURE 9.20   (a) Schematic illustrations of (i) wall pinning in soft and hard materials and (ii) nucleation-controlled coercivity in modern, hard magnetic materials. (b) (i) Nucleation-controlled coercivity in a ferrite (left) and (ii) pinning-controlled coercivity in Ce–Co–Cu–Fe magnet (right). (From du Trémolet de Lacheisserie, E., D. Gignoux, and M. Schlenker. 2002. *Magnetism: Fundamentals.* Berlin: Springer. With permission.)

Another way to enhance coercivity is to pin down the domain walls of the existing domains as much as possible. This is known as *pinning-controlled coercivity*. In this case, multiple domains already exist. Typically, microstructural defects, especially grain boundaries, can accomplish this. If the range of fields required for proceeding from the demagnetized state (origin of the hysteresis loop curve) to saturation (see the dotted curve in Figure 9.20a (i)) is similar to that for moving from $-M_r$ to $+M_s$, then the coercivity is controlled by the domain-pinning process. An example of pinning-controlled coercivity is shown in Figure 9.20b for a Ce–Cu–Co–Fe magnetic material.

This is in contrast with the hysteresis loop for a Ce–Co–Cu–Fe magnet, in which defects that pin the domains control the coercivity. An understanding of the microstructural features that control the coercivity is important for the development of magnetic, especially those used as permanent magnets (see Section 9.9) and magnetic recording materials (see Section 9.12).

9.6.5 MAGNETIC ANISOTROPY

The term "magnetic anisotropy" means that magnetic properties are dependent on the crystallographic direction. This property is critical for the development of permanent or hard magnetic materials. The coercivity of magnetic materials is related to the magnetic anisotropy in two ways. First, there is the *magnetocrystalline anisotropy*. Simply stated, this means that for a given single crystal material, magnetic properties, such as the coercive field, change depending on the crystallographic direction. The second type of magnetic anisotropy is known as *magnetoshape anisotropy* or *shape anisotropy*. This anisotropy means that if we have particles of two identical materials, with one particle being needle-like (acicular) and the other one nearly spherical, then the acicular particle will have a higher coercivity because of its shape. The cause of shape anisotropy can be traced to the difference in the demagnetization factors (due to differences in geometry; Table 9.3b). For an elongated particle, the hard axis tends to line up along the longer axis of the particles. This effect is used in magnetic recording materials wherein elongated or needle-like particles of materials, such as iron or barium ferrites, are used (see Section 9.12). The elongated shape of these particles causes them to have higher coercivity, which helps with the retention of information stored.

The magnetocrystalline anisotropy is important in many applications of hard and soft materials and is discussed here in detail. For example, in magnetic materials used for data storage, we take advantage of the magnetocrystalline anisotropy by using oriented or textured thin films that possess higher coercivity and hence show better retention of data. On the contrary, we use grain-oriented steels for transformer cores so that the steel magnetizes easily along the easy directions.

Owing to the magnetocrystalline anisotropy present in an iron single crystal, the magnetization develops most easily along the [100] direction. The crystallographic direction in which the magnetization develops with a smaller coercivity is known as the easy magnetization direction or the *easy axis*. This represents the direction(s) in which the average magnetic moment will be directed for a material with no application of an external magnetic field (Bloor et al. 1994). Alignment of magnetic moments along the easy-axis directions would minimize the free energy of the material. On the contrary, for a BCC-iron single crystal, the [111] direction that represents a body diagonal will be the hard magnetization direction or *hard axis*. All the body diagonals, that is, the <111> family of directions, and not just the [111] direction, will represent the hard axes for BCC iron.

For cobalt (Co), the easy axis is the direction perpendicular to the hexagonal planes. The axes in the basal plane represent the hard directions (Chapter 2). The energy needed to rotate from the easy-magnetization direction is known as the magnetocrystalline anisotropy energy. This energy can be written as

$$E_a = K_1 \sin^2\theta + K_2 \sin^4\theta + K_3' \sin^6\theta \cos 6\phi \quad \text{(for hexagonal crystals)} \quad (9.16)$$

For tetragonal crystals, the magnetocrystalline anisotropy energy is written as

$$E_a = K_1 \sin^2\theta + K_2 \sin^4\theta + K_3 \sin^4\theta \cos 4\phi \quad \text{(for tetragonal crystals)} \quad (9.17)$$

In these equations, K_1, K_2, K_3, and K_3' are the coefficients of anisotropy. Usually, the coefficient K_1 dominates. The angle θ is that formed between the magnetization vector and the c axis. The angle ϕ is that between the projection of the magnetization vector on the basal plane and one of the a axes (Bloor et al. 1994). The coefficients of anisotropy are expressed in units of Joule/cubic meter or Gauss–Oersteds (G·Oe). The conversion factor is 1 mega G·Oe = 7.9577 kJ/m^3.

The anisotropy field, K_1 values, and other properties of some hard magnetic materials are listed in Tables 9.7a and b.

Note that the values of K_1 are listed as Megajoule/cubic meter, that is, the value of SmCo$_5$ is 17 MJ/m^3 or 17×10^6 J/m^3. The K_1 values of some materials are listed in Table 9.7b.

In a ferromagnetic or ferrimagnetic material, the long-range spin ordering occurs because of what is called the *exchange interaction*, that is, the combination of the electrostatic coupling between electron orbitals and Pauli's exclusion principle from Chapter 2. In ferromagnetic materials, this interaction aligns the spins of neighboring atoms in the same direction (Figure 9.3). In ferrimagnetic materials, this exchange interaction orients the spins of ions on different crystallographic sites in an antiparallel arrangement (Figure 9.15). In ceramic ferrites, this is known as the superexchange

TABLE 9.7A
Values of T_C, Anisotropy Field, K_1, and Saturation Magnetization for Some Hard Magnets

Compound	T_c(C)	$\mu_0 H_a$(T)	K_1 (MJ m^{-3})	$\mu_0 M_s$(T)
SmCo$_5$	720	40	17	1.05
Sm$_2$Co$_{17}$	823	6.5	3.3	1.30
Nd$_2$Fe$_{14}$B	312	6.7	5	1.60

Note: 1 MG.Oe = 7.9577 kJ/m^3.

Source: Jakubowicz, J., et al. 2000. *J Magn Magn Mater* 208(3):163–8; and Bloor, D. et al., eds. 1994. *Encyclopedia of Advanced Materials*, vol. 4. Oxford, UK: Pergamon Press.

TABLE 9.7B
Approximate Values of Anisotropy Coefficient (K_1) for Some Magnetic Materials

Material	K_1 (kJ/m^3)	Material	K_1 (kJ/m^3)
Iron	48	SmCo$_5$	17,000
Nickel	−4.5	Sm$_2$Co$_{17}$	3300
Cobalt	530	Nd$_2$Fe$_{14}$B	4900
Fe$_3$O$_4$	−13	BaFe$_{12}$O$_{19}$	250
CoFe$_2$O$_4$	180–200	Cu$_2$MnAl	−0.47
NiFe$_2$O$_4$	−6.9	Y$_3$Fe$_5$O$_{12}$	−2.5
MgFe$_2$O$_4$	−4		
MnFe$_2$O$_4$	−4		

Note: 1 MG.Oe = 7.9577 kJ/m^3.

Source: Fiorillo, F. 2004. *Measurement and Characterization of Magnetic Materials*. Oxford: Elsevier; and du Trémolet de Lacheisserie, E., D. Gignoux, and M. Schlenker. 2002. *Magnetism: Fundamentals*. Berlin: Springer. With permission.

interaction because it is mediated by oxygen ions. The anisotropy, however, comes about because of the interactions between the electron orbitals and the potential associated with the atoms in the crystal (Bloor et al. 1994).

9.6.6 MAGNETIC DOMAIN WALLS

Normally, we would expect the magnetic moments to reverse their direction only when the applied magnetic field is equal to or greater than the so-called anisotropy field (H_a), which is given by

$$H_a = \frac{2K_1}{M_s} \tag{9.18}$$

This means that for an ideal magnetic material, domain-switching must occur only at one value of the applied field, that is, the hysteresis loop should be rectangular for the M (or J)–H loop and a parallelogram for the B–H loop (Figure 9.21a).

It has been experimentally observed that most permanent or hard magnetic materials (see Section 9.8) show magnetization reversal at fields that are only 10%–15% of the value of anisotropy field (H_a) (Buschow and De Boer 2003).

The reason for this phenomenon is that magnetic materials consist of magnetic domains, which are separated by domain walls known as *Bloch walls* (Figure 9.21b). The magnetization direction for adjacent domains is in opposite directions. This helps reduce the magnetostatic energy. The magnetization direction changes gradually from one direction to an opposite direction through the thickness of the domain wall.

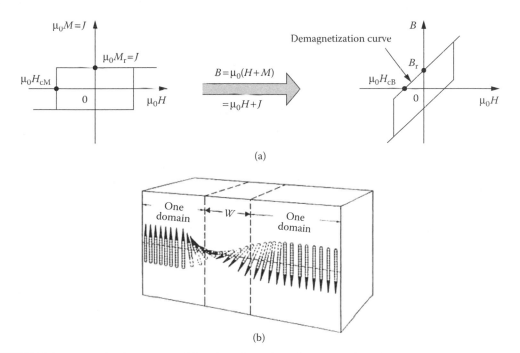

(a)

(b)

FIGURE 9.21   (a) M (or J)–H loop and B–H loop for an ideal ferromagnetic or ferrimagnetic material. (From du Trémolet de Lacheisserie, E., D. Gignoux, and M. Schlenker. 2002. *Magnetism: Fundamentals*. Berlin: Springer. With permission.) (b) Illustration of change in magnetization across an 180° Bloch wall. (From Buschow, K. H., and F. R. De Boer. 2003. *Physics of Magnetism and Magnetic Materials*. Boston: Kluwer. With permission.)

The thickness of the domain wall (W) in magnetic materials is often much greater than that of the ferroelectric domains in ferroelectric materials. The domain-wall thickness depends on the relative strengths of the anisotropy energy and the exchange energy. If the exchange energy is large, it tends to maintain the magnetic moment in the given direction, that is, the domain wall would be thicker.

If A is the average exchange energy and K_1 is the anisotropy energy constant, then it can be shown that the domain-wall thickness (W) is given by

$$W = \pi \sqrt{\frac{A}{K_1}} \tag{9.19}$$

and the domain-wall energy per unit area of a domain wall of width W is given by

$$E_{\text{wall}} = 2\pi\sqrt{K_1 A} \tag{9.20}$$

In deriving these equations, it is assumed that the effect of the demagnetization energy can be neglected.

If the anisotropy energy is large, then the domain walls would be thinner because the material prefers to undergo magnetization in the easy direction of magnetization. For example, in highly anisotropic materials such as tetragonal $Nd_2Fe_{14}B$ with high values of K_1, reaching ~4900 × 10^3 J/m^3 (Table 9.7a and b), the domain-wall thickness is only ~5 nm. In materials such as BCC iron, the domain-wall thickness may be ~50 nm because of the relatively smaller anisotropy energy (smaller value of K_1, ~4.8 × 10^4 J/m^3). Because the lattice constant (a_0) of iron is ~0.3 nm (Example 9.4), the domain-wall thickness will be greater than ~200 lattice spacings.

Thus, in many permanent magnetic materials (see Section 9.9), the observed value of coercivity is *significantly lower* than that expected from the value of the anisotropy field (H_a) because magnetization reversal occurs by the nucleation of Bloch walls and growth of reversed domains. This situation is somewhat similar to the yield stress (τ_{ys}) of metals and alloys. We know that based on the strengths of metallic and covalent bonds, most metallic materials should be extremely strong and brittle. However, in reality, their strength is much lower (almost 1000 times lower). Also, metals and alloys are relatively ductile, because mechanical deformation occurs by the motion of dislocations (known as slip) at much lower levels of stress and not by breaking of all the strong bonds (Chapter 2).

In some materials, such as thin films, magnetic domains can extend across the entire width of a sample. If there are Bloch walls between two domains that run the entire width of the sample, then in the wall region, the magnetization will be perpendicular to the plane of the film. This causes a considerable increase in the demagnetization energy. In this case, a *Néel wall* forms, in which the magnetic moment rotates within the plane of the film (Jiles 1991). This is shown in Figure 9.22. The formation of a Néel wall is seen in thin films because it occurs with lower energy. Néel walls do not appear in bulk magnetic materials because they actually cause a higher demagnetization energy. The formation of Néel walls is favored in thin films when the thickness falls below a certain critical value.

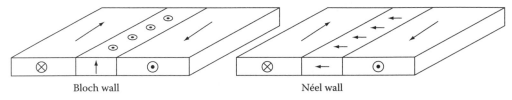

Bloch wall   Néel wall

FIGURE 9.22 Formation of Bloch and Néel walls in a magnetic thin film. (From Jiles, D. C. 1991. *Introduction to Magnetism and Magnetic Materials*. London: Chapman and Hall. With permission.)

9.6.7 180° AND NON-180° DOMAIN WALLS

Both Bloch and Néel walls are examples of what we call 180° domain walls. This means that the direction of magnetization in the domains separated by the walls is antiparallel. Accordingly, by definition, for these walls, the crystallographic directions will be equivalent (e.g., [100] and [$\bar{1}$00]). It is possible to have domain walls that are not 180°. For example, in cubic materials with the anisotropy energy constant (K_1) greater than zero, it is possible to have 90° domain walls. An example of this is BCC iron, in which it is possible to get magnetization in adjacent domains in the [100] and [010] directions (Jiles 1991). In nickel (Ni), the anisotropy constant (K_1) is less than zero ($K_1 = -4.5 \times 10^3$ J/m³), and the result is domain walls of 71° and 109°. Note that for nickel, the easy-magnetization directions are all body diagonals (<111>; Figures 9.23 and 9.24).

Such 90° domains, shown in Figure 9.23, appear as closure domains in materials such as grain-oriented silicon containing steel used in transformer cores. It is difficult to form 90° domain walls

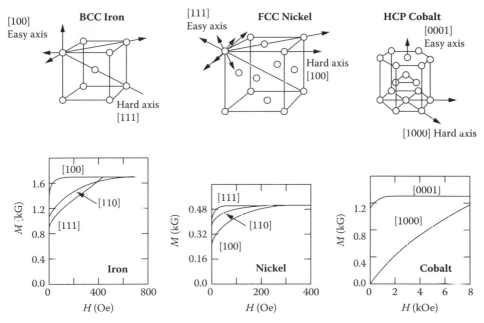

FIGURE 9.23 Crystal structures and magnetocrystalline anisotropy for Fe, Ni, and Co single crystals. (From O'Handley, R. C. 1999. *Modern Magnetic Materials, Principles, and Applications.* New York: Wiley. With permission.)

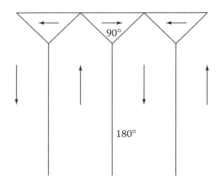

FIGURE 9.24 Illustration showing closure domains and 90° and 180° domain walls. (From Jiles, D. C. 1991. *Introduction to Magnetism and Magnetic Materials.* Boca Raton, FL: Chapman and Hall. With permission.)

in materials such as cobalt (Co) because the directions in the basal plane are the hard magnetic directions, as shown in Figure 9.23.

One important significance of 90° walls is that they are stress-sensitive. For example, consider a domain wall between the [100] and [010] directions. If a tensile stress is applied to this material in the [100] direction, then the [100] domain is energetically favored. This causes the growth of [100] domains at the expense of the [010] domain. Consequently, 90° domain walls will move under a stress. This contributes to the strain produced by *magnetostriction* (see Section 9.7). However, if there are two 180° domains (say in the [100] and [$\bar{1}$00] directions), the energies of both will be reduced equally by the application of stress in the [100] direction. Thus, 180° domains will be insensitive to stress.

9.6.8 Maximum Energy Products for Magnets

We saw the nature of the hysteresis loops for both ferromagnetic and ferrimagnetic materials in Figure 9.18. The first quadrant of the loop is important because it shows the dependence of permeability on the applied field. Once the material is saturated and the magnetic field is removed, the material is left in a state with a remnant magnetic induction (B_r; Figure 9.25), which makes the magnet permanent. The part of the loop in the second quadrant where the magnetic material is demagnetized from B_r to 0 is known as the *demagnetization curve* (Figure 9.25b). Magnetic circuit designs involving permanent magnets make use of this curve.

If we have a permanent magnet and we create a magnetic circuit with an air gap, then the magnetic induction decreases from B_r to some lower value, say to point P (Figure 9.26). The actual level of decrease depends on the geometry of the magnetic circuit and the demagnetization curve. The ratio of B to H at a point of operation along this curve is known as the permeance coefficient. The line drawn from the origin to the operating point (P) is known as the shearing line.

For permanent magnets, a frequently reported value is the so-called $(BH)_{max}$ *product,* also known as the *maximum energy product*. This is the maximum value of the product obtained by multiplying the corresponding B and H values on the demagnetization curve (Figure 9.26), which represents the maximum energy that can be stored within a given volume of a magnetic material.

The improvements in energy product and coercivity of permanent magnets are shown in Figure 9.27a. A useful conversion between different units for expressing the energy product is:

$$1 \text{ MG Oe} = \frac{10^6}{4\pi} \text{ erg/cc} = 7.96 \text{ kJ/m}^3 \tag{9.21}$$

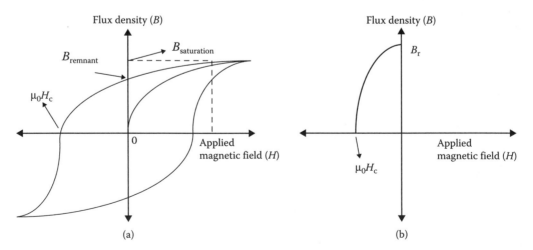

(a)   (b)

FIGURE 9.25   (a) Hysteresis loop at saturation and (b) demagnetization curve (second quadrant).

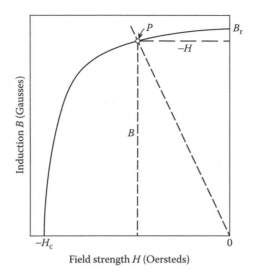

FIGURE 9.26 Demagnetization curve and a shearing line for $(BH)_{max}$. (From Bozroth, R. M. 1955. *Ferromagnetic Materials*. Van Nostrand: IEEE Press. [Reprinted 1993]. With permission.)

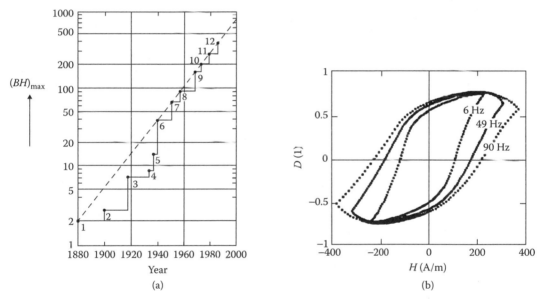

FIGURE 9.27 (a) Improvements in maximum energy product $(BH)_{max}$ in kJ/m³ over the years 1880 to 1980: 1, carbon steel; 2, tungsten steel; 3, cobalt steel; 4, Fe–Ni–Al alloy; 5, Ticonal II; 6, Ticonal G; 7, Ticonal GG; 8, Ticonal XX; 9, $SmCo_5$; 10, $(SmPr)Co_5$; 11, $Sm_2(Co_{0.85} F_{0.11} Mn_{0.04})_{17}$; and 12, $Nd_2Fe_{14}B$. (From Jiles, D. C. 1991. *Introduction to Magnetism and Magnetic Materials*. London: Chapman and Hall. With permission.) (b) Hysteresis loops for soft iron at different frequencies for induction up to 0.75 T. (From du Trémolet de Lacheisserie, E., D. Gignoux, and M. Schlenker. 2002. *Magnetism: Fundamentals*. Berlin: Springer. With permission.)

9.6.9 Magnetic Losses

Magnetic losses refer to the energy dissipation that occurs when a magnetic material is subjected to a time-varying external magnetic field. They are similar to the dielectric losses occurring in dielectric materials (originating from different polarization mechanisms such as ferroelectric polarization). Magnetic losses are especially important in some applications, such as the use of different materials (e.g., iron–silicon steels) for transformers. The energy dissipation that occurs in a magnetic material can be seen from the area under the B–H loop, represented as $\int H \cdot dB$. In materials with higher electrical resistivity (e.g., ceramic ferrites), losses do not increase much with increase in the frequency of applied field. However, in magnetic materials that are good conductors (e.g., iron and steel), losses increase with increasing frequency. This is seen from the increase in the B–H loop area (Figure 9.27b).

The magnetic loss component can be treated mathematically in a way similar to that for dielectric losses using a complex dielectric constant. Magnetic permeability (μ) can be written as a complex number (μ^*) that has a real part (μ') and an imaginary part (μ''). The imaginary part is related to the magnetic losses that occur. The ratio of the imaginary and the real parts of magnetic permeability is defined as the loss tangent (tan δ). In some applications of ceramic ferrites used in high-frequency applications for microwave devices, characterization of such properties becomes very important. Magnetic losses due to eddy currents induced by the applied magnetic field are known as *hysteresis losses*. Other components include excess losses due to domain motion and classical losses. For a detailed description of both origin and types of magnetic losses, the reader is directed to the work by Bloor et al. (1994). Losses due to eddy currents can be reduced by increasing the resistivity. This is the reason why silicon is added (up to ~3.5%) to iron to make the steel used in transformer cores. Addition of silicon also reduces the coercivity of the steel.

9.7   MAGNETOSTRICTION

We saw in Chapter 8 that every material responds to an electric field and shows electrostriction to some extent. The electrostriction strain is larger and more useful in materials that are more polarizable (e.g., relaxor ferroelectrics). Similarly, *magnetostriction* is seen in ferromagnetic and ferrimagnetic materials. Magnetostriction is defined as the development of strain in a material subjected to a magnetic field. A familiar example of the magnetostriction phenomenon is seen in magnetic steels used in transformers, which causes a "transformer hum." This occurs as the magnetic materials in the transformer expand and contract to their original dimensions.

Magnetostriction in ferromagnetic and ferrimagnetic materials can come about spontaneously when there is a spontaneous alignment of magnetic moments at the Curie temperature. This is known as *spontaneous magnetostriction*.

Another form of magnetostriction is encountered when domains become reoriented in the presence of an external magnetic field. This is known as *Joule magnetostriction*. The extent of the magnetostriction effect is indicated by measuring the magnetostriction coefficient (λ). A material known as Terfenol-D ($Tb_{0.3}Dy_{0.7}Fe_{1.9}$) has a magnetostriction coefficient of ~1500 ppm, that is, 1500×10^{-6}, at room temperature. Other materials include $TbFe_2$, known as Terfenol, with a value of $\lambda_{111} = +2460 \times 10^{-6}$ (the highest reported so far), and $SmFe_2$ with $\lambda_{111} = -2100 \times 10^{-6}$ (Table 9.8).

TABLE 9.8

Room-Temperature Magnetostriction Coefficients of Some Materials

Material	$\lambda_{100}(10^{-6})$	$\lambda_{111}(10^{-6})$
Fe	24	−22
Ni	−51	−23
$TbFe_2$	—	2460
$SmFe_2$	—	−2100

Source: Buschow, K. H., and F. R. De Boer. 2003. *Physics of Magnetism and Magnetic Materials.* Boston: Kluwer. With permission.

The magnetostriction coefficient λ_{100} represents a change in length or saturation magnetostriction in the [100] direction when the magnetization is also along the [100] direction (after the material is cooled through the Curie temperature). The λ_{111} is defined similarly.

The acronym Terfenol comes from "Ter" for Terbium, Fe for iron, and NOL for the Naval Ordnance Laboratory that developed this material. Terfenol-D means that it has dysprosium (Dy) in it. Addition of Dy lowers the magnetocrystalline anisotropy, and this leads to better magnetostriction properties. These materials are being developed for a number of actuator applications, similar to the applications of piezoelectric materials. The cost of the elements such as Tb and Dy is high, making it difficult to create widespread applications of these materials. Current efforts are being directed to the development of relatively low-cost materials that can show large magnetostriction coefficients. Metglas™ ($Fe_{81}B_{13.5}Si_{3.5}C_2$) is one of the best isotropic magnetostrictive materials. Because magnetostriction basically involves causing strain using magnetic fields, many applications are similar to those of piezoelectrics (e.g., force sensors and sonar).

For polycrystalline materials with no texture or preferred grain orientation, the saturation magnetostriction coefficient (λ_s) is given by

$$\lambda_s = \frac{2}{5}\lambda_{100} + \frac{3}{5}\lambda_{111} \tag{9.22}$$

In some materials such as Invar (an iron–nickel alloy), the thermal expansion effect is neutralized by the magnetostriction-induced contraction. This yields a material whose dimensions are essentially invariable with temperature, and hence the name Invar.

Equations and coefficients used are similar to those used for piezoelectrics (e.g., d_{33} and k_{33}). For example, the d_{33} coefficient of magnetostriction refers to the amount of strain produced by the application of unit magnetic field.

The following example illustrates the use of such coefficients.

EXAMPLE 9.6: TERFENOL-D MAGNETOSTRICTION

The maximum d_{33} magnetostriction coefficient for Terfenol-D is 57×10^{-9} m/A (du Trémolet de Lacheisserie et al. 2002). A bar made from this material is 10 cm long and is exposed to a magnetic field (H) of 5 A/m. What will be the elongation produced in this bar?

SOLUTION

The d_{33} coefficient represents the strain produced per magnetic field. Thus, in this case, the strain produced $\frac{\Delta l}{l}$ will be given by

$$\lambda = 57 \times 10^{-9} = \frac{\text{strain}}{\text{magnetic field}} = \frac{\Delta l / l}{5 \text{ A/m}} = \frac{\Delta l}{5 \text{ A/m} \times 0.1\,\text{m}}$$

Therefore, the increase in length will be = $(57 \times 10^{-9}) \times (5 \text{ A/m}) \times (0.1 \text{ m}) = 28.5$ nm.

9.8 SOFT AND HARD MAGNETIC MATERIALS

As seen in Figure 9.19, materials with $H_c \sim 5000$ A/m are considered soft magnetic materials. Those with $H_c \sim >10^4$ A/m are considered hard or permanent magnets. A summary of some of these materials is shown in Figure 9.19.

The following example illustrates the coercivity magnitudes associated with soft and hard magnetic materials.

EXAMPLE 9.7: SHAPES OF HYSTERESIS LOOPS: SOFT AND HARD MAGNETIC MATERIALS

The coercivity (H_c) of a sample of the material known as grain-oriented steel is 300 A/m. The saturation magnetization ($\mu_0 M_s$) of this material is 2.1 T. A hysteresis loop for this material was measured to determine the properties.

    a. If the coercivity of this material is also expressed in Tesla, how does the y-axis value of saturation magnetization compare with the x-axis value of coercivity?
    b. What is the value of the coercivity of this material in Oersteds?
    c. What is the value of the saturation magnetization of this grain-oriented steel in Gauss?
    d. Repeat this calculation for a sample of a neodymium–iron–boron ($Nd_2Fe_{14}B$) permanent magnet, whose coercive field is 700,000 A/m, if the saturation magnetization is 1.2 T.

SOLUTION

    a. The coercivity is 300 A/m. To convert this value into Tesla, we multiply it by $\mu_0 = 4\pi \times 10^{-7}$ Wb/m·A.
    The coercivity value expressed in Tesla will be $\mu_0 H_c = (300 \text{ A/m})(4\pi \times 10^{-7} \text{ Wb/m} \cdot \text{A}) = 3.78 \times 10^{-4}$ Wb/m² or T. Compare this with $\mu_0 M_s = 2.1$ T. If we were to plot the hysteresis loop as $\mu_0 M$ versus $\mu_0 H$, the loop will look extremely slim.
    b. The coercive field is expressed as ampere/meter or Oersteds (Oe). From Section 9.6.3 and Table 9.5, we know that 1 A/m = $4\pi \times 10^{-3}$ Oe. Thus, the coercivity of 300 A/m for this soft-magnetic grain-oriented steel will be

$$= (300 \text{ A/m}) \times \left(4\pi \times 10^{-3} \frac{\text{Oe}}{\text{A/m}} \right) = 3.78 \text{ Oe}$$

    c. The saturation magnetization of this grain-oriented steel is 2.1 T. One Tesla is 10,000 Gauss; thus, the saturation magnetization will be 21,000 Gauss.
    d. For the $Nd_2Fe_{14}B$ magnet, the saturation magnetization is listed as 1.2 T. The coercivity of 700,000 A/m will be = (700,000 A/m)($4\pi \times 10^{-7}$ Wb/m·A) = 0.88 Wb/m² or T. Thus, for this permanent magnet, the hysteresis loop will appear square in shape. Also, because this material is stated to be a permanent magnet, we expect the remnant magnetization ($\mu_0 M_r$) and saturation magnetization ($\mu_0 M_s$) to be similar.

In the following sections, we discuss hard and soft magnetic materials in detail.

9.9 HARD MAGNETIC MATERIALS

Hard or permanent magnets are used in many applications to supply a permanent magnetic flux and to generate a static magnetic field. They are used in simple applications such as holding objects, electrical motors, and so on. Permanent magnets are also used for mechanical systems, such as magnetic gears, bearings, shock absorbers, and so on.

    The properties of some of the commercially available types of hard magnetic materials are summarized in Table 9.9. These include the H_c values for the J–H and B–H loops and the variations of B_r with temperature.

    The anisotropy field, magnetocrystalline anisotropy coefficient (K_1), Curie temperature (T_c), and saturation magnetization for some hard magnetic materials were listed in Table 9.7b.

    The relative costs of some permanent magnet materials, expressed in U.S. dollars/Joule, versus the maximum energy per unit volume are shown in Figure 9.28.

    From this diagram, it can be seen that the so-called *rare-earth magnets* (often based on NdFeB and SmCo) offer products with the highest energy. One problem with the use of rare-earth magnets is that their Curie temperature is low ($T_c \sim 300°C$). These materials also undergo surface oxidation, and this leads to a decrease in the coercivity. Magnets based on Neodymium (e.g., $Nd_2Fe_{14}B$) are commonly known as *neomagnets*.

TABLE 9.9
Properties of Some Hard Magnetic Materials

Material	T_c (°C)	$(BH)_{max}$ (kJ m^{-3})	B_r (T)	dB_r/dT (%/degr.)	$_JH_c$ (kA m^{-1})	$_BH_c$ (kA · m^{-1})
Ferroxdure (SrFe$_{12}$O$_{19}$)	450	28	0.39	−0.2	275	265
Alnico$_4$	850	72	1.04	−0.015	−	124
SmCo$_5$	720	130–180	0.8–0.9	−0.01	1100–1500	600–670
Sm(Co, Fe, Cu, Zr)$_7$	800	200–240	0.95–1.15	−0.03	600–1300	600–900
Nd-Fe-B (sintered magnet)	310	200–280	1.0–1.2	−0.13	750–1500	600–850

Source: Buschow, K. H., and F. R. De Boer. 2003. *Physics of Magnetism and Magnetic Materials*. Boston: Kluwer. With permission.

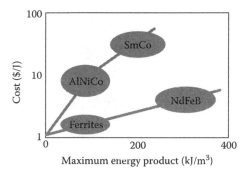

FIGURE 9.28 Relative cost of permanent magnets per unit energy versus maximum energy product. (From du Trémolet de Lacheisserie, E., D. Gignoux, and M. Schlenker. 2002. *Magnetism: Fundamentals*. Berlin: Springer. With permission.)

The AlNiCo (pronounced al-knee-ko) magnets have lower coercivity. However, they have very good temperature-stability, and because of their high Curie temperatures (~857°C), they can be used up to 500°C.

Ceramic ferrites offer the lowest cost per unit energy stored. Thus, ceramic ferrites are preferred for low-cost applications. The negative aspect of ceramic ferrites is that their saturation magnetization is low (~0.3–0.6 T, Table 9.6a), and hence the energy products are relatively low (Figure 9.28). Ceramic ferrite compositions can belong to both soft and hard categories.

The samarium–cobalt magnets (SmCo$_5$ and Sm$_2$Co$_{17}$) are expensive but high-performance magnets. Their Curie temperatures are high (727°C for SmCo$_5$; and 827°C for Sm$_2$Co$_{17}$).

9.10   ISOTROPIC, TEXTURED (ORIENTED), AND BONDED MAGNETS

Permanent magnets are sometimes classified based on the orientation of magnetic domains. An *isotropic magnet* is a permanent magnet in which the spatial distribution of the easy directions of magnetizations is random. An isotropic magnet, in principle, is similar to an unpoled ferroelectric material. In *textured magnets*, also known as *oriented magnets*, most of the easy directions of magnetizations are oriented in a particular direction. Therefore, the distribution of orientations of the easy direction of magnetism inside the material is not random. This can be achieved by a process similar to the poling of ferroelectrics. In some cases, processing of a material (e.g., using a rolling process or extrusion) will cause orientation of different grains in a particular direction.

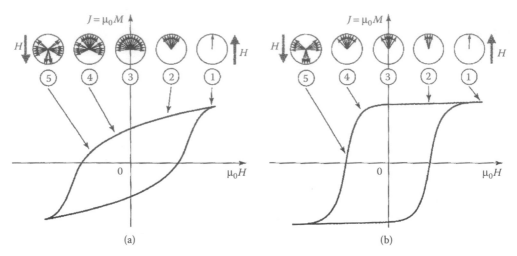

FIGURE 9.29   Hysteresis loops for (a) oriented and (b) isotropic magnets. (From du Trémolet de Lacheisserie, E., D. Gignoux, and M. Schlenker. 2002. *Magnetism: Fundamentals*. Berlin: Springer. With permission.)

Compared to isotropic magnets, the hysteresis loop of textured magnets is square (Figure 9.29), because after saturation, most domains prefer to remain aligned in specific directions. In Figure 9.29, the distribution of the easy axes is shown in the circles drawn for different stages of magnetization. For example, under saturation conditions, for both oriented and isotropic magnets, the easy axes are aligned along only one direction, which will be the direction of the applied magnetic field.

From the distribution of the directions of easy axes, we can see that for textured or oriented magnets, the remnant magnetization is much higher, typically ~88%–97% of the saturation magnetization (du Trémolet de Lacheisserie et al. 2002). For isotropic magnets, the remnant magnetization is much lower (~50% of the saturation magnetization).

Bonded magnets are essentially composites made by dispersing magnetic powders in a polymer or metallic matrix. These are usually isotropic in nature because the easy axes within each particle (small crystal) are randomly arranged in the composite material. It is possible to orient the magnetic particles, for example, ferrite in a polymer, using a special processing technique. For example, oriented-bonded magnets are made with ceramic ferrites in a plastic using the calendaring process. Such materials, known as magnetic rubbers, have a saturation magnetization of ~0.25 T. In refrigerators, for example, we make use of these magnetic rubbers as door gaskets. Compared to the so-called *sintered magnets*, obtained by high-temperature sintering of metal or ceramic powders, the energy product and coercive fields of bonded magnets are lower. The sintered materials also could be oriented or isotropic.

The demagnetization curves for some ceramic hard ferrites and NdFeB magnets, which are (1) isotropic bonded, (2) oriented bonded, and (3) oriented and sintered, are shown in Figure 9.30.

9.11   SOFT MAGNETIC MATERIALS

Ceramic ferrites, which are alloys of iron containing nickel, cobalt, silicon and some of the amorphous materials known as metallic glasses, constitute the most important classes of soft magnetic materials (Table 9.10a and b). The market share is dominated by electrical steels (especially nonoriented) as shown in Figure 9.31. The ranges of the properties of some technologically important magnetic materials are shown in Figure 9.32. These materials are used in transformers, electrical motors, generators, electromagnetic switches, and so on.

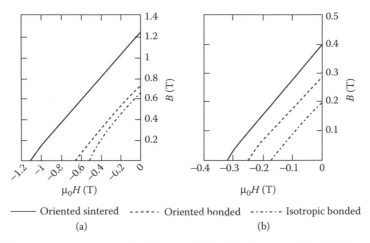

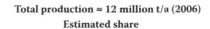

—— Oriented sintered   - - - - - Oriented bonded   - - - - - - Isotropic bonded

(a)   (b)

FIGURE 9.30 Demagnetization curves for (a) ceramic hard ferrites and (b) NdFeB magnets. (From du Trémolet de Lacheisserie, E., D. Gignoux, and M. Schlenker. 2002. *Magnetism: Fundamentals*. Berlin: Springer. With permission.)

Total production ≈ 12 million t/a (2006)
Estimated share

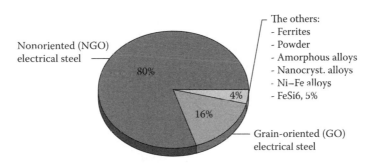

FIGURE 9.31 Market share of soft magnetic materials. (From Varga, L. K., and H. A. Davies. 2008. *J Magn Magn Mater* 320(20):2411–22. With permission.)

For many applications of soft magnetic materials, it is desirable to have materials that have high saturation magnetization (for lightweightness and compactness), very low coercivity and high resistivity (to minimize eddy-current losses), and high permeability (to minimize the magnetic reluctance—equivalent of electrical resistance). No single material meets all these requirements. A choice is made depending on the technical specifications needed and cost.

In grain-oriented steel, silicon is added (up to ~3.25%) to enhance resistivity and to reduce coercivity (Figure 9.32b). This reduces transformer-core losses. Permeability of this steel depends on the crystallographic directions. As shown in Figure 9.23, for iron, the main ingredient of steels, magnetization is preferred along the [100] and [110] directions. As a result, the permeability of steel is higher in these directions. This is the reason why grain-oriented steels are preferred for making transformer cores; however, they also cost more.

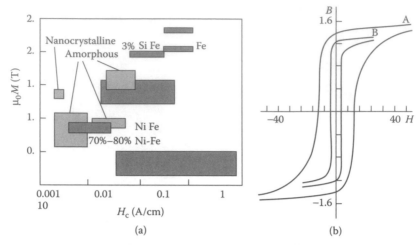

FIGURE 9.32 (a) Coercivity and saturation magnetization for some soft magnetic materials. (From Bloor, D. et al., eds. 1994. *Encyclopedia of Advanced Materials*, vol. 4. Oxford, UK: Pergamon Press. With permission.) (b) Hysteresis loops for pure iron and a grain-oriented steel with 97% Fe and 3% Si. (From Buschow, K. H., and F. R. De Boer. 2003. *Physics of Magnetism and Magnetic Materials*. Boston: Kluwer. With permission.)

TABLE 9.10A
Properties of Some Soft Magnetic Materials

	Composition	μ_{max}	H_c (A/m)	J_s (T)
Fe	Fe 100	$3\text{--}50 \times 10^3$	1–100	2.16
NO Fe–Si	Fe(>96)–Si(<4)	$3\text{--}10 \times 10^3$	30–80	1.98–2.12
GO Fe–Si	Fe 97–Si 3	$20\text{--}80 \times 10^3$	4–15	2.03
Fe–Si 6.5%	Fe 93.5–Si 6.5	$5\text{--}30 \times 10^3$	10–40	1.80
Sintered powders	Fe 99.5–P 0.5	$0.2\text{--}2 \times 10^3$	100–500	1.65–1.95
Permalloy	Fe 16–Ni 79–Mo 5	5×10^5	0.4	0.80
Permendur	Fe 49–Co 49–V 2	2×10^3	100	2.4
Ferrites	$(Mn, Zn)O.Fe_2O_3$	3×10^3	20–80	0.2–0.5
Sendust	Fe 85–Si 95–Al 5.5	50×10^3	5	1.70
Amorphous (Fe-based)	$Fe_{78}B_{13}Si_{10}$	10^5	2	1.56
Amorphous (Co-based)	$Co_{17}Fe_4B_{15}Si_{10}$	5×10^5	0.5	0.86
Nanocrystalline	$Fe_{73.5}Cu_1Nb_3Si_{13.5}B_9$	10^5	0.5	1.2

The composition is given in weight %; for the amorphous and nanocrystalline alloys, it is expressed in atomic %. μ_{max}, maximum DC relative permeability; H_c, coercive field; J_s, saturation polarization at room temperature.

Data from: Fiorillo, F. 2004. *Measurement and Characterization of Magnetic Materials*. Oxford: Elsevier; and du Trémolet de Lacheisserie, E., D. Gignoux, and M. Schlenker. 2002. *Magnetism: Fundamentals*. Berlin: Springer.

The properties of some commercially important soft magnetic materials are shown in Tables 9.10a (Fiorillo 2004) and 9.10b. In Table 9.10a, the maximum value of permeability (see Figure 9.16), coercive field (H_c in ampere/meter), and saturation polarization ($J_s = \mu_0 M_s$ in Tesla) are presented. All compositions are in weight %, except those for amorphous and nanocrystalline materials, which are in atomic %.

As discussed in Section 9.6.4, the coercivity of a magnetic material is controlled by the difficulty in nucleation or by the pinning of domains. In most hard magnetic materials, the nucleation of new domains controls the coercivity. In soft magnetic materials, defects such as grain boundaries affect the coercivity. Introduction of microstructural imperfections, which are of the order of domain-wall

TABLE 9.10B

Properties of Some Commonly Used Soft Magnetic Materials

Name	Composition	Permeability (μ_r)		Coercivity (H_c) (A·m⁻¹)	Retentivity (B_r) (T)	B_{max}	Resistivity (μΩ m)
		Initial	Maximum				
Ingot iron	99.8% Fe	150	5,000	80	0.77	2.14	0.10
Low-carbon steel	99.5% Fe	200	4,000	100		2.14	1.12
Silicon iron, unoriented	Fe–3% Si	270	8,000	60		2.01	0.47
Silicon iron, grain-oriented	Fe–3% Si	1400	50,000	7	1.20	2.01	0.50
4750 alloy	Fe–48% Ni	11,000	80,000	2		1.55	0.48
4-79 permalloy	Fe–4% Mo–79% Ni	40,000	200,000	1		0.80	0.58
Superalloy	Fe–5% Mo–80% Ni	80,000	450,000	0.4		0.78	0.65
2V-permendur	Fe–2% V–49% Co	800	450,000	0.4		0.78	0.65
Supermendur	Fe–2% V–49% Co		100,000	16	2.00	2.30	0.40
Metglas[a]2650SC	$Fe_{81}B_{13.5}Si_{3.5}C_2$		300,000	3	1.46	1.61	1.35
Metglas[a]2650S-2	$Be_{78}B_{13}S_9$		600,000	2	1.35	1.56	1.37
MnZn ferrite	H5C2[b]	10,000		7	0.09	0.40	1.5×10^5
MnZn ferrite	H5E[b]	18,000		3	0.12	0.44	5×10^4
NiZn ferrite	K5[b]	290		80	0.25	0.33	2×10^{12}

[a] Allied corporation trademark.

[b] TDK ferrite code.

Source: Chin, G. Y., et al. 1994. Magnetic materials: An overview, basic concepts, magnetic measurements, magnetostrictive materials. In *Encyclopedia of Advanced Materials*, vol. 1, eds. D. Bloor et al., 1424. Oxford, UK: Pergamon Press; and Askeland, D., and P. Fulay. 2006. *The Science and Engineering of Materials*. Washington, DC: Thomson. With permission.

TABLE 9.10C

Properties of Some Materials Used in Magnetic Tape Media

	Particle Length (μm)	Aspect Ratio	Magnetization (B_r)		Coercivity (H_c)		Surface Area (m^2/g)	Curie Temperature (T_c) (°C)
			Wb/m²	emu/cc	kA/m	Oe		
γ-Fe_2O_3	0.20	5:1	0.44	350	22–34	420	15–30	600
Co-γ-Fe_2O_3	0.20	6:1	0.48	380	30–75	940	20–35	700
CrO_{-2}	0.20	10:1	0.50	400	30–75	950	18–55	125
Fe	0.15	10:1	1.40[a]	1100[a]	56–176	2200	20–60	770
Barium ferrite	0.05	0.02 μm thick	0.40	320	56–240	3000	20–25	350

[a] For overcoated, stable particles, use only 50%–80% of these values due to reduced volume of magnetic particles.

Source: Jorgensen, F. 1996. *Complete Handbook of Magnetic Recording*, 4th ed. New York: McGraw Hill. With permission.

thickness, usually results in an increase in coercivity. The coercivity (H_c) of a magnetic material scales with the average grain size (d), as shown by the following equation:

$$H_c = H_{c,0} + \frac{\text{constant}}{d} \tag{9.23}$$

In this equation, $H_{c,0}$ is a constant that defines the baseline coercivity of a material and depends on factors such as intrinsic properties, stress, other defects such as inclusions, and so on (Bloor et al. 1994). In fact, a class of soft magnetic materials is made from amorphous materials. These materials are made in the form of metallic ribbons obtained by the rapid solidification of alloys. These materials are known as metallic glasses. Because there are no grain boundaries in these materials, the coercivity of these materials is very low (Table 9.10c).

9.12 MAGNETIC DATA STORAGE MATERIALS

For storage of information using magnetic materials, the main idea has been to use the remnant magnetization as the basis for storing data. For this application, magnetically semihard materials are needed because the information must be written, and we must be able to erase the information and rewrite it. There are two major types of magnetic media (Figure 9.33). One is in the form of thin films, used for making magnetic hard disks for personal computers and other electronic gadgets. The second technology involves magnetic tape media, such as audio and video tapes and computer disks. The magnetic tape technology, although available and useful, has encountered stiff competition with the advent of optically written media, such as DVDs and CDs, in addition to semiconductor-based flash memories widely used in digital cameras and other electronic hardware.

For magnetic tapes, cobalt-modified metallic iron, cobalt-modified gamma iron oxide (γ-Fe_2O_3), and barium ferrite ($BaFe_{12}O_{19}$) are some of the commonly used materials. Cobalt modification of gamma iron oxide leads to higher coercivity values that enhance data retention. Most particles used in this application are elongated to effectively take advantage of the magnetoshape anisotropy (Figure 9.34). The important properties of some of the magnetic particles used for magnetic tapes are included in Table 9.10c.

The technology of magnetic hard-disk manufacture has advanced at a very rapid pace. Significant advances in the area of development of nanoscale materials, perpendicular recording, advanced read-and-write heads, devices based on giant magnetoresistance, spin electronics, and bit-patterned media have been made. These have led to remarkable advances related to the amount of information that can be stored on a one-inch square of a hard disk. The goal in the year 2008 was to be able to

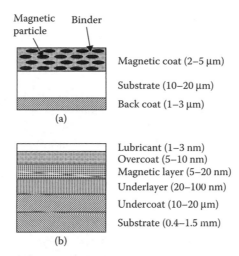

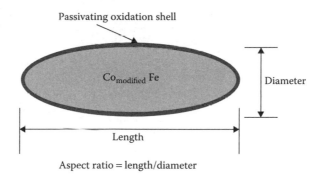

FIGURE 9.33 Schematic representation of (a) magnetic tape and (b) thin-film hard disk. (From Li., Y., and A. K. Menon. 2008. Magnetic recording technologies: Overview. In *Encyclopedia of Materials: Science and Technology*, eds. D. Bloor et al. Amsterdam: Elsevier. With permission.)

FIGURE 9.34 Schematic illustration of an advanced acicular particle of cobalt-modified iron. (From Bloor, D. et al., eds. 1994. *Encyclopedia of Advanced Materials*, vol. 4. Oxford, UK: Pergamon Press. With permission.)

store an ultra-high density of 1 terabits in a one-inch square area (Tbit/in^2). A terabit is one trillion bits. The hard disks today can store up to 200 Gbit/in^2. A gigabit is one billion bits.

There are also many other technologies such as magnetic random-access memories (MRAMs) that are being developed for information storage. Similarly, researchers have also been developing different classes of devices based on multiferroics for information storage. There are many rapidly evolving and very exciting developments occurring in the field of magnetic materials and devices.

9.13 PROBLEMS

9.1 Comment on the following statement: "All materials in this world are magnetic."

9.2 What is the difference between a diamagnetic and paramagnetic material? Give an example of each type of material.

9.3 What is an antiferromagnetic material? Are there examples of antiferromagnetic coupling in ferromagnetic materials?

9.4 What is the difference between a ferromagnetic and a ferrimagnetic material?

9.5 Is it necessary for ferromagnetic or ferrimagnetic materials to contain ferromagnetic elements? Explain.

9.6   What is a superparamagnetic material?

9.7   What is a ferrofluid? Where are ferrofluids used commercially?

9.8   What is the basis of magnetic levitation using superconductors?

9.9   The coercive field for a neodymium iron boron–based magnet is reported to be 1.25 T; how much is this value in Oersted?

9.10 Bohr magneton (μ_B) is the basic unit for magnetic moment. It is based on the spin component of the magnetic moment of an electron that is aligned with an external magnetic field, and its value is given by $\mu_B = \hbar q_e / 2m_e = 9.274 \times 10^{-24}$ A $\cdot$ m^2. Sometimes, it is necessary or useful to express the number n, which is the number of Bohr magnetons per mole of a material. Show that if MW is the molecular weight of a material and if x is the number of Bohr magnetons per atom, the value of Bohr magnetons per mole of a material is $(5.585 \times x)/(\text{MW})$.

9.11 The magnetic moment of a nickel atom is listed as 0.606 Bohr magneton at 0 K (shown in Table 9.6a). The value of saturation magnetization at 290 K is listed as 0.61 T. Nickel has a face-centered cubic (FCC) structure, and its lattice constant at 290 K is 3.5167 Å. What will be the effective magnetic moment of nickel atoms in FCC nickel at $T = 290$ K?

9.12 In Table 9.6a, why are the values of saturation magnetization not listed for gadolinium (Gd) and dysprosium (Dy) at 290 K?

9.13 The exchange-energy (A) values for iron and nickel are 2.5×10^{-21} and 3.2×10^{-21} J, respectively. The anisotropy energies (K_1) for iron and nickel are 4.8×10^4 J/m^3 and -0.5×10^4 J/m^3, respectively (Jiles 1991). In which material would you expect the domain walls to be thicker? Why? How will the domain energies compare for iron and nickel?

9.14 What is a Bloch wall? How is it different from the Néel wall?

9.15 Why is the coercivity of permanent magnetic materials significantly lower than that expected from the value of the anisotropy field?

9.16 How is the coercivity of a polycrystalline soft magnetic material related to the grain size?

9.17 Why are 180° domain walls not stress-sensitive, whereas 90° domain walls are stress-sensitive?

9.18 What will be the flux density (B) created for a current of 1 A if a toroidal solenoid has an air core? Assume that the average circumference of the solenoid is 50 cm and that there are 1000 turns.

9.19 How much will the flux density (B) be if the core is filled with ceramic ferrite of permeability 300? Assume that the average circumference of the solenoid is 50 cm and that there are 1000 turns. If the diameter of the solenoid core is 5 cm, what will be the flux in Wb?

9.20 The demagnetization curves for different commercially available permanent magnets are shown in Figure 9.35.

   (a) Which magnets have the lowest and the highest coercivity values?

   (b) Calculate the coercivity of the material Sm_2Co_{17} in Oersteds.

   (c) Calculate the maximum energy product for $SmCo_5$ and $Nd_2Fe_{14}B$ (shown in Figure 9.35).

9.21 The properties of isotropic-bonded, oriented-bonded, and oriented-sintered magnets made from hard ceramic ferrites and NdFeB are shown in Figure 9.30.

   (a) Why is the energy product for the oriented-sintered material (ferrite or NdFeB) better than that of either of the bonded magnets?

   (b) Calculate the energy products for the oriented-sintered NdFeB and compare the value with those for oriented-bonded and isotropic-bonded NdFeB materials.

9.22 The lifting force (F) of a permanent magnet is given by the following equation:

$$F = \frac{(\mu_0 M)^2 A}{2\mu_0} \tag{9.24}$$

where A is the area of the magnet. Calculate the lifting force, in kN, for a permanent magnet with saturation magnetization of 1.5 T. Assume that the area is 0.25 m^2.

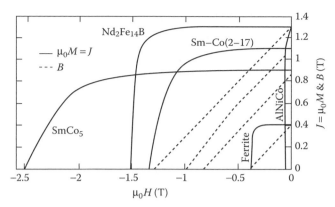

FIGURE 9.35 Demagnetization curves for some permanent magnets. (From du Trémolet de Lacheisserie, E., D. Gignoux, and M. Schlenker. 2002. *Magnetism: Fundamentals*. Berlin: Springer. With permission.)

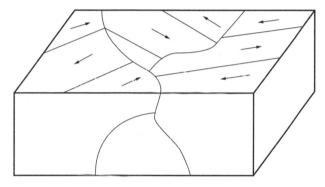

FIGURE 9.36 Domain pattern in rolled Ni–Fe alloy. (From Buschow, K. H., and F. R. De Boer. 2003. *Physics of Magnetism and Magnetic Materials*. Boston: Kluwer. With permission.)

9.23 What types of materials are used for magnetic tape data storage? Why?

9.24 Permanent magnets have high coercivity; why can these not be used for magnetic data storage?

9.25 In Figure 9.32b, why is the maximum magnetization for the grain-oriented steel lower than that for pure iron?

9.26 What is the material used for the magnetic stripe on credit cards and automated teller machine cards used for banking?

9.27 What magnetic properties are important for a material to be used for transformer cores?

9.28 What is grain-oriented steel? What magnetic properties of steel are improved by producing grain-oriented steel? In what crystallographic directions are the grains oriented? Why?

9.29 Why is silicon added to grain-oriented steels?

9.30 The domain pattern in a rolled polycrystalline Ni–Fe alloy is shown in Figure 9.36.
    (a) Are the grains oriented in this sample?
    (b) Based on the domain geometry shown, is the material likely to have any net remnant magnetization?
    (c) What will happen if this material is heated above its T_c and then "field-cooled," that is, annealed at temperatures below T_c by placing it in a magnetic field parallel to the rolling direction.

9.31 What are the different smart materials based on the use of magnetic materials?

9.32 What is a ferrofluid? How is it different from a magnetorheological (MR) fluid?

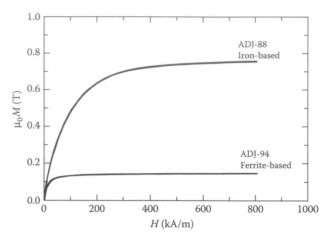

FIGURE 9.37 Magnetization curves for magnetorheological fluids. (From Fulay, P. P., A. D. Jatkar, and J. M. Ginder. 1997. Synthesis and properties of magnetorheological fluids for active vibration control. In *Proceeding of MRS Symposium on Materials for Smart Systems* 459:99. With permission.)

9.33 How does an MR fluid work in a magnetic shock absorber?

9.34 The magnetization curves for two MR fluids, one based on iron and another based on a ceramic ferrite, are shown in Figure 9.37.

    (a) Calculate the initial permeabilities of both these MR fluids from the magnetization curves.

    (b) Calculate the volume fraction of iron in the iron-based fluid. Assume that the saturation magnetization of iron is 2.1 T and that the MR fluid is made using a dispersion of iron particles in synthetic oil.

    (c) Assuming that the ferrite-based fluid is made using iron oxide (Fe_3O_4), what is the volume fraction of this material in this MR fluid made using synthetic oil?

9.35 If nickel ferrite has an inverse spinel structure, what will be the net magnetic moment of this structure per unit formula? Assume that the lattice constant of nickel ferrite is 8.37 Å.

9.36 In Table 9.6b, zinc ferrite is shown to have zero magnetic moment. The magnetic moment per formula for $MnFe_2O_4$ is 5 μ_B and that of $NiFe_2O_4$ is 2 μ_B. Why does the initial addition of zinc ferrite to $MnFe_2O_4$ and $NiFe_2O_4$ *increase* the total magnetic moments of these materials? However, as increasing quantities of zinc ferrite are added, the magnetization eventually decreases. Explain.

9.37 What is magnetostriction? Explain how this phenomenon can be used to create ultrasonic waves and a nanopositioning device.

9.38 What phenomenon causes an electrical transformer to "hum"?

GLOSSARY

Antiferromagnetic material: A material that shows zero net magnetic moment because the magnetic moments of different ions or atoms cancel each other out completely.

$(BH)_{max}$ product: See **Maximum energy product**.

Bloch wall: A wall between two magnetic domains across which the magnetization direction changes by 180°. The thickness of a Bloch wall increases with increase in exchange energy (as the material tries to maintain the magnetization in the already-existing direction) and decreases with increased anisotropy (where the material prefers to magnetize in the easy direction of magnetization).

Bohr magneton: The unit in which the magnetic dipole moment of an electron, atom, or an ion is measured. A Bohr magneton is also equal to the component of the spin magnetic moment of an electron that is aligned with the applied magnetic field

$$\mu_B = \hbar \frac{q_e}{2m_e} = 9.274 \times 10^{-24} \ A \cdot m^2$$

Bonded magnets: Magnetic composites made by dispersing magnetic particles in a polymeric or metallic matrix, which are low-cost magnetic materials that could be either isotropic or oriented.

Curie temperature (T_c): Temperature at which a ferromagnetic or ferrimagnetic material becomes paramagnetic. Note that this term is also used for ferroelectrics (Chapter 8).

Demagnetization curve: The second quadrant of the hysteresis loop in which a magnetic material moves from flux density B_r to zero. This curve is important for designing circuits involving permanent magnets.

Demagnetizing factor (N_d): A numerical factor that shows the extent of internal field created in a magnetic material. This field opposes the effect of the applied magnetic field.

Diamagnetism: Exclusion of the applied magnetic field from a material. Diamagnetic materials are defined by negative values of magnetic susceptibility, ranging from -10^{-6} to -1 (for superconductors). Most materials (e.g., most metals, inert gases, and organic compounds) that are classified as "nonmagnetic" are actually diamagnetic.

Domain: See **Magnetic domain**.

Domain wall: A surface across which the magnetization direction rotates. If the magnetization rotates about the normal to a domain wall, the domain wall is known as a Bloch wall (Figure 9.21), and if it rotates such that it has a component along the wall normal, then it is called a Néel wall.

Easy axis: The crystallographic directions in which the magnetic moments in a material tend to line up naturally in the absence of any magnetic field, thus minimizing the free energy.

Exchange interaction: The long-range spin ordering occurring as a result of a quantum mechanical interaction, which is due to the combination of electrostatic coupling between electron orbitals and the Pauli's exclusion principle. In ferromagnetic materials, this interaction causes the spin to be aligned in the same direction; and in ferrimagnetic materials, the magnetic moments due to the electron spin of ions at different crystallographic locations are antiparallel.

Ferrofluid: A stable dispersion of superparamagnetic particles, used as heat-transfer fluid; the carrier is typically an oil or organic fluid, but it could be water.

Ferrimagnetic materials: Materials in which the magnetic moments of some of the atoms or ions are antiparallel but do not completely cancel out, thereby creating a net magnetic moment.

Ferromagnetic materials: Materials in which all the magnetic moments of atoms or ions remain in the same direction (e.g., Fe, Ni, Co, Gd, and alloy of Fe–Co).

Flux density (B): See **Magnetic flux density**.

Hard axis: Crystallographic directions along which it is difficult to reorient the magnetization direction.

Hard magnetic materials: Materials that have a high ($\sim >10^4$ A/m) coercivity, also known as permanent magnets.

Hysteresis loop: See **Magnetic hysteresis loop**.

Hysteresis losses: Power loss in a magnetic material because of the occurrence of eddy currents induced by an applied magnetic field.

Induction (B): See **Magnetic flux density**.

Isotropic magnet: A magnet in which the spatial distribution of the magnetization directions is random.

Joule magnetostriction: The magnetostriction strain component caused by rotation of magnetic domains, which is different from the spontaneous magnetostriction that sets in when magnetic dipoles undergo a spontaneous alignment at temperatures near the T_c.

Liquid magnets: Materials showing an entire body motion when placed next to a permanent magnet. See also **Ferrofluids**.

Magnetic domain: Region of a ferromagnetic or ferrimagnetic material in which the magnetization (i.e., net magnetic moment due to different atoms or ions) is in the same direction.

Magnetic flux density (B): The magnetic flux created per unit area; unit is Weber/square meter or Tesla, which represents the strength of the magnetic field created inside a magnetic material; its level depends on the applied external field and the magnetic permeability of the magnetic material.

Magnetic induction: See **Magnetic flux density**.

Magnetic hysteresis loop: The trace of magnetization (M) developed in magnetic materials as a function of the applied magnetic field (H), plotted as $\mu_0 M$ versus H or flux density B versus H.

Magnetic levitation: See **Meissner effect**.

Magnetic losses: Energy dissipation occurring when a magnetic material is subjected to a time-varying external magnetic field.

Magnetization (M): The total magnetic moment per unit volume; unit is ampere/meter.

Magnetic permeability of free space (μ_0): A constant that relates the flux density B created by a magnetic field H in space. Its value is $\mu_0 = 4\pi \times 10^{-7}$ H/m or Wb/A · m.

Magnetic permeability (μ): A property that describes the ease with which magnetic flux lines are carried through a material. Ferromagnetic and ferrimagnetic materials have large magnetic permeability values, which are field-dependent. The units are Henry/meter, which is the same as Weber/ampere · meter.

Magnetic polarization (J): The quantity $\mu_0 M$ that is shown using the symbol J.

Magnetocrystalline anisotropy: The dependence of magnetic properties, such as coercive field, on the crystallographic direction.

Magnetocrystalline anisotropy energy: The energy that is needed to rotate from the easy-magnetization direction to a given direction.

Magnetorheological (MR) fluids: Dispersions of noncolloidal and magnetically multidomain soft ferromagnetic or ferrimagnetic particles. Applications of MR fluids include vibration control.

Magnetoshape anisotropy: Anisotropy created by the shape of magnetic grains or particles; same as shape anisotropy; causes needle-like magnetic particles to exhibit higher coercivity, which is used in magnetic recording media.

Magnetostriction: A change in the dimension of a magnetic material after magnetization caused by either the development of spontaneous magnetization as the material goes through the Curie temperature or by application of a magnetic field, which causes changes in the domain configuration.

Maximum energy product: The maximum value of a product obtained by multiplying the corresponding B and H values on the demagnetization curve (Figure 9.26); also known as the $(BH)_{max}$ product; is a measure of the energy stored in a magnet of a given size.

Meissner effect: Expulsion of magnetic flux lines due to the diamagnetic behavior of a superconductor in a superconducting state, which is the basis of magnetic levitation using superconductors.

Minor loop: A magnetic hysteresis loop in which the applied fields do not reach saturation levels.

Multiferroics: Materials that possess two or more switchable properties, such as spontaneous polarization, spontaneous magnetization, or spontaneous strain.

Néel temperature (T_N or O_N): Temperature at which antiferromagnetic exchange-coupling interactions develop spontaneously in the sublattice (Figure 9.12c). In an antiferromagnetic material, these are overcome by *increasing* the temperature, due to which the material becomes paramagnetic. Note that below the T_N, because the antiferromagnetic coupling is reduced, the *susceptibility increases* (i.e., $1/\chi_m$ decreases).

Néel wall: A domain wall in which the magnetization rotates in the plane of the sample as we move from one direction of the magnetic moment to another (Figure 9.22); occurs in thin films and not in bulk magnetic materials.

Neomagnets: Permanent magnets based on neodymium (e.g., $Nd_2Fe_{14}B$).

Nonlinear magnetic materials: Ferromagnetic and ferrimagnetic materials below their Curie temperature.

Nucleation-controlled coercivity: When control of coercivity of a magnetic material is accomplished by making it difficult to nucleate new domain walls, the coercivity is said to be nucleation-controlled. See also **Pinning-controlled coercivity**.

Oriented magnets: Magnetic materials in which the distribution of the easy axes of magnetization is not random. The hysteresis loops of these materials are squarer, and the remnant magnetization is higher; the same as textured magnets.

Paramagnetic atoms: Atoms that have a net magnetic moment (e.g., Al, Cu, Fe).

Paramagnetic ions: Ions that have a net magnetic moment (e.g., Fe^{2+}).

Paramagnetic materials: Materials with relatively small, but positive, susceptibility ($\sim+10^{-6}$–10^{-3}); for example, liquid oxygen, copper, and aluminum. Ferromagnetic and ferrimagnetic materials become paramagnetic above their Curie temperatures.

Permanent magnets: Materials that have a large coercivity ($\sim>10^4$ A/m); also known as hard magnetic materials.

Pinning-controlled coercivity: When coercivity values are controlled by controlling the concentration of defects or grain boundaries that can pin domains. In this case, domains already exist, and the focus is on pinning the domain walls. See also **Nucleation-controlled coercivity**.

Rare-earth magnets: Permanent magnets based on rare-earth elements (e.g., Sm, Co, and Nd).

Relative magnetic permeability (μ_r): Ratio of the magnetic permeability of a material to that of magnetic permeability of free space; has no units and is related to the susceptibility as shown by the following equation: $\mu_r = 1 + \chi_m$.

Saturation magnetization ($\mu_0 M_s$): The maximum possible magnetization in a ferromagnetic or ferrimagnetic material; the unit is Tesla.

Semihard magnetic materials: Materials whose coercivity values are between a few hundred and $\sim10^4$ A/m (Figure 9.19).

Shape anisotropy: See **Magnetoshape anisotropy**.

Sintered magnets: Solid and relatively dense magnetic materials obtained by the compaction of metal or ceramic powders of a magnetic material, followed by high-temperature sintering. These materials could be either oriented or isotropic.

Smart material: A material whose properties are controllable using an external field or stimulus (e.g., piezoelectrics and magnetostrictive materials).

Soft magnetic materials: In general, materials with a coercivity less than ~5000 A/m are considered magnetically soft materials.

Spontaneous magnetostriction: Development of strain in ferromagnetic and ferrimagnetic materials, which occurs spontaneously when there is a spontaneous alignment of magnetic moments at the Curie temperature.

Superparamagnetism: A behavior in which a material that is originally either ferromagnetic or ferrimagnetic in its bulk form behaves as a paramagnetic material on a nanoscale.

Superparamagnetic materials: Nanoscale particles, grains, or structures made from materials that are ferromagnetic or ferrimagnetic in their bulk form but behave as paramagnetic materials because of randomization of magnetic interactions by thermal energy.

Textured magnets: See **Oriented magnets**.

REFERENCES

Askeland, D., and P. Fulay. 2006. *The Science and Engineering of Materials*. Washington, DC: Thomson.

Bloor, D., M. C. Flemings, R. J. Brook, S. Mahajan, and R. W. Cahn, eds. 1994. *Encyclopedia of Advanced Materials*, vol. 4. Oxford, UK: Pergamon Press.

Bozroth, R. M. 1955 (Reprinted 1993). *Ferromagnetic Materials*. Van Nostrand: IEEE Press.

Buschow, K. H., and F. R. De Boer. 2003. *Physics of Magnetism and Magnetic Materials*. Boston: Kluwer.

Chin, G.Y., et al. 1994. Magnetic materials: An overview, basic concepts, magnetic measurements, and magnetostrictive materials. In *Encyclopedia of Advanced Materials*, vol. 1. eds. D. Bloor, M. C. Flemings, R. J. Brook, S. Mahajan, and R. W. Cahn, 1424. Oxford, UK: Pergamon Press.

Chu, Y.-H., L. W. Martin, M. B. Holcomb, and R. Ramesh. 2007. Controlling magnetism with multiferroics. *Mater Today* 10(10):16–23.

Fiorillo, F. 2004. *Measurement and Characterization of Magnetic Materials*. Oxford: Elsevier.

Featonby, D. 2005. Experiments with neodymium magnets. *Physics Educ* 40(6): 505–8.

Fulay, P. P., A. D. Jatkar, and J. M. Ginder. 1997. Synthesis and properties of magnetorheological fluids for active vibration control. In *Proceedings of MRS Symposium on Materials for Smart Systems*. 459:99.

Goldman, A. 1999. *Handbook of Modern Ferromagnetic Materials*. Boston: Kluwer.

Hadjipanayis, G. C. 1999. *J Magn Magn Mater* 200:373.

Jakubowicz, J., M. Jurczyk, A. Handstein, D. Hinz, O. Gutfleisch, and K. -H. Müller. 2000. Temperature dependence of magnetic properties for nanocomposite $Nd_2(Fe,Co,M)_{14}B/\alpha$-Fe magnets. *J Magn Magn Mater* 208(3):163–8.

Jiles, D. C. 1991. *Introduction to Magnetism and Magnetic Materials*. London: Chapman and Hall.

Jiles, D. C. 2001. *Introduction to the Electronic Properties of Materials*. Cheltenham, UK: Nelson Thornes.

Jorgensen, F. 1996. *Complete Handbook of Magnetic Recording*. 4th ed. New York: McGraw Hill.

Kasap, S. O. 2002. *Principles of Electronic Materials and Devices*. New York: McGraw Hill.

Li., Y., and A. K. Menon. 2008. Magnetic recording technologies: Overview. In *Encyclopedia of Materials: Science and Technology*, eds. D. Bloor et al. Amsterdam: Elsevier.

O'Handley, R. C. 1999. *Modern Magnetic Materials, Principles, and Applications*. New York: Wiley.

Rosensweig, R. E., Y. Hirota, S. Tsuda, and K. Raj. 2008. Study of audio speakers containing ferrofluid. *J Phys Condens Matter* 20(204):147.

Scott, W. T. 1959. *Physics of Electricity and Magnetism*. New York: Wiley.

Skomski, R., and J. M. D. Coey, eds. 1999. *Permanent Magnetism*. Bristol, UK: Institute of Physics.

du Trémolet de Lacheisserie, E., D. Gignoux, and M. Schlenker. 2002. *Magnetism: Fundamentals*. Berlin: Springer.

Varga, L. K., and H. A. Davies. 2008. Challenges in optimizing the magnetic properties of bulk soft magnetic materials. *J Magn Magn Mater* 320(20): 2411–22.

Index